Annals of Mathematics Studies

Number 102

SEMINAR ON DIFFERENTIAL GEOMETRY

EDITED BY

SHING-TUNG YAU

PRINCETON UNIVERSITY PRESS

AND

UNIVERSITY OF TOKYO PRESS

PRINCETON, NEW JERSEY

1982

Published in Japan exclusively by
University of Tokyo Press;
In other parts of the world by
Princeton University Press

Printed in the United States of America
by Princeton University Press, 41 William Street
Princeton, New Jersey

Library of Congress Cataloging in Publication data will
be found on the last printed page of this book

TABLE OF CONTENTS

INTRODUCTION ix

SURVEY ON PARTIAL DIFFERENTIAL EQUATIONS IN DIFFERENTIAL GEOMETRY 3
by S.-T. Yau

POINCARÉ INEQUALITIES ON RIEMANNIAN MANIFOLDS 73
by P. Li

BONNESEN-TYPE INEQUALITIES IN ALGEBRAIC GEOMETRY, I. INTRODUCTION TO THE PROBLEM 85
by B. Teissier

LIOUVILLE THEOREMS FOR HARMONIC MAPPINGS, AND AN APPROACH TO BERNSTEIN THEOREMS 107
by S. Hildebrandt

SUBHARMONIC FUNCTIONS, HARMONIC MAPPINGS AND ISOMETRIC IMMERSIONS 133
by L. Karp

AN ISOPERIMETRIC INEQUALITY AND WIEDERSEHEN MANIFOLDS 143
by J. L. Kazdan

ON THE BLASCHKE CONJECTURE 159
by C. T. Yang

BEST CONSTANTS IN THE SOBOLEV IMBEDDING THEOREM: THE YAMABE PROBLEM 173
by T. Aubin

GAUSSIAN AND SCALAR CURVATURE, AN UPDATE 185
by J. L. Kazdan

CONFORMAL METRICS WITH ZERO SCALAR CURVATURE AND A SYMMETRIZATION PROCESS VIA MAXIMUM PRINCIPLE 193
by W.-M. Ni

RIGIDITY OF POSITIVELY CURVED MANIFOLDS WITH LARGE DIAMETER 203
by D. Gromoll and K. Grove

COMPLETE THREE DIMENSIONAL MANIFOLDS WITH POSITIVE RICCI CURVATURE AND SCALAR CURVATURE 209
by R. Schoen and S.-T. Yau

ENTIRE SPACELIKE HYPERSURFACES OF CONSTANT MEAN CURVATURE IN MINKOWSKI SPACE 229
by A. Treibergs

APPLICATIONS OF THE MONGE-AMPERE OPERATORS TO THE DIRICHLET PROBLEM FOR QUASILINEAR ELLIPTIC EQUATIONS 239
by I. Bakelman

EXTREMAL KÄHLER METRICS 259
by E. Calabi

L^2 HARMONIC FORMS ON COMPLETE MANIFOLDS 291
by J. Dodziuk

L^2-COHOMOLOGY AND INTERSECTION HOMOLOGY OF SINGULAR ALGEBRAIC VARIETIES 303
by J. Cheeger, M. Goresky, and R. MacPherson

FUNCTION THEORY OF NONCOMPACT KÄHLER MANIFOLDS OF NONPOSITIVE CURVATURE 341
by R. Greene

KÄHLER MANIFOLDS WITH VANISHING FIRST CHERN CLASS 359
by M. L. Michelsohn

COMPACTIFICATION OF NEGATIVELY CURVED COMPLETE KÄHLER MANIFOLDS OF FINITE VOLUME 363
by Y.-T. Siu and S.-T. Yau

LOCAL ISOMETRIC EMBEDDINGS 381
by H. Jacobowitz

YANG-MILLS THEORY: ITS PHYSICAL ORIGINS AND DIFFERENTIAL GEOMETRIC ASPECTS 395
by J. P. Bourguignon and H. B. Lawson, Jr.

SYMMETRY AND ISOLATED SINGULARITIES OF CONFORMALLY FLAT METRICS AND OF SOLUTIONS OF THE YANG-MILLS EQUATIONS 423
by B. Gidas

ON PARALLEL YANG-MILLS FIELDS 443
by Gu Chaohao

VARIATIONAL PROBLEMS FOR GAUGE FIELDS 455
by K. Uhlenbeck

SOME GEOMETRICAL ASPECTS OF INTEGRABLE NONLINEAR EVOLUTION EQUATIONS 465
by H.-H. Chen and Y.-C. Lee

THE CONFORMALLY INVARIANT LAPLACIAN AND THE INSTANTON VANISHING THEOREM 483
by R. O. Wells, Jr.

CAUSALLY DISCONNECTING SETS, MAXIMAL GEODESICS AND GEODESIC INCOMPLETENESS FOR STRONGLY CAUSAL SPACE-TIMES 499
by J. Beem and P. Ehrlich

RENORMALIZATION 507
by A. Jaffe

METRICS WITH PRESCRIBED RICCI CURVATURE 525
by D. DeTurck

BLACK HOLE UNIQUENESS THEOREMS IN CLASSICAL AND QUANTUM GRAVITY 539
by A. Lapedes

GRAVITATIONAL INSTANTONS 603
by M. Perry

SOME UNSOLVED PROBLEMS IN CLASSICAL GENERAL RELATIVITY 631
by R. Penrose

PROBLEM SECTION 669
by S.-T. Yau

INTRODUCTION

In the academic year 1979-80, the Institute for Advanced Study and the National Science Foundation sponsored special activities in differential geometry, with particular emphasis on partial differential equations. In this volume, we collect all the papers which were presented in the seminars of that special program. Since there were many papers presented in the areas of closed geodesics and minimal surfaces, all the papers in these subjects have been collected in a separate volume. We would like to thank all the speakers for their enthusiastic participation and their cooperation in writing up their talks. We would also like to thank the National Science Foundation for supporting this special year.

SHING-TUNG YAU

Seminar on Differential Geometry

SURVEY ON PARTIAL DIFFERENTIAL EQUATIONS IN DIFFERENTIAL GEOMETRY

Shing-Tung Yau*

In these talks, we are going to survey some analytic methods in differential geometry. The basic tools will be partial differential equations while the basic motivation is to settle problems in geometry or subjects related to geometry such as topology and physics. We shall order our exposition according to the nonlinearity of the partial differential equations that are involved in the geometric problems. It should be emphasized that these equations are related to each other in an intriguing manner, the major reason being that all these equations serve the same purpose of understanding geometric phenomena.

It is obvious that nonlinear equations are more complicated than linear equations and coupled systems of equations are more complicated than scalar valued equations. However, we should bear in mind that the understanding of linear equations is of fundamental importance in understanding nonlinear equations.

(I) Scalar Equations.

(A) Linear equations.

The basic linear operator in differential geometry is the Laplace-Beltrami operator $\Delta = \frac{1}{\sqrt{g}} \sum_{i,j} \frac{\partial}{\partial x^i}\left(\sqrt{g}\, g^{ij} \frac{\partial}{\partial x^j}\right)$ where $\sum g_{ij}\, dx^i dx^j$ is the

*I would like to thank the typing staff at the Institute for Advanced Study for their usual excellent work, and Robert Bartnik for his assistance in compiling the bibliography.

Seminar on Differential Geometry
0-691-08268-5/82/000003-69$03.45/0 (cloth)
0-691-08296-0/82/000003-69$03.45/0 (paperback)

metric and $g = \det(g_{ij})$. Associated with this operator, we have the Laplace equation, the heat equation and the wave equation. All these linear operators are connected with the eigenfunctions. These are functions u so that $\Delta u = -\lambda u$ where λ is a constant. Besides these linear operators, we also have the linear operator associated with the bending of the surface. If we consider a motion of a surface in three space which preserves the metric up to the first order, the field of motion satisfies a linear equation. If the surface is a graph of some function, this equation can be interpreted as the linearized equation of the Monge-Ampère operator which will be discussed later. The linear equation arising in this way is rather complicated because it is of mixed type in general.

(B) Equations whose highest order term is linear.

The typical equation that appears has the form $\Delta u = F(x, u)$ where F is a given smooth function. When we deform a metric conformally, the equation has either the form $\Delta u = k_1 e^u + k_2$ or $\Delta u = k_1 u^p + k_2 u$ where p is a constant and k_1, k_2 are given functions.

(C) Quasilinear equations.

The most important quasilinear equation in geometry is the minimal surface equation which has the form

$$\sum_i \frac{\partial}{\partial x^i}\left(\frac{\frac{\partial u}{\partial x^i}}{\sqrt{1+|\nabla u|^2}}\right) = 0 .$$

Notice that the coefficients of the highest order term involve the first derivatives of the unknown. This is what happens for quasilinear equations in general.

(D) The Monge Ampère equation.

This equation is nonlinear even in the highest order term. It has the form $\det\left(\frac{\partial^2 u}{\partial x^i \partial x^j}\right) = F$ for u defined on a domain in R^n. If one studies complex analysis, one will study the equation $\det\left(\frac{\partial^2 u}{\partial z^i \partial \bar{z}^j}\right) = F$. These equations are closely related to the study of the curvature of a manifold.

(II) Systems.

(A) Linear systems.

The most important linear systems are the systems of harmonic forms, the Dirac equation, and the $\bar{\partial}$-equation. The last is overdetermined when the dimension of the complex space is greater than one. It makes the system more rigid. The study of these systems is related to harmonic theory.

(B) Linear systems whose highest order term is linear.

The typical system is the system of harmonic maps between Riemannian manifolds. The other system is the Yang-Mills equation when we choose a suitable gauge.

(C) Quasilinear systems.

As in the scalar case, the most important quasilinear system is the system corresponding to minimal submanifolds.

(D) Systems associated to the isometric immersion of a Riemannian manifold into another manifold.

This is an underdetermined system and the most celebrated work was done by Nash [N1].

(E) Systems associated with a prescribed curvature tensor.

This may be considered as a generalization of the Monge-Ampère equation to systems. The most important system is the Einstein Field equation. The question is that given a tensor on the manifold, how do we find a metric on the manifold so that some part of the curvature tensor is the given tensor? In the case of the Einstein Field equation, we are given the energy-stress tensor and we are asked to find a metric whose normalized

Ricci tensor is this energy-stress tensor. If we are looking in the category of Lorentz metrics, the system is hyperbolic.

§1. *The isoperimetric, Poincaré and Sobolev inequalities*

We start with the most basic inequalities in analysis. These are the Poincaré and Sobolev inequalities. The Poincaré inequality can be derived from the Sobolev inequality while the Sobolev and isoperimetric inequalities are essentially equivalent.

The Poincaré inequality states that for any compact manifold M with boundary ∂M, there exists a constant $c > 0$ such that for any smooth function f which vanishes on ∂M,

$$c \int_M f^2 \le \int_M |\nabla f|^2 . \tag{1.1}$$

The Sobolev inequality states that there exists a constant $c' > 0$ such that for any smooth function f which vanishes on ∂M,

$$c' \left(\int_M f^{\frac{n}{n-1}} \right)^{\frac{n-1}{n}} \le \int_M |\nabla f| . \tag{1.2}$$

Here n is the dimension of M.

These inequalities are for functions satisfying Dirichlet boundary conditions. If we assume $\int_M f = 0$ instead of $f = 0$ on ∂M, then the inequalities (1.1) and (1.2) are still valid with different constants c and c' (which are independent of f). The condition $\int_M f = 0$ is usually called the Neumann condition.

It should be noted that the inequality (1.2) implies that for all $n > p > 1$:

$$\frac{(n-p)c'}{p(n-1)} \left(\int_M |f|^{\frac{np}{n-p}} \right)^{\frac{n-p}{np}} \le \left(\int_M |\nabla f|^p \right)^{\frac{1}{p}} . \tag{1.3}$$

This is a simple application of the Hölder inequality.

The Poincaré inequality and the Sobolev inequality are basic tools in the theory of partial differential equations. It turns out that the largest positive constant c in the Poincaré inequality is the smallest eigenvalue for the Laplacian acting on functions satisfying the Dirichlet boundary condition or the Neumann condition depending on the assumption $f = 0$ on ∂M or $\int_M f = 0$. This is a consequence of the mini-max principle (see Courant-Hilbert [CH] Vol. I, page 399).

In fact, let H be the Hilbert space of functions f on M so that $\int_M |\nabla f|^2 < \infty$ and $f = 0$ on ∂M in case we are dealing with the Dirichlet boundary condition and $\frac{\partial f}{\partial \nu} = 0$ on ∂M in case we are dealing with the Neumann condition. Then the spectral theory says that we can find a countable orthonormal basis of H consisting of eigenfunctions f_i with

$$(1.4) \qquad \Delta f_i = -\lambda_i f_i$$

and

$$0 < \lambda_1 \leq \lambda_2 \leq \cdots .$$

The mini-max principle asserts that

$$(1.5) \qquad \lambda_1 = \inf \left\{ \left(\int_M |\nabla f|^2 \right) \left(\int_M f^2 \right)^{-1} : f \in H \right\}$$

$$\lambda_i = \inf \left\{ \left(\int_M |\nabla f|^2 \right) \left(\int_M f^2 \right)^{-1} : f \in H \right.$$

and

$$\left. \int_M f f_j = 0 \quad \text{for} \quad j = 1, \cdots, i-1 \right\} .$$

This characterization of the eigenvalues of the Laplacian has been the basic tool to estimate the eigenvalues. It is especially effective for the first few eigenvalues.

It is also well known (see [Ch2]) that λ_1 is closely related to some constants arising in the isoperimetric inequality. Let

$$h_D(M) = \inf\left\{\frac{\mathrm{Vol}(\partial\Omega)}{\mathrm{Vol}(\Omega)} : \Omega \text{ is a compact subdomain of } M\right\} \tag{1.6}$$

and

$$h_N(M) = \inf\left\{\frac{\mathrm{Vol}(H)}{\min(\mathrm{Vol}(M_1), \mathrm{Vol}(M_2))} : H \text{ is a hypersurface of } M \text{ which decomposes } M \text{ into } M_1 \text{ and } M_2\right\}. \tag{1.7}$$

Then by studying the level set of the first eigenfunction [Ch1], one can prove that

$$\lambda_1 \geq \frac{1}{4} h_D(M)^2 \tag{1.8}$$

for the first eigenvalue of the Dirichlet problem and

$$\lambda_1 \geq \frac{1}{4} h_N(M)^2 \tag{1.9}$$

for the first eigenvalue of the Neumann problem.

In this explicit form, (1.8) and (1.9) are due to Cheeger [Ch1]. A formula of this sort is still lacking for the Laplacian acting on differential forms. Such a formula will give a better understanding of nonlinear elliptic systems on a Riemannian manifold.

Inequalities (1.8) and (1.9) were used in [Y2] to give a lower estimate of λ_1 in terms of some more precise geometric data of M. C. Croke [Cr] was able to push [Y2] further by using an idea of Berger and Kazdan [Bes]. He was able to estimate the Sobolev constant for a compact Riemannian manifold by the same geometric data as in [Y2].

The Sobolev inequality is equivalent to the isoperimetric inequality. In fact, if Ω is a compact subdomain in M and if we choose f in (1.2) to be a function which approximates the characteristic function of Ω, then taking the limit one obtains

$$(1.10) \qquad c'(\mathrm{Vol}\,\Omega)^{\frac{n-1}{n}} \le \mathrm{Vol}(\partial\Omega)$$

which is the isoperimetric inequality for domains in M.

One can prove (see [FF]) that if (1.10) holds for all compact subdomains of M, then (1.2) holds. Hence a demonstration of (1.2) is reduced to a demonstration of (1.10).

To state the isoperimetric inequality of Croke, we define a constant $\omega(\Omega)$. For each point $x \in \Omega$, let ω_x be the Lebesgue measure of the set of all unit vectors $v \in T_x(\Omega)$ such that the geodesic issuing from x with tangent v intersects $\partial\Omega$ and the geodesic segment from x to its first point of intersection with $\partial\Omega$ has minimal distance. The quantity $\omega(\Omega)$ is defined to be $\min_{x\in\Omega} \omega_x$.

In [Y2], we estimate $h(M)$ in terms of $\omega(M)$ and the diameter of M. Croke then proved

$$(1.11) \qquad \mathrm{Vol}(\partial\Omega)^n \ge c_n\,\omega(\Omega)^{n+1}\,\mathrm{Vol}(\Omega)^{n-1}$$

where the equality holds if and only if $\omega(\Omega) = 1$ and Ω is isometric to a hemisphere of constant curvature.

If Ω is a subdomain of a simply connected complete manifold without conjugate points, then $\omega(\Omega) = 1$. If Ω is a compact subdomain of a general complete manifold M, then $\omega(\Omega)$ can be estimated as follows:

Suppose Ω is a subset of some geodesic ball $B(r)$ of radius r. For each $x \in \Omega$, we can consider the exponential map $\exp_x$ at x and its Jacobian $\sqrt{g}(x,y)$ at $y \in T_x(\Omega)$. As is well known ([BC]), an upper bound of $\sqrt{g}$ can be estimated in terms of the lower bound of the Ricci curvature of M. Using this function, we can estimate $\omega(\Omega)$ as follows:

$$(1.11) \quad \omega(\Omega) \ge \mathrm{Vol}(B(kr)-B(r))\left(\sup_{x\in\Omega}\int_0^{(k+1)r} \sup_{\overline{x,y}=t} \sqrt{g}(x,y)\,t^{n-1}\,dt\right)^{-1}$$

where $\overline{x,y}$ denotes the distance between x and y and k is any number greater than one.

The proof of (1.11) is rather easy. One forms a cone by joining every point in $B(kr)-B(r)$ to $x \in \Omega$ by a shortest geodesic. These geodesics must intersect $\partial\Omega$. Hence ω_x must be not less than the solid angle formed by the cone. By computing the volume of this cone in terms of $\sqrt{g}$, we obtain the formula (1.11).

If we have information about the Ricci curvature of M, we can estimate $\sqrt{g}$ as follows: Let $(n-1)K$ be the lower bound of the Ricci curvature of M where $K \leq 0$. Then

$$\sqrt{g}(x,y) \leq \left(\frac{\sinh\sqrt{-K}\,r}{\sqrt{-K}\,r}\right)^{n-1} . \tag{1.12}$$

where $r = \overline{x,y}$.

In particular, if the Ricci curvature of M is nonnegative, then $\sqrt{g}(x,y) \leq 1$. Hence it follows easily from (1.11), and (1.12) that *for a complete manifold with nonnegative Ricci curvature such that* $\lim_{r\to\infty} \mathrm{Vol}(B(r))r^{-n} > 0$, *the isoperimetic inequality holds for all compact subdomains* Ω *of* M *where the constant is independent of* Ω. *As was pointed out before, this means that the Sobolev inequality holds for smooth functions with compact support for this class of manifolds.*

The Sobolev inequality for functions with compact support is, of course, very important. However, in applications, it is also very important to prove a Poincaré inequality or Sobolev inequality for functions without compact support. This type of inequality is much more subtle and is much more sensitive to the boundary of the domain under consideration. We mention an inequality of this type in the following.

Let $B(r)$ be the geodesic ball with radius r, and with fixed centre. Let $\beta > 0$ be chosen so that

$$\mathrm{Vol}((1-\beta)r) = \frac{3}{4}\,\mathrm{Vol}(B(r)) . \tag{1.13}$$

Then using the method of [Y2], one can prove that for $p \geq 1$,

$$\inf_a \int_{B((1-\beta)r)} |f-a|^p \le cr^p \int_{B(r)} |\nabla f|^p \tag{1.14}$$

where c is a constant depending on p and

$$\operatorname{Vol}(B(r)) \left[\sup_{x \in B((1-\beta)r)} \int_0^r \left(\sup_{\overline{x,y} \le \beta t} \sqrt{g}(x,y) \right) t^{n-1}\, dt \right]^{-1} .$$

As was mentioned earlier, this last constant can be estimated in terms of the lower bound of the Ricci curvature of M and the lower bound of the volume of B(r). In particular, *if the Ricci curvature of* M *is nonnegative and if* $\liminf \operatorname{Vol}(B(r))\, r^{-n} > 0$, (1.14) *holds with* c *and* β *independent of* r.

Up to now, we have mainly surveyed results which apply for a general Riemannian manifold. For special classes of manifolds, more precise information can be obtained. For example, a Sobolev inequality for functions with compact support is known for minimal submanifolds of Euclidean space due to the works of Federer-Fleming [FF], Bombieri-De Giorgi-Miranda [BDM], Miranda [MM], Allard [Ald], and Michael-Simon [MS]. In the last two works, a suitable form of the inequality was also proved for submanifolds with bounded mean curvature. Since it is possible that complete manifolds with bounded curvature can be embedded into Euclidean space with bounded mean curvature, there may be a link between the two different approaches. Finally, it should be mentioned that the best constant in the Sobolev inequality for minimal submanifolds is not known. For the literature see [O2].

§2. *Harmonic functions and eigenfunctions on a complete Riemannian manifold*

One of the most interesting linear equations in geometry is the eigenfunction equation:

$$\Delta u = -\lambda u \tag{2.1}$$

where λ is a constant.

For reasons from Physics, people also study equations similar to (2.1) where a potential function V is added:

$$\Delta u + Vu = -\lambda u . \tag{2.2}$$

As was mentioned in Section 1, for compact manifolds M with boundary ∂M, we impose either the Dirichlet condition or the Neumann condition. We begin by studying the theory of harmonic functions, i.e., we assume $\lambda = 0$ in (2.1).

One of the most outstanding global theorems in the theory of harmonic functions is the Liouville theorem which states that the only positive harmonic function defined on the entire space R^n is constant. The generalization of this theorem to other Riemannian manifolds is very interesting. Besides its own beauty, the proof usually requires sharp estimates. These estimates give deeper understanding of the Laplacian and hence provide broad applications to problems in geometry.

The basic problem that we would like to discuss is to give a geometric condition on a complete manifold M so that the Liouville theorem holds. In case of two-dimensional surfaces, we have the uniformization theorem that every non-compact surface is either conformally covered by $\mathbb{C}$ or the unit disk. Since the property of harmonicity is invariant under conformal change, the study of the Liouville theorem is relatively easy. In fact, it was a theorem of Blanc-Fiala-Huber [Hu] that every complete surface with nonnegative curvature admits no nonconstant positive harmonic function. On the other hand, it follows from a theorem of Ahlfors [Ah] that a simply connected complete surface with curvature less than a negative constant is conformally the unit disk. In particular, these manifolds admit a lot of nontrivial bounded harmonic functions.

A natural generalization of the above two-dimensional results is the following. If M is a complete manifold with nonnegative Ricci curvature, then M admits no nonconstant positive harmonic function. If M is complete simply connected with sectional curvature less than a negative constant, then M admits a nonconstant bounded harmonic function. While the last statement is still not known, the first statement was solved in [Y1]. We shall describe various methods of approaching this problem.

Classically the Liouville theorem is proved by mean-value theorems. The uniformization theorem for surfaces allows this type of method to be generalized to curved surfaces. However, for higher-dimensional curved manifolds, these methods do not work and we need different arguments. For a metric defined on R^n which is uniformly equivalent to the Euclidean metric, one can apply the regularity theory of second order elliptic operators as developed by Bernstein, Leray, Morrey, Schauder, Nirenberg, Ladyzhenskaya, etc. However, except for two dimensions, the classical theory assumes the regularity of the coefficients and the method is too restrictive to generalize to curved manifolds. It was not until De Giorgi [DeG], Nash [N2], Moser [Ms1] developed a theory which assumes no regularity of the coefficients, that these methods were used extensively in nonlinear analysis. (See also Ladyzhenskaya [LU] and Morrey [M2] for generalizations.)

Bombieri and Giusti [BGi] found that by improving Moser's method, the proof of the Harnack inequality depends only on the validity of the Sobolev inequality for functions with compact support and the Poincaré inequality for functions satisfying the Neumann condition. In particular, for this class of manifolds, Liouville's theorem, which follows from Harnack's inequality, is valid.

To illustrate an application of the Liouville theorem in this form, Bombieri and Giusti proved that for a properly embedded hypersurface which is area-minimizing in Euclidean space, the above two inequalities hold and hence Liouville's theorem holds. On the other hand, it is well known that the coordinate functions of Euclidean space restricted to the hypersurface are harmonic. This proves that an area-minimizing properly embedded hypersurface cannot be a subset of a half space.

Let us now examine this method in connection with the problem of proving Liouville's theorem for complete manifolds with nonnegative Ricci curvature. Unfortunately, the Sobolev inequality is not valid for such manifolds. This is easily illustrated by the cylinder because the Sobolev inequality implies the isoperimetric inequality which in turn implies that the volume of geodesic balls of radius r grows like r^n where n is the dimension of the manifold. Conversely, if we assume that the volume of the geodesic balls of a manifold with nonnegative Ricci curvature grows like r^n, then the Sobolev inequality does hold. In fact, Section 1 also shows that the Poincaré inequality for functions satisfying the Neumann condition also holds. Therefore, Liouville's theorem holds for complete manifolds with nonnegative Ricci curvature whose geodesic balls have volume growth like r^n. The question is then whether one can drop the latter condition. It was proved by Calabi [Ca5] and the author [Y3] that the volume of the geodesic ball must grow at least like r. Therefore, it seems that suitable modification of the arguments of De Giorgi-Nash-Moser, as modified by Bombieri-Giusti, might lead to a proof of the Liouville theorem without assumption on the growth of the manifold.

In any case, many years ago the author [Y1] was able to devise a maximum principle to settle the question completely. The method was used successively to deal with other problems in geometry. If u is an harmonic function which is bounded from below by a constant a, then the theorem of [Y1] gives an estimate of $|\nabla u|^2(u-a)^{-2}$ in terms of the lower bound of the Ricci curvature of the manifold. This number turns out to be zero if the Ricci curvature is nonnegative and this of course implies that u is constant.

Recently R. Schoen pointed out that by modifying the arguments of [Y1] by the method of integration, one can simplify and improve the estimate in some cases. The result may be stated as follows.

(∗) Let $B(R)$ be a geodesic ball in a complete Riemannian manifold M with center x_0. Let u be a harmonic function defined on $B(R)$ which is bounded from below by a constant a. Let K be a nonnegative function so that $-K$ is the lower bound of the Ricci curvature of M. Then for any $0 < \beta < 1$ and $p \geq n$, there is an estimate of the following form:

$$(2.3)\quad \left[\int_{B((1-\beta)R)} \frac{|\nabla u|^{2p}}{(u-a)^{2p}}\right]^{1/p} \leq c\left(\int_{B(R)} K^p\right)^{1/p} + cp^{1/2}\left(\int_{B(R)} \left(\frac{\Delta u}{u-a}\right)^p\right)^{1/p}$$
$$+ c\beta^{-2}\cdot p\cdot R^{-2}\cdot [\mathrm{Vol}\, B(R)]^{1/p},$$

where c is a constant depending only on n.

If $\Delta u = 0$ and u is globally defined on M, then we let R and p tend to infinity together such that R/p is bounded. Then, since

$$\mathrm{Vol}\, B(R) \leq c_n R^n \exp(\sqrt{n-1}\, KR),$$

we obtain an estimate of $\sup \dfrac{|\nabla u|^2}{(u-a)^2}$ in terms of $\sup K$. In particular, if $K = 0$, u is constant and we have proved the Liouville theorem for complete manifolds with nonnegative Ricci curvature. The estimate of $\sup \dfrac{|\nabla u|^2}{(u-a)^2}$ was proved in [Y1] using a maximum principle. The estimate of $\sup\limits_{B(R)} \dfrac{|\nabla u|^2}{(u-a)^2}$ was carried out in [CY2].

Since the method is useful for many nonlinear problems, we sketch it as follows.

Let $F = |\nabla \log(u-a)|^2$. Then by the computations performed in [Y1], we have

$$(2.4)\quad \Delta F \geq \frac{1}{n-1} F^2 + \left(\frac{1}{n-1} - 2\right)\frac{\sum_i F_i u_i}{u-a} - \frac{2}{n-1}\left(\frac{\Delta u}{u-a}\right)^2$$
$$+ \frac{2\sum_i u_i(\Delta u)_i}{(u-a)^2} + \frac{2\sum_{i,j} R_{ij}u_iu_j}{(u-a)^2} - \frac{2|\nabla u|^2}{(u-a)^2}\left(\frac{\Delta u}{u-a}\right)$$

where $n = \dim M$ and R_{ij} is the Ricci tensor of M.

Let η be any Lipschitz function with compact support. Then we can multiply (2.4) by $\eta^{2p}F^{p-2}$ and integrate by parts. For $p \geq n$, we can choose a positive constant c_1 depending only on n so that

$$(2.5) \quad \int_M \eta^{2p}F^p \leq c_1 \int_M \eta^{2p}KF^{p-1} + c_1 p \int_M \eta^{2p}F^{p-2}\left(\frac{\Delta u}{u-a}\right)^2$$

$$+ c_1 \int_M \eta^{2p}F^{p-1}\left|\frac{\Delta u}{u-a}\right| + c_1 p \int_M \eta^{2p-2}F^{p-1}|\nabla\eta|^2$$

where $-K$ is a lower bound of the Ricci curvature. (We may assume $K \geq 0$.)

By applying Hölder's inequality, it follows that

$$\|F\eta^2\|_{L^p} \leq c_1\|K\eta^2\|_{L^p} + c_1 p\,\|\nabla\eta\|^2_{L^p} + c_1\sqrt{p}\,\left\|\left(\frac{\Delta u}{u-a}\right)\eta^2\right\|_{L^p}.$$

For any $0 < \beta < 1$, we choose η so that $\eta = 1$ for $r \leq (1-\beta)R$, $\eta = 0$ for $r \geq R$, and η is linear elsewhere. This gives the inequality (2.3).

In [Y1] we provided an estimate of $\sup \frac{|\nabla u|}{u-a}$ in terms of K, $\sup \frac{|\Delta u|}{u-a}$ and $\sup \frac{|\nabla(\Delta u)|}{u-a}$. This follows by noting that at the point where F achieves its maximum, the left-hand side of (2.2) is nonpositive and hence we can use (2.2) to estimate $\sup F$. This kind of estimate turns out to be useful in estimating the first eigenvalue of the Laplacian, as was demonstrated by P. Li in [Li]. To illustrate the idea, let us assume that M is a compact manifold without boundary. If $\Delta u = -\lambda_1 u$ and $a = \inf u - \delta$, then we have an estimate of $\frac{|\nabla u|}{u-a}$ in terms of K and δ. Let σ be the shortest geodesic joining a point where $u = 0$ and a point $u = \inf u$. Then we can integrate $\frac{|\nabla u|}{u-a}$ along σ and obtain an upper bound of $\log\left(1 + \frac{1}{\delta}\right)$ in terms of K, λ_1 and the length of σ. By suitably choosing δ, one obtains a lower estimate of λ_1.

In the original paper of P. Li, he assumed that the Ricci curvature cannot be too negative. This assumption was later removed in [LiY] where the best estimate for the lower bound of λ_1 for a general manifold was obtained. (When $\partial M = \emptyset$, Gromov [G] was able to find a similar bound but with a worse constant.) The results can be stated as follows.

For the Dirichlet problem, λ_1 can be estimated from below by an (explicit) positive constant depending only on the lower bound of the Ricci curvature of M, the upper bound of the diameter of the largest geodesic ball inscribed in M and the lower bound of the mean curvature of ∂M.

For the Neumann problem, λ_1 can be estimated from below by the lower bound of the Ricci curvature of M and the upper bound of the diameter of the largest geodesic ball inscribed in M if ∂M is convex or empty.

Qualitatively speaking, this result is rather satisfactory if one compares it with the following result of S. Y. Cheng [Cn1]. Cheng proved that for the Dirichlet problem or for the Neumann problem when $\partial M = \emptyset$, λ_1 is estimated from above by a constant depending only on the lower bound of the Ricci curvature and the diameter of the largest geodesic ball inscribed in M.

The above results can be best illustrated by inspecting a special case. When $\partial M = \emptyset$ and Ricci curvature is nonnegative, the above results say that

$$\frac{1}{4}\pi^2 \leq \lambda_1 d^2 \leq n\pi^2$$

where d is the diameter of M.

Historically speaking, Faber and Krahn [Fa], [Kr] were the first to obtain information about the lower estimates of the first eigenvalue of a compact domain in Euclidean space with Dirichlet condition. Their work was then followed by the extensive works of Polyà-Szegö [PS], Payne, Weinberger [PW], etc. However, the work of Lichnerowicz [Lc1] and Obata [Ob] appeared to be the first nontrivial work on estimating λ_1 on a compact curved manifold. They estimated λ_1 for compact manifolds whose Ricci curvature is bounded from below by a positive constant. Reilly [Re] later extended the results of Lichnerowicz and Obata by allowing the manifold to have convex boundary. One should also mention the works of Hayman [Hay], Osserman [O3], Taylor [Tay], Cheng [Cn3], and Protter [Pro] on improving the inequality of Faber-Krahn for Euclidean domains.

The mini-max principle provides an easy way to give an upper estimate of λ_1. However, a good upper estimate can be intricate. The above-mentioned comparison theorem of Cheng is of this nature. It provides the best estimate for a general compact manifold. The work of Cheng was motivated by the work of Cheeger [Ch1] who obtained a rough estimate for compact manifolds with nonnegative curvature.

For manifolds with special properties, the above-mentioned estimates can sometimes be improved. For example, R. Schoen [Sch] proved that for a compact three-dimensional manifold M with sectional curvature equal to -1, λ_1 is bounded from below by $c\,\mathrm{Vol}(M)^{-2}$ where c is a universal constant.

It seems that eigenvalues behave quite differently for two-dimensional surfaces. To mention a few results, Cheng [Cn1] proved a very beautiful theorem that for a compact surface of genus g, the multiplicity of the n-th eigenfunction is bounded by a constant depending only on n and g. (Cheng's estimate was improved later by Besson [Be] quantitatively.) A corollary of Cheng's theorem is that the multiplicity of the first eigenvalue of any metric over S^2 is not greater than three. Note that this bound is sharp and is achieved by the standard metric on S^2.

A natural generalization of Cheng's theorem to higher dimension would be to find an upper bound for the multiplicity in terms of n and the topology of the manifold. Cheng's proof cannot be generalized readily because it depends to a great extent on the analysis of the topology of the zero set of the eigenfunction. In dimension two, the latter set is a graph and the analysis can be carried out. In any case, Urakawa [Ur] demonstrated that the upper bound of the multiplicity of the first eigenfunction of the metrics on S^n is not achieved by the standard metric of S^n.

The other special feature of the eigenvalues of a two-dimensional surface can be described as follows. Adapting a method of Szegö [Sz] from a simply-connected noncompact surface to the two-dimensional sphere, Hersch [Her] proved that the first eigenvalue for any metric on S^2 is bounded by $8\pi/A$ where A is the area of this metric. A natural higher-dimensional analogue of this inequality turns out to be false [Ur]. For compact orientable surfaces of genus g, Yang and the author [YY] proved that the upper bound of λ_1 can be chosen to be $8\pi(g+1)/A$. At this moment, it is not known how to obtain a sharp upper estimate for $\lambda_1 A$ when $g \geq 1$. (When the surface is hyperelliptic, one can prove that $\lambda_1 A \leq 16\pi$.) Recently, P. Li and the author were able to obtain an upper estimate for $\lambda_1 A$ for nonorientable surfaces also.

Based on these results and a conjecture of Polya [Pol] for planar domains, it seems reasonable to believe that for a compact surface with genus g, $\lambda_n A(n+1)^{-1}$ has an upper bound depending only on g. It may be possible that for the sphere, the upper estimate of $\sup_n \lambda_n A(n+1)^{-1}$ is achieved by the standard metric of the sphere.

If we consider the Moduli space of genus g as the space of metrics with curvature -1 over a compact surface of genus g, then λ_i can be considered as a function defined on this space. A qualitative description of these functions can be found in the works of Schoen-Wolpert-Yau [SWY].

Finally, we would like to mention an interesting property of the eigenfunctions of a two-dimensional surface. Let ϕ_n be the n-th eigenfunction of the surface. Then the zero set of ϕ_n forms a graph with finite length L_n. It is not hard to show that $\underline{\lim}\, L_n(\sqrt{\lambda_n})^{-1}$ has a positive lower bound depending on the area of the surface. (This was independently proved by Brüning [Brü].) It also seems probable that $\overline{\lim}\, L_n(\sqrt{\lambda_n})^{-1}$ has an upper bound, but this appears to be a nontrivial problem.

§3. *The heat equation and the wave equation*

The heat equation is given by

$$\frac{\partial u}{\partial t} = \Delta u \,, \tag{3.1}$$

and the wave equation is given by

$$\frac{\partial^2 u}{\partial t^2} = \Delta u \,. \tag{3.2}$$

These two equations are very useful in studying the eigenvalues of Riemannian manifolds. In studying these equations, one imposes suitable boundary conditions if the manifold has boundary.

To illustrate the relation of these equations to the distribution of the eigenvalues of the Laplacian, we start by discussing the kernel function associated to the heat equation. Given any L^2 function ϕ on the compact manifold M, one can uniquely solve (3.1) for all $t > 0$ so that at $t = 0$, $u = \phi$. Hence one can define a semigroup $H(t)$ of operators from $L^2(M)$ into itself by mapping ϕ to u at time $t > 0$. These operators are integral operators given by kernel functions $K(x, y, t)$ defined on $M \times M \times R^+$. If $\{\phi_i\}$ form the complete set of normalized eigenfunctions with eigenvalues $\{\lambda_i\}$, then one can prove that (see [BGM])

$$K(x, y, t) = \sum_i e^{-\lambda_i t} \phi_i(x) \phi_i(y) \,. \tag{3.3}$$

Letting $x = y$ and integrating, one obtains the following important identity:

$$\int_M K(x, x, t)\, dx = \sum_i e^{-t\lambda_i} \,. \tag{3.4}$$

Hence if we know the "trace" of the heat operator, we can determine the sum $\sum_i e^{-t\lambda_i}$ which, as a function of t, is the Laplace transform of the measure defined by taking the sum of the point masses at $-\lambda_i$. If we know the behaviour of the Laplace transform when $t \to 0$, then Tauberian theorems will tell us the behaviour of λ_i when $i \to \infty$.

To compute the asymptotic behaviour of the left-hand side of (3.4), one constructs "explicitly" another kernel $H_k(x, y, t)$ so that $|K(x, x, t) - H_k(x, x, t)| = 0(t^k)$ when $t \to 0$. The asymptotic behaviour of $\int_M K(x,x,t)\,dx$ is then reduced to the study of the asymptotic behaviour of $\int_M H_k(x,x,t)\,dx$ by choosing k large enough. In this way, one proves that

$$(3.5) \qquad \sum_i e^{-\lambda_i t} = (4\pi t)^{-\frac{n}{2}}(a_0 + a_1 t + \cdots + a_k t^k + \cdots)\,,$$

where $a_0 = \mathrm{Vol}(M)$.

From (3.5) and the Tauberian theorems, one can then derive the following asymptotic formula of H. Weyl.

$$(3.6) \qquad \lim_{i\to\infty} \frac{\lambda_i}{i^{\frac{2}{n}}} = c_n \mathrm{Vol}(M)^{\frac{2}{n}}\,,$$

where c_n is a constant depending only on $\dim M$.

In connection with the formula (3.6), Polyà [Pol] made the remarkable conjecture that for a bounded domain with the Dirichlet condition,

$$(3.7) \qquad \lambda_i \geq c_n \mathrm{Vol}(M)^{\frac{2}{n}} i^{\frac{2}{n}}\,.$$

For the Neumann condition, one should reverse this inequality.

For a restrictive class of domains, Polyà showed the validity of (3.7). Lieb [Lb] has shown that (3.7) is true when c_n is replaced by a smaller constant.

Let us now indicate a method of Cheng-Li [CLi] to obtain a lower estimate of λ_i when M need not be a domain.

The uniqueness of the heat equation implies the following semigroup property:

$$(3.8) \qquad K(x, y, t) = \int_M K(x, z, s)\, K(z, y, t-s)\, dz$$

for $0 < s < t$.

Hence

$$K(x,x,2t) = \int_M K(x,z,t)^2\,dz\,. \tag{3.9}$$

On the other hand,

$$\begin{aligned} \frac{\partial}{\partial t}\int_M K(x,z,t)^2\,dz &= 2\int_M K(x,z,t)\Delta K(x,z,t)\,dz \\ &= -2\int_M |\nabla_z K(x,z,t)|^2\,dz \\ &\le -2c\left(\int_M |K(x,z,t)|^{\frac{2n}{n-2}}\,dz\right)^{\frac{n-2}{n}}\,, \end{aligned} \tag{3.10}$$

where c is the Sobolev constant of the manifold M.

Since $\int_M K(x,z,t)\,dz \le 1$ for all $t > 0$,

$$\left(\int_M |K(x,z,t)|^{\frac{2n}{n-2}}\right)^{\frac{n-2}{n}} \ge \left(\int_M K(x,z,t)^2\,dz\right)^{\frac{2+n}{n}}\,. \tag{3.11}$$

Noting that $\lim_{t\to 0} K(x,x,t) = \infty$, one derives from (3.9), (3.10) and (3.11) that

$$K(x,x,2t) \le \left(\tfrac{4}{n}\,ct\right)^{-\frac{n}{2}}\,. \tag{3.12}$$

Integrating (3.12) over M, one obtains

$$\sum_i e^{-\lambda_i t} \le \left(\tfrac{2c}{n}\,t\right)^{-\frac{n}{2}} \mathrm{Vol}(M)\,. \tag{3.13}$$

Putting $t = \frac{1}{\lambda_m}$ in (3.13), one finds

$$\lambda_m \ge (m\,\mathrm{Vol}(M)\,e^{-1})^{\frac{2}{n}}\left(\tfrac{2c}{n}\right)\,. \tag{3.14}$$

If M is a compact minimal submanifold (with boundary) in Euclidean space, and if we impose the Dirichlet boundary condition, then as was mentioned in Section 1, the Sobolev constant c can be chosen to depend only on n. The estimate of c for general manifolds was discussed in Section 1. In general, it depends on the geometry of the manifold. It should be mentioned that the above argument can be used to estimate eigenvalues of operators besides the Laplacian.

For a compact manifold without boundary, the number $\inf_i \lambda_i (i \operatorname{Vol}(M))^{-\frac{2}{n}}$ certainly depends on some other geometric invariants of M. Since eigenvalues for these manifolds behave like eigenvalues for domains with Neumann condition, it is more natural to look at $\sup_i \lambda_i (i \operatorname{Vol}(M))^{-\frac{2}{n}}$. When $\dim M > 2$, one cannot expect this number to depend only on n (see [Ur]). However, when $\dim M = 2$, this number may have an absolute upper bound. This is true at least for the number $\lambda_1 (\operatorname{Vol}(M))^{-1}$. (See Hersch [Her], Yang-Yau [YY], Li-Yau [LiY].)

The method of using the heat kernel to study eigenvalues is very old. The major works include Weyl [W], Carleman [Cl], Minakshisundaram [Mn], Kac [Ka], McKean-Singer [MKS], Berger [BGM], Patodi [Pt], Atiyah-Bott-Patodi [ABP], Gilkey [Gi], etc. The coefficients a_i in (3.5) were also studied extensively by these authors. They are invariants of the manifold M. The formula (3.5) says that the λ_i's determine these invariants a_i. For quite a long time, the computation of a_i has been the main tool with which to study the following important problem.

(†) Let M_1 and M_2 be two compact manifolds with the same set of eigenvalues counting multiplicities. (If ∂M_1 and ∂M_2 are both non-empty, one imposes the same boundary conditions.) Then are M_1 and M_2 isometric to each other?

The answer to this question turns out to be negative in general. (See Milnor [Mℓ1], Vignerás [V], Ikeda [Ik].) However, the counterexamples still exhibit enough similarities between M_1 and M_2, and it is believable that (†) is true in most cases. Hence one asks the following question.

(††) Let ds_t^2 be a one-parameter family of metrics on a compact manifold so that the spectrum of ds_t^2 is independent of t. Then are the metrics ds_t^2 isometric to each other?

Guillemin and Kazdhan [GK] have made major progress on this problem. They introduced the generic condition of having simple length spectrum, and proved (††) for such manifolds with curvature close to -1. For surfaces, they only need the curvature to be negative. Note that the above counterexamples are all highly non-generic.

Their method comes from study of asymptotic distribution of eigenvalues by using the wave equation. This type of method was initiated by Hörmander [Hö1]. Instead of studying the Laplace transform of the measure defined by the point mass located at λ_i, one studies the Fourier transform of the measure defined by the point mass located at $\sqrt{\lambda_i}$. This method was pursued by Colin de Verdière [CdV], Chazarain [Chz] and Duistermaat and Guillemin [DG]. They showed that for a generic Riemannian manifold, the spectrum of the Laplacian determines the lengths of the closed geodesics and their Morse indices modulo 4. This provides a lot of information about the manifold, especially when it has negative curvature. More recently, Berard [Ber] and Melrose [Mr] were able to use this method to obtain more precise asymptotic information on the eigenvalues. The best result is recently obtained by Ivrii [Iv] who estimated the second term in the asymptotic behavior of λ_i for compact manifolds with boundary.

Finally, it should be mentioned that very little is known about the spectrum of a complete noncompact manifold. When the manifold has finite volume and is locally symmetric, people have tried to use the method of Selberg [Se] to study the spectrum. It would be interesting to understand the discrete spectrum of a general manifold with finite volume.

§4. *The isometric deformation of surfaces in* R^3

Let $X_t : M \to R^3$ be a one-parameter family of immersions of M into R^3. Then $V = \frac{\partial X_t}{\partial t}\Big|_{t=0}$ is a deformation vector field defined on M. We want to study deformation vector fields which leave the metrics $|dX_t|^2$ invariant up to the first order at $t = 0$. The equation has the form

$$dX_0 \cdot dV = 0 . \tag{4.1}$$

The *rotation vector field* R associated to such a deformation is defined by

$$dV = R \times dX_0 . \tag{4.2}$$

From (4.2), one derives $dR \times dX_0 = 0$. Hence if $\{e_1, e_2\}$ form a local orthonormal frame field for X_0, and if

$$\begin{aligned} R_i &= e_i(R) \\ &= \sum_j a_{ij} e_j , \end{aligned} \tag{4.3}$$

then

$$a_{11} + a_{22} = 0 . \tag{4.4}$$

Differentiating (4.3), one derives the following equations:

$$\sum_j a_{ij} h_{jk} = \sum_j a_{kj} h_{ji} \tag{4.5}$$

$$a_{ij,k} = a_{kj,i} \tag{4.6}$$

where h_{ij} is the second fundamental form of the surface, and $a_{ij,k}$ denotes the covariant derivatives of a_{ij} in direction e_k.

The equations (4.5) and (4.6) are the fundamental equations for isometric deformation. The deformation is trivial iff $a_{ij} = 0$. When M is the graph of a function f defined over a domain in the (x, y) plane, the equations can be reduced to the single equation

$$f_{xx}\zeta_{yy} - 2f_{xy}\zeta_{xy} + f_{yy}\zeta_{xx} = 0 , \tag{4.7}$$

where ζ is the vertical component of the vector field V.

From (4.7), one concludes that the equation for infinitesimal isometric deformation is elliptic if the curvature is positive and hyperbolic if the curvature is negative. In general, the equation is a mixed type equation if the surface is flat somewhere. Infinitesimal isometric deformation is not well understood if the curvature is negative. The difficulty comes from the understanding of the characteristics of the equation.

In a manner similar to the above, one can derive differential equations for deformations which preserve the metrics up to higher order. It is not hard (see [E1]) to prove that if a surface has no isometric deformations up to first or second order, then it has no analytic isometric deformations except the Euclidean motions. A compact surface is said to be infinitesimal rigid of order n if there is no nonzero deformation vector field of order n which vanishes on the boundary of the surface (if the boundary exists). It is easy to see that a piece of the plane is infinitesimal rigid of order two but not infinitesimal rigid of order one.

The first major theorem proved in the theory of infinitesimal isometric deformation was due to Blaschke [Bl]. It says that a closed surface whose curvature is nonnegative and is not equal to zero on any region is infinitesimal rigid of order one. The proof is based on the following integral formula:

$$\int_M \det(a_{ij}) X_0 \cdot dX_0 \times dX_0 = \int_{\partial M} X_0 \cdot R \times dR \,, \tag{4.8}$$

which follows easily from (4.3) and Stokes' theorem.

On the other hand, equations (4.4) and (4.5) show that when (h_{ij}) is a definite quadratic form, then $\det(a_{ij}) < 0$ unless $a_{ij} = 0$ for all i, j. If M is a closed convex ovaloid, we can place the origin in the interior of M. Then the integrand in the left-hand side of (4.8) is nonpositive unless $a_{ij} = 0$ for all i, j. As $\partial M = \emptyset$, the right-hand side of (4.8) is zero, and Blaschke's theorem follows readily from this information.

The generalization of Blaschke's theorem to nonsmooth convex ovaloids was studied extensively by the Russian geometers including

Alexandrov [AZ], Efimov [E1], and Pogorelov [P5]. They have basically answered all the questions that one would like to know. However, the study of the rigidity of nonconvex surfaces is still in its infancy. There are the following important questions to be answered.

Do there exist closed surfaces which have nontrivial infinitesimal deformations of all orders? Cohn-Vossen [CV1] found surfaces of revolution which admit infinitesimal isometric deformations of order one and two.

It seems that among closed surfaces of a fixed genus, the rigid surfaces are "more" dense than those nonrigid surfaces. Can one make this statement more precise and demonstrate it? In the works of Cohn-Vossen [CV2], Rembs [Rm] and Lyukshin [Ly], surfaces of revolution were studied, and it was found that in very special cases the class of nonrigid surfaces is related to the Sturm-Liouville property of a second order ordinary differential operator. The question of density of nonrigid surfaces is not known even if we restrict our category of surfaces to surfaces of rotation.

Finally, we would like to point out that generalization of infinitesimal isometric deformations to hypersurfaces in R^n is possible. However, it would take more work to find the correct generalization for higher codimension.

§5. *Scalar equations which are linear in the highest order term*

A class of nonlinear equations which occurs frequently in geometry is obtained by adding a nonlinear first order term to a linear equation. The simplest example is the equation for the scalar curvature under a conformal change of the Riemannian metric. The equation for conformal change is somewhat different for two-dimensional manifolds and higher-dimensional manifolds.

For two-dimensional manifolds, if we multiply the metric ds^2 by a function $e^{2\rho}$, the curvature of the metric $e^{2\rho}ds^2$ is given by

$$-\Delta\rho + K = e^{2\rho}\tilde{K} . \tag{5.1}$$

In higher-dimensional manifolds, if we multiply the metric ds^2 by the positive function $u^{4/(n-2)}$, the scalar curvature of the deformed metric is given by

$$-\frac{4(n-1)}{(n-2)}\Delta u + uR = u^{\frac{n+2}{n-2}}\tilde{R}\,. \tag{5.2}$$

In the above two equations, if we consider K, $\tilde{K}$, R, $\tilde{R}$ as known functions defined on the manifold and ρ, u as unknown functions, we obtain quasilinear equations.

These equations have had long histories. In connection with the conformal structures on Riemann surfaces, the first equation was extensively studied from a different point of view. Poincaré, for example, made a great many contributions to this equation and constructed the Poincaré metric. In the last twenty years the equation was studied by Berger, Kazdan-Warner, Moser, Nirenberg, etc. We refer the reader to the article by Kazdan in these Proceedings. Perhaps the most interesting problem remaining to be solved is to find a condition on the function defined on S^2 so that it can be achieved as the curvature function of a metric conformally related to the standard metric of S^2. Kazdan-Warner found a necessary condition in [KW]. J. Moser [Ms2] made the best contribution in providing a sufficient condition that the function is invariant under the antipodal map of S^2.

For higher-dimensional manifolds, the equation was studied by Yamabe [Yam] who asserted that every metric is conformally equivalent to one with constant scalar curvature. After the death of Yamabe, a gap in [Yam] was found by Trudinger [T]. Trudinger pointed out that Yamabe's proof can still be pushed through to cover the case when the scalar curvature of the manifold is not too positive. (See the later works of Eliasson [El] and Aubin [Au].) Aubin pushed the argument further to prove that if the manifold is not conformally flat and has dimension greater than five, then Yamabe's assertion is true.

§6. *The minimal surface equation*

Perhaps the most important quasilinear equation in geometry is the minimal surface equation. Besides their own beauty, minimal surfaces provide a powerful tool for us to understand the topology and the geometry of the manifold. Since there is an excellent survey paper written by Simon [SL], we will just discuss the part of the theory which is more related to geometry.

The first problem in minimal surface theory is the Plateau problem, which can be described as follows. Let σ be a smooth Jordan curve in a manifold. Let $\mathcal{C}$ be a class of surfaces in the manifold whose boundary is given by σ. Then we want to know the possibility of finding a surface Σ in $\mathcal{C}$ so that Σ has least area among all the surfaces in $\mathcal{C}$. When the manifold is the Euclidean space or a "homogeneously regular" manifold (see [M1]) and $\mathcal{C}$ is the class of branched immersed disks, the problem was solved by Douglas-Rado [D], [R] and Morrey [M1]. Their solution was shown to have no branched point in the interior by Osserman [O1] and Gulliver [Gu]. It is not known whether boundary branched points exist or not. For the case when σ is real analytic, Gulliver-Lesley [GL] proved that they do not exist. If $\mathcal{C}$ is the class of surfaces with all possible genus, the Plateau problem was solved successfully by the methods of geometric measure theory. (See [HS] and the references therein.) In this latter case the minimal surface is automatically an embedded surface. The advantages of these two different categories are complementary to each other. The minimal surface constructed by Douglas-Rado-Morrey always has genus zero, while the minimal surface constructed by geometric measure theory is always embedded.

It was an important problem (see [MY1] to find a condition on σ to guarantee that the Douglas-Rado-Morrey solution is embedded. Osserman conjectured that this would be true if σ is in R^3 and lies on the boundary of its convex hull. The problem was first attacked by Gulliver-Spruck [GS] and then more successfully by Tomi-Tromba [TT] and Almgren-Simon [AS]. While these methods are of great interest, the final solution of Osserman's conjecture was produced by Meeks and the author [MY1] using a topological approach. We used the method provided by the topologists Papakyriakopoulos [Pp] and Whitehead-Shapiro [SW] in proving Dehn's lemma. Because of the nature of the minimal surfaces, the classical proof of Dehn's lemma can in fact be simplified and generalized.

Osserman's conjecture can be generalized to the following form. Let σ be a Jordan curve on the boundary of a compact three-dimensional manifold M. Suppose ∂M has nonnegative mean curvature with respect to the outward normal, and σ is contractible in M. Then σ bounds an embedded minimal disk in M, and every disk with least area bounded by σ is embedded. We may call this statement the geometric Dehn's lemma. As we shall see later, the geometric Dehn's lemma can be exploited to study the actions of finite groups on a three-dimensional manifold.

When the ambient manifold is compact, it is natural to hunt for minimal surfaces which have no boundary. In this regard, important progress was made by Sacks-Uhlenbeck [SU]. They proved that every compact manifold with finite fundamental group admits a branched minimal immersion of the two-dimensional sphere. More importantly, they proved that one can minimize area among homotopically nontrivial maps from S^2 into any compact manifold. When the manifold is a three-dimensional compact manifold, it was also demonstrated by Meeks-Yau [MY2] that this area-minimizing immersed sphere must be either embedded or a two-to-one covering of an embedded RP^2. The method here is similar to the one used in Dehn's lemma. This theorem can be considered as a geometric analogue of the sphere theorem [MY2] in the theory of three-dimensional manifolds.

The geometric sphere theorem turns out to be very useful in studying finite groups acting on a three-dimensional manifold. The reason can be explained as follows. Any compact group acting on a manifold can be realized as the group of isometries of some metric on the manifold. If g is an isometry and S^2 is a minimal sphere minimizing area, then $g(S^2)$

still minimizes area. The area-minimizing property of S^2 and $g(S^2)$ can be used to prove that either $S^2 = g(S^2)$ or $S^2 \cap g(S^2) = \emptyset$. This last property is purely a topological statement and can be used to split the group action. The point is that any three-dimensional manifold can be decomposed as the connected sum of "prime" manifolds, and the group action has to respect this decomposition. Essentially the splitting theorem for group actions says that the group is partly permuting these factors and partly acting on each separate factor. The factors are joined together by taking an equivariant connected sum. (For precise statements, see [MY2].) By using the theory of minimal surfaces, one can also produce the geometric loop theorem. Combining the geometric loop theorem, a theorem of Thurston, a theorem of Bass, and an observation of Gordon, one can solve the classical Smith conjecture (see [MY3] and the Proceedings of the Smith Conference at Columbia University).

Minimal surfaces also play an important role in studying the curvature of a manifold. In fact, Sacks-Uhlenbeck [SU] and Schoen-Yau [SY2] found that if Σ is a compact surface with genus ≥ 1 and M is any compact manifold such that for some continuous map f from Σ into M the induced map f_* on $\pi_1(\Sigma)$ is injective, then there is a conformal branched minimal immersion $g : \Sigma \to M$ so that $g_* = f_*$. This theorem was used by Schoen-Yau [SY3] to study the topology of manifolds which admit metrics with positive scalar curvature. We proved that for such manifolds the fundamental group does not contain a nontrivial subgroup isomorphic to the fundamental group of a surface with genus ≥ 1. The idea is to use the subgroup and the above theorem to produce a stable minimal surface of higher genus. By looking at the formula for stability carefully, we were able to prove such surfaces do not exist. This argument was later generalized to prove the positive mass theorem in general relativity [SY4].

The embedding problem for minimal surfaces of higher genus is more difficult. However, in Meeks-Simon-Yau [MSY], it was demonstrated that any embedded compact surface in a three-dimensional manifold is isotopic to a sum of embedded stable minimal surfaces. This was used to solve a classical problem that a three-dimensional manifold is prime iff its universal cover is prime.

Very recently, Uhlenbeck [U2] studied embedded minimal surfaces in a hyperbolic manifold. Part of her argument was extended by Freedman, Hass and Scott [FHS] to prove that if an embedded surface is incompressible, then the minimal surface that minimizes area in the homotopic class of this embedded surface is also embedded. (When the surface is a torus, a special case was established earlier by Schoen and Shalen [SS] in their classification of three-dimensional manifolds with nonpositive curvature.)

Two years ago, Fischer-Colbrie and Schoen [FS] and doCarmo-Peng [dCP] independently generalized the Berstein theorem to prove that the only complete stable minimal surface in R^3 is the plane. Actually the theorem of Fischer-Colbrie and Schoen goes beyond this. They obtained very strong theorems concerning stable minimal surfaces in a three-dimensional manifold with nonnegative scalar curvature or nonnegative Ricci curvature. For example, they proved that no complete stable minimal surface exists in a complete three-dimensional manifold with positive Ricci curvature. These theorems were used by Schoen-Yau [SY7] to find topological obstructions for complete three-dimensional manifolds with nonnegative scalar curvature. It is hopeful that this type of manifold can eventually be classified topologically. In any case, Schoen-Yau [SY7], using the theorem of Meeks-Simon-Yau and Fischer-Colbrie and Schoen, were able to prove that a complete three-dimensional manifold with positive Ricci curvature is diffeomorphic to R^3.

So far, all the minimal surfaces that we have discussed are stable. For unstable surfaces, both existence and embeddedness are much harder to understand. The first existence theorem for unstable surface was due to Morse-Tompkins-Shiffman ([MT], [Sh]), where they assumed the existence of two distinct strictly stable minimal disks bounding the Jordan curve in R^3. The best generalization of this theorem to compact manifolds is not known, although the assertion can be proved if the compact manifold admits no minimal sphere. Even when the three-dimensional manifold is the standard three sphere, our knowledge of closed minimal surfaces is not good yet.

Using an argument of Hopf, Almgren [Alm1] proved the only closed minimal surface of genus zero in S^3 is totally geodesic. Lawson [L] constructed examples of closed minimal surfaces in S^3 with arbitrary genus, except RP^2 which cannot be minimally immersed into S^3 [Alm1]. Lawson also demonstrated that embedded closed minimal surfaces in S^3 must be standardly embedded. It is interesting to understand the modulus space of minimal surfaces in S^3. In connection with this question, the following two questions seem to be important.

Which conformal structure over a closed surface can be conformally and minimally immersed into S^3?

Can one bound the area of a closed embedded minimal surface of a fixed genus in terms of the genus alone?

The last question has an affirmative answer if we can prove that the first eigenvalue of the surface is equal to two. This follows from the result of [YY].

Lawson also asked whether the standard Clifford torus is the only embedded minimal torus in S^3. It should also be noted that the only known complete embedded minimal surfaces with finite genus in R^3 are the plane, the heloid and the catenoid. Are these the only embedded minimal surfaces of finite genus? In higher dimension, one does not even have a good way to construct examples of complete immersed minimal hypersurfaces.

From geometric measure theory, one expects that an isolated singularity of a minimal hypersurface behaves like the cone over a closed minimal hypersurface in S^n. In this connection, a question of Chern is very important. It says that the only embedded minimal hypersphere in the sphere is totally geodesic. An affirmative answer to this question would be interesting even if we assume the cone is stable. In this special form, the question was answered by Simons [SJ] when the hypersurface has dimension less than six.

§7. *Equations which are nonlinear in the highest order term*

When one studies the Gauss-Kronecker curvature or the Ricci curvature of a Kähler manifold, the real Monge-Ampère equations or the complex Monge-Ampère equations appear naturally. By definition, the real Monge-Ampère equation has the following form:

$$\det\left(\frac{\partial^2 u}{\partial x_i \partial x_j} + f_{ij}\right) = F \tag{7.1}$$

where f_{ij} and F are given functions of x, u and ∇u.

The complex Monge-Ampère equation has the same form except that u is a function of complex variables $z_1, \cdots, z_n$, $\bar{z}_1, \cdots, \bar{z}_n$, and $\frac{\partial^2 u}{\partial x_i \partial x_j}$ is replaced by $\frac{\partial^2 u}{\partial z_i \partial \bar{z}_j}$.

For many problems in geometry, $f_{ij} = 0$, and we shall restrict our attention to this case. The Monge-Ampère equation was studied extensively when the equation is elliptic. For the real Monge-Ampère, this means that u is strictly convex. For the complex Monge-Ampère equation, this means that u is strictly plurisubharmonic. We shall discuss the real Monge-Ampère equation first.

The real Monge-Ampère equation of two variables is closely related to the Minkowski problem and the Weyl embedding problem. In these problems, one tries to construct closed surfaces with positive curvature in R^3. The positivity of the curvature is used to guarantee the ellipticity of the Monge-Ampère equation associated to the problem. The theory of generalized solutions for the Monge-Ampère equation was developed by A. D. Alexandroff [AZ] and his followers. However, the difficult task of constructing classical solutions (or proving the generalized solution to be smooth) began only in the nineteen thirties when Lewy [Lw] studied the elliptic Monge-Ampère equation of two variables. He assumed the coefficients of this equation to be real analytic and proved the solution to be real analytic also. Around the year 1952, Nirenberg [NL] and Pogorelov [P1] independently solved the elliptic Monge-Ampère equation of two variables with smooth coefficients. They also solved the Minkowski problem and the Weyl's embedding problem at the same time. While Nirenberg proved a smooth solution to the Monge-Ampère equation directly, Pogorelov proved that the generalized solution of Alexandrov is smooth.

Even in two variables, the knowledge of the nonelliptic Monge-Ampère equation is much less complete. For example, one does not even know whether a local solution of a general Monge-Ampère equation of two variables exists or not. (It exists if the equation is real analytic or if the equation is elliptic or hyperbolic.) This problem is closely related to the problem of whether one can locally embed an abstract surface into R^3 isometrically or not. The problem is unsolved even if we assume the right-hand side of (7.1) is nonnegative. On the other hand, the hyperbolic Monge-Ampère [E3] equation was studied quite extensively by the Russian geometers. The best result in this direction was due to Efimov [E2] who generalized the result of Hilbert [H] and proved that no complete surface with curvature ≤ -1 exists in R^3. One should note that Heinz [Hnz] studied the same theorem when the surface is a graph over a disk. He gave a quantitative statement; i.e., he estimated the curvature in terms of the radius of the disk. It would be interesting to find a quantitative statement for Efimov's theorem when the surface is not a graph.

The generalization of the theory of the two-dimensional Monge-Ampère equation to higher dimension is not trivial. The major difficulty comes from the lack of an *a priori* estimate for the third derivative of the solution.

The first progress was made when Calabi [Cal] was trying to generalize the theorem of Jörgens [JK] to higher dimension. The theorem of Jörgens says that any convex solution u of the equation

$$\frac{\partial^2 u}{\partial x^2}\frac{\partial^2 u}{\partial y^2} - \left(\frac{\partial^2 u}{\partial x\, \partial y}\right)^2 = 1$$

defined on R^2 is a convex quadratic polynomial.

Calabi studied global convex solutions of the equation

$$\det\left(\frac{\partial^2 u}{\partial x_i\, \partial x_j}\right) = 1\ . \tag{7.2}$$

He considered the geometric meaning of such equations. It turns out that the graph of these equations defines a *parabolic affine sphere*. In affine geometry, one is interested in the invariants of a submanifold that are preserved by the affine transformation group. One can canonically define the affine normal by moving the tangent space parallelly and tracing the center of gravity of the body cut by these parallel spaces. One calls a convex hypersurface an affine sphere if the affine normals meet at one point which is called the center of the affine sphere. When this point is located at infinity, one calls the affine sphere a parabolic affine sphere.

In the study of the Monge-Ampère equation, it is natural to introduce the concept of affine geometry because the Monge-Ampère equation is clearly invariant under the special linear group. For a convex hypersurface, one can introduce the concept of affine metric. When the hypersurface is defined by the graph of the solution u of (7.1), the metric is simply given by $\sum_{i,j} \frac{\partial^2 u}{\partial x^i \partial x^j} dx^i \otimes dx^j$. In affine geometry, one has the concept of Fubini-Pick form. It is a cubic form, and in our case it is simply $\sum \frac{\partial^3 u}{\partial x^i \partial x^j \partial x^k} dx^i \otimes dx^j \otimes dx^k$. This cubic form measures the deviation of a solution u of (7.2) from a quadratic polynomial. Following the familiar method of Bochner, one computes the (affine) Laplacian of the length of the cubic form $\sum u^{im}u^{jn}u^{kp}u_{ijk}u_{mnp}$ where (u^{ij}) denotes the inverse matrix of $\left(\frac{\partial^2 u}{\partial x^i \partial x^j}\right)$, and u_{ijk} is the abbreviation of $\frac{\partial^3 u}{\partial x^i \partial x^j \partial x^k}$. This computation was carried out by Calabi [Cal] in 1957 in his attempt to generalize Jörgen's result. The method of Jörgen depends on function theory of one complex variable and hence cannot be generalized to higher dimension. Using the computation mentioned above, Calabi [Cal] was able to prove the assertion up to dimension five. It was not until 1971 that Pogorelov [P1] gave the first proof which works for all dimensions. Pogorelov's proof depends heavily on the theory of convex bodies. Subsequently, Cheng and the author were able to give a simpler and more analytic proof along the lines of affine geometry. In any case, the proof of the generalization of Jörgen's theorem implies that any parabolic affine sphere which is complete in Euclidean topology is a paraboloid.

From the point of view of affine geometry, it is equally interesting to study elliptic and hyperbolic affine spheres. Using the method mentioned above, Calabi-Cheng-Nirenberg-Yau (see [CY1]) were able to solve some problems of Calabi. The first problem says that an elliptic affine sphere is an ellipsoid, while the second says that a hyperbolic affine sphere properly embedded in affine space is asymptotic to a unique convex cone, and conversely, for any nontrivial convex cone, there is a unique hyperbolic affine sphere asymptotic to the convex cone with a given mean curvature.

The existence of the hyperbolic affine sphere is equivalent to the solution of the following Monge-Ampère equation

$$\det\left(\frac{\partial^2 u}{\partial x^i \partial x^j}\right) = \left(-\frac{1}{u}\right)^{n+2}$$

where u is defined on a bounded convex domain with zero boundary data.

When u is a function of two variables, this equation was solved by Loewner and Nirenberg [LN]. They viewed the equation from a different point of view. For higher dimension, the solution depends on the theory developed by Pogorelov [P5] and Cheng-Yau [CY1]. In [CY3] and [P3], the solution depends on the solution of the higher-dimensional generalization of the Minkowski problem (see [P2] and [CY3]). This solution also depends crucially on the computation of Calabi of the length of the Fubini-Pick form mentioned above. Recently, Cheng-Yau [CY4] were able to solve the Monge-Ampère equation by a method avoiding the use of the theory of generalized solutions and the Minkowski problem. The method resembles the way that we constructed a complete Kähler Einstein metric on a bounded pseudoconvex domain.

The theory of the complex Monge-Ampère equation and its application to Kähler geometry is more lengthy; we hope to discuss it in a separate survey paper.

§8. *Harmonic forms and the Dirac equation*

The most classical and the most useful system of linear partial differential equations in Riemannian geometry is the system of Harmonic forms. The Hodge theory tells us that for a compact manifold, we can represent the deRham cohomology by harmonic forms. The harmonic form is unique in each cohomology class and hence enjoys many properties which relate the geometry and the topology of the manifold. Later when Kodaira-Spencer [KS] and then Calabi-Vesentini [Ca2], [CV] studied the deformation theory of complex structures and Weil [We] and other people generalized to the study of deformation theory of discrete groups in a Lie group, harmonic forms with coefficients in a bundle became important. When one combines the Hodge theory with the complex structure, one has the Hodge decomposition for complex manifolds which gives a lot of information for the cohomological structure of algebraic manifolds.

One of the major applications of harmonic forms comes from the Bochner method. A typical way is to compute the Laplacian of the length of the harmonic form and integrate the resulting formula over the manifold. Since the curvature appears in the formula, one can usually draw conclusions from the curvature of the manifold about the topology or the rigidity of the manifold. The first "vanishing" theorem was due to Bochner [Bo2] who proved that the first Betti number of a compact manifold with positive Ricci curvature is zero. (Actually, in this special case, this theorem can be derived from a theorem of Bonnet-Myers [Bn], [My].) Since the vanishing of cohomology groups means that one can prove the existence of solutions of certain systems under some compatibility conditions, the Bochner method is also useful in establishing theorems for the existence of differential equations.

The harmonic theory became very powerful after the appearance of the index theorem of Atiyah-Singer. The index of a linear elliptic operator A is defined to be the difference of the dimension of the kernel and the cokernel of A. It is not hard to show that the index of A depends only on the symbol of A. The index theorem tells us how to compute the index by topological means. Using the topological information, we can then gain knowledge about the existence or the nonexistence of solutions of a globally defined elliptic system. Conversely, if we know the existence or nonexistence of solutions of the elliptic system has topological meaning, we obtain identities involving topological quantities only. Hence the index theorem provides a link between geometry and topology.

In the last decade, a local index theorem was developed through the efforts of Atiyah, Bott, Patodi, Gilkey and Singer ([ABP], [Gi], [APS]). The basic idea can be described as follows. One looks at the semigroups e^{-tA^*A} and e^{-tAA^*} where A^* is the adjoint of A. As in Section 3, the asymptotic behaviour of the trace of these operators as t tends to zero can be computed by integrating some kernel functions whose asymptotic behaviour can be calculated by invariant theory. The invariant theory is based on some obvious scaling effects of the coefficients that appear in expanding the kernel functions in increasing powers of t. On the other hand, the trace of $e^{-tA^*A} - e^{-tAA^*}$ is equal to $\dim(\ker A) - \dim(\ker A^*) + \sum_i e^{-t\lambda_i} - \sum_{i'} e^{-t\lambda_{i'}}$ where the λ_i's and the $\lambda_{i'}$'s are the nonzero eigenvalues of A^*A and AA^* respectively. If ϕ is a nonzero eigenvector for A^*A, then $A\phi$ is also a nonzero eigenvector for AA^* with the same eigenvalue. When the eigenvalue is not zero, the map $\phi \to A\phi$ gives rise to an isomorphism between the eigenspaces of A^*A and AA^*. Therefore, $\lim_{t\to 0} \operatorname{tr}[\exp(-tA^*A) - \exp(-tAA^*)]$ is equal to the index of A. In this way, the index of A is found to be expressible in terms of the integral of some functions which can be computed, via the invariant theory, in terms of the symbols of A. For almost all the operators arising in geometry, one can express the integrand in terms of some combinations of the curvature tensor of the manifold and the bundles on which the operators act. They can also be considered as forms representing the Chern classes or the Pontryagin classes through the fundamental theorem of Chern [C1]. In this way, one achieves a "local" index formula.

There are three classes of "classical" operators that are commonly used as the index operator; the exterior derivative, the "$\bar{\partial}$" operator and the Dirac operator. If the structures involved here are also invariant under some compact group action, one gets corresponding formulae also. From the first operator, one obtains the Hirzebruch signature formula and the Lefschetz formula. From the second operator, one obtains the Riemann-Roch-Hirzebruch formula for complex manifolds. This formula has been the key source of information for non-Kähler complex manifolds. It plays a crucial role in Kodaira's classification of complex surfaces. The third operator gives the $\hat{A}$-genus of the manifold.

Since we shall discuss the $\bar{\partial}$ operator in the next section, we shall discuss the Dirac operator here. An oriented n-dimensional compact Riemannian manifold M is said to admit a spin structure if we can find a principal Spin(n) bundle P over M and an isomorphism of the oriented bundles $PX_{\mathrm{spin}(n)}R^n$ to the tangent bundle of M. It is well known [M12] that the necessary and the sufficient condition for the existence of a spin structure is the vanishing of the second Stiefel-Whitney class of M.

The group Spin(n) has a complex representation space Δ of dimension 2^n. The space Δ can be considered as the module for the Clifford algebra of R^n for the form $-\sum x_i^2$. Spin(n) can be considered as a subgroup of the group of the units of the Clifford algebra and the action of spin(n) is induced by the Clifford multiplication. If n is equal to an even integer 2ℓ, the representation Δ splits into two irreducible representations Δ^+ and Δ^- of dimension 2^{n-1}. The decomposition $\Delta = \Delta^+ \oplus \Delta^-$ is interchanged by Clifford multiplication.

Hence if M is a Spin(2ℓ) manifold with principal bundle P, we can form the associated complex vector bundles $E^+ = Px_{\mathrm{spin}(2\ell)}\Delta^+$ and $E^- = Px_{\mathrm{spin}(2\ell)}\Delta^-$. The Dirac operator is a first order differential operator mapping sections of E to sections of E. It can be defined as follows. Since the Riemannian connection defines a connection for P, we have the concept of covariant differentiation for sections of E. In terms of a local orthonormal frame field e_i of M, the Dirac operator D is defined by $Ds = \sum e_i(\nabla_{e_i} s)$ where s is a section of E and $\nabla_{e_i} s$ is the covariant derivative of s in the direction e_i. The Dirac operator is self-adjoint. However, because of the properties of Clifford multiplication, $D = D^+ + D^-$ where D^+ maps sections of E^+ to sections of E^- and D^- maps sections of E^- to sections of E^+. The operator D^+ is the formal adjoint of D^-. Its kernel is the subspace of the kernel of D consisting of the harmonic spinors. The Atiyah-Singer index formula shows that the index of D^+ can be expressed as a combination of Pontryagin numbers of M. It is called the $\hat{A}$-genus of M. (Note that it is zero if $\dim M \not\equiv 0 \bmod 4$.)

The first application of the Atiyah-Singer index theorem to differential geometry comes in the following way. If M is a compact spin manifold with positive scalar curvature, then Lichnerowicz [Lc2] has shown that no harmonic spinors exist. This follows from a Bochner-type argument. Hence the Atiyah-Singer index theorem shows that the $\hat{A}$-genus of such manifolds must be zero.

When the dimension of the manifold is not even, the representation for spin(n) is different. However, if $n = 8k+1$, one can define a skew-adjoint Dirac operator. The dimension of the kernel of such an operator modulo the even integers is also a topological invariant. The Atiyah-Singer index theorem has identified it as a KO-characteristic number. When $n = 8k+2$, one can define a similar skew-adjoint operator which gives a mod 2 KO-characteristic number.

It was observed later by Hitchin [Ht1] that the theorem of Lichnerowicz and the above mod 2 invariant can be used to prove that many exotic spheres do not admit a metric with positive scalar curvature.

The index theorem for the Dirac operator can also be used to study compact groups acting on manifolds. In fact, Atiyah-Hirzebruch [AHz] used the equivariant index theorem for the Dirac operator to prove that a compact 4k-dimensional spin manifold admits no effective circle group action unless the $\hat{A}$-genus of the manifold is equal to zero. Later, Lawson-Yau [LY] proved that if a compact manifold admits a smooth effective action of a compact connected nonabelian group, then this manifold also admits a metric with positive scalar curvature. This theorem gives a relation between the theory of group actions and the curvature of the manifold. In particular, it shows that if a compact manifold admits a smooth effective action of a compact connected nonabelian group, then all the KO-characteristic numbers mentioned above must be equal to zero. In particular, for exotic spheres which do not bound spin manifolds, the only possible connected group action is given by the torus. It is not clear how to obtain a sharp upper bound of the dimension of such a torus. Bredon [Br] has shown that the dimension is at least one for many cases. Incidentally, this means that we cannot replace nonabelian by abelian in the above-mentioned theorem.

§9. *The* $\bar{\partial}$ *operator*

The most important system of linear partial differential equations in complex geometry is the $\bar{\partial}$-equation. When the (complex) dimension of the manifold is greater than one, the system is overdetermined. However, there is the obvious compatibility condition which comes from $\bar{\partial} \circ \bar{\partial} = 0$. For the analytic aspects of the equation, we refer the reader to Hörmander [Hö2], Kohn-Folland [FK], Siu [S1] and the references mentioned there. We shall report on the more geometric aspect of the $\bar{\partial}$-equation.

The basic problem that we want to study is the geometric aspects of the uniformization problem. The classical uniformization theorem says that any one dimensional complex manifold is covered by S^2, the complex line $\mathbb{C}$ or the unit disk $\mathbb{D}$. There are several ways to generalize this theorem to higher dimensional complex manifolds. One generalization is based on the observation that each one-dimensional complex manifold admits a complete conformal metric with constant curvature. A natural generalization of this is to try to find a necessary and sufficient condition for a complex manifold to admit a complete Kähler metric whose Ricci form is proportional to the Kähler form. Such a metric is called a Kähler-Einstein metric and can be formulated in terms of the complex Monge-Ampère equation. This point of view will be reported elsewhere. Another generalization of the uniformization theorem depends on the observation that the sign of the curvature has important implications for the complex structure of the manifold. This point of view has a long history, but the emphasis due to Greene-Wu [GW2] has stirred up a lot of activities in the last decade.

In the following, we divide the study of the higher dimensional Kähler manifolds from this point of view into three cases.

a. Elliptic manifolds:

We want to study the complex structure of a compact Kähler manifold whose curvature is positive in a suitable sense. The most outstanding problem was the Frankel conjecture [Fr1] which says that every compact Kähler manifold with positive bisectional curvature is biholomorphic to CP^n. There were works by Andreotti-Frankel [Fr1], Kobayashi-Ochiai [KO], Iitaka [I], Mabuchi [Mb] and Hartshorne [Hts1] for Kähler manifolds with dimension not greater than three. The final solution was achieved by Mori [Mo] and Siu-Yau [SiuY2]. Mori's method is purely algebraic and uses methods of algebraic geometry of characteristic p. In fact, Mori proved the Hartshorne conjecture which is stronger than the Frankel conjecture, i.e., Mori has only to assume the tangent bundle of the Kähler manifold is positive. The method that Siu and the author used depends on minimal surface theory and can be described as follows.

For a compact Kähler manifold with positive bisectional curvature, it was proved by Bishop and Goldberg [BG] that the second Betti number is equal to one. Hartshorne had already observed that the key point in establishing the conjecture is to prove that the primitive generator of $H_2(M, Z)$ can be represented by a rational curve. (One can then compute the first Chern class $c_1(M)$ by restricting the tangent bundle of M to the rational curve. The positivity of the tangent bundle implies the positivity of the normal bundle and hence $c_1(M)$ is $n+1$ times the generator of $H^2(M, Z)$ where $n = \dim M$. By applying the Riemann-Roch theorem, one can then produce enough sections of the anticanonical bundle to map M onto CP^n.) In order to establish Harthorne's hypothesis, we represent the primitive generator by a map from S^2 into M. (Note that it is easy to prove M is simply connected and hence, the Hurewicz theorem shows that one can represent the second homology classes of M by spheres. We shall identify the second homology with second cohomology mod torsion.) By using the argument of Sacks-Uhlenbeck, we deform this sphere into sums of stable minimal spheres S_i. By applying the second variational formula to each stable sphere S_i, we prove that each S_i is either antiholomorphic or holomorphic.

We observe that the S_i's must be antiholomorphic or holomorphic in the same way. In fact, by restricting the (positive) tangent bundle to each of these holomorphic curves, we find that the normal bundle has enough sections and we can deform one curve to touch the other one. If the orientations of these curves are opposite to each other, we can decrease area of the sum of these and arrive at a contradiction. Hence all spheres are either holomorphic or antiholomorphic in the same way. Hence we have represented the primitive generator of $H_2(M, Z)$ by the sum of holomorphic curves or the sum of antiholomorphic curves. Hence there can be at most one summand and we have proved Hartshorne's hypothesis and hence the Frankel conjecture.

After the Frankel conjecture is proved, one may be interested in weakening the hypothesis on the curvature. For example, what kind of Kähler manifolds admit a complete Kähler metric with positive Ricci curvature? If the manifold is compact, can one prove that they are rationally connected, i.e., can every two points be joined by a chain of rational curves? Mori's argument produced at least one rational curve for this class of manifolds.

b. Parabolic manifolds:

If a complete Kähler manifold is contractible and if the bisectional curvature of the manifold is asymptotic to zero in a suitable rate, can one prove that M is biholomorphic to $\mathbb{C}^n$? If M is a complete noncompact Kähler manifold with positive bisectional curvature, is M biholomorphic to $\mathbb{C}^n$?

Greene-Wu [GW1] asked the weaker question that if M is a complete simply-connected Kähler manifold with nonpositive curvature which decays like $r^{-2-\varepsilon}$, then M is biholomorphic to $\mathbb{C}^n$. This conjecture of Greene-Wu is relatively easier than the above one because in this case, we know the geometry quite well and the metric on M is uniformly equivalent to the Euclidean metric and the analysis is easier. It was established by Siu and the author in [SiuY1] and can be described as follows.

We tried to produce holomorphic functions on M which are similar to linear functions on $\mathbb{C}^n$. It turns out that these functions are characterized by their growth rate. We first produce a non-vanishing holomorphic n-form by solving the $\bar{\partial}$-equation by the L^2-method. We find a solution u to the equation $\bar{\partial}u = f$ so that $\int_M |u|^2 e^{-\phi}$ is finite where ϕ is a certain plurisubharmonic function. Typically, f is obtained in the following manner. Let σ be a function with compact support in a complex coordinate chart so that $\sigma = 1$ in a neighborhood of a preassigned point x. Then we set $f = \bar{\partial}(\sigma dz^1 \wedge \cdots \wedge dz^n)$. If we solve the equation $\bar{\partial}u = f$, we will obtain a holomorphic n-form $(u-\sigma)\, dz^1 \wedge \cdots \wedge dz^n$. However, to guarantee that this holomorphic n-form is nontrivial, we must require that our plurisubharmonic function grows like $\frac{n}{2} \log \left(\sum_i |z_i|^2\right)$ near x. This condition ensures that u is zero at x. (Otherwise $\int_M |u|^2 e^{-\phi} = \infty$.) In this way we obtain a globally defined holomorphic n-form. By suitably choosing ϕ, we can control the growth of this n-form also. On the other hand, the growth of the n-form is related to the growth of the volume of its zero locus. Since the volume of a complex subvariety must grow suitably fast, we can prove that the n-form does not vanish if it does not grow fast. In this way we produce a nonvanishing holomorphic n-form for which we can find both an upper estimate and the lower estimate on its norm. By using a similar method, we produce holomorphic n-forms of nearly linear growth which have the form $z_i dz_1 \wedge \cdots \wedge dz_n$ near x. Dividing these holomorphic n-forms by the previous one, we obtain n holomorphic functions with linear growth. It turns out that they give a biholomorphic map of M onto $\mathbb{C}^n$.

It does not seem that the method outlined above can be readily generalized to manifolds whose geometry is not close to Euclidean geometry. For example, if the manifold is simply connected and is asymptotic to a cylinder near infinity, every nontrivial holomorphic function must grow at least exponentially. Since it is rather difficult to make use of holomorphic functions with exponential growth in our problem, one needs some more ideas besides the one mentioned above.

There is, however, a case which can be treated in a similar manner. If our complete Kähler manifold M has nonnegative bisectional curvature and if the volume of the geodesic balls in M grows at least as fast as r^{2n} where $n = \dim_{\mathbb{C}} M$, then it is reasonable to expect that M is biholomorphic to $\mathbb{C}^n$. The basic reason is the following. The hypothesis of the growth of the volume of the geodesic balls guarantees the validity of the Sobolev and the Poincare inequalities, thanks to the results of Section 1. These inequalities should enable one to find a function u which grows like r^2 and which satisfies the equation $\partial\bar{\partial}u = \omega$ where ω is the Kähler form of M. One should be able to make use of u and the methods of [SiuY1] to prove that M is biholomorphic to $\mathbb{C}^n$. This will be discussed in a joint paper with Mok and Siu [MkSY].

In that paper, we discuss the following results. We prove that if a complete Kähler manifold with a pole has curvature decay faster than $r^{-2-\varepsilon}$ and if the volume of the geodesic ball grows like r^{2n}, then it is biholomorphic to $\mathbb{C}^n$. We also prove that if the curvature has the same sign, then the manifold is in fact isometric to $\mathbb{C}^n$.

Let us now discuss the second question where we assume the bisectional curvature is nonnegative. It is likely that the question can be generalized to the following one: A complete Kähler manifold with nonnegative bisectional curvature and positive Ricci curvature is biholomorphic to the total space of a holomorphic fiber bundle whose base is a compact locally symmetric space and whose fiber is biholomorphic to $\mathbb{C}^n$. The motivation of this problem comes from the works of Cheeger-Gromoll-Meyer [CG], [GM] who showed that every complete noncompact manifold with nonnegative curvature is the total space of a vector bundle over a compact manifold with nonnegative curvature. If the manifold has positive curvature, then the manifold is in fact diffeomorphic to R^n. The method that Cheeger-Gromoll [CG] used is also useful in studying the above-mentioned problem. In fact, Greene and Wu [GW1] were able to use the construction of Cheeger-Gromoll to demonstrate that complete noncompact Kähler manifolds with positive curvature are Stein.

Besides this result, the only information comes from the Liouville theorem discussed in Section 2. Namely, we know that such a complete manifold does not admit any bounded holomorphic function. It should be noted that if we merely assume the bisectional curvature to be positive, we do not even know that the manifold is Stein. However, Siu and the author did use the method of [SiuY1] to prove that there are globally defined holomorphic functions on the manifold which separate points and define local coordinate systems.

If one wants to weaken the hypothesis on the curvature further, one may ask the following question. If a complete Kähler manifold has nonnegative Ricci curvature or if the Ricci curvature decays fast enough, can one prove that the manifold is a Zariski open set of some compact complex manifold?

c. Hyperbolic manifolds:

A complete simply connected Kähler manifold with bisectional curvature less than a negative constant is biholomorphic to a proper subvariety of a bounded domain in $\mathbb{C}^n$.

Very little is known about this problem. We do not even know whether such a manifold is noncompact or not. Thus we shall restrict our attention to treat a simpler problem first. Namely we replace "bisectional curvature" by "sectional curvature" in the above problem. In this case, we know the geometry of the manifold quite well. For example, the distance function provides a convex exhaustion of the manifold and using this exhaustion Wu [Wu] observed that the manifold is Stein. By the Cartan-Hadamard theorem, the manifold is diffeomorphic to $\mathbb{C}^n$ and we are tempted to believe that the manifold is a domain in $\mathbb{C}^n$. However, we do not know how to construct a nontrivial bounded holomorphic function on M. Hence even if we assume the manifold is a Stein domain, we cannot tell whether it is a bounded domain or not.

If one attempts to produce a bounded holomorphic function by classical methods, the first difficulty one encounters is that the volume element of the manifold is not necessary uniform at infinity when one views the infinity in every direction. Hence the problem may be more tractable if the manifold covers a compact Kähler manifold with negative curvature.

When the dimension of a compact Kähler manifold of negative curvature is greater than one, the complex structure tends to be rigid. In fact, one expects that there is only one complex structure over such a manifold. In [Y4], the author proved that any compact complex surface which is homotopic equivalent to an algebraic surface covered by the ball is in fact biholomorphic to such a surface. (The proof partly depends on the work of Mostow on the rigidity of rank one locally symmetric spaces with dimension greater than two.) Then Siu [S2] was able to demonstrate that harmonic maps from a compact Kähler manifold to another Kähler manifold with *strongly negative* curvature are either holomorphic or antiholomorphic if the rank of the map is not less than four. Since Eells and Sampson [ES] proved that any map from a compact manifold to a compact negatively curved manifold is homotopic to a harmonic map, this shows that there is only one Kähler structure over any compact strongly negatively curved Kähler manifold. The method of Siu will be discussed in the next section. Siu's theorem was generalized by Siu and the author [SiuY3] to cover the case when the strongly negatively curved manifold is complete noncompact but has finite volume and bounded curvature.

In connection with the last statement, Siu and the author proved that such a manifold can be compactified as a compact complex variety. It will be interesting to show that any real analytic complete Kähler manifold with curvature bounded between zero and a negative constant can be compactified complex analytically if its volume is finite. If the manifold is locally symmetric, this was done by Satake [St] and Borel-Baily [BB].

§10. *Harmonic maps*

The second class of quasilinear equation is the equation for harmonic mapping. It was first introduced and studied by Bochner [Bo1] as a generalization of minimal surfaces. They are interesting primarily because they tell us some geometric properties of mappings between Riemannian manifolds. They are canonical in a certain sense and therefore should reflect the geometry of the manifolds. They are very closely related to minimal submanifolds because if the harmonic mapping is given by an isometric immersion, then it is a minimal immersion. However, except in dimension two where the conformal structure plays a special role, the theory of harmonic maps is quite different from the theory of minimal submanifolds. Even in dimension two the theory is sufficiently different that we report on theorems more related to minimal surfaces in the other sections.

Let M and N be two compact Riemannian manifolds. Then the energy of a map $f = M \to N$ is defined to be $\frac{1}{2}\int_M |df|^2$. A harmonic map is a map which is a critical point of this functional. It satisfies the following equation

$$\Delta f^{\alpha} + \sum g^{ij}\Gamma^{\alpha}_{\beta\gamma}\frac{\partial f^{\beta}}{\partial x^{i}}\frac{\partial f^{\gamma}}{\partial x^{j}} = 0 \tag{9.1}$$

where g_{ij} is the metric tensor of M and $\Gamma^{\alpha}_{\beta\gamma}$ is the Christoffel symbol of the image manifold.

The equation that governs harmonicity of maps is perhaps the simplest nonlinear equation that one can pose on maps. As there is no global coordinate system in the image in general, one has to be careful in applying the classical methods. It was observed very early by Morrey (see [ES]) that if one does not impose any geometric restriction on the image manifold, a global minimum for the energy need not exist in a fixed homotopy class of maps between manifolds. (A few years ago, it was demonstrated by Eells and Wood [EW] that a smooth map need not be homotopic to any harmonic map (see also [EL]).) The first existence theorem was established by Eells and Sampson [ES]. It says that if N has nonpositive curvature, then every smooth map from M into N is homotopic to a harmonic map with minimal energy. Later Lemaire [Le] established the same existence theorem if M is a two-dimensional surface with genus ≥ 1 and $\pi_2(N) = 0$. (Later, in the course of proving existence theorems for minimal surfaces, Sacks-Uhlenbeck [SU] and Schoen-Yau [SY2] gave different proofs of this fact.) Most recently, Schoen and Uhlenbeck [SU] gave a rather satisfactory regularity theorem for energy minimizing maps. They proved the codimension of the singular set is not less than three. These theorems can also be established if M has boundary and if a suitable boundary condition is imposed. When N has nonpositive curvature, this boundary-value problem was solved by R. Hamilton [Ham].

Besides the existence, a very interesting question about harmonic maps is their qualitative behaviour. One potential use of the theory of harmonic maps is to study the geometry of negatively curved manifolds. In fact, when N has negative curvature, it was proved by Hartman [Htm] that there is only one harmonic map in each homotopy class of maps from M to N if the image of $\pi_1(M)$ in $\pi_1(N)$ is not a cyclic group. The uniqueness certainly indicates that the harmonic map must enjoy a lot of nice properties. If both M and N are negatively curved Einstein manifolds with dimension ≥ 3 and if the harmonic map is a homotopy equivalence, there may be a chance that the map is in fact an isometry. This will give a different proof of Mostow's rigidity theorem in the class of rank one symmetric spaces. It is also a conjecture that if both M and N are negatively curved compact manifolds which are homotopically equivalent to each other, then any harmonic map from M to N which is a homotopy equivalence is also a diffeomorphism. The only progress on this problem was due to Schoen-Yau [SY1] and Sampson [Sa]. They proved the assertion when $\dim M = \dim N = 2$. (The more difficult case when ∂N is not empty was also treated in [SY1]. Jost [JJ] also studied the case when the curvature may be positive.)

The most successful progress on the properties of harmonic maps is due to Siu [S2] who proved that a harmonic map f from M into N is either holomorphic or antiholomorphic if N is a strongly negatively curved Kähler manifold and if the rank of f is not less than four at some point. While the proof is ingenious, it can be outlined in a few lines. One considers $\partial\bar{\partial}(\partial f \wedge \overline{\bar{\partial} f}) \wedge \omega^{n-2}$ where n is the complex dimension of M and ω is the Kähler form of M. By applying the divergence theorem, the integral of this form is zero. On the other hand, by using the harmonicity of f and computing as in the Bochner type argument, the integral can be expressed in a different manner which involves the curvature tensor of N. When N is strongly negatively curved, the formula shows that f is either holomorphic or antiholomorphic.

Another potential application of harmonic maps is to exploit their uniqueness to study finite groups acting smoothly on the manifold. For manifolds with nonpositive curvature, this application was carried out by Schoen-Yau [SY5].

§11. *The Yang-Mills equations*

This is a subject of interest to both physicists and mathematicians. It became a popular subject to mathematicians after the works of Atiyah-Hitchin-Singer [AHS1], Atiyah-Hitchin-Drinfield-Manin [AHDM], etc.

In general, it studies the following variational problem. Let P be a principle bundle over a compact Riemannian manifold with compact structure group. Then for any connection A defined on P, we can consider its curvature Ω_A. The Yang-Mills functionals are defined by

$$\int_M \|\Omega_A\|^2 .$$

The critical points of this functional in the space of connections are called Yang-Mills fields. The equation of the Yang-Mills field depends how we choose a section of the Lie algebra valued bundle. The choice of such a section is called the choice of a gauge. In a suitable gauge, the Yang-Mills equations become a quasilinear elliptic system whose highest order term is linear.

The physicists are mostly interested in Yang-Mills field over R^4. However, Uhlenbeck [U1] proved the fundamental theorem that a L^2 Yang-Mills field on R^4 can be extended smoothly to a L^2 Yang-Mills field on S^4. (Note that the Yang-Mills equation is invariant under conformal change of the metric.) Hence one can reduce the study of L^2 Yang-Mills fields to smooth Yang-Mills fields over S^4.

If one fixes the Pontryagin number of the connection, then it is easy to see that the Yang-Mills functional has an absolute lower bound depending only on the Pontryagin number. If the Yang-Mills field is a global minimum, it is a self-dual or anti-self-dual field. (This means that the curvature is invariant or anti-invariant under the $*$ operator of S^4.) The self-dual and anti-self-dual fields have been studied extensively in recent years. Physicists, especially t'Hooft, constructed many examples of these fields. Atiyah-Hitchin-Singer [AHS1] were able to compute the number of parameters of these fields. Ward [Wa] and Atiyah-Ward [AW] found that these fields can be studied, via the Penrose transform [Pn], by algebraic geometric methods. Finally Atiyah-Hitchin [At], Drinfield-Manin [DM] were able to classify them completely. However, many important questions remained unsolved. The first one is concerning the topology of the space of self-dual fields with the same Pontryagin number. One does not know whether this is connected or not. There is work done by Hartshorne [Hts2], Atiyah [At], etc. The second one is the fundamental question whether every SU(2) Yang-Mills field is self-dual or anti-self-dual. The only progress on the second question was due to Bourguignon-Lawson-Simons [BLS] who proved that stable solutions are either self-dual or anti-self-dual. The third question is how to prove existence of

Yang-Mills fields by methods of partial differential equations. These three questions are perhaps related to each other.

Finally one should also mention that Yang-Mills theory is a very natural theory from a purely mathematical point of view. Therefore it is natural that Atiyah-Bott [AB] found applications of this theory in the study of vector bundles over Riemann surfaces. We should expect more applications in the near future. Besides the Yang-Mills field, physicists also consider the more general functional which includes the sections of a vector bundle associated to the principal bundle. These functionals are very important in physics. Recently, Parker [Pr] was able to generalize the fundamental theorem of Uhlenbeck to these fields.

§12. *Higher codimension minimal submanifolds*

The higher codimension minimal submanifolds are defined by a system of quasilinear partial differential equations. At this moment, our knowledge of this class of submanifolds is not enough for us to put them into applications. (The only exceptional case is the use of minimal spheres in the Frankel conjecture.) However, it should be emphasized that eventually they will be very useful in understanding the geometry of higher dimensional manifolds.

For minimal surfaces in R^n, one can define the Gauss map in a natural manner. The image of the map is the Grassmannian of two-planes in R^n. It has a complex structure and Chern [C2] was the first one to observe that the Gauss map is antiholomorphic. This enables him and Osserman [CO] to generalize the theorem of Bernstein and some theorems of Osserman to higher codimension. Basically, they prove that the image of the Gauss map cannot stay in a "small" set. Because of the simplicity of the Grassmannian of two-planes in R^4, it seems that minimal surfaces in R^4 should have special properties.

One important question here is to study area minimizing surfaces in R^4. It is important for both geometric and topological applications. It is well known (see Federer [F] §5.4.19) that complex curves in $C^2 = R^4$ are always area minimizing. It would be interesting to know the truth of the converse of this statement, i.e., given a properly embedded area minimizing surface in R^4, can one find a linear complex structure in R^4 so that the surface is complex analytic? (The paper of F. Morgan [Mg] on the union of linear planes may be relevant here.) It would also be interesting to find conditions on a Jordan curve in R^4 so that the area minimizing disk bounded by this Jordan curve is a nonsingular embedded disk. Any theorems of this sort will be important in understanding the topology of four dimensional manifolds. If the area minimizing surface has singularities, it would be interesting to compare these singularities with the singularities arising in complex varieties. An understanding of these singularities will provide understandings of the topology of surfaces in a four-dimensional manifold, and will lead to resolution of classical problems.

When the ambient manifold is compact, there has been more work done. For example, Calabi [Ca4] and Chern [C3] have studied minimal two-dimensional spheres in higher dimensional spheres. Simons [SJ] obtained integral formula for higher dimensional minimal submanifolds of S^n. The work of Simons was pursued by Chern-doCarmo-Kobayashi [CCK]. Hsiang [Hs], Hsiang-Lawson [HL] and doCarmo-Wallach [dCW] studied minimal immersions of homogeneous spaces into S^n. Lawson-Simons [LS] proved that no stable minimal current exists in S^n and that stable minimal currents in the standard CP^n consist of complex subvarieties.

It seems that if a compact Kähler manifold is rigid in the complex analytic sense, then the area minimizing subvarieties tend to be closer to being complex analytic. This is apparent in the work of Siu and the author [SiuY2] and Siu [S2]. Perhaps one can exploit this phenomenon to prove the Hodge conjecture for some special class of algebraic manifolds. (By the way, the attempt of proving the Hodge conjecture by minimal subvarieties is not new. For example, it motivated the paper of Lawson-Simons.)

Recently Almgren [Alm2] has announced that the set of singularities of an area minimizing current has measure zero. The details of his proof should give good understanding of the singularities also.

§13. *The Einstein equation*

The Einstein equation is another important system of quasilinear equations. The unknown for this system is the metric tensor g_{ij} and it is required that $R_{ij} - \frac{1}{2} R g_{ij}$ be equal to a given symmetric tensor T_{ij}. Here R_{ij} is the Ricci tensor of g_{ij} and R is its trace.

If the metric tensor g_{ij} is required to be positive definite, the above system is an elliptic system. If the metric tensor is required to be Lorentzian, the above system is a hyperbolic system. The Einstein equation is invariant under coordinate change. The choice of the coordinate system complicates the way to solve the equations. The physicists have been interested in the Cauchy problem for the Einstein equation for a long time. The works of Hawking and Penrose demonstrated that under physically reasonable conditions, the Cauchy problem cannot have a global solution. Hence a singularity must occur. The most important problem in classical general relativity is to study the nature of this singularity. At this moment, one does not have good ways to formulate the concept of singularity. However, Penrose proposed the cosmic censorship hypothesis which roughly states that no naked singularity will develop from nonsingular Cauchy data. This means that the singularity that will appear must hide itself behind some surface in a way similar to the black hole singularity. Although the formulation of the cosmic censorship hypothesis is not clear yet, it is certainly one of the newest important problems in classical relativity. Because of the nonlinear nature of the problem, the classical energy method is not readily applicable. However, one does have a concept of total energy in relativity. Schoen and the author [SY4] did prove that it is positive unless the spacetime is trivial. One hopes that this statement will be useful. The methods of [SY6] can be used to deal with spacetimes where singularities of the type of black hole are allowed. More interestingly, even when we are dealing with a nonsingular Cauchy surface, the concept of apparent horizon naturally comes into our proof. This may indicate the possible relevance of the proof with the cosmic censorship.

For a long time, differential geometers have studied Riemannian manifolds whose Ricci curvature is proportional to the metric tensor. (This is the Einstein equation for the vacuum spacetime.) At this moment there are two major categories of examples of these manifolds which are called Einstein manifolds. One category comes from the quotient spaces of irreducible symmetric spaces. The other one comes from Kähler geometry which was constructed via the Monge-Ampère equation (see [Y4]). However, in general, we do not know whether every manifold with dimension ≥ 5 admits an Einstein metric or not. In four dimension, the existence of such a metric on a compact manifold implies the Thorpe-Hitchin [Ht2] inequality

$$|\tau(M)| \leq \frac{2}{3}\chi(M)$$

where $\tau(M)$ is the index and $\chi(M)$ is the Euler number of the manifold. It is not known whether every noncompact four-dimensional manifold admits an Einstein metric or not. So far, no method has been devised to produce an Einstein metric on a general compact manifold except when the manifold is Kähler.

Many of the difficulties that we encounter come from our ignorance of the implications of a compact manifold admitting a metric with negative Ricci curvature. We do not even know whether S^3 or T^3 admits such a metric or not. Any topological information that we may obtain will not only be important for differential geometry but also for algebraic geometry, in view of the solution of the Calabi conjecture ([Y5]).

§14. *Isometric immersion*

The equation for the isometric immersion of a Riemannian manifold M is a first order nonlinear system. If we want to isometrically immerse M into R^N, then we are looking for N functions f^i on M so that the metric tensor of M can be expressed as $\sum_{i=1}^{N} (df^i)^2$. The fundamental theorem on isometric immersions was proved by J. Nash [N1] which says that any complete Riemannian manifold can be isometrically embedded into the Euclidean space. More recently, Gromov [GR] has improved Nash's theorem so that the manifold can be isometrically immersed in a Euclidean space of much lower dimension than the one given by Nash.

The question of whether we can always locally isometrically immerse an n-dimensional manifold into $\frac{n(n+1)}{2}$ dimensional Euclidean space is still not known. The Cartan-Janet [CÉ] theorem answers the question affirmatively if the metric is real analytic. However, if we merely assume the metric is smooth, we do not even know how to locally isometrically immerse a two-dimensional surface into R^3.

For two-dimensional surfaces in R^3, our understanding is considerably better than for higher dimensions. For example, the Hadamard theorem and the Cohn-Vossen theorem say that isometric immersions of a compact surface with nonnegative curvature are unique up to rigid motion. The Weyl embedding problem (which was solved by Nirenberg [NL] and Pogorelov [P1]) says that every compact surface with positive curvature can be isometric embedded into R^3. The Hilbert-Efimov [E2] theorem says that no complete surface of curvature ≤ -1 can be isometrically immersed into R^3. The study of complete surfaces with negative curvature in R^3 is far from being exhausted.

The difficulty of generalizing the theory of surfaces to higher dimension comes from the lack of suitable understanding of the rigidity of higher dimensional manifolds in Euclidean space. Based on algebraic calculations, one knows that a generic submanifold M^n with codimension not greater than $\frac{n}{3}$ in the Euclidean space is rigid, (see Allendoerfer [All]). The first global rigidity theorem for higher dimensional manifolds is due to Moore [Mr] who proved that the isometric immersion of a product manifold $M^m \times N^n$ into R^{m+n+2} splits as a product immersion if $m, n \geq 2$. One can define an isometric embedding of M^n into R^N to be elliptic if the second fundamental form associated to each normal vector has two nonzero eigenvalues of the same sign. One may expect that if M is compact, then the embedding is rigid. For some special elliptic embeddings of symmetric spaces, Tanaka [Ta] proved a local rigidity theorem. However, in order that M^n be elliptic, one requires $N \leq 2n-1$ and this is not a generic dimension if $n \geq 3$.

Finally we mention some interesting global problems of isometric immersions of Riemannian manifolds here. The first one says that every compact surface can be isometrically immersed into R^4. The second one says that every compact n-dimensional manifold with positive curvature can be isometrically immersed into the Euclidean space of dimension $\frac{n(n+1)}{2}$. This can be considered as a higher dimensional generalization of the Weyl embedding problem. The third one says that the complete n-dimensional hyperbolic space cannot be isometrically immersed into R^{2n-1}. The fourth one is to find an intrinsic condition on a complete non-compact manifold which admits a proper isometric immersion into R^n with bounded mean curvature.

Finally we recall the most fundamental problem of the isometric immersion of a compact surface Σ into R^3. It is an old problem and can be described as follows. Let f_t be a smooth family of isometric immersion of Σ into R^3. Then for each t, does there exist a rigid motion E_t of R^3 so that $f_t = E_t(f_0)$?

After these notes were compiled, Phillip Griffiths sent the author a manuscript in which he studies submanifolds M^n of codimension $\frac{1}{2}n(n-1)$ in Euclidean space. He considers the complicated algebraic problems involved in such isometric embeddings, and can deal with these problems locally. These are the first results in such high codimension and should lead to many important new developments.

REFERENCES

[Ah] L. V. Ahlfors: An extension of Schwarz's lemma, Trans. Amer. Math. Soc. *43* (1938), 359-364.

[AZ] A.D. Alexandrov and V.A. Zalgaller: Intrinsic geometry of convex surfaces, Moscow, 1968 (Russian) AMS Translations *15*, 1967.

[Ald] W. Allard: On the first variation of a varifold, Ann. of Math. *95* (1972), 417-491.

[All] C.B. Allendoerfer: Global theorems in Riemannian geometry, Bull. Amer. Math. Soc. *54* (1948), 249-259.

[Alm1] F.J. Almgren, Jr.: Some interior regularity theorems for minimal surfaces and an extension of Bernstein's theorem, Ann. of Math. *84* (1966), 277-292.

[Alm2] ————: Dirichlet's problem for multiple-valued functions and the regularity of mass-minimizing integral currents, in Minimal Submanifolds and Geodesics (M. Obata, editor) Kaigai, Tokyo (1978), 1-6.

[AS] F.J. Almgren, Jr. and L.M. Simon: Existence of embedded solutions of Plateau's problem, Ann. Scuola Norm. Sup. Pisa *6* (1979), 447-495.

[At] M.F. Atiyah: Geometry of Yang-Mills fields, Proc. Intern. Conf. Math. Phys., Rome 1977.

[AB] M.F. Atiyah and R. Bott: On the Yang-Mills equation and Riemann surfaces, preprint.

[ABP] M.F. Atiyah, R. Bott and V.K. Patodi: On the heat equation and the index theorem, Invent. Math. *19* (1973), 279-330.

[AHDM] M.F. Atiyah, N. Hitchin, U. Drinfield and Yu Manin: Construction of instantons, Phys. Lett. *65A* (1978), 185-187.

[AHS1] M.F. Atiyah, N. Hitchin and I.M. Singer: Deformations of instantons, Proc. Nat. Acad. Sci. USA *74* (1977), 2662-2663.

[AHS2] ————: Self-duality in four-dimensional Riemannian geometry, Proc. Roy. Soc. London *362A* (1978), 425-461.

[AHz] M.F. Atiyah and F. Hirzebruch: Riemann-Roch theorems for differentiable manifolds, Bull. Amer. Math. Soc. *65* (1959), 276-281.

[APS] M.F. Atiyah, V.K. Patodi and I.M. Singer: Spectral asymmetry and Riemannian geometry, Bull. London Math. Soc. *5* (1973), 229-234.

[AW] M.F. Atiyah and R. Ward: Instantons and algebraic geometry, Comm. Math. Phys. *55* (1977), 111-124.

[Au] T. Aubin: The scalar curvature, in Differential Geometry and Relativity, Holland 1976, 5-18.

[BB] W.L. Baily and A. Borel: Compactification of arithmetic quotients of bounded symmetric domains, Ann. of Math. *84* (1966), 442-528.

[Ber] P. Berard: On the wave equation on a compact Riemannian manifold without conjugate points, Math. Zeit. *155* (1977), 249-276.

[BGM] M. Berger, P. Gauduchon and E. Mazet: Le Spectre d'une Variété Riemannienne, Lecture Notes in Math. 194, Springer 1972.

[Bes] A. Besse: Manifolds All of Whose Geodesics are Closed, Springer 1978.

[Be] G. Besson: Sur la Multiplicité de la Premiere Valeur Propre des Surfaces Riemanniennes, Thesis at Université de Paris VII, 1979.

[BC] R. L. Bishop and R. J. Crittenden: Geometry of Manifolds, Academic Press, New York, 1964.

[BG] R. L. Bishop and S. I. Goldberg: On the second cohomology group of a Kähler manifold of positive curvature, Proc. Amer. Math. Soc. *16* (1965), 119-122.

[Bl] W. Blaschke: Vorlesungen über Differentialgeometrie, vol. 1, Springer 1924.

[Bo1] S. Bochner: Harmonic surfaces in Riemann metric, Trans. Amer. Math. Soc. *47* (1940), 146-154.

[Bo2] ———: Vector fields and Ricci curvature, Bull. Amer. Math. Soc. *52* (1946), 776-797.

[BDM] E. Bombieri, E. De Giorgi and M. Miranda: Una maggiorazione a priori relativa alle ipersuperfici minimali non parametriche, Arch. Rat. Mech. Anal. *32* (1969), 255-267.

[BGi] E. Bombieri and E. Giusti: Harnack's inequality for elliptic differential equations on minimal surfaces, Invent. Math. *15* (1972), 24-46.

[Bn] O. Bonnet: Sur quelques propriétés des lignes géodésiques, C. R. Acad. Sci. Paris *40* (1855), 1311-1313.

[BLS] J. P. Bourguinon, H. B. Lawson, Jr. and J. Simons: Stability and gap phenomena in Yang-Mills fields, Proc. Nat. Acad. Sci. USA *76* (1979), 1550-1553.

[Br] G. E. Bredon: A π^*-module structure for Θ_* and applications to transformation groups, Ann. of Math. *86* (1967), 434-448.

[Brü] J. Brüning: Über knoten von eigenfunktionen des Laplace-Beltrami operators, Math. Zeit. *158* (1978), 15-21.

[Ca1] E. Calabi: Improper affine hypersurfaces of convex type and a generalization of a theorem by K. Jörgens, Mich. Math. J. *5* (1958), 105-126.

[Ca2] ———: On compact Riemannian manifolds with constant curvature I, AMS Symposium on Differential Geometry III, Arizona (1960), 155-180.

[Ca3] ———: Examples of Bernstein problems for some nonlinear equations, AMS Symposium on Global Analysis XV, Berkeley (1960), 223-230.

[Ca4] E. Calabi: Quelques applications de l'analyse complex aux surfaces d'aire minima, Topics in Complex Analysis, Univ. of Montreal, 1968, 59-81.

[Ca5] ________: On manifolds with non-negative Ricci curvature II, Notices Amer. Math. Soc. *22* (1975), A205.

[CaV] E. Calabi and E. Vesentini: On compact, locally symmetric Kähler manifolds, Ann. of Math. *71* (1960), 472-507.

[Cl] T. Carleman: Propriétés asymptotiques des fonctions fondamentales des membranes vibrantes, C.R. 8ème Congr. des Math. Scand., Stockholm 1934, 34-41.

[CÉ] É. Cartan: Sur la possibilité de plonger un espace riemannien donné dans un espace euclidien, Ann. Soc. Math. Pol. *6* (1927), 1-17.

[Chz] J. Chazarain: Formule de Poisson pour les variétés riemanniennes, Invent. Math. *24* (1974), 65-82.

[Ch1] J. Cheeger: The relation between the Laplacian and the diameter for manifolds of non-negative curvature, Arch. der Math. *19* (1968), 558-560.

[Ch2] ________: A lower bound for the smallest eigenvalue of the Laplacian, in Problems in Analysis, Princeton University Press, 1970, 195-199.

[CG] J. Cheeger and D. Gromoll: The splitting theorem for manifolds of non-negative Ricci curvature, J. Diff. Geom. *6* (1971), 119-128.

[Cn1] S.Y. Cheng: Eigenvalue comparison theorems and its geometric applications, Math. Zeit. *143* (1975), 289-297.

[Cn2] ________: Eigenfunctions and nodal sets, Comm. Math. Helv. *51* (1976), 43-55.

[Cn3] ________: To appear.

[CLi] S.Y. Cheng and P. Li: To appear.

[CY1] S.Y. Cheng and S-T. Yau: Complete affine hypersurfaces I, preprint.

[CY2] ________: Differential equations on Riemannian manifolds, Comm. Pure Appl. Math. *28* (1975), 333-354.

[CY3] ________: On the regularity of the solution of the n-dimensional Minkowski problem, Comm. Pure Appl. Math. *29* (1976), 495-515.

[CY4] ________: On the regularity of the Monge-Ampère equation $\det(\partial^2 u/\partial x^i \partial x^j) = F(x,u)$, Comm. Pure Appl. Math. *30* (1977), 41-68.

[C1] S.S. Chern: Characteristic classes of Hermitian manifolds, Ann. of Math. *47* (1946), 85-121.

[C2] S.S. Chern: Minimal surfaces in an Euclidean space of N dimensions, in Differential and Combinatorial Topology, Princeton University Press (1965), 187-198.

[C3] ———: On minimal surfaces in the four-sphere, in Studies and Essays presented to Y.W. Chen, Taiwan (1970), 137-150.

[CCK] S.S. Chern, M. do Carmo and S. Kobayashi: Minimal submanifolds of a sphere with second fundamental form of constant length, in Functional Analysis and Applications, Springer 1970.

[CO] S.S. Chern and R. Osserman: Complete minimal surfaces in Euclidean n-space, J. d'Analyse Math. *19* (1967), 15-34.

[CV1] S. Cohn-Vossen: Unstarre geschlossene Fläche, Math. Ann. *102* (1929), 159-184.

[CV2] ———: The isometric deformability of surfaces in the large, Uspehi Mat. Nauk *1* (1936), 33-76.

[CdV] Y. Colin de Verdière: Spectre du Laplacien et longuers des géodésiques périodiques II, Comp. Math. *27* (1973), 159-184.

[CH] R. Courant and D. Hilbert: Methods of Mathematical Physics, Interscience 1962.

[Cr] C. Croke: Some isoperimetric inequalities and consequences, Ann. Scient. Éc. Norm. Sup. Paris *13* (1980).

[DeG] E. De Giorgi: Sulla differenziabilità e l'analiticità delle estremali degli integrali multipli regolari, Mem. Accad. Sci. Torino (3) *3* (1957), 25-43.

[dCP] M. do Carmo and C.K. Peng: Stable complete minimal surfaces in $\mathbf{R}^3$ are planes, Bull. Amer. Math. Soc. *1* (1979), 903-905.

[dCW] M. do Carmo and N. Wallach: Minimal immersions of spheres into spheres, Ann. of Math. *93* (1971), 43-62.

[D] J. Douglas: Solution of the problem of Plateau, Trans. Amer. Math. Soc. *33* (1931), 263-321.

[DM] V.G. Drinfield and Yu I. Manin: A description of instantons, Comm. Math. Phys. *63* (1977), 177-192.

[DG] J.J. Duistermaat and V.W. Guillemin: The spectrum of positive elliptic operators and periodic bicharacteristics, Invent. Math. *29* (1975), 39-79.

[EL] J. Eells, Jr. and L. Lemaire: On the construction of harmonic and holomorphic maps between surfaces, Math. Ann. *252* (1980), 27-52.

[ES] J. Eells, Jr. and J.H. Sampson: Harmonic mappings of Riemannian manifolds, Amer. J. Math. *86* (1964), 109-160.

[EW] J. Eells, Jr. and J.C. Wood: Restrictions on harmonic maps of surfaces, Topology *15*(1976), 263-266.

[E1] N.V. Efimov: Qualitive problems of the theory of deformations, Uspehi Mat. Nauk, *24*(1948), 47-158 (Russian) AMS Translations *37*(1951).

[E2] ________: Generation of singularities on surfaces of negative curvature, Mat. Sbornik *64*(1964), 286-320.

[E3] ________: Hyperbolic problems in the theory of surfaces, Proc. Intern. Congr. Math., Moscow (1966), AMS Translation *70*(1968), 26-38.

[El] H. Eliasson: On variation of metrics, Math. Scand. *29*(1971), 317-327.

[Fa] C. Faber: Beweis, dass unter allen homogenen Membranen von gleicher Fläche und gleicher Spannung die Kreisförmige den teifsten Grundton gibt, Sizungsber. Bayer. Akad. der Wiss. Math. Phys. Munich (1923), 169-172.

[F] H. Federer: Some theorems on integral currents, Trans. Amer. Math. Soc. *117*(1965), 43-67.

[FF] H. Federer and W. Fleming: Normal and integral currents, Ann. of Math. *72* (1960), 458-520.

[FS] D. Fischer-Colbrie and R. Schoen: The structure of complete stable minimal surfaces in 3-manifolds of non-negative scalar curvature, Comm. Pure Appl. Math. *33* (1980), 199-211.

[FK] G.B. Folland and J.J. Kohn: The Neumann problem for the Cauchy-Riemann complex, Ann. of Math. Studies *75*, Princeton University Press (1972).

[Fr1] T. Frankel: Manifolds with positive curvature, Pacific J. Math. *11* (1964), 165-174.

[Fr2] ________: On the fundamental group of a compact minimal submanifold, Ann. of Math. *83* (1966), 68-73.

[FHS] M. Freedman, J. Hass and P.J. Scott: Least area incompressible surfaces, to appear.

[Gi] P. Gilkey: The spectral geometry of a Riemannian manifold, J. Diff. Geom. *10*(1975), 601-618.

[GW1] R.E. Greene and H. Wu: Some function-theoretic properties of non-compact Kähler manifolds, AMS Symposium on Differential Geometry XXVII, Stanford (1973), 33-42.

[GW2] ________: Function theory on manifolds which possess a pole, Lecture Notes in Math. 699, Springer 1979.

[GM] D. Gromoll and W. Meyer: On complete open manifolds of non-negative curvature, Ann. of Math. *90*(1969), 75-90.

[G] M. Gromov: Paul Levy's isoperimetric inequality, I.H.E.S. preprint, 1980.

[GR] M. Gromov and V. Rohlin: Embeddings and immersions in Riemannian geometry, Uspehi Mat. Nauk *25*(3) (1970), 3-62; Russian Math. Surveys *25*(3) (1970), 1-57.

[GK] V. Guillemin and D. Kazhdan: Some inverse spectral results for negatively curved n-manifolds, AMS Symposium on Geometry of the Laplace Operator XXXVI, Hawaii (1979), 153-180.

[Gu] R. Gulliver: Regularity of minimizing surfaces of prescribed mean curvature, Ann. of Math. *97*(1973), 275-305.

[GL] R. Gulliver and J. Leslie: On boundary branch points of minimizing surfaces, Arch. Rat. Mech. Anal. *52*(1973), 20-25.

[GS] R. Gulliver and J. Spruck: On embedded minimal surfaces, Ann. of Math. *103*(1976), 331-347.

[Ham] R.S. Hamilton: Harmonic maps of manifolds with boundary, Lecture Notes in Math. *471*, Springer 1975.

[Htm] P. Hartman: On homotopic harmonic maps, Canadian J. Math. *19* (1967), 673-687.

[Hts1] R. Hartshorne: Ample subvarieties of algebraic varieties, Lecture Notes in Math. *159*, Springer 1970.

[Hts2] ________: Stable vector bundles and instantons, Comm. Math. Phys. *59*(1978), 1-15.

[Hay] W.K. Hayman: Some bounds for principal frequency, Applicable Analysis *7*(1978), 247-254.

[Hnz] E. Heinz: Über gewisse elliptische Systeme von Differentialgleichungen zweiter Ordnung mit Anwendung auf die Monge-Ampèresche gleichung, Math. Ann. *131* (1956), 411-428.

[Her] J. Hersch: Quatre propriétés isopérimétriques de membranes sphériques homogènes, C.R. Acad. Sci. Paris, *270*(1970), 1645-1648.

[H] D. Hilbert: Über Flächen von constanter Gausscher Krümmung, Trans. Amer. Math. Soc. *2*(1901), 87-99.

[Ht1] N. Hitchin: Harmonic spinors, Advances in Math. *14*(1974), 1-55.

[Ht2] ________: Compact four-dimensional Einstein manifolds, J. Diff. Geom. *9*(1974), 435-441.

[Hö1] L. Hörmander: The spectral function of an elliptic operator, Acta Math. *121* (1968), 193-218.

[Hö2] ________: An Introduction to Complex Analysis in Several Variables, North-Holland, 2nd edition, 1973.

[Hs] W-Y. Hsiang: On compact homogeneous minimal submanifolds, Proc. Nat. Acad. Sci. USA *56* (1966), 5-6.

[HL] W-Y. Hsiang and H.B. Lawson, Jr.: Minimal submanifolds of low cohomogeneity, J. Diff. Geom. *5* (1970), 1-37.

[Hu] A. Huber: On subharmonic functions and differential geometry in the large, Comm. Math. Helv. *32* (1957), 13-72.

[I] S. Iitaka: On some new birational invariants of algebraic varieties and their application to rationality of certain algebraic varieties of dimension three, J. Math. Soc. Japan *24* (1972), 384-396.

[Ik] A. Ikeda: On lens spaces which are isospectral but not isometric, Ann. Sci. École Norm. Sup. *13* (1980), 303-315.

[Iv] V. Ya. Ivrii: Second term of the Spectral asymptotic expansion of the Laplace Beltrami operator on manifolds with boundary. Funct. Analy. Appl. 14 (2) (1980), 98-105.

[JK] K. Jörgens: Über die Lösungen der Differentialgleichung $rt - s^2 = 1$, Math. Ann. *127* (1954), 130-134.

[JJ] J. Jost: Univalency of harmonic mappings between surfaces, J. Reine Angew. Math. *324* (1981), 141-153.

[Ka] M. Kac: Can one hear the shape of a drum?, Amer. Math. Monthly *73* (1966), 1-23.

[KW] J. Kazdan and F. Warner: Integrability conditions for $\Delta u = k - Ke^{\alpha u}$ with applications to Riemannian geometry, Bull. Amer. Math. Soc. *77* (1971), 819-821.

[KO] S. Kobayashi and T. Ochiai: Three-dimensional compact Kähler manifolds with positive holomorphic bisectional curvature, J. Math. Soc. Japan *24* (1972), 465-480.

[KS] K. Kodaira and D.C. Spencer: On deformations of complex analytic structures I and II, Ann. of Math. *67* (1958), 328-466.

[Kr] E. Krahn: Über eine von Rayleigh formulierte Minimaleigenschaft des Kreises, Math. Ann. *94* (1925), 97-100.

[LU] O.A. Ladyzhenskaya and N.N. Ural'tseva: Quasilinear elliptic equations and variational problems with several independent variables, Uspehi. Mat. Nauk *16* (1), (1961), 19-90 (Russian) Russian Math. Surveys *16* (1), (1961), 17-92.

[L] H.B. Lawson, Jr.: Complete minimal surfaces in S^3, Ann. of Math. *92* (1970), 335-374.

[LS] H.B. Lawson, Jr. and J. Simons: On stable currents and their application to global problems in real and complex geometry, Ann. of Math. *98* (1973), 427-450.

[LY] H.B. Lawson, Jr. and S-T. Yau: Scalar curvature, non-Abelian group actions and the degree of symmetry of exotic spheres, Comm. Math. Helv. *49* (1974), 232-244.

[Le] L. Lemaire: Applications harmoniques de surfaces riemanniennes, J. Diff. Geom. *13* (1978), 51-78.

[Lw] H. Lewy: On the existence of a closed convex surface realizing a given Riemannian metric, Proc. Nat. Acad. Sci. USA *24* (1938), 104-106.

[Li] P. Li: A lower bound for the first eigenvalue of the Laplacian on a compact Riemannian manifold, Indiana J. Math. *28* (1979), 1013-1019.

[LiY] P. Li and S-T. Yau: Eigenvalues of a compact Riemannian manifold, AMS Symposium on Geometry of the Laplace Operator XXXVI, Hawaii (1979), 205-240.

[Lc1] A. Lichnerowicz: Geometrie des Groupes de Transformations, Dunod 1958.

[Lc2] ————: Spineurs harmoniques, C.R. Acad. Sci. Paris, AB *257* (1963), 7-9.

[Lb] E. Lieb: The number of bound states of one-body Schrödinger operators and the Weyl problem, AMS Symposium on Geometry of the Laplace Operator XXXVI, Hawaii (1979), 241-252.

[LN] C. Loewner and L. Nirenberg: Partial Differential Equations Invariant under Conformal or Projective Transformations, in Contributions to Analysis, Academic Press 1974.

[Ly] V.S. Lyukshin: Zur Theorie der Verbeigung der Rotationsflächen negativer Krümmung, Mat. Sbornik (1937), 557-565.

[Mb] T. Mabuchi: $\mathbb{C}$-actions and algebraic three-folds with ample tangent bundle, Nagoya Math. J. *69* (1978), 33-64.

[MKS] H. P. McKean and I. M. Singer: Curvature and the eigenvalues of the Laplacian, J. Diff. Geom. *1* (1967), 43-69.

[MY1] W. Meeks, III and S-T. Yau: The classical Plateau problem and the topology of three-dimensional manifolds, Arch. Rat. Mech. Analysis, to appear.

[MY2] ————: Topology of three-dimensional manifolds and the embedding problem in minimal surface theory, Ann. of Math. *112* (1980), 441-484.

[MY3] ————: Conference on the Smith Conjecture, Columbia University 1979, to appear.

[MSY] W. Meeks, III, L. M. Simon and S-T. Yau: To appear.

[Mr] R. Melrose: Weyl's conjecture for manifolds with concave boundary, AMS Symposium on Geometry of the Laplace Operator XXXVI, Hawaii (1979), 257-274.

[MS] J. Michael and L. M. Simon: Sobolev and mean-value inequalities on generalized submanifolds of $\mathbf{R}^n$, Comm. Pure Appl. Math. *26* (1973), 361-379.

[Ml1] J. Milnor: Eigenvalues of the Laplace operator on certain manifolds, Proc. Nat. Acad. Sci. USA *51* (1964), 542.

[Ml2] ———: Remarks concerning spin manifolds, in Differential and Combinatorial Topology, Princeton University Press (1965), 55-62.

[Mn] S. Minakshisundaram: Eigenfunctions on Riemannian manifolds, J. Indian Math. Soc. *17* (1953), 158-165.

[MM] M. Miranda: Diseguaglianze di Sobolev sulle ipersuperfici minimali, Rend. Sem. Math. Univ. Padova *38* (1967), 69-79.

[MkSY] N. Mok, Y-T. Siu and S-T. Yau: To appear.

[Mr] J.D. Moore: Isometric immersions of Riemannian products, J. Diff. Geom. *5* (1971), 159-168.

[Mg] F. Morgan: On the singular structure of two-dimensional area-minimizing surfaces in R^n, preprint.

[Mo] S. Mori: Projective manifolds with ample tangent bundles, Ann. of Math. *110* (1979), 593-606.

[M1] C.B. Morrey, Jr.: The problem of Plateau in a Riemannian manifold, Ann. of Math. *49* (1948), 29-37.

[M2] ———: Second order elliptic equations in several variables and Hölder continuity, Math. Zeit. *72* (1959), 146-164.

[MT] M. Morse and C.B. Tompkins: On the existence of minimal surfaces of critical types, Ann. of Math. *40* (1939), 443-472.

[Ms1] J. Moser: A new proof of De Giorgi's theorem concerning the regularity problem for elliptic differential equations, Comm. Pure Appl. Math. *13* (1960), 457-468.

[Ms2] ———: On a nonlinear problem in differential geometry, in Dynamical Systems (M. Peixoto, editor) Academic Press, New York (1973).

[My] S.B. Myers: Riemann manifolds with positive mean curvature, Duke Math. J. *8* (1941), 401-404.

[N1] J. Nash: The embedding problem for Riemannian manifolds, Ann. of Math. *63* (1956), 20-63.

[N2] ———: Continuity of solutions of parabolic and elliptic equations, Amer. J. Math. *80* (1958), 931-954.

[NL] L. Nirenberg: The Weyl and Minkowski problems in differential geometry in the large, Comm. Pure Appl. Math. *6* (1953), 337-394.

[Ob] M. Obata: Certain conditions for a Riemannian manifold to be isometric with a sphere, J. Math. Soc. Japan *14* (1962), 333-340.

[O1] R. Osserman: A proof of the regularity everywhere of the classical solution of Plateau's problem, Ann. of Math. *91* (1970), 550-569.

[O2] R. Osserman: A note on Hayman's Theorem on the bass note of a drum, Comment. Math. Helv. *52* (1977), 545-555.

[Pp] C.D. Papakyriakopolous: On Dehn's lemma and the asphericity of knots, Ann. of Math. *66* (1957), 1-26.

[Pr] T. Parker: Ph.D. dissertation, Stanford 1980.

[Pt] V.K. Patodi: Curvatures and eigenforms of the Laplace operator, J. Diff. Geom. *5* (1971), 233-249.

[PW] L.E. Payne and H.F. Weinberger: Some isoperimetric inequalities for membrane frequencies and torsional rigidity, J. Math. Anal. Appl. *2* (1961), 210-216.

[Pn] R. Penrose: The twistor program, Reports in Math. Phys. *12* (1977), 65-76.

[P1] A.V. Pogorelov: The regularity of a convex surface with given Gauss curvature, Mat. Sbornik *31* (1952), 88-103 (Russian).

[P2] ————: The Dirichlet problem for the n-dimensional analogue of the Monge-Ampère equation, Soviet Math. Dokl. *12* (1971), 1192-1196.

[P3] ————: On the regularity of generalized solutions of the equation $\det(\partial^2 u/\partial x_i \partial x_j) = \phi(x_1,\cdots,x_n) > 0$, Soviet Math. Dokl. *12* (1971), 1436-1440.

[P4] ————: Improper convex affine hypersurfaces, Soviet Math. Dokl. *13* (1972), 240-244.

[P5] ————: Extrinsic geometry of convex surfaces, Nauka, Moscow 1969 (Russian) AMS Translations *35* (1973).

[Pol] G. Pólya: On the eigenvalues of a vibrating membrane, London Math. Soc. Proc. Ser. 3 *11* (1961), 419-433.

[PS] G. Pólya and G. Szegö: Isoperimetric inequalities in Mathematical Physics, Ann. of Math. Studies 27, Princeton University Press (1951).

[Pro] M.H. Protter: To appear.

[R] T. Rado: On Plateau's problem, Ann. of Math. *31* (1930), 457-469.

[Re] R.C. Reilly: Applications of the integral of an invariant of the Hessian, Bull. Amer. Math. Soc. *82* (1976), 579-580.

[Rm] E. Rembs: Über Gleitverbiegungen, Math. Ann. *111* (1935), 587-595.

[SU] J. Sacks and K. Uhlenbeck: The existence of minimal immersions of 2-spheres, Ann. of Math. *113* (1981), 1-24.

[Sa] J.H. Sampson: Some properties and applications of harmonic mappings, Ann. Sci. École Norm. Sup. *11* (1978), 201-218.

[St] I. Satake: On compactifications of the quotient spaces for arithmetically defined discontinuous groups, Ann. of Math. *72* (1960), 555-580.

[Sch] R. Schoen: To appear.

[SS] R. Schoen and P. Shalen: In preparation.

[SU] R. Schoen and K. Uhlenbeck: To appear.

[SY1] R. Schoen and S-T. Yau: On univalent harmonic maps between surfaces, Invent. Math. *44* (1978), 265-278.

[SY2] ________: Existence of incompressible minimal surfaces and the topology of manifolds with non-negative scalar curvature, Ann. of Math. *110* (1979), 127-142.

[SY3] ________: On the structure of manifolds with positive scalar curvature, Man. Math. *28* (1979), 159-183.

[SY4] ________: Positivity of the total mass of a general space-time, Phys. Rev. Lett. *43* (1979), 1457-1459.

[SY5] ________: Compact group actions and the topology of manifolds with non-positive curvature, Topology *18* (1979), 361-380.

[SY6] ________: The energy and linear momentum of space-times in general relativity, Comm. Math. Phys. *79* (1981), 47-51.

[SY7] ________: Complete three-dimensional manifolds with positive Ricci curvature and scalar curvature, these proceedings.

[Se] A. Selberg: Harmonic analysis and discontinuous groups in weakly symmetric Riemannian spaces with applications to Dirichlet series, J. Indian Math. Soc. *20* (1956), 47-87.

[SW] A. Shapiro and J.H.C. Whitehead: A proof and extension of Dehn's lemma, Bull. Amer. Math. Soc. *64* (1958), 174-178.

[Sh] M. Shiffman: The Plateau problem for non-relative minima, Ann. of Math. *40* (1939), 834-854.

[SL] L.M. Simon: These proceedings.

[SJ] J. Simons: Minimal varieties in Riemannian manifolds, Ann. of Math. *88* (1968), 62-105.

[S1] Y-T. Siu: Pseudo-convexity and the problem of Levi, Bull. Amer. Math. Soc. *84* (1978), 481-512.

[S2] ________: The complex-analyticity of harmonic maps and the strong rigidity of compact Kähler manifolds, Ann. of Math. *112* (1980), 73-111.

[SiuY1] Y-T. Siu and S-T. Yau: Complete Kähler manifolds with non-positive curvature of faster than quadratic decay, Ann. of Math. *105* (1977), 225-264.

[SiuY2] Y-T. Siu and S-T. Yau: Compact Kähler manifolds of positive bisectional curvature, Invent. Math. *59*(1980), 189-204.

[SiuY3] ———: These proceedings.

[Sz] G. Szegö: Inequalities for certain eigenvalues of a membrane of given area, J. Rat. Mech. Anal. *3* (1954), 343-356.

[Ta] N. Tanaka: Rigidity for elliptic isometric embeddings, Nagoya Math. J. *51* (1973), 137-160.

[Tay] M. Taylor: Estimates on the fundamental frequency of a drum, Duke Math. J. *46*(1979), 447-453.

[TT] F. Tomi and A. Tromba: Extreme curves bound an embedded minimal surface of the type of a disc, Math. Zeit. *158*(1978), 137-145.

[T] N. Trudinger: Remarks concerning the conformal deformation of Riemannian structures on compact manifolds, Ann. Scuola Norm. Sup. Pisa *3* (1968), 265-274.

[U1] K. Uhlenbeck: Removable singularities in Yang-Mills fields, Bull. Amer. Math. Soc. *1* (1979), 579-581.

[U2] ———: Closed minimal hypersurfaces in hyperbolic manifolds, to appear.

[Ur] H. Urakawa: On the least eigenvalue of the Laplacian for compact group manifolds, J. Math. Soc. Japan *31* (1979), 209-226.

[V] M. Vignerás: Variétés riemanniennes isospectrales et non-isométriques, Ann. of Math. *112* (1980), 21-32.

[Wa] R. Ward: On self-dual gauge fields, Phys. Lett. *61A* (1977), 81-82.

[We] A. Weil: Remarks on the cohomology of groups, Ann. of Math. *80* (1964), 149-157.

[W] H. Weyl: Das asymptotische Verteilungsgesetz der Eigenwerte linearer partieller Differentialgleichungen, Math. Ann. *71* (1912), 441-479.

[Wu] H. Wu: Negatively curved Kähler manifolds, Notices Amer. Math. Soc. *14* (1967), 515.

[Yam] H. Yamabe: On a deformation of Riemannian structures on compact manifolds, Osaka Math. J. *12* (1960), 21-37.

[YY] P. Yang and S-T. Yau: Eigenvalues of the Laplacian of compact Riemannian surfaces and minimal submanifolds, Ann. Scuola Norm. Sup. Pisa *7*(1980), 55-63.

[Y1] S-T. Yau: Harmonic functions on complete Riemannian manifolds, Comm. Pure Appl. Math. *28*(1975), 201-228.

[Y2] ———: Isoperimetric inequalities and the first eigenvalue of a compact Riemannian manifold, Ann. Sci. École Norm. Sup. Paris *8*(1975), 487-507.

[Y3] S-T. Yau: Some function-theoretic properties of complete Riemannian manifolds and their applications to geometry, Indiana Math. J. *25* (1976), 659-670.

[Y4] ________: Calabi's conjecture and some new results in algebraic geometry, Proc. Nat. Acad. Sci. USA *74* (1977), 1798-1799.

[Y5] ________: On the Ricci curvature of a compact Kähler manifold and the complex Monge-Ampère equation, I, Comm. Pure Appl. Math. *31* (1978), 339-411.

POINCARÉ INEQUALITIES ON RIEMANNIAN MANIFOLDS

Peter Li

§1. Let M^n be an n-dimensional compact Riemannian manifold with (possibly empty) boundary ∂M. In local coordinates $(x^1, x^2, \cdots, x^n)$, the Riemannian metric is given by

$$ds^2 = \sum_{i,j} g_{ij} dx^i dx^j . \tag{1.1}$$

One defines on M a second ordered elliptic differential operator by

$$\Delta = \frac{1}{\sqrt{g}} \sum_{i,j} \frac{\partial}{\partial x^i} \sqrt{g}\, g^{ij} \frac{\partial}{\partial x^j} , \tag{1.2}$$

where $(g^{ij}) = (g_{ij})^{-1}$ and $g = \det(g_{ij})$, which is known as the Laplace operator.

(A) When $\partial M = \emptyset$, the Laplacian Δ is a self-adjoint operator acting on $H_1^2(M)$. Its spectrum consists of discrete eigenvalues $\{0 = \lambda_0 < \lambda_1 \leq \lambda_2 \leq \cdots \leq \lambda_k \leq \cdots\}$, and the set of eigenfunctions $\{\phi_i\}$ satisfying

$$\Delta \phi_i = -\lambda_i \phi_i , \quad \phi_i \neq 0 , \tag{1.3}$$

form an orthonormal basis for the Hilbert space $H_1^2(M)$.

(B) When $\partial M \neq \emptyset$, it becomes necessary to adopt one of the following boundary conditions to guarantee the self-adjointness of Δ.

Seminar on Differential Geometry
0-691-08268-5/82/000073-11 $00.55/0 (cloth)
0-691-08296-0/82/000073-11 $00.55/0 (paperback)

(i) Dirichlet boundary condition: The domain of Δ is restricted to

$$H^2_{1,D}(M) = \{f \in H^2_1(M) | f|_{\partial M} \equiv 0\} .$$

with spectrum in $H^2_{1,D}(M)$ of the form

$$\{0 < \mu_1 < \mu_2 \leq \mu_3 \leq \cdots \leq \mu_k \leq \cdots\} .$$

The eigenfunctions $\{\phi_i\}$ satisfying

$$\Delta\phi_i = -\mu_i\phi_i$$

(1.4) and

$$\phi_i|_{\partial M} \equiv 0$$

form an orthonormal basis for $H^2_{1,D}(M)$.

(ii) Neumann boundary condition: The Laplacian Δ is now acting on the space

$$H^2_{1,N}(M) = \left\{f \in H^2_1(M) | \frac{\partial f}{\partial \nu}|_{\partial M} \equiv 0\right\}$$

where $\partial/\partial\nu$ is the outward normal vector to ∂M. Similar to that of (A), the spectrum of Δ is again of the form

$$\{0 = \eta_0 < \eta_1 \leq \eta_2 \leq \cdots \leq \eta_k \leq \cdots\}$$

with eigenfunctions $\{\phi_i\}$ satisfying

$$\Delta\phi_i = -\eta_i\phi_i$$

(1.5) and

$$\partial\phi_i/\partial\nu|_{\partial M} \equiv 0 .$$

By the compactness of M, it is known that the set of functions which satisfied (1.3) and (1.5) with $i = 0$ (i.e., $\lambda_i = 0$) are constants. The first nonzero eigenvalues λ_1 and η_i in problems (A) and (Bii) are hence characterized as the optimal constants in the Poincaré inequalities:

(A) When $\partial M = \emptyset$, then

$$\int |\nabla f|^2 \geq \lambda_1 \int f^2 \tag{1.6}$$

for all $f \in H_1^2(M)$ such that $\int_M f = 0$.

(Bii) When $\partial M \neq \emptyset$, then

$$\int |\nabla f|^2 \geq \eta_1 \int f^2 \tag{1.7}$$

for all $f \in H_1^2(M)$ such that $\int_M f = 0$.

Analogously, the first eigenvalue μ_1 of problem (Bi) is characterized as the Poincaré constant for Dirichlet boundary condition, namely,

$$\int |\nabla f|^2 \geq \mu_1 \int f^2 \tag{1.8}$$

for all $f \in H_{1,D}^2(M)$.

§2. Due to the importance of Poincaré inequalities for analysis on a manifold, it is hence much desirable to obtain optimal quantitative estimates for the first eigenvalues λ_1, μ_1 and η_1 from below in terms of geometric elements. Classically, for domains in $\mathbf{R}^n$, lower bounds for μ_1 and η_1 were established by: Faber-Krahn, Hayman, Payne, Polyá, Szegö, Weinberger, etc.

In the year 1958, Lichnerowicz [7] initiated the task of estimating λ_1 from below. He proved that: If the Ricci curvature of M satisfies

$$\mathrm{Ric} \geq K > 0,$$

then

$$\lambda_1 \geq \frac{nK}{n-1}.$$

Four years later, Obata [8] showed that under the same hypothesis, $\lambda_1 = \frac{nK}{n-1}$ iff M^n is isometric to the standard sphere S^n with constant curvature $\frac{K}{n-1}$. Reilly [10] later generalized these results of Lichnerowicz and Obata to problem (Bi): If the Ricci curvature of M satisfies $\mathrm{Ric} \geq K > 0$, and if the mean curvature H of ∂M is nonnegative, then

$$\mu_1 \geq \frac{nK}{n-1} .$$

Equality holds iff M is the upper hemisphere of the standard sphere of curvature $\frac{K}{n-1}$.

In 1970, Cheeger [2] showed that λ_1 and μ_1 have lower bounds in the form of some isoperimetric constants. By defining

$$h = \inf_N \frac{A(N)}{\min\{V(M_1), V(M_2)\}} ,$$

where inf is taken over all codimension 1 submanifolds N which divides M into two parts M_1 and M_2; and

$$\overline{h} = \inf \frac{A(N)}{V(M_1)} ,$$

where N is any codimension 1 submanifold such that $N \cap \partial M = \emptyset$, and M_1 is the part which does not intersect ∂M. He proved that:

$$\lambda_1 \geq \frac{h^2}{4}$$

and

$$\mu_1 \geq \frac{\overline{h}^2}{4} .$$

In 1975, by studying a similar isoperimetric constant, Yau [11] demonstrated that λ_1 has a lower bound in terms of: diameter of M, d; volume of M, V; and lower bound of the Ricci curvature, K. In that paper, he conjectured that λ_1 should have a lower bound depending only on d and K.

Meanwhile, upper estimates of λ_1 were obtained by Chavel-Feldman, Cheeger, Cheng and Mazet. In particular, Cheng [3] proved that

$$\lambda_k \leq C_1 \max\{0, -K\} + C_2 \frac{k^2}{d^2}$$

where C_1 and C_2 are constants depending on n alone. Examples also showed that this estimate could not be improved quantitatively without special assumption on M. Hence, for general manifolds, Yau's conjecture will provide optimal estimate for λ_1.

§3. In his thesis and [4], the special case when $K \geq 0$ was proved by the author. A few months later, Yau and the author gave a complete proof for the general case [6].

THEOREM 1. *Let* M *be a compact manifold without boundary. Suppose* $\mathrm{Ric} \geq K$ *with* $K \leq 0$; *then*

$$\lambda_1 \geq \frac{\exp - [1 + (1-2cd^2K)^{1/2}]}{cd^2}$$

where c = *constant depending only on* n, *and* d = *diameter of* M.*

Lower bounds for μ_1 and η_1 were also established.

THEOREM 2. *Let* M *be a compact manifold with boundary* ∂M. *Suppose* $\mathrm{Ric} \geq K$ *with* $K \leq 0$.

(i) *If* H *is a lower bound for the mean curvature of* ∂M *with respect to* $\partial/\partial\nu$, *then*

$$\mu_1 \geq \frac{1}{a}\left[\frac{(\log a)^2}{4(n-1)\rho^2} + K\right]$$

where $a = \max\{\exp[1 + (1 - c\rho^2K)^{1/2}], \exp[-cH\rho]\}$, *with* c = *constant depending on* n, *and* ρ = *inscribe radius.*

(ii) *If* ∂M *is convex, i.e., the second fundamental form of* ∂M *is non-negative, then*

*Gromov has announced a different proof to this theorem.

$$\eta_1 \geq \frac{\exp - [1 + (1-2cd^2K)^{1/2}]}{cd^2}$$

where c = *constant depending only on* n, *and* d = *diameter of* M.

The proofs of Theorems 1 and 2 depended on some estimates for the first eigenfunctions. In fact, gradient estimates for functions satisfying equations of the form

$$\Delta u = F(u) \tag{3.1}$$

was developed. This method would also yield lower bounds for eigenvalues of Δ with boundary conditions of the type

$$\partial u/\partial \nu = -\alpha u \tag{3.2}$$

with $\alpha \geq 0$ being a fixed constant.

In the same paper, sharp results for the case $K \geq 0$ were also derived.

THEOREM 3. *Let* M *be a compact manifold without boundary. If* $\mathrm{Ric} \geq 0$, *then*

$$\lambda_1 \geq \frac{\pi^2}{2d^2} .$$

In addition, if the multiplicity of λ_1 *is greater than* 1, *then*

$$\lambda_1 \geq \frac{\pi^2}{d^2} .$$

THEOREM 4. *Let* M *be a manifold with* $\partial M \neq \emptyset$. *Suppose* $\mathrm{Ric} \geq 0$, *and*

(i) *if the mean curvature of* ∂M *is nonnegative, then*

$$\mu_1 \geq \frac{\pi^2}{4\rho^2}$$

where ρ = *inscribe radius of* M;

(ii) *if* ∂M *is convex, then*

$$\eta_1 \geq \frac{\pi^2}{2d^2} .$$

Also if the multiplicity of η_1 *is greater than* 1 , *then*

$$\eta_1 \geq \frac{\pi^2}{d^2} .$$

The proofs of Theorems 3 and 4 are based on inequalities for the first eigenfunctions of the type

$$\frac{|\nabla f|}{\sqrt{(\sup f)^2 - f^2}} \leq \lambda_1^{1/2} \text{ (respectively } \mu_1^{1/2} \text{ or } \eta_1^{1/2}) \tag{3.3}$$

By completeness of M, let γ be the shortest geodesic joining the supremum point x_1 of f and the infimum point x_2 of f. For Theorem 4(i), we may assume $f \geq 0$, hence $\inf f = 0$. Also since the length of γ, $\ell(\gamma)$, is at most ρ, it is clear that the desired estimate can be obtained by integrating (3.3) along γ. For the cases of Theorems 3 and 4(ii), since $\int_M f = 0$, $\inf f < 0$ and $\ell(\gamma) \leq d$. Integrating (3.3) along γ gives

$$\frac{\pi}{2} + \sin^{-1}\left(\frac{|\inf f|}{\sup f}\right) \leq d\lambda_1^{1/2} \text{ (or } d\eta_1^{1/2}) . \tag{3.4}$$

If the multiplicity of λ_1 (or η_1) ≥ 2, then we claim that there exists an eigenfunction with eigenvalue λ_1 (η_1 respectively) such that

$$\sup f = -\inf f . \tag{3.5}$$

If not, we may assume $\sup f_i > -\inf f_i$ for any two linearly independent eigenfunctions f_1 and f_2. Consider the 1-parameter of eigenfunctions

$$f_t = tf_1 - \sqrt{1-t^2}\, f_2$$

for $0 \leq t \leq 1$. Clearly $f_t \not\equiv 0$ is an eigenfunction with eigenvalue λ_1 (or η_1). Define $g(t) = \sup f_t + \inf f_t$. Then

$$\begin{aligned} g(0) &= \sup(-f_2) + \inf(-f_2) \\ &< 0 \end{aligned}$$

and

$$g(1) = \sup f_1 + \inf f_1$$
$$> 0 .$$

Hence by continuity of g, there exists $0 < t_0 < 1$ such that $g(t_0) = 0$. Thus f_{t_0} is the required eigenfunction which satisfies (3.5). Substituting f_{t_0} into (3.4) yields

(3.6) $$\lambda_1 \geq \frac{\pi^2}{d^2} \quad \left(\eta_1 \geq \frac{\pi^2}{d^2}\right).$$

When without the multiplicity assumption, one only has to consider the product manifold $M \times M$. One verifies that the hypothesis on the Ricci curvature is preserved on $M \times M$. Moreover, $\lambda_1(M \times M) = \lambda_1(M)$ with multiplicity ≥ 2. Applying (3.6), we have

$$\lambda_1(M) = \lambda_1(M \times M)$$
$$\geq \frac{\pi^2}{d^2(M \times M)} = \frac{\pi^2}{2d^2(M)}$$

which is to be proved (similarly for $\eta_1(M)$).

It is worth pointing out that Payne and Weinberger [9] showed that

(3.7) $$\eta_1 \geq \frac{\pi^2}{d^2}$$

for convex Euclidean domains without any assumption on the multiplicity of η_1. Later in their paper [1], Chavel and Feldman pointed out that the proof of Payne and Weinberger appeared to be incomplete. They then proceeded to give a complete proof of (3.7) for 2-dimensional manifolds with convex boundaries which have curvature conditions

$$0 \leq \text{curvature of } M \leq \beta < \infty .$$

Although the objection of Chavel and Feldman has not yet been overcome for dimension ≥ 3, the multiplicity assumption on Theorems 3 and 4(ii) appears to be a purely technical one.

COROLLARY 5. *Let* M *be a compact homogeneous Riemannian manifold with* $\mathrm{Ric} \geq 0$. *Then*

$$\lambda_1 \geq \frac{\pi^2}{d^2}.$$

This follows directly from the fact that the eigenvalues on a homogeneous manifold have multiplicities at least 2 [5]. Moreover, if the multiplicity of any eigenvalue is exactly 2, one concludes that the universal covering $\tilde{M}$ of M isometrically splits into $\mathbf{R} \times N$. In fact, if $M = G/H$ where G is the isometry group of M, then G acts on any eigenspace E_λ orthogonally. Hence E_λ has the following irreducible decomposition:

$$E_\lambda = E_1 \oplus E_2 \oplus \cdots \oplus E_k .$$

PROPOSITION 6. *Let* M *be a compact homogeneous Riemannian manifold. There exists an eigenvalue* λ *such that one of the irreducible components, say* E_1, *of* E_λ *is of dimension* 2 *iff* $\tilde{M}$ *isometrically splits into* $\mathbf{R} \times N$.

Under special curvature assumption on M, we demonstrated that λ_1, μ_1 and η_1 possess sharper lower bounds than that obtained in Theorems 1 and 2. It is important to point out that in the case of Riemann surfaces Σ_g^2, eigenvalue estimates have been obtained by Buser, Huber, McKean, Randol, Schoen-Wolpert-Yau, etc.

§4. It has been a general belief that without any curvature restriction, λ_1 and d will not satisfy any direct relation. Indeed, the example in [2] clearly confirmed the above observation. However, this was disproved among the class of compact homogeneous manifolds.

THEOREM 7 [5]. *Let* M *be a compact homogeneous Riemannian manifold. Then*

$$\lambda_1 \geq \frac{\pi^2}{4d^2} .$$

If M *is an irreducible homogeneous space, then*

$$\lambda_1 \geq \frac{n\pi^2}{4d^2} .$$

Due to the presence of a transitive isometry group, gradient estimates on zonal functions (see [5] for definition) were derived. This enabled us to apply similar technique as mentioned above in order to yield lower bounds for λ_1. One also observes that a generalized version of Lichnerowicz's theorem also holds in the category of irreducible homogeneous manifolds. More specifically, let R be the scalar curvature of M, which under the above hypothesis is known to be a positive constant, then

$$\lambda_1 \geq R .$$

STANFORD UNIVERSITY
STANFORD, CALIFORNIA 94305

BIBLIOGRAPHY

[1] Chavel, I., and Feldman, E., An optimal Poincaré inequality for convex domains of nonnegative curvature, *Arch. Rat. Mech. Anal.*, vol. 65, 3 (1977), 263-273.

[2] Cheeger, J., A lower bound for the smallest eigenvalue of the Laplacian, in *Problems in Analysis, a symposium in honor of S. Bochner*, Princeton University Press (1970).

[3] Cheng, S. Y., Eigenvalue comparison theorems and its geometric applications, *Math. Z.*, vol. 143 (1975), 289-297.

[4] Li, P., A lower bound for the first eigenvalue of the Laplacian on a compact Riemannian manifold, *Indiana U. Math. J.*, Vol. 28 (1979), 1013-1019.

[5] ———, Eigenvalue estimates on homogeneous manifolds, *Comm. Math. Helv.*, Vol. 55 (1980), 347-363.

[6] Li, P., and Yau, S. T., Estimates of eigenvalues of a compact Riemannian manifold, *Proc. Sym. Pure Math.*, Vol. 36 (1980), 205-239.

[7] Lichnerowicz, A., Géométrie des groupes de transformations, Dunod (1958).

[8] Obata, M., Certain conditions for a Riemannian manifold to be isometric with a sphere, *J. Math. Soc. Japan*, vol. 14 (1962), 333-340.

[9] Payne, L. E., and Weinberger, H., The optimal Poincaré inequality for convex domains, *Arch. Rat. Mech. Anal.*, vol. 5, 4 (1960), 282-292.

[10] Reilly, R. C., Applications of the Hessian operator in a Riemannian manifold, *Indiana U. Math. J.*, vol. 26, 3 (1977), 459-472.

[11] Yau, S. T., Isoperimetric constants and the first eigenvalue of a compact Riemannian manifold, *Ann. Scient. Éc. Norm. Sup.* 4^c série, t. 8 (1975), 487-507.

BONNESEN-TYPE INEQUALITIES IN ALGEBRAIC GEOMETRY, I: INTRODUCTION TO THE PROBLEM

B. Teissier*

Introduction

Let $K \subset \mathbf{R}^2$ be a compact convex domain bounded by a curve of length L. The area S of K is subjected to the isoperimetric inequality $L^2 - 4\pi S \geq 0$, which implies that among all such domains with a given perimeter L, the disk maximizes the area. The most constructive proof of the fact that the disk is the only domain with this property is in my opinion that given by T. Bonnesen in ([2], p. 69): Bonnesen shows that if we consider the greatest radius r of a disk contained in K, and the smallest radius R of a disk containing K, we have the inequality

$$L^2 - 4\pi S \geq \pi^2(R-r)^2 \tag{1}$$

and this settles immediately the equality case of the isoperimetric inequality. The proof of (1) in fact contains the proof of stronger inequalities (see [2], p. 60, and compare with p. 123 of L. A. Santaló's magnificent book [14], and R. Osserman's beautiful paper [12]):

$$\frac{L + \sqrt{L^2 - 4\pi S}}{2\pi} \geq R \geq r \geq \frac{L - \sqrt{L^2 - 4\pi S}}{2\pi} . \tag{2}$$

*Partially supported by NSF Grant MCS 7715524.

Seminar on Differential Geometry
0-691-08268-5/82/000085-21 $01.05/0 (cloth)
0-691-08296-0/82/000085-21 $01.05/0 (paperback)

Our aim is to investigate the generalization of these inequalities (2) to compact convex domains in $\mathbf{R}^d$ for $d \geq 2$ in a much broader context.

More precisely, let d be an integer, $d \geq 2$, and let K_1 and K_2 be two compact convex subsets of $\mathbf{R}^d$. Following Minkowski (see [2], p. 105, [$2\frac{1}{2}$], p. 60) one defines the *mixed volumes* of K_1 and K_2 as the coefficients in the following expression for the volume of the Minkowski sum $\nu_1 K_1 + \nu_2 K_2$ of the homothetics $\nu_1 K_1$, $\nu_2 K_2$ of K_1 and K_2, as ν_1 and ν_2 range over the nonnegative real numbers $\mathbf{R}_+$. (By definition $\nu_1 K_1 + \nu_2 K_2 = \{y_1 + y_2, y_1 \in \nu_1 K_1, y_2 \in \nu_2 K_2\}$.) One has then the expression (see [$2\frac{1}{2}$], p. 40, [2], p. 106),

$$\mathrm{Vol}(\nu_1 K_1 + \nu_2 K_2) = \sum_{i=0}^{d} \binom{d}{i} v_i \nu_1^i \nu_2^{d-i},$$

and the $v_i \in \mathbf{R}$, sometimes written $v(K_1^{[i]}, K_2^{[d-i]})$ $(0 \leq i \leq d)$, are the mixed volumes of K_1 and K_2. Note that $v_0 = \mathrm{Vol}(K_2)$ and $v_d = \mathrm{Vol}(K_1)$, and that if $K_1 = K_2$ up to translation, then $v_0 = \cdots = v_d$.

In the special case where $K_1 = \mathbf{B}$, the unit ball in $\mathbf{R}^d$, one finds that $d \cdot v_1$ is equal to $\mathrm{Vol}(\partial K_2)$, where $\mathrm{Vol}(\partial K_2)$ is the $(d-1)$-dimensional volume. Taking in particular $d = 2$, we find that

$$\mathrm{Vol}(\nu_1 \mathbf{B} + \nu_2 K) = S\nu_2^2 + L\nu_1\nu_2 + \pi\nu_1^2 \qquad (\text{for } \nu_1 \geq 0,\ \nu_2 \geq 0).$$

Going back to K_1, K_2 in $\mathbf{R}^d$ we define two real numbers, the inradius $r(K_2; K_1)$ and the outradius $R(K_2; K_1)$ of K_2 with respect to K_1, by

$$r(K_2; K_1) = \mathrm{Sup}\{ r \in \mathbf{R}_+ \mid rK_1 \subseteq K_2 \ , \text{ up to translation}\}$$

$$R(K_2; K_1) = \mathrm{Inf}\{ R \in \mathbf{R}_+ \mid K_2 \subseteq RK_1, \text{ up to translation}\}.$$

In the case where $d = 2$ and $K_1 = \mathbf{B}$ we find r and R of the inequalities (1) and (2).

PROBLEM A. Given two compact convex subsets K_1 and K_2 in $\mathbf{R}^d$, give bounds for $r(K_2;K_1)$ and $R(K_2;K_1)$ in terms of the mixed volumes v_i $(0 \le i \le d)$ of K_1 and K_2. These bounds should imply that when $v_0 = \cdots = v_d$, then $r(K_2;K_1) = R(K_2;K_1) = 1$, and therefore $K_1 = K_2$ up to translation.

Note that the inequalities (2) answer that problem for $d = 2$ and $K_1 = \mathbf{B}$. There is an answer for arbitrary $d \ge 2$ and $K_1 = \mathbf{B}$, due to Hadwiger and Dinghas (see [12], IIIC). There is also a splendid proof by A.D. Alexandrov, of the fact that the equalities $v_0 = \cdots = v_d$ imply that $K_1 = K_2$, up to translation. I give below a proof of the answer, originally due to Flanders [$6\frac{1}{2}$], to Problem A for $d = 2$, but my main interest, and the motivation for this paper, lies in the problems suggested in algebraic geometry by Problem A when one "embeds" a part of the theory of convex sets into algebraic geometry as explained below. In other words, I am interested in the problem in algebraic geometry of which Problem A is an avatar!

The principal new fact contained in this paper is that Problem A can be deemed to be a rather special case of a problem of algebraic geometry: to find sufficient numerical conditions for an invertible sheaf to have sections. This fact is established by associating to integral polyhedra K_1 and K_2 in $\mathbf{R}^d$ algebraic varieties of dimension d with two invertible sheaves L_1 and L_2, the properties of which reflect very well the properties of K_1 and K_2. Using approximation of arbitrary compact convex sets by integral ones, we obtain a dictionary which translates problems on compact convex subsets of $\mathbf{R}^d$ into problems on invertible sheaves on some very special algebraic varieties of dimension d (Demazure varieties, or torus embeddings). This dictionary was used in [19] to show that the basic inequalities of the isoperimetric problem in dimension $d \ge 2$, namely the Minkowski-Alexandrov-Fenchel inequalities between the mixed volumes of two compact convex subset of $\mathbf{R}^d$:

$$v_{i-1}^2 - v_i v_{i-2} \ge 0 \qquad (2 \le i \le d) \tag{3}$$

were consequences of the *Hodge index theorem* in the theory of algebraic surfaces. In this paper we show that our viewpoint is operative also for Problem A by giving a proof of the solution for $d = 2$. This proof relies on an easy (nowadays) but rather deep result on the geometry of invertible sheaves on a projective algebraic surface.

A point of interest is that once again this is a quite general result, the validity of which is not restricted to the Demazure surfaces which we encounter when starting from convex sets.

On the way, we recall in §1 the construction of our Note [19] and make precise the general inequalities of algebraic geometry which imply the inequalities (3). Thus this paper, although it is a continuation of [19], can be read without prerequisites. In this connection, after the Note [19] went to press, I learned that Mr. A.G. Hovanski, of Moscow, has independently given a construction analogous to that of §§1 and 2 of that Note, and, inspired like myself by my inductive proof in [16], has given in a Note [8] a sketch of a proof of the Alexandrov-Fenchel inequalities which is very similar to that of [19]. I also point to a recent paper [13] of D. Rees and R.Y. Sharp showing that the "Minkowski-type inequalities" for multiplicities of [16] were valid for arbitrary noetherian local rings. The range of validity of this type of inequality is therefore large.

I should like to thank Tadao Oda, whose paper [11] influenced the construction in §1 (or [19]) and Professor Rolf Schneider who, by the material he has sent to me, has helped me very much to study the beautiful theory of mixed volumes.

§1. *Alexandrov-Fenchel type inequalities in algebraic geometry*

1.1. The construction (after [4], [9], [19]). Let d be an integer, $d \geq 2$, and let $\mathcal{K}$ (resp. $\mathcal{K}_M$) be the set of compact convex subsets of $\mathbf{R}^d$ (resp. of those which are the convex hull of a finite number of points in the integral lattice $M \simeq \mathbf{Z}^d \subset \mathbf{R}^d$).

Given $K_1, \cdots, K_r$ in $\mathcal{K}$, each K_i has a support function $H_i : \mathbf{R}^{d^*} \to \mathbf{R}$ defined by:

$$H_i(u) = \min\{u(m)/m \in K_i\}; \quad H_i \text{ is a convex function.}$$

If the K_i are in $\mathcal{K}_M$, the H_i are piecewise-linear and it is not hard to see that there exists a decomposition $\Sigma = (\sigma_\alpha)_{\alpha \in A}$ of $\mathbf{R}^{d^*}$ into rational convex polyhedral cones σ_α such that:

1) Each face of a σ_α is a σ_β for some $\beta \in A$.
2) $\sigma_\alpha \cap \sigma_\beta$ $(\alpha, \beta \in A)$ is a face of σ_α and of σ_β.
3) For each i, $1 \le i \le r$, H_i is linear on σ_α for all $\alpha \in A$.

Now to each σ_α, associate its convex-dual $\check{\sigma}_\alpha = \{x \in \mathbf{R}^d \mid u(x) \ge 0, \forall u \in \sigma_\alpha\}$. The subset $\check{\sigma}_\alpha \cap M$ of $\mathbf{R}^d$ is a submonoid of M, and hence one can define for any given field, say the field $\mathbf{C}$ of complex numbers, the "algebra of the monoid $\check{\sigma}_\alpha \cap M$," which is the subalgebra $\mathbf{C}[\check{\sigma}_\alpha \cap M]$ of the algebra $\mathbf{C}[M] \simeq \mathbf{C}[X_1, X_1^{-1}, \cdots, X_d, X_d^{-1}]$ generated by those monomials which have their exponents in $\check{\sigma}_\alpha \cap M$.

We can glue up the affine varieties Spec $\mathbf{C}[\check{\sigma}_\alpha \cap M]$ along the Spec $\mathbf{C}[\widecheck{\sigma_\alpha \cap \sigma_\beta} \cap M]$ to obtain a compact algebraic variety $X = X(\Sigma)$, the Demazure variety associated to the decomposition Σ. (For all this, see [4], §4 and [9].) X is a normal, integral and rational variety of dimension d. The field of fractions of each $\mathbf{C}[\check{\sigma}_\alpha \cap M]$ is $\mathbf{C}(M)$ and we are going to recall how to associate to each support function H_i a line bundle L_i on X: For each $\alpha \in A$ and i, $1 \le i \le r$, consider the sub $\mathbf{C}[\check{\sigma}_\alpha \cap M]$-module of $\mathbf{C}(M)$ generated by $\{m \in M \mid u(m) \ge H_i(u)$ for all $u \in \sigma_\alpha\}$ and denote it by $L_{i,\alpha}$. Since H_i is linear on σ_α, there exists $m_{i,\alpha} \in M$ such that $H_i(u) = u(m_{i,\alpha})$ on σ_α, and therefore $L_{i,\alpha}$ is generated by $m_{i,\alpha}$. Now clearly the $L_{i,\alpha}$ glue up together into an invertible sheaf of fractional ideals L_i, and a basis of $H^0(X, L_i)$ is given by $\{m \in M \mid u(m) \ge H_i(u)$ for all $u \in \mathbf{R}^{d^*}\}$ which is exactly $K_i \cap M$ in view of the convexity of K_i. Furthermore, it is proved in ([9], pp. 42-44) that for all invertible sheaves L_i obtained in this way from $K_i \in \mathcal{K}_M$, not only is L_i generated by its global sections (because necessarily $m_{i,\alpha} \in K_i \cap M$) but also we have

$$H^j(X, L_i) = 0 \quad \text{for} \quad j \ge 1 .$$

We recall also that there is an operation on $\mathcal{K}$, called the Minkowski sum: $K_1 + K_2$ is by definition $\{k_1+k_2;\ k_1 \in K_1, k_2 \in K_2\}$; a special case of it is the translation $K+m$ by the vector $\vec{Om}$. The result is independent of the choice of origin $O \in \mathbf{R}^d$, up to translation. The support function of $K_1 + K_2$ is $H_1 + H_2$ and if we took K_1, K_2 in $\mathcal{K}_M$, and Σ as above, then $H_1 + H_2$ is again linear in each σ_α, hence we can associate to $K_1 + K_2$ a line bundle on X, which is nothing but the tensor product $L_1 \otimes L_2$. In particular to the homothetic $\nu \cdot K_i$ for $\nu \in N$ we associate L_i^ν, where L_i is associated to K_i as above. (Note that if K is a point, e.g., $O \in \mathbf{R}^d$, then $L \simeq \mathcal{O}_X$, and therefore more generally the sheaf associated to $K_i + m$, for $m \in M$, is isomorphic to L_i.)

1.2. Given $K_1, \cdots, K_r$ in $\mathcal{K}$, consider for all $\nu_i \geq 0$ the homothetic convex sets $\nu_i K_i$ and their Minkowski sum $\nu_1 \cdot K_1 + \cdots + \nu_r \cdot K_r = \{x_1 + \cdots + x_r \in \mathbf{R}^d / x_1 \in \nu_1 K_1, \cdots, x_r \in \nu_r \cdot K_r\}$. By a result of Minkowski-Steiner (cf. [2], [$10\frac{1}{2}$]) the volume of this set has an expression in terms of the ν_i which is a homogeneous polynomial of degree d:

$$\mathrm{Vol}(\nu_1 \cdot K_1 + \cdots + \nu_r \cdot K_r) = \sum_{\substack{\alpha \in \mathbf{N}^r \\ |\alpha| = d}} \frac{d!}{\alpha!} v_\alpha \nu_1^{\alpha_1} \cdots \nu_r^{\alpha_r} .$$

The v_α which are defined by this expression and sometimes written $v(K_1^{[\alpha_1]}, \cdots, K_r^{[\alpha_r]})$ are called the mixed volumes of the K_i, and they have the following elementary properties:

1) All the v_α are left unchanged by translations on the K_i (indeed, translations on the K_i induce translations on the $\sum \nu_i K_i$).
2) The v_α are increasing functions of the K_i, i.e., if $K_i \subset K'_i$ (up to translation) we have:

$$v(K_1^{[\alpha_1]}, \cdots, K_i^{[\alpha_i]}, \cdots, K_r^{[\alpha_r]}) \leq v(K_1^{[\alpha_1]}, \cdots, K_i'^{[\alpha_i]}, \cdots, K_r^{[\alpha_r]})$$

and if $\alpha_i = 0$, we have equality.

In particular, all the v_α are ≥ 0.

3) For positive λ_i, we have

$$v((\lambda_1 K_1)^{[\alpha_1]}, \cdots, (\lambda_r K_r)^{[\alpha_r]}) = \lambda_1^{\alpha_1}, \cdots, \lambda_r^{\alpha_r} v(K_1^{[\alpha_1]}, \cdots, K_r^{[\alpha_r]}) .$$

4) Giving $\mathcal{K}$ its natural Hausdorff topology (cf. [10bis]) we have that the v_α are continuous functions on $\mathcal{K}^r$.

1.3. Now given invertible sheaves $L_1, \cdots, L_r$ on an algebraic variety X of dimension d, Snapper (see [10], Chap. I, §1) has shown that there is a polynomial expression for the coherent Euler-characteristic of $L_1^{\nu_1} \otimes \cdots \otimes L_r^{\nu_r}$, as follows: for $\nu_1, \cdots, \nu_r$ in $\mathbf{Z}$, we have

$$\chi(X, L_1^{\nu_1} \otimes \cdots \otimes L_r^{\nu_r}) = \sum_{\substack{\alpha \in \mathbf{N}^r \\ |\alpha| = d}} \frac{1}{\alpha!} s_\alpha \nu_1^{\alpha_1} \cdots \nu_r^{\alpha_r} \tag{1.3.1}$$

$$+ \text{ polynomial of degree } \leq d-1$$

and this expression defines integers s_α which we may call the *mixed degrees* of the invertible sheaves L_i (sometimes written $s_\alpha = \deg(L_1^{[\alpha_1]}, \cdots, L_r^{[\alpha_r]})$).

If we start with $K_1, \cdots, K_r$ in $\mathcal{K}_M$, and associate to them an "eventail" Σ, X and the L_i as above, we saw that the groups $H^j(X, L_1^{\nu_1} \otimes \cdots \otimes L_r^{\nu_r})$ are 0 for $j \geq 1$ and $\nu_i \geq 0$ and hence we have, for $\nu_i \geq 0$

$$\chi(X, L_1^{\nu_1} \otimes \cdots \otimes L_r^{\nu_r}) = h^0(X, L_1^{\nu_1} \otimes \cdots \otimes L_r^{\nu_r}) = \#(M \cap (\nu_1 K_1 + \cdots + \nu_r K_r))$$

as we saw, and therefore letting the ν_i tend to $+\infty$ and remembering that for any $K \in \mathcal{K}$, we have

$$\mathrm{Vol}(K) = \lim_{\nu \to +\infty} \nu^{-d} \cdot \#(M \cap \nu \cdot K)$$

we deduce that *the mixed degrees of the* L_i *are linked to the mixed volumes of the* K_i *by the equality*:

(1.3.2) $$s_\alpha = d!\,v_\alpha .$$

1.4. Now let us fix an integer t, $2 \le t \le d$, and set $v_i = v_{(i,t-i,1,\cdots,1)}$ and $s_i = s_{(i,t-i,1,\cdots,1)}$ where the sequence $1,\cdots,1$ has $d-t$ terms. In the note [19] we showed how to use the Hodge index theorem to prove the quadratic inequalities

(1.4.1) $$s_i \cdot s_{i-2} \le (s_{i-1})^2 \qquad (2 \le i \le t)$$

which imply the *Alexandrov-Fenchel inequalities*

(1.4.2) $$v_i \cdot v_{i-2} \le (v_{i-1})^2 \ \text{ for } K_1,\cdots,K_r \in \mathcal{K}_M \quad \text{(by 1.3.2)}$$

which in turn imply the same Alexandrov-Fenchel inequalities for compact convex sets $K_1,\cdots,K_r$ in $\mathcal{K}$, by using the continuity of the mixed volumes, their homogeneity, and a standard approximation procedure using the following:

1.4.2.1. FACT. For $K \in \mathcal{K}$, let us denote by $[K] \in \mathcal{K}_M$ the convex hull of $M \cap K$, where M is the integral lattice of $\mathbf{R}^d$. Then, for any $K \in \mathcal{K}$ we have for large ν:

$$0 \le \mathrm{Vol}\,(K - \tfrac{1}{\nu}[\nu \cdot K]) < \frac{c(K)}{\nu} \ \text{ where } c(K) \text{ is a constant.}$$

1.4.3. We emphasize that the result in algebraic geometry is much more general than what is needed for the Alexandrov-Fenchel inequalities:

Let us define the degree of an invertible sheaf on a complete algebraic variety X of dimension d by the equality (see [10], Ch. I)

$$\chi(X, L^\nu) = \deg L \cdot \frac{\nu^d}{d!} + \text{polynomial of degree} \le d-1 \ \text{ in } \nu .$$

Then we may define the mixed degrees s_α of $L_1,\cdots,L_r$ by the equality

$$\deg(L_1^{\nu_1} \otimes \cdots \otimes L_r^{\nu_r}) = \sum_{\substack{\alpha \in \mathbf{N}^r \\ |\alpha| = d}} \frac{d!}{\alpha!} s_\alpha \nu_1^{\alpha_1} \cdots \nu_r^{\alpha_r}$$

as a glance at 1.3.1 will show.

Then, choosing an integer t, $0 \leq t \leq d$ as in 1.4 and with the same notations, we have:

1.4.3.1. PROPOSITION. *Let* X *be a proper integral algebraic variety and let* $L_1, \cdots, L_r$ *be invertible sheaves on* X, *generated by their sections, with* $\deg L_i > 0$ $(1 \leq i \leq r)$.

Then we have

1) $s_\alpha \geq 0$

2) *For each* i, $2 \leq i \leq t$, we have the *Alexandrov-Fenchel-type inequalities*

$$s_{i-1}^2 - s_i s_{i-2} \geq 0 .$$

COROLLARY. *Take now* $r = 2$; *then*

1.4.3.2
$$\deg L_1^{\nu_1} \otimes L_2^{\nu_2} = \sum_{i=0}^{d} \binom{d}{i} s_i \nu_1^i \nu_2^{d-i}$$

and the inequalities (2) above imply easily (see [16]) *the inequalities*

1.4.3.3
$$\underline{s_i^d \geq s_0^{d-i} \cdot s_d^i}, \qquad 0 \leq i \leq d .$$

(The analogue of the classical isoperimetric inequality $\mathrm{Vol}(\partial \check{K})^d \geq d^d \mathrm{Vol}(B) \mathrm{Vol}(K)^{d-1}$ is the case $i = 1$.) In turn 1.4.3.3 implies, in view of 1.4.3.2

1.4.3.4
$$\deg(L_1 \otimes L_2)^{1/d} \geq (\deg L_1)^{1/d} + (\deg L_2)^{1/d}$$

(compare [16]): this is the analogue of the Brünn-Minkowski inequality (see [2], [5]) and at least if L_1 is ample, say, we have equality if and only if there exist positive integers a, b such that L_1^a and L_2^b are numerically equivalent (see [3], exp. XIII, and compare [17]).

1.5. REMARK (ubiquitous inequalities). All these analogies between isoperimetric inequalities and algebraic ones first appeared (see [18]) in the case of multiplicities of primary ideals n_1, n_2 in a complex analytic algebra $\mathcal{O}$, the mixed multiplicities of n_1 and n_2 being defined by

$$e(n_1^{\nu_1} n_1^{\nu_2}) = \sum_{i=0}^{d} \binom{d}{i} e_i \nu_1^i \nu_2^{d-i} \qquad \text{where } d = \dim \mathcal{O} \geq 1 .$$

and the inequalities which were proved were: (see [16], [17])

1) $e_i \cdot e_{i-2} \geq e_{i-1}^2$
2) $e_i^d \leq e_0^{d-i} e_d^i$
3) $e(n_1 \cdot n_2)^{1/d} \leq e(n_1)^{1/d} + e(n_2)^{1/d}$.

Provided $\mathcal{O}$ is normal, there is equality in 3) if and only if there exist positive integers a and b such that n_1^a and n_2^b have the same integral closure. These results on multiplicities correspond to the negative-definiteness of the intersection matrix of the components of the exceptional divisor in a resolution of singularities of a germ of a normal surface, and the results on degrees, inspired by these, correspond, as already explained, to the Hodge Index theorem. It is interesting to note that the same year 1937 saw the publication by A. D. Alexandrov of his inequalities (cf. [1]) and the publication by Hodge of his Index theorem (cf. [7]). Furthermore, Hodge thanks DuVal for the formulation of the Index theorem, and DuVal himself had a few years before discovered the negative-definiteness of the intersection matrix (cf. [6]), a fact which is the local version of the Hodge Index theorem; and is equivalent to the following statement: Given two curves A and B on a germ of a normal surface $(S, 0)$, their intersection multiplicity defined by Mumford, which is a rational number, must satisfy the inequality $(A, B)_0 \geq \dfrac{m_0(A) \cdot m_0(B)}{m_0(S)}$ where m_0 is the multiplicity, and equality must hold if and only if the two curves have no common tangent at 0. See [16].

§2. *The translation*

Recall from the Introduction that given K_1 and K_2 in $\mathcal{K}$ we defined the inradius of K_2 with respect to K_1

$$r(K_2; K_1) = \operatorname{Sup}\{r \in R_+ / rK_1 \subseteq K_2 \text{ up to translation}\} .$$

Our problem as set in the introduction is so symmetric that it is enough to study one-half of it:

2.0. PROBLEM A′. Give bounds for $r(K_2; K_1)$ in terms of the mixed volumes v_i of K_1 and K_2 defined by

$$\operatorname{Vol}(\nu_1 K_1 + \nu_2 K_2) = \sum_{i=0}^{d} \binom{d}{i} v_i \nu_1^i \nu_2^{d-i} \qquad (\nu_1, \nu_2 \in R^+)$$

$$(v_d = \operatorname{Vol}(K_1), v_0 = \operatorname{Vol}(K_2)) .$$

We are going to translate this problem into algebraic geometry; we use the constructions and notations of §1. Here we have

2.1. KEY LEMMA. *Let* d *be an integer,* $d \geq 2$, *let* K_1 *and* K_2 *be in* $\mathcal{K}_M$ *and let the compact algebraic variety* X *and the line bundles* L_1 *and* L_2 *be associated to them as in paragraph 1. Then given integers* a, b, $a \geq 0$, $b \geq 0$, *for every integer* $\nu > 0$ *there is a bijection*

$$\{m \in \tfrac{1}{\nu} M \mid a \cdot K_1 + m \subseteq b \cdot K_2\} \xrightarrow{\sim} \textit{basis of } H^0(X, (L_1^{-a} \otimes L_2^{b})^{\nu}) .$$

Proof. The inclusion $a \cdot K_1 + m \subseteq b \cdot K_2$ is equivalent to the inequalities $a \cdot H_1(u) + u(m) \geq b \cdot H_2(u)$ for all $u \in R^{d^*}$.

That is: $u(m) \geq b \cdot H_2(u) - a \cdot H_1(u)$ and this precisely means that if we have $\nu \cdot m \in M$, then $\nu \cdot m$ corresponds to a global section of the invertible sheaf $(L_1^{-a} \otimes L_2^{b})^{\nu}$. The converse is obtained by reading backwards, in view of the fact that $H^0(X, (L_1^{-a} \otimes L_2^{b})^{\nu})$ has a natural M-grading for which all non-zero homogeneous components are of dimension 1 .

2.2. Let X be a complete integral algebraic variety and let L_1 and L_2 be two invertible sheaves on X, of positive degree. Define the *inradius* of L_2 with respect to L_1:

$$r(L_2; L_1) = \operatorname*{Sup}_{(a,b)\in N} \left\{ \frac{a}{b} \,\middle|\, H^0(X, L_1^{-a} \otimes L_2^{b}) \neq 0 \right\}$$

and remark that for all integers $\nu > 0$,

$$r(L_2^{\nu}; L_1^{\nu}) = r(L_2; L_1) .$$

It follows immediately from the equality 1.3.2, the Key Lemma 2.1 and 1.4.2.1 that Problem A′ is a special case of the following:

2.2.1. PROBLEM B. Given two invertible sheaves of positive degree L_1 and L_2, generated by their sections, on a complete algebraic variety X, give bounds for $r(L_2; L_1)$ in terms of the mixed degrees s_i of L_1 and L_2, defined by

$$\deg(L_1^{\nu_1} \otimes L_2^{\nu_2}) = \sum_{i=0}^{d} \binom{d}{i} s_i \nu_1^i \nu_2^{d-i}$$

$$(s_d = \deg L_1 ,\ s_0 = \deg L_2) .$$

2.3. Here we are going to study this problem only in the case where X is a projective variety, and we assume that $L_1 \otimes L_2$ is very ample. Indeed, this is sufficient for the applications to the problem on convex set which is outlined in the introduction, for the following reason: if we take for Σ the *coarsest* subdivision of R^{d^*} which makes H_1 and H_2 linear in each piece, then by the amplitude criterion given by Demazure (cf. [4], §4, and also [9], p. 48) we have precisely that $L_1 \otimes L_2$ is ample and since we may freely replace L_1 and L_2 by L_1^{ν} and L_2^{ν} $(\nu > 0)$, we may assume that $L_1 \otimes L_2$ is very ample.

§3. *An example: the case* $d=2$ *(and a proof of Bonnesen's inequality)*

We shall use the following

3.1. LEMMA (see [10], p. 3.8 and [3], Exp XIII, Appendice, Lemma 7.1.2, and Exp. X). *Let* S *be an integral projective surface with a very ample invertible sheaf* $H = \mathcal{O}_X(D)$. *Let* L *be an invertible sheaf on* X. *The following conditions are equivalent:*

i) $\deg L > 0$ *and* $\deg L_D > 0$ ($L_D = L \otimes \mathcal{O}_D$, *an invertible sheaf on* D)

ii) *For sufficiently large* ν *we have*

$$\dim H^0(X, L^\nu) > \varepsilon \cdot \nu^2 \quad \textit{with} \quad \varepsilon > 0 .$$

(Note: Use [3], Exp. X, to check that $\deg L = c_1(L)^2$, $\deg L_D = c_1(L) \cdot c_1(H)$).

Let us apply this lemma with $L = L_1^{-a} \otimes L_2^b$, $a, b \in \mathbb{N}$, where L_1 and L_2 are two invertible sheaves of positive degree such that $H = L_1 \otimes L_2$ is very ample. We have, with the notations introduced above

$$\deg L = s_0 b^2 - 2 s_1 ab + s_2 a^2$$

and using a result in ([10], Chap. I), we can compute $\deg L_D$ as the coefficient of $2\mu\nu$ in the homogeneous expression of $\deg(L^\nu \otimes H^\mu)$, that is:

$$\deg(L_1^{-a\nu+\mu} \otimes L_2^{b\nu+\mu}) = s_0(b\nu+\mu)^2 + 2s_1(b\nu+\mu)(-a\nu+\mu) + s_2(-a\nu+\mu)^2$$

where the coefficient of $2\mu\nu$ is

$$(b(s_0+s_1) - a(s_1+s_2)) .$$

Therefore, we have $\deg L > 0$ and $\deg L_D > 0$ if and only if $a, b \in \mathbb{N}$ satisfy:

$$s_0 b^2 - 2s_1 ab + s_2 a^2 > 0$$

$$b(s_0+s_1) - a(s_1+s_2) > 0$$

but we can remark that

$$\frac{s_1 - \sqrt{s_1^2 - s_0 s_2}}{s_2} \leq \frac{s_0 + s_1}{s_1 + s_2}$$

and therefore, when we let a/b increase starting from -1, as long as we have

$$\frac{a}{b} < \frac{s_1 - \sqrt{s_1^2 - s_0 s_2}}{s_2}$$

which is the smallest root of the polynomial $s_0 - 2s_1 T + s_2 T^2 = 0$, we are sure that $H^0(X, (L_1^{-a} \otimes L_2^{b})^{\nu}) \neq 0$ for large enough ν.

In other words we have just proved the inequality

$$r(L_2; L_1) \geq \frac{s_1 - \sqrt{s_1^2 - s_0 s_2}}{s_2}$$

(we note that the Hodge Index theorem has been used to get that $s_1^2 - s_0 s_2 \geq 0$).

On the other hand, using the additivity of Euler characteristics we see that the mixed degrees are increasing functions of the sheaves, i.e., if $L_1 \subseteq L_1'$ and $L_2 \subseteq L_2'$, then we have

$$\deg(L_1^{[i]}, L_2^{[d-i]}) \leq \deg(L_1'^{[i]}, L_2'^{[d-i]}),$$

and using this and the homogeneity, we obtain immediately an upper bound for $r(L_2; L_1)$, namely

$$r(L_2; L_1) \leq \frac{s_0}{s_1} \leq \frac{s_1}{s_2}.$$

(The first inequality comes from writing that if $H^0(X, L_1^{-a} \otimes L_2^{b}) \neq 0$, i.e., $L_1^a \subset L_2^b$, we must have:

$$as_1 = a \deg(L_1^{[1]}, L_2^{[1]}) = \deg((L_1^a)^{[1]}, L_2^{[1]}) \leq \deg((L_2^b)^{[1]}, L_2^{[1]}) = bs_0$$

i.e., $\frac{a}{b} \leq \frac{s_0}{s_1}$.)

The second inequality is once again $s_0 s_2 \leq s_1^2$.

Finally we have proved:

3.2. PROPOSITION. *Let* L_1 *and* L_2 *be two invertible sheaves on a projective integral surface* X, *such that* $\deg L_1 > 0$, $\deg L_2 > 0$ *and* $L_1 \otimes L_2$ *is ample. Define the mixed degrees* s_0, s_1, s_2 *by*

$$\deg L_1^{\nu_1} \otimes L_2^{\nu_2} = s_2 \nu_1^2 + 2 s_1 \nu_1 \nu_2 + s_0 \nu_2^2 .$$

Then

1) $s_2 = \deg L_1$, $S_0 = \deg L_2$ *and* $s_1 = \deg(L_1^{[1]}, L_2^{[1]})$ *are positive.*

2) *The roots of the polynomial*

$$s_0 + 2 s_1 T + s_2 T^2$$

are real and negative, hence the roots of the polynomial $s_0 - 2s_1 T + s_2 T^2$ *are real and positive.*

3) *We have the inequalities*

$$\frac{s_1}{s_2} \geq \frac{s_0}{s_1} \geq r(L_2; L_1) \geq \frac{s_1 - \sqrt{s_1^2 - s_0 s_2}}{s_2} .$$

A symmetric proof (considering $L = L_1^a \otimes L_2^{-b}$ and letting $\frac{a}{b}$ decrease) gives symmetric inequalities for

$$R(L_2; L_1) = \inf\left\{\frac{a}{b} / H^0(X, L_1^a \otimes L_2^{-b}) \neq 0\right\}$$

namely:

$$\frac{s_0}{s_1} \leq \frac{s_1}{s_2} \leq R(L_2; L_1) \leq \frac{s_1 + \sqrt{s_1^2 - s_0 s_2}}{s_2} .$$

Using what we have seen so far, and the easily established fact that the inradius $r(K_2; K_1)$ and the outradius $R(K_2; K_1)$ are continuous functions on the product space $\mathcal{K} \times \mathcal{K}$ endowed with the product of Hausdorff topologies, we obtain:

3.2.1 COROLLARY (Flanders [$6\frac{1}{2}$]). *Let* K_1 *and* K_2 *be two compact convex domains in* $\mathbf{R}^2$, *with mixed volumes* $v_0 = \mathrm{Vol}(K_2)$, $v_1 = \mathrm{Vol}(K_2^{[1]}, K_1^{[1]})$ *and* $v_2 = \mathrm{Vol}(K_1)$. *The inradius* $r = r(K_2; K_1)$ *and the outradius* $R = R(K_2; K_1)$ *of* K_2 *with respect to* K_1 *satisfy the following inequalities*:

$$\frac{v_1 + \sqrt{v_1^2 - v_0 v_2}}{v_2} \geq R \geq \frac{v_1}{v_2} \geq \frac{v_0}{v_1} \geq r \geq \frac{v_1 - \sqrt{v_1^2 - v_0 v_2}}{v_2}$$

and by difference, we get Bonnesen's inequality

$$v_1^2 - v_0 v_2 \geq \frac{v_2^2}{4}(R-r)^2 \qquad \text{3.2.2}$$

and when we specialize to the case where K_1 *is the unit disk, we get the inequalities (1) and (2) of the introduction.*

3.3. REMARK. Going back to the case of arbitrary $d \geq 2$, one can also use the construction of §1 to obtain results on the measure of the set of translations sending K_1 into K_2, in the special case where the difference of the support functions $H_2 - H_1$ is again a convex function: in this case, thanks to the theorem of ([9], pp. 42-44) quoted in §1, we have $H^j(X, L_1^{-1} \otimes L_2) = 0$ for $j \geq 1$, and therefore

$$\chi(X, (L_1^{-1} \otimes L_2)^{\nu}) = h^0(X, (L_1^{-1} \otimes L_2)^{\nu})$$

from this equality, the Key Lemma 2.1 and approximation, we obtain that:

3.3.1. PROPOSITION (for $d \geq 1$, and K_1, K_2 in $\mathcal{K}$). *If* $H_2 - H_1$ *is convex, the measure of the set of translations sending* K_1 *into* K_2 *is given by*

$$m(K_1; K_1 \subset K_2) = \sum_{i=0}^{d} \binom{d}{i} (-1)^i v_i \quad (\textit{or } 0 \textit{ if this is } \leq 0)$$

(*compare with* [14], p. 95).

§4. *A problem*

I propose to study the following precise form of Problem B:

4.1. PROBLEM C. Determine for which algebraic varieties X of dimension $d \geq 1$ we have that, given any two invertible sheaves L_1 and L_2 on X which satisfy the conditions:

i) L_i is generated by its global sections, $i = 1, 2$

ii) $\deg L_i > 0$, $i = 1, 2$

iii) $L_1 \otimes L_2$ is ample.

Then:

1) The roots of the polynomial $R(T) = \sum_{i=0}^{d} (-1)^i \binom{d}{i} s_i T^i \in Z[T]$ where

the s_i are defined by: $\deg L_1^{\nu_1} \otimes L_2^{\nu_2} = \sum_{i=0}^{d} \binom{d}{i} s_i \nu_1^i \nu_2^{d-i}$, all have

positive real parts, say $0 < \rho_1 \leq \rho_2 \leq \cdots \leq \rho_d$.

2) The following inequality holds (notation of 2.2):

$$r(L_2; L_1) \geq \rho_1 .$$

As we have seen above, when $d \leq 2$, the answer is: all algebraic varieties. (For $d = 2$ we used the Index theorem and Lemma 3.1, which is essentially a consequence of Riemann's theorem on curves; the case $d = 1$ follows directly from Riemann's theorem.) Now we show that:

4.2. PROPOSITION. *For any dimension* d, *the class of algebraic varieties defined in Problem C contains all abelian varieties.*

This is an almost immediate consequence of a remarkable theorem of Mumford and Kempf:

THEOREM (Kempf-Mumford, see $[9\frac{1}{2}]$). *Let* L *and* M *be invertible sheaves on an abelian variety* X, *with* L *ample. Let* $P_{L,M}(n) = \chi(X, L^n \otimes M)$. *Then*:

i) *All the* d *roots of the polynomial* $P_{L,M}$ *are real* ($d = \dim X$).

ii) *Counting roots with multiplicities*:

$$H^k(X, M) = 0 \quad \textit{for} \quad 0 \leq k < \textit{number of positive roots};$$

$$H^{d-k}(X, M) = 0 \quad \textit{for} \quad 0 \leq k < \textit{number of negative roots};$$

and if 0 *is not a root, exactly one cohomology group is not zero.*

We apply this result to our problem, taking $L = L_1 \otimes L_2$ and $M = L_1^{-a} \otimes L_2^b$. Then, by the properties of line bundles on abelian varieties (cf., loc. cit.) and the definition of the polynomial $R(T)$, we have the equality:

$$P_{L,M}(T) = R\left(\frac{a-T}{b+T}\right) \cdot (b+T)^d .$$

Therefore, all the roots of $R(T)$ are real, and are of the form $\rho_j = \dfrac{a-r_j}{b+r_j}$ where $r_1, r_2, \cdots, r_d$ are the roots of $P_{L,M}(T)$. Hence we have $r_j = \dfrac{a-b\cdot\rho_j}{1+\rho_j}$. Since the ρ_i depend only upon L_1 and L_2, we see that taking $b = 0$ and applying the theorem gives $1 + \rho_j > 0$ for $j=1,\cdots,d$, and taking $a = 0$ gives $\rho_j > 0$, $j=1,\cdots,d$. Furthermore, as long as $\frac{a}{b} < \rho_1$, the smallest of the ρ_j, we have that all the r_j are < 0, and hence, applying the theorem again, $H^0(X, L_1^{-a} \otimes L_2^b) \neq 0$. This shows that $r(L_2; L_1) \geq \rho_1$, as desired.

4.3. Remark. We emphasize that in general the cohomology of line bundles is much more complicated than in the case $d \leq 2$ or in the case of abelian varieties, so that other methods must be used to study Problem C.

4.4. Remarks. If a polynomial such as $R(T)$ has all its roots real, then, the coefficients s_i must satisfy the inequalities

$$s_{i-1}^2 - s_i \cdot s_{i-2} \geq 0 \qquad (2 \leq i \leq d)$$

as we know (compare with 1.4.3.1).

One might think that these inequalities and the positivity of the s_i suffice to imply that all the roots of the polynomial $R(T)$ have positive real parts; this is easily checked for $d \leq 5$, using the Routh-Hurwitz criterion. However, it is not true for $d \geq 10$: According to a random search programmed by G. Wanner at the Math. Inst. of the University of Geneva, it seems that the inequalities imply the positivity of the real parts of the roots for $d \leq 9$, but he found a counterexample with $d = 10$. I am also very grateful to Douady who suggested a construction of a counterexample by hand, and to Coray who made the first search of counterexamples on the Geneva computer, and found one of degree $d = 16$.

David Mumford has shown to me that the three-dimensional variety obtained by blowing up a point in $P^2 \times P^1$ does not belong to the class defined in Problem C. [The inequality 2) is not satisfied on some lines in the affine subspace of $NS(X)$ consisting of classes of divisors of the form $xH + yE + K$ where H is the total transform of $P^2 \times \{a\}$, E is the exceptional divisor, and K is the total transform of $\ell \times P^1$, ℓ being a line which contains the point b, and we blow up the point (b,a).] Also, there are examples showing that if $d \geq 3$, the roots of the polynomial $R(T)$ need not be all real; the first such example was given to me by Mr. L. Brown of Purdue University: Take $K_1 = B^3 \subset R^3$ and for K_2 a very close approximation, of positive volume, of $B^2 \subset R^2 \subset R^3$.

CENTRE DE MATHS.
ÉCOLE POLYTECHNIQUE (CNRS)

AND

HARVARD UNIVERSITY

REFERENCES

[1] A.D. Alexandrov: Theory of mixed volumes. (4 papers in Mat. Sbornik. *44* (N.S. 2), pp. 947-972 and 1205-1238, and *45* (N.S. 2), pp. 27-46 and 227-251, in 1937.) I have used the translation made by Prof. J. Firey in 1966-1967 (Dept. of Math., Oregon State University, Corvallis, Oregon 97331) kindly sent to me by Prof. R. Schneider.

[2] Bonnesen: Sur le problème des isopérimètres et des isépiphanes. Gauthier-Villars, Paris 1929.

[$2\frac{1}{2}$] Bonnesen-Fenchel: Theorie der Konvexen körper, Ergebnisse der. Math. Berlin, Verlag von Julius Springer, 1934.

[3] Berthelot-Grothendieck – Illusie, SGA 6. Springer Lecture Notes No. 225.

[4] M. Demazure: Sous-groupes algebriques de rang maximum du groupe de Cremona. Ann. Sci. E.N.S. 4e serie t. 3, Fasc. 4 (1970).

[5] H. G. Eggleston: Convexity. Cambridge University Press 1958.

[6] P. DuVal: On the isolated singularities of surfaces which do not affect the condition of adjunction. Part I. Proc. Camb. Phil. Soc. *30* (1934).

[$6\frac{1}{2}$] H. Flanders: A proof of Minkowski's inequality for convex curves. American Math. Monthly *75* (1968), p. 581.

[7] W. V. D. Hodge: Note on the theory of the base for curves on an algebraic surface. Journ. London Math. Soc. *12* (1937), p. 58.

[8] A. G. Hovanski: Newton Polyhedra and Alexandrov-Fenchel inequalities. Uspekhi Ak. Nauk, 34, *4* (208), pp. 160-161.

[9] G. Kempf, F. Knudsen, D. Mumford, B. St. Donat: Toroidal embeddings. Springer Lecture Notes No. 339.

[$9\frac{1}{2}$] G. Kempf: Appendix to D. Mumford's course, in: C.I.M.E. 1969: Questions on algebraic varieties. Edizioni Cremonese, Roma 1970.

[10] S. Kleiman: Toward a numerical theory of ampleness. Annals of Math., Vol. 84, No. 2, Sept. 1966, pp. 293-344.

[$10\frac{1}{2}$] C. Matheron: Random sets and integral geometry. Wiley, New York, 1975.

[11] T. Oda: A dual formulation for Hironaka's game problem. Preprint, Institute of Math., Tohoku University, Sendai, Japan.

[12] R. Osserman: Bonnesen-style isoperimetric inequalities. American Math. Monthly, Vol. 86, No. 1 (1979).

[13] D. Rees and R. Y. Sharp: On a theorem of B. Teissier on multiplicities of ideals in local rings. Journ. London Math. Soc. 2nd series, Vol. 18, part 3, 1978, p. 449.

[14] Luis A. Santaló: Integral geometry and geometric probability. Encyclopedia of Mathematics and its applications. Addison-Wesley 1976.

[15] R. Schneider: On A. D. Alexandrov's inequalities for mixed discriminants. Journ. of Math. and Mech. Vol. 15, No. 2, p. 285 (1966).

[16] B. Teissier: Sur une inégalité à la Minkowski pour les multiplicités. Annals of Math. Vol. 106 (1977), pp. 38-44.

[17] B. Teissier: On a Minkowski-type inequality for multiplicities II in: C. P. Ramanujam: a Tribute, Tata Institute for Fundamental Research, Bombay 1978.

[18] ________: Jacobian Newton Polyhedra and equisingularity. Proc. R.I.M.S. Conference on singularities. April 1978. R.I.M.S. Kyoto 1978.

[19] ________: Du Théorème de l'Index de Hodge aux inégalités isopérimetriques. Note C.R.A.S. Paris, tome 288 (29 Jan. 1979), pp. 287-289.

LIOUVILLE THEOREMS FOR HARMONIC MAPPINGS, AND AN APPROACH TO BERNSTEIN THEOREMS

Stefan Hildebrandt*

1. *The theorem of Liouville for harmonic mappings*

The classical theorem of Liouville states that a nonconstant holomorphic function on the complex plane cannot be bounded. Actually, Liouville formulated a somewhat weaker assertion concerning elliptic functions. At a session of the French Academy of Sciences in 1844, Liouville announced that there exist no holomorphic doubly periodic functions besides the trivial ones, the constants [44], [45]. Cauchy recognized at once the importance of this result and presented already at the next meeting a note [5] to the Academy containing the general theorem mentioned at the beginning which carries Liouville's name up to our days.[1] A well-known generalization states that a real-valued nonconstant harmonic function on $\mathbb{R}^n$ has to be unbounded. Let us consider the analogous question for harmonic functions which are defined on an n-dimensional Riemannian manifold $\mathcal{X}$. That is, we consider solutions $u \in C^2(\mathcal{X}, \mathbb{R})$ of the equation

$$\Delta_{\mathcal{X}} u = 0 \quad \text{on} \quad \mathcal{X}$$

where $\Delta_{\mathcal{X}}$ denotes the Laplace-Beltrami operator on $\mathcal{X}$.

*Supported in part by NSF grant MCS-77-18723(02).

[1]Cf. E. Neuenschwander [53] for a detailed historical account. We wish to thank Prof. Kuhlmann for drawing our attention to this paper.

Seminar on Differential Geometry
0-691-08268-5/82/000107-25$01.25/0 (cloth)
0-691-08296-0/82/000107-25$01.25/0 (paperback)

We infer from E. Hopf's maximum principle that the constants are the only harmonic as well as the only subharmonic functions on a compact Riemannian manifold.

A function of class $C^2(\mathcal{X}, \mathbf{R})$ is said to be subharmonic if $\Delta_{\mathcal{X}} u \geq 0$ holds.

The matter becomes more difficult if we consider harmonic functions on complete but noncompact manifolds $\mathcal{X}$. Yau [71], p. 217, Cor. 1, has found the following:

THEOREM 1. *A bounded harmonic function on a complete Riemannian manifold with nonnegative Ricci curvature has to be a constant.*

The proof follows from gradient estimates for solutions of partial differential equations on $\mathcal{X}$. Moreover, Yau has proved that the assertion remains true for harmonic functions which are bounded from one side only, say, $u \leq \text{const.}$

Let us now mention some further results about the growth of harmonic or subharmonic functions on a complete but noncompact Riemannian manifold $\mathcal{X}$. Firstly we state a theorem by Greene and Wu (cf. [24], p. 231, and [22], p. 270):

THEOREM 2. *Let $\mathcal{X}$ be a complete noncompact Riemannian manifold whose sectional curvature is positive outside some compact set. Then, for any $p \geq 1$ and for any nonnegative, not identically vanishing subharmonic function u of class $C^0(\mathcal{X}, \mathbf{R})$, there exists a number $C(p, u) > 0$ such that for every $x_0 \in \mathcal{X}$,*

$$\int_{B_r(x_0)} u^p \, d\,\text{vol} \geq C \cdot r$$

for all sufficiently large r. Here $B_r(x_0)$ denotes the geodesic ball in $\mathcal{X}$ with center x_0 and radius r.

Continuous subharmonic functions are defined by the principle of harmonic dominance, cf. [22], pp. 267-268. We infer by taking $u \equiv 1$ that $\mathcal{X}$ must have infinite volume under the assumptions of Theorem 2. Yau [72], p. 667, and Calabi [4] have found the following generalization:

THEOREM 3. *Every complete noncompact Riemannian manifold with non-negative Ricci curvature has infinite volume.*

Next we mention a theorem on the integrability of harmonic and non-negative subharmonic functions due to Yau [72], p. 664.

THEOREM 4. *Let* $\mathcal{X}$ *be a complete Riemannian manifold,* $0 < p < \infty$, $p \neq 1$, *and suppose that* $u \in C^2(\mathcal{X}, \mathbf{R})$ *satisfies* $u\Delta_{\mathcal{X}} u \geq 0$. *Then either* $u \equiv \text{const}$ *or*

$$\int_{\mathcal{X}} |u|^p d\ \text{vol} = \infty .$$

This result has been sharpened by L. Karp [40] and [41]. Since there will appear a paper by Karp in this issue of Annals of Mathematical Studies we omit a description of his interesting results, except for the following:

LEMMA 1. *A complete noncompact Riemannian manifold* $\mathcal{X}$ *is strongly parabolic if it has moderate volume growth.*

Following [40], we call a noncompact Riemannian manifold $\mathcal{X}$ *strongly parabolic* if it admits no nonconstant negative subharmonic function. Moreover a complete noncompact Riemannian manifold $\mathcal{X}$ is said to have *moderate volume growth* if there exist a point x_0 in $\mathcal{X}$ and a positive nondecreasing function $F(r)$ such that $\int_a^\infty \frac{dr}{rF(r)} = \infty$ for some $a > 0$ and

$$\limsup_{r\to\infty} \frac{\text{vol}\, B_r(x_0)}{r^2 F(r)} < \infty.$$

where $B_r(x_0)$ is the geodesic ball of radius r around x_0.

Proof of Lemma 1. Suppose that $v \le 0$ and $\Delta v \ge 0$. Then $u = e^v$ satisfies $0 < u \le 1$ and $\Delta u \ge 0$. Therefore,

$$\limsup_{r\to\infty} \frac{1}{r^2F(r)} \int_{B_r(x_0)} |u|^2 \, d\,\mathrm{vol} \le \limsup_{r\to\infty} \frac{\mathrm{vol}\, B_r(x_0)}{r^2F(r)}$$

for all admissible functions F. But this is a contradiction to the moderate volume growth of $\mathcal{X}$ since, by Theorem 2.2 in [40], the limes superior on the left-hand side is $+\infty$ unless v is a constant.

Now we consider the general case of harmonic mappings. A mapping $U: \mathcal{X} \to \mathfrak{M}$ from an n-dimensional Riemannian manifold $\mathcal{X}$ into an N-dimensional Riemannian manifold $\mathfrak{M}$ is said to be *harmonic* if it is of class $C^2(\mathcal{X}, \mathfrak{M})$ and satisfies the Euler equations of the energy integral

$$E(U) = \int_{\mathcal{X}} e(U)\, d\,\mathrm{vol}_{\mathcal{X}}$$

where the energy density

$$e(U) = \frac{1}{2} \operatorname{tr}_{\mathcal{X}} \langle U_*, U_* \rangle_{\mathfrak{M}}$$

of a map $U: \mathcal{X} \to \mathfrak{M}$ is the trace with respect to the metric tensor of $\mathcal{X}$ of the pull-back of the metric tensor of $\mathfrak{M}$ under the mapping U.

By a well-known formula (cf. [14], p. 14) we obtain (cf. [21], [40]):

LEMMA 2. *If $U: \mathcal{X} \to \mathfrak{M}$ is a harmonic map and if ψ is a convex function on $U(\mathcal{X})$ of class $C^2(U(\mathcal{X}), \mathbb{R})$, then the composition $v = \psi \circ U$ is subharmonic on $\mathcal{X}$.*

If, in addition, ψ is bounded on $U(\mathcal{X})$, and $\mathcal{X}$ is either compact or strongly parabolic then $\psi \circ U$ is a constant, and

$$\operatorname{tr}_{\mathcal{X}} U^*(\nabla^2 \psi) = 0 .$$

Finally, if also ψ is strictly convex, then U is a constant map.

Combining Lemma 2 with estimates for the Christoffel symbols proved in [26] and [29], we obtain:

LEMMA 3. *Let* $B_M(Q)$ *be a closed geodesic ball in* $\mathfrak{M}$ *with center* Q *and radius* M *such that* $B_M(Q)$ *does not meet the cut locus* C(Q) *of* Q. *Suppose also that the sectional curvature* $K_{\mathfrak{M}}$ *of* $\mathfrak{M}$ *is bounded on* $B_M(Q)$ *from above by a number* $\kappa \geq 0$ *such that* $M\sqrt{\kappa} < \pi/2$. *Then the function* ψ, *defined by*

$$\psi(P) = \mathrm{dist}^2_{\mathfrak{M}}(P, Q) ,$$

is a smooth strictly convex function on $B_M(Q)$.

THEOREM 5. *Let* $\mathfrak{X}$ *and* $\mathfrak{M}$ *be complete Riemannian manifolds, where* $K_{\mathfrak{M}}$ *denotes the sectional curvature of* $\mathfrak{M}$. *Suppose that* U *is a harmonic map from* $\mathfrak{X}$ *into* $\mathfrak{M}$. *Then* U *is a constant map if either one of the following conditions holds:*

(i) $\mathfrak{X}$ *is compact and* $\pi_1(\mathfrak{X})$ *is finite,* $K_{\mathfrak{M}} \leq 0$;

(ii) $\mathfrak{X}$ *is compact or strongly parabolic (for instance, of moderate volume growth),* $\mathfrak{M}$ *is simply connected and* $K_{\mathfrak{M}} \leq 0$, *and* $U(\mathfrak{X})$ *is bounded in* $\mathfrak{M}$;

(iii) $\mathfrak{X}$ *is compact or strongly parabolic,* $K_{\mathfrak{M}} \geq 0$, $K_{\mathfrak{M}} > 0$ *outside a compact set,* $\mathfrak{M}$ *is noncompact, and* $U(\mathfrak{X})$ *is bounded in* $\mathfrak{M}$;

(iv) $\mathfrak{X}$ *is compact or strongly parabolic,* $U(\mathfrak{X})$ *is contained in a geodesic ball* $B_M(Q)$ *which does not meet the cut locus of its center, and there is a number* $\kappa \geq 0$ *such that* $M\sqrt{\kappa} < \pi/2$ *and* $K_{\mathfrak{M}} \leq \kappa$ *on* $B_M(Q)$.

Proof. The assertion (iv) follows immediately from the Lemmata 2 and 3.

Under the assumptions of (ii) C(Q) is empty in virtue of the Theorem of Hadamard-Cartan. Therefore (ii) is a consequence of (iv).

The assumptions of (iii) imply that $\mathfrak{M}$ supports a strictly convex smooth function (cf. Wu [68], and Greene-Wu [24]). Thus we can infer (iii) from Lemma 2. In order to prove (i), we consider the universal covering $p : \tilde{\mathfrak{X}} \to \mathfrak{X}$ and $q : \tilde{\mathfrak{M}} \to \mathfrak{M}$ of $\mathfrak{X}$ and $\mathfrak{M}$, respectively. The manifolds $\tilde{\mathfrak{X}}$ and $\tilde{\mathfrak{M}}$ are simply connected, and the mappings p and q are isometric and locally invertible. Thus $U \circ p : \mathcal{X} \to \mathfrak{M}$ is harmonic, and $U \circ p$ can be lifted to a harmonic map $\tilde{U} : \tilde{\mathcal{X}} \to \tilde{\mathfrak{M}}$ such that $q \circ \tilde{U} = U \circ p$. Since $\pi_1(\mathcal{X})$ is finite and $\mathcal{X}$ is compact, the universal covering $\tilde{\mathcal{X}}$ has also to be compact. Since $\tilde{\mathfrak{M}}$ is simply connected and $K_{\tilde{\mathfrak{M}}} \leq 0$, (ii) implies that $\tilde{U}$ and therefore U is a constant.

The results of Theorem 7 are taken from Gordon [21], Karp [40], and Hildebrandt-Kaul [26].

The next theorem has been proved by S.-Y. Cheng [7]. It generalizes Yau's Liouville theorem for harmonic functions which we have formulated as Theorem 1.

THEOREM 6. *Suppose that* $\mathcal{X}$ *is a complete Riemannian manifold of nonnegative Ricci curvature, and that* $\mathfrak{M}$ *is a simply connected complete Riemannian manifold of nonpositive sectional curvature. Let* U *be a harmonic map from* $\mathcal{X}$ *into* $\mathfrak{M}$ *such that* $U(\mathcal{X})$ *is bounded in* $\mathfrak{M}$. *Then* U *is a constant mapping.*

For compact $\mathcal{X}$ this result is contained in Theorem 5, (ii). If $\mathcal{X}$ is noncompact, Cheng derives his Liouville theorem from the following *gradient estimate*:

There is a number $c > 0$ depending only on $n = \dim \mathcal{X}$ such that, on every geodesic ball $B_R(x_0)$ in $\mathcal{X}$, the following estimate holds:

$$(1.1) \qquad (R^2 - r^2) \inf_{B_R(x_0)} \{A^2 - \rho^2\}^2 e(U) \leq cA^2R^2 \sup_{B_R(x_0)} \{A^2 - \rho^2\}^2 .$$

Here we have set

$$r(x) = \mathrm{dist}_{\mathcal{X}}(x, x_0) \quad \text{and} \quad \rho(x) = \mathrm{dist}_{\mathfrak{M}}(U(x), P) ,$$

where P is a point of $\mathfrak{M}$ which is not contained in the closure of $U(B_R(x_0))$, and A is a number which satisfies $A > \sup \{\rho(x) : X \in B_R(x_0)\}$.

If we choose $P \notin \mathrm{clos}\, U(\mathcal{X})$ and $A > \sup \{\rho(x) : x \in \mathcal{X}\}$ then there exists a number $C > 0$ depending on x_0 but not on R such that

$$e(U)(x_0) \leq CR^{-2} \quad \text{for all} \quad R > 0$$

whence $e(U)(x_0) = 0$ for every $x_0 \in \mathcal{X}$, and therefore $U \equiv \text{const}$.

Actually, the estimate (1.1) yields a somewhat stronger result. We can infer that a harmonic map has to be a constant if its growth in $B_R(x_0)$ is of order $o(R)$.

Next we shall describe a Liouville theorem by Hildebrandt-Jost-Widman [32] which applies to harmonic mappings $U : \mathcal{X} \to \mathfrak{M}$ where $\mathcal{X}$ is neither compact nor necessarily strongly parabolic and where the sectional curvature $K_{\mathfrak{M}}$ of $\mathfrak{M}$ may assume positive values. For this purpose, we introduce the notion of a simple manifold.

DEFINITION. A Riemannian manifold $\mathcal{X}$ is said to be *simple* if the following holds:

(i) $\mathcal{X}$ is homeomorphic to $\mathbf{R}^n$;

(ii) there exist two positive numbers λ and μ as well as a chart $\phi : \mathcal{X} \to \mathbf{R}^n$ mapping $\mathcal{X}$ topologically onto $\mathbf{R}^n$ such that the line element

$$d\sigma^2 = \gamma_{\alpha\beta}(x)\,dx^\alpha dx^\beta$$

of $\mathcal{X}$ with respect to the coordinates $x = \phi(p)$, $p \in \mathcal{X}$, satisfies

$$\lambda|\xi|^2 \leq \gamma_{\alpha\beta}(x)\,\xi^\alpha \xi^\beta \leq \mu|\xi|^2$$

for all x and ξ in $\mathbf{R}^n$. In other words, $\mathcal{X}$ is topologically $\mathbf{R}^n$ furnished with a metric for which the associated Laplace-Beltrami operator $\Delta_{\mathcal{X}}$ defined by (1.3) is uniformly elliptic on $\mathbf{R}^n$.

Moser [52], p. 588, has proved:

THEOREM 7. *A bounded harmonic function on a simple Riemannian manifold has to be a constant.*

This result is an immediate consequence of Moser's well-known Harnack principle.

Clearly, a simple manifold $\mathcal{X}$ is a complete noncompact simply connected Riemannian manifold. However it is not yet known which natural assumptions on $\mathcal{X}$ are needed in addition to guarantee that the converse holds although simple Riemannian manifolds should be an interesting geometrical topic. This question can be subsumed under the general problem which manifolds are quasi-isometric to each other, the study of which has been started by F. John [36I], [36II]. Fortunately, we can provide at least one sufficient condition due to Greene and Wu [25], pp. 56-57:

LEMMA 4. *Let* $\mathcal{X}$ *be a complete simply-connected Riemannian manifold with nonpositive sectional curvature* $K_{\mathcal{X}}$. *Then* $\mathcal{X}$ *is simple provided that there are numbers* $A > 0$, $\varepsilon > 0$, *and a point* $x_0 \in \mathcal{X}$ *such that*

$$\frac{-A}{r^2(\log r)^{1+\varepsilon}} \leq K_{\mathcal{X}} \quad \textit{for} \quad r > 1$$

where $r(x) = \mathrm{dist}_{\mathcal{X}}(x, x_0)$.

In fact, Greene and Wu prove a somewhat stronger result involving only an integral condition about the so-called radial curvature of $\mathcal{X}$.

Moreover we wish to mention that simple manifolds arise naturally if we wish to prove "Bernstein theorems" for manifolds which are minimally immersed into Euclidean space. For details, we refer the reader to Section 2.

Now we shall state the Liouville theorem proved in [32] which generalizes Moser's result for harmonic functions to harmonic mappings.

THEOREM 8. *Let* U *be a harmonic map of a simple Riemannian manifold* $\mathcal{X}$ *into a complete Riemannian manifold* $\mathfrak{M}$ *the sectional curvature of which is bounded from above by a constant* $\kappa > 0$. *Denote by* $B_M(Q)$ *a closed geodesic ball in* $\mathfrak{M}$ *with radius* $M < \pi/(2\sqrt{\kappa})$ *which does not meet the cut locus of its center* Q. *Assume also that the range* $U(\mathcal{X})$ *of the map* U *is contained in* $B_M(Q)$. *Then* U *is a constant map.*

The proof follows from an a priori estimate for the Hölder seminorm of harmonic maps which has been derived in [32]. We formulate this estimate as

LEMMA 5. *Let* $B_M(Q)$ *be a geodesic ball in* $\mathfrak{M}$ *which does not intersect the cut locus of* Q. *Suppose that the sectional curvature* K *of* $\mathfrak{M}$ *satisfies* $K \leq \kappa$ *on* $B_M(Q)$, *where* $\kappa \geq 0$ *and* $M\sqrt{\kappa} < \pi/2$. *Let* $u = (u^1, \cdots, u^N)$ *be normal coordinates of* $\mathfrak{M}$ *on* $B_M(Q)$, *and assume that* $\Omega_{2d} = \{x \in R^n : |x| < 2d\}$ *is a coordinate patch for* $\mathfrak{X}$ *such that the components of the metric tensor of* $\mathfrak{X}$ *on* Ω_{2d} *satisfy*

$$\lambda|\xi|^2 \leq \gamma_{\alpha\beta}(x)\xi^\alpha\xi^\beta \leq \mu|\xi|^2, \quad 0 < \lambda \leq \mu,$$

for all $\xi \in R^n$ *and all* $x \in \Omega_{2d}$.

Assume that $U : \mathfrak{X} \to \mathfrak{M}$ *is a harmonic map with range* $U(\Omega_{2d})$ *contained in* $B_M(Q)$, *and let* $u = u(x)$ *be the representation of* U *with respect to the normal coordinates* $u^1, \cdots, u^N$. *Then there exist numbers* σ, $0 < \sigma < 1$, *and* $c > 0$ *depending only on* n, N, M, λ, μ, ω, κ *but not on* $d > 0$ *such that the* σ*-Hölder seminorm of* u *on* $\overline{\Omega}_d$ *is estimated by*

(1.2) $$[u]_{C^\sigma(\overline{\Omega}_d)} \leq cd^{-\sigma}.$$

The proof of this lemma is rather complicated. It is based on the differential geometric estimates stated in [29] and on appropriate bounds for the Green function of $\Delta_{\mathfrak{X}}$ where the Harnack inequality of Moser enters. Therefore Theorem 8 does not yield a conceptually new proof for Theorem 7 although this result is a special case of Theorem 8.

Now we turn to harmonic mappings which are not necessarily bounded but have finite energy. We begin with a result due to Schoen and Yau [62], pp. 336-337.

THEOREM 9. *Let* $\mathfrak{X}$ *be a complete noncompact Riemannian manifold with nonnegative Ricci curvature and* $\mathfrak{M}$ *be a Riemannian manifold with nonpositive sectional curvature. Then each harmonic map* U *from* $\mathfrak{X}$ *to* $\mathfrak{M}$ *with finite energy* E(U) *has to be a constant map.*

In the original paper [62] the manifold $\mathfrak{M}$ was assumed to be compact; cf. also [14], p. 11. However, this assumption seems to be superfluous. The proof was carried out by considering the continuous subharmonic function $\sqrt{e}$. Yet it is not a priori evident that $\sqrt{e}$ is of class C^1 or C^2 while the theorems 2 and 3 in [72] require the smoothness of the investigated function u, and we do not see how the approximation procedure of Greene and Wu can be utilized. However, it turns out that analogous estimates can be obtained for $\sqrt{e+\varepsilon}$, $\varepsilon > 0$, and the final inequality follows after the passage to the limit $\varepsilon \to 0$.

THEOREM 10. *Let* $U : \mathfrak{X} \to \mathfrak{M}$ *be a harmonic map with finite energy. Suppose that* $\mathfrak{X}$ *is conformally equivalent to* $\mathbf{R}^n$, *and that, for appropriate coordinates* $x \in \mathbf{R}^n$, *the fundamental tensor* $(\gamma_{\alpha\beta})$ *of* $\mathfrak{X}$ *is of the form*

$$\gamma_{\alpha\beta}(x) = f(x)\delta_{\alpha\beta} \tag{1.3}$$

where $f(x) > 0$ *satisfies*

$$(2-n-\sigma)f(x) \le \nu x^\alpha D_\alpha f(x) \quad \text{on} \quad \mathbf{R}^n \tag{1.4}$$

for some number $\sigma < 0$ *and with* $\nu = \frac{n-2}{2}$. *Then* U *is a constant map.*

REMARKS: 1) This result is a straightforward generalization of a theorem due to Garber, Ruijsenaars, Seiler, and Burns [19]. These authors proved that each harmonic map $U : \mathbf{R}^n \to S^n$ with finite Dirichlet integral has to be a constant if $n > 2$. Eells and Lemaire observed[2] that the conclusion holds as well if one replaces S^N by an arbitrary N-dimensional Riemannian manifold $\mathfrak{M}$.

2) Added in proof (June 21, 1980): During the Arbeitstagung in Bonn (June 13-19, 1980) J. Eells and H. Sealey have kindly pointed out to us the following additional information concerning Theorem 10.

[2] This has kindly been pointed out to us by J. Eells in a letter dated Nov. 7, 1978.

(i) H. Sealey (Warwick Thesis 1980) has constructed a harmonic map $\phi_1 : \mathcal{X}_1 \to S^2$ from a paraboloid of revolution $\mathcal{X}_1$ which has finite energy and is nonconstant. The manifold $\mathcal{X}_1$ is conformally flat and has positive Ricci curvature. For any integer $k \geq 1$, we define $\mathcal{X} = \mathcal{X}_1 \times \mathcal{T}^k$ and $\phi : \mathcal{X} \to S^2$ by $\phi(x, y) = \phi(x_1)$. Then ϕ is a nonconstant harmonic map with $E(\phi) < \infty$, and the Ricci curvature of $\mathcal{X}$ is nonnegative.

(ii) If $n \geq 3$, then Theorem 10 remains true under the assumption $\sigma \leq 0$ instead of $\sigma < 0$. A somewhat more general theorem can be found in *H. C. J. Sealey, Some conditions ensuring the vanishing of harmonic differential forms*, Preprint, pp. 1-10. Another interesting result of this paper is the following:

Let $\mathcal{X} = H^n$ denote the simply connected hyperbolic space form. If $U : H^n \to \mathfrak{M}$ is harmonic and $E(U) < \infty$, then U is a constant provided that $n \geq 3$.

Proof of Theorem 10. Define a family of maps $V_t : \mathbf{R}^n \to \mathfrak{M}$, $t > 0$, by

$$V_t(x) = U(tx) \quad \text{for} \quad x \in \mathbf{R}^n ,$$

and set

$$\Phi(R, t) = E_{B_R}(V_t) = \int_{B_R} e(V_t)\, dx ,$$

where $B_R = \{x \in \mathbf{R}^n : |x| < R\}$. Then

$$\frac{\partial}{\partial t} \Phi(R, t)\Big|_{t=1} = \delta E_{B_R}(U, \zeta) = \int_{B_R} \delta e(U, \zeta)\, dx$$

is the first variation of E_{B_R} at U in direction of some vector field ζ. Let $u = u(x)$ be a representation of U with respect to local coordinates $u^1, \cdots, u^N$ on $\mathfrak{M}$ and to standard cartesian coordinates $x^1, \cdots, x^n$ on $\mathbf{R}^n$. Then, ζ can be expressed as

$$\zeta(x) = (\zeta^1(x), \cdots, \zeta^N(x)), \zeta^i(x) = x^\alpha D_\alpha u^i(x) .$$

Since U is harmonic, we infer

$$\delta E_{B_R}(U,\zeta) = \frac{1}{R}\int_{S_R} g_{ik}(u)\zeta^i\zeta^k\, d\omega_R$$

where $S_R = \{x \in \mathbf{R}^n : |x| = R\}$, and $d\omega_R$ is the area element on S_R. Therefore,

$$\frac{\partial}{\partial t}\Phi(R,t)\Big|_{t=1} \geq 0 .$$

On the other hand, we can write

$$\Phi(R,t) = \int_{B_R} f^\nu(x)\, g_{ik}(u(tx))\, u^i_{x^\alpha}(tx)\, u^k_{x^\alpha}(tx)\, dx$$

$$= t^{2-n}\int_{B_{Rt}} f^\nu\!\left(\frac{x}{t}\right) g_{ik}(u(x))\, u^i_{x^\alpha}(x)\, u^k_{x^\alpha}(x)\, dx .$$

Then we infer from assumption (1.4) that

$$\frac{\partial}{\partial t}\Phi(R,t)\Big|_{t=1} \leq \sigma\phi(R) + R\frac{d}{dR}\phi(R)$$

where we have set

(1.5) $$\phi(R) = \Phi(R,1) .$$

Thus we obtain

(1.6) $$\frac{d}{dR}\{R^\sigma\phi(R)\} \geq 0 \quad \text{for all} \quad R > 0 .$$

If $U \not\equiv$ const, then there is a number $\rho > 0$ such that $\phi(\rho) > 0$. We then conclude from (1.5) and (1.6) that

$$\left(\frac{R}{\rho}\right)^{-\sigma} E_{B_\rho}(U) \leq E_{B_R}(U)$$

whence

$$E(U) = \lim_{R\to\infty} E_{B_R}(U) = \infty$$

which contradicts the assumption of our theorem.

COROLLARY. *Suppose that there exists a* C^1*-function* $g(r)$ *of the real variable* $r \geq 0$ *such that* $f(x) = g(|x|)$. *Then (1.4) is equivalent to*

$$2\left(1 + \frac{\sigma}{n-2}\right)g(r) \geq -rg'(r) \ . \tag{1.7}$$

For instance, $g(r) = (1+r)^{-1}$ *and* $g(r) = 1 + h(r)$ *with* $h(r) = 0$ *for* $0 \leq r \leq 1$, $h(r) = (r-1)^{\lambda}$ *for* $r > 1$, *with* $\lambda > 1$ *are solutions of (1.7).*

In the last part of this section, we consider harmonic mappings with finite Dirichlet integral which, in addition, satisfy a minimum property. Firstly, we mention a result by Xin [70]:

THEOREM 11. *If* $n > 2$, *there is no nonconstant stable harmonic map from* S^n *to any complete Riemannian manifold* $\mathfrak{M}$.

Note that a harmonic map $U : \mathfrak{X} \to \mathfrak{M}$ is said to be *stable* if the second variation of the energy integral at U, taken on an arbitrary compact subset of $\mathfrak{X}$, is nonnegative. It is well known that any harmonic map $U : \mathfrak{X} \to \mathfrak{M}$ is stable if the target manifold $\mathfrak{M}$ has nonpositive curvature.

The result of Xin corresponds to the observation by Morrey [15], p. 130, that the energy functional does not assume its absolute minimum on the class $H_k(S^n, S^n)$ of C^∞-mappings from S^n onto itself with degree $k \neq 0$ if $n \geq 3$.

The proof of Theorem 11 is carried out as follows. Let $I(V)$ be the second variation of E at U in direction of a vector field V on $U(S^n)$. Choose $V = U_*\nu$ where ν is a vector field on S^n, i.e., $\nu(x) \in T_xS^n$ for $x \in S^n$. By an elementary but tricky computation, one obtains that

$$I(U_*\nu) = (2-n)\int_{S^n} \langle U_*\nu, U_*\nu\rangle_{\mathfrak{M}}\, d\,\mathrm{vol}$$

holds for vector fields ν which are the projection of a constant vector field in R^{n+1} onto TS^n. The stability of U implies that $U_* \equiv 0$, or $U \equiv \text{const}$.

Probably the next result has already been known to Morrey. We provide a simple proof.

THEOREM 12. *Let* U *be a harmonic map with finite energy from* $\mathcal{X}$ *into* $\mathfrak{M}$ *where* $\mathcal{X}$ *and* $\mathfrak{M}$ *are simple Riemannian manifolds. Moreover, suppose that* U *is strictly stable, i.e.,* $E(U) \le E(V)$ *for each* C^2*-map* V *from* $\mathcal{X}$ *into* $\mathfrak{M}$ *which agrees with* U *except on a compact subset of* $\mathcal{X}$. *Then* U *is a constant map.*

Proof. Since $\mathcal{X}$ and $\mathfrak{M}$ are simple we can introduce local coordinates $x = (x^1, \cdots, x^n)$ and $u = (u^1, \cdots, u^N)$ such that, for any map $U \in C^2(\mathcal{X}, \mathfrak{M})$ with representation $u = u(x)$, the energy integral is of the form

$$E(U) = \int_{R^n} f(x, u(x), \nabla u(x))\, dx$$

where there exist two numbers m_1 and m_2 with $0 < m_1 \le m_2$ such that

$$m_1 |p|^2 \le f(x, u, p) \le m_2 |p|^2 \tag{1.8}$$

for all $(x, u, p) \in R^n \times R^N \times R^{nN}$.

Set $B_R = \{x \in R^n : |x| < R\}$. Then, for each $R > 0$, there is a function $\eta \in C_c^\infty(B_{2R}, R)$ with $0 \le \eta \le 1$ and $\eta(x) \equiv 1$ on B_R such that $|\nabla \eta| \le C \cdot R^{-1}$ where $C > 0$ does not depend on R. Let

$$\omega = \frac{1}{\text{vol } T_{2R}} \int_{T_{2R}} u(x)\, dx\,, \quad \text{where} \quad T_{2R} = B_{2R} - B_R\,,$$

and set $v(x) = u(x) - \eta(x)\{u(x) - \omega\}$. Let V be the map from $\mathcal{X}$ into $\mathfrak{M}$ with the representation $v = v(x)$. Since

$$E(U) \leq E(V)$$

we infer from (1.8) that

$$m_1 \int_{B_{2R}} |\nabla u|^2\, dx \leq 2m_2 \left\{ \int_{T_{2R}} (1-\eta)^2\, |\nabla u|^2\, dx + C^2 R^{-2} \int_{T_{2R}} |u-\omega|^2\, dx \right\} .$$

By Poincaré's inequality, there is a number C^* independent of R such that

$$\int_{T_{2R}} |u-\omega|^2\, dx \leq C^* R^2 \int_{T_{2R}} |\nabla u|^2\, dx .$$

Thus, for $c = 2m_1^{-1} m_2 \{1 + C^* C^2\}$, we obtain that

$$\int_{B_R} |\nabla u|^2\, dx \leq c \int_{B_{2R}-B_R} |\nabla u|^2\, dx . \tag{1.9}$$

However, $E(U) < \infty$ implies that $\int_{R^n} |\nabla u|^2\, dx < \infty$, on account of (1.8). Then, $\lim_{R\to\infty} \int_{B_{2R}-B_R} |\nabla u|^2\, dx = 0$, and we infer from (1.9) that $\int_{R^n} |\nabla u|^2\, dx = 0$. Thus $u(x) \equiv \text{const.}$

Finally we state a theorem by J. C. Wood [67], p. 1338, which poses a different type of restriction to the size of the image of harmonic maps.

THEOREM 13. *Let* $\mathcal{X}$ *and* $\mathcal{M}$ *be two-dimensional compact and real analytic Riemannian manifolds, and let* $\mathcal{M}$ *be orientable. Suppose that* $U: \mathcal{X} \to \mathcal{M}$ *is a nonconstant harmonic map. Then the total curvature* $\mathcal{K}$ *and the Euler characteristic* χ *of the image* $U(\mathcal{X})$ *of* $\mathcal{X}$ *under* U *are related by* $\mathcal{K} \geq 2\pi\chi$.

2. *Bernstein theorems for minimal submanifolds*

We shall not try to give a description of the origin of Bernstein's theorem and of the subsequent developments. We shall also refrain from exhibiting the various ties between minimal surfaces and the theory of functions of a complex variable although it is very tempting to consider the close relations between the theorems of Liouville, Picard, Bernstein, Osserman, and Chern-Osserman. Instead we refer to the books of Osserman [55] and Nitsche [54], to the recent reports by Osserman [56] and Do Carmo [11], and to the papers by Hoffman and Osserman [33], [34]. Moreover, we mention the remarkable papers by Do Carmo and Peng [12], [13], Fischer-Colbrie and Schoen [17], de M. Jorge and Xavier [38], Jones [37], and Xavier [69] which are relevant in this context.

We restrict ourselves to the relation between minimal submanifolds in $\mathbb{R}^n$ and harmonic mappings into Grassmannian manifolds provided by the Ruh-Vilms theorem. Since the sectional curvature of Grassmannians is nonnegative but, in general, not positive, only Theorem 5, (iv), and Theorem 8 are Liouville results which are applicable to the present situation.

We begin with a brief survey on results about the Grassmannian manifolds which will be used in the following. For detailed references we refer to [32].

By $G(n,p)$ we denote the (real) Grassmann manifold consisting of all n-dimensional oriented subspaces (called n-planes) of $\mathbb{R}^{n+p}$. $G(n,p)$ carries a Riemannian metric which is invariant under all transformations of the group of motions induced in $G(n,p)$ by $O(n+p)$. Except for $n = p = 2$, this metric is, up to a constant positive factor, the only invariant metric of this kind on $G(n,p)$. Therefore, $G(n,p)$ is in a uniquely defined sense a Riemannian manifold (except for $n = p = 2$). This manifold forms a symmetric, homogeneous, complete space. Moreover $G(1,n)$ and $G(n,1)$ can be identified with S^n. Except for $G(1,1)$, $G(n,p)$ is simply connected, and $\dim G(n,p) = np$.

The minimum locus of a point P_0 of $G(n, p)$ coincides with the first conjugate locus of P_0. Let $\kappa \geq 0$ be an upper bound for the sectional curvature K of $G(n, p)$, and let $B_M(Q)$ be a closed geodesic ball in $G(n, p)$ with center Q and radius M. If $M < \pi/\sqrt{\kappa}$, then the first conjugate locus of Q does not meet $B_M(Q)$. If $M < \pi/(2\sqrt{\kappa})$, then we can introduce normal coordinates on $B_M(Q)$ around each point P of $B_M(Q)$ as center. Let us now fix an invariant metric on $G(n, p)$. For this purpose, we note that

$$G(n, p) = SO(n+p)/SO(n) \times SO(p) .$$

Then it is easy to see that the tangent space to $G(n, p)$ at some point P_0 can be described by the $(n \times p)$-matrices $X = (x^i_\alpha)$, $Y = (y^i_\alpha), \cdots,$ and that the inner product of two tangent vectors X and Y at P_0 can be chosen as

$$<X, Y> = \operatorname{tr}(XY^*) \tag{2.1}$$

where Y^* is the transpose of Y, and tr denotes the trace. Let $n + p \geq 3$. With respect to the normalized metric (2.1), the sectional curvature $K(X, Y)$ of $G(n, p)$ satisfies

$$K(X, Y) = 1 \quad \text{if} \quad n = 1 \quad \text{or} \quad p = 1$$

and

$$0 \leq K(X, Y) \leq 2 \quad \text{if} \quad \min\{n, p\} \geq 2 .$$

Thus we may choose κ as

$$\kappa = \left\{\begin{matrix} 1 & & m = 1 \\ & \text{if} & \\ 2 & & m \geq 2 \end{matrix}\right\}, \text{ where } m = \min\{n, p\} . \tag{2.2}$$

Suppose that $\{\dot{e}_1, \cdots, \dot{e}_n\}$ and $\{\dot{f}_1, \cdots, \dot{f}_n\}$ form an orthonormal basis of the two n-planes P_0 and P, respectively. Then we define the inner product $<P_0, P>$ of P_0 and P by

$$<P_0, P> = \det(<\dot{e}_\alpha, \dot{f}_\beta>) \tag{2.3}$$

where $\langle \dot{e}_\alpha, \dot{f}_\beta \rangle$ denotes the Euclidean inner product of the vectors $\dot{e}_\alpha$ and $\dot{f}_\beta$. It follows from a computation by Fischer-Colbrie that

$$B^*_M(P_0) \subset B_M(P_0) \tag{2.4}$$

for $0 \le M < \pi/2$ where $B^*_M(P_0)$ is defined by

$$B^*_M(P_0) = \{P \in G(n,p) : \langle P_0, P \rangle \ge \cos^m(M/\sqrt{m})\} . \tag{2.5}$$

By virtue of the Theorems 5 and 8, and by the previous results, we obtain the following Liouville theorem for harmonic mappings into Grassmannian manifolds:

THEOREM 14. *Let* $U : \mathcal{X} \to G(n,p)$ *be a harmonic map of a compact or a simple Riemannian manifold* $\mathcal{X}$ *into a Grassmannian manifold* $G(n,p)$ *such that the range* $U(\mathcal{X})$ *of* U *is contained in a closed ball* $B_M(P_0)$ *in* $G(n,p)$ *with center* P_0 *and radius* $M < \pi/(2\sqrt{\kappa})$ *where* $\kappa = 1$ *for* $m = 1$, $\kappa = 2$ *for* $m \ge 2$, *and* $m = \min\{n,p\}$. *Then* U *is a constant map.*

The link between minimal submanifolds in the Euclidean space E^{n+p} and harmonic mappings is formed by an important result due to Ruh and Vilms [61].

THEOREM 15. *The Gauss map* $\mathcal{G} : \mathcal{X} \to G(n,p)$ *of a* C^∞*-immersion*[3] $F : \mathcal{X} \to E^{n+p}$ *of an* n*-dimensional manifold* $\mathcal{X}$ *into* E^{n+p} *is harmonic if and only if* $\mathcal{X}$ *is immersed with parallel mean curvature field. In particular,* $\mathcal{G}$ *is harmonic if* $F(\mathcal{X})$ *is a minimal* n*-dimensional submanifold in* E^{n+p}.

Now we consider a minimal, n-dimensional submanifold $M = F(\mathbf{R}^n)$ which is given by a nonparametric representation

$$F(x) = (x, f(x)) = (x^\alpha, f^i(x))_{1 \le \alpha \le n,\, n+1 \le i \le n+p} \tag{2.6}$$

where $x = (x^1, \cdots, x^n) \in \mathbf{R}^n$ and $f \in C^\infty(\mathbf{R}^n, \mathbf{R}^p)$. Clearly, F is an embedding.

[3]In fact, $F \in C^3$ would be sufficient.

Osserman [57] has proved that a manifold $\mathfrak{M}$, given in the nonparametric form (2.6), is a minimal submanifold of E^{n+p} if and only if the functions $f^i(x)$, $i = n+1, \cdots, n+p$, are solutions of the system

$$D_\beta\{\sqrt{\gamma}\,\gamma^{\alpha\beta} D_\alpha f^i\} = 0 \quad \text{on} \quad \mathbf{R}^n, \; i = n+1, \cdots, n+p, \tag{2.7}$$

where

$$\gamma_{\alpha\beta}(x) = \delta_{\alpha\beta} + \langle D_\alpha f(x), D_\alpha f(x)\rangle = \delta_{\alpha\beta} + f^i_{x^\alpha}(x) f^i_{x^\beta}(x)$$

(summation with respect to i from $n+1$ to $n+p$),

$$\gamma = \det(\gamma_{\alpha\beta}), \qquad (\gamma^{\alpha\beta}) = (\gamma_{\alpha\beta})^{-1}.$$

Morrey has proved that weak solutions f of (2.7), which are of class $C^1(R^n, R^p)$, are real analytic. On the other hand, Lawson and Osserman [43] have shown that there exist Lipschitz continuous weak solutions of (2.7) which are not of class C^1.

From the previous results of this section we can readily derive the following "Bernstein theorem" which has been proved by Hildebrandt-Jost-Widman [32]:

THEOREM 16. *Let* $z^i = f^i(x)$, $i = n+1, \cdots, n+p$, $x = (x^1, \cdots, x^n) \in \mathbf{R}^n$, *be a nonparametric* C^2*-solution of the minimal surface system (2.7). Suppose that there exists a number* β_0 *with*

$$\beta_0 < \cos^{-m}(\pi/(2\sqrt{\kappa m})), \; m = \min\{n, p\}, \; \kappa = \begin{Bmatrix} 1 & & m = 1 \\ & \text{if} & \\ 2 & & m \geq 2 \end{Bmatrix}, \tag{2.8}$$

such that

$$\Delta_f(x) \leq \beta_0 \quad \textit{for all} \quad x \in \mathbf{R}^n, \tag{2.9}$$

where we have set

$$\Delta_f(x) = \{\det(\delta_{\alpha\beta} + f^i_{x^\alpha}(x) f^i_{x^\beta}(x))\}^{1/2} = \sqrt{\gamma(x)}. \tag{2.10}$$

Then, $f^1(x), f^2(x), \cdots, f^n(x)$ *are linear functions on* $\mathbf{R}^n$ *representing an affine* n*-plane in* E^{n+p}.

If $p = 1$, then $m = \kappa = 1$. Therefore, (2.8) implies no restriction for β_0, and condition (2.9) becomes $|\nabla f(x)| \leq$ const. Thus, Theorem 16 furnishes for $p = 1$ the following result: An entire solution $f(x)$ of the minimal surface equation

$$\operatorname{div}\left\{\frac{\nabla f}{\sqrt{1+|\nabla f|^2}}\right\} = 0$$

with $\sup_{\mathbb{R}^n}|\nabla f(x)| < \infty$ is necessarily a linear function. This is nothing but Moser's weak Bernstein theorem cf. [52], pp. 590-591. In the classical Bernstein theorem which has been proved up to $n \leq 7$ by Bernstein $(n=2)$, De Giorgi $(n=3)$, Almgren $(n=4)$, Simons $(n=5,6,7)$, no uniform boundedness of $|\nabla f(x)|$ on $\mathbb{R}^n$ is required. This is equivalent to replacing $<$ in (2.8) by $\leq$, or to replacing the condition $M < \pi/(2\sqrt{\kappa})$ in Theorem 14 by $M \leq \pi(2\sqrt{\kappa})$. We are unable to handle this situation. In fact, many examples for analogous situations show that, in general, the Liouville property breaks down if the equality sign is admitted in the decisive criterion, cf. [30], [31], [46]. Also, it is known that Bernstein's theorem does not hold for $p = 1$, $n \geq 8$, because of the example in [3].

Thus, our Theorem 16 can be considered as a generalization of Moser's theorem to higher codimension p. It is not clear in which cases our quantitative assumption is sharp. But it is very likely that quantitative conditions of the kind (2.8), (2.9) are necessary for a Bernstein theorem to hold for C^2-solutions on $\mathbb{R}^n$, since Lawson and Osserman have found Lipschitz solutions on $\mathbb{R}^n$ of (2.7) which are not linear, provided that $n \geq 4$. For instance, the Lipschitz function $f: \mathbb{R}^4 \to \mathbb{R}^3$ given by

$$f(x) = \frac{\sqrt{5}}{2}|x|\,\eta\left(\frac{x}{|x|}\right) \quad \text{for} \quad x \neq 0\,,$$

where

$$\eta(z_1, z_2) = (|z_1|^2 - |z_2|^2, 2z_1\bar{z}_2)$$

denotes the Hopf map $\eta: S^3 \to S^2$, turns out to be a solution of (2.7), cf. [43], p. 15. On the other hand, using an idea of Barbosa as well as results from geometric measure theory, Fischer-Colbrie [16] Theorem 5.4, has sketched a proof that each entire solution $f: \mathbb{R}^3 \to \mathbb{R}^p$ of the minimal surface system has to be linear if it satisfies $\sup_{\mathbb{R}^3} |\nabla f| < \infty$. Her approach can also be used to give another proof of Theorem 16. For $n = 2$, Chern and Osserman found an even better result. In [9] they proved that a complete, two-dimensional regular minimal surface in E^{2+p} has to be a 2-plane if its image under the Gauss map does not intersect a dense set of hyperplanes.

3. *Remarks on Liouville theorems for more general elliptic systems*

During the last years, various authors have proved Liouville theorems for solutions of nonlinear elliptic systems. To our knowledge, the first paper dealing with systems has been written by Frehse [18]. His results have been known since 1974. Hildebrandt and Widman [31] have dealt with a class of nonlinear systems which contain the harmonic maps as subclass. Ivert [35] has pointed out that several of the results of [31] can be derived from the a priori estimates established in [65] and [28] by using a blow down procedure. Meier [46] has extended the results of [31]. In addition, Section 3 of his paper contains remarkable examples which show that the results of [31] and [46] are optimal. We observe that similar examples exist for harmonic mappings. For this reason, Theorem 9 of the present paper cannot be improved. A survey on the results mentioned before can be found in the lecture notes [30] by Hildebrandt. While most of these results are concerned with systems the principal part of which is in diagonal form, Uhlenbeck [64] has proved a Liouville theorem for more general systems. Recently, further results in this direction have been found by Meier [47], Kawohl [42], and Wiegner [66]. Kawohl and Wiegner investigate also the question as to whether the "Liouville property" and the existence of a priori estimates are equivalent problems. This question had been raised in a paper of Giaquinta and Necas [20], and earlier by Frehse [18], pp. 98-99.

WEGELERSTRASSE 10
D-5300 BONN
MATHEMATISCHES INSTITUT
der UNIVERSITÄT BONN
FEDERAL REPUBLIC OF GERMANY

REFERENCES

[1] Almgren, F. J., Some interior regularity theorems for minimal surfaces and an extension of Bernstein's theorem, Annals of Math. (2), 84 (1966), 277-292.

[2] Bernstein, S. N., Über ein geometrisches Theorem und seine Anwendung auf die partiellen Differentialgleichungen vom elliptischen Typus, Math. Z. 26 (1927), 551-558.

[3] Bombieri, E., E. De Giorgi and E. Giusti, Minimal cones and the Bernstein theorem, Inventiones Math. 7 (1969), 243-269.

[4] Calabi, E., On manifolds with non-negative Ricci curvature, II, Notices Amer. Math. Soc. 22 (1975), A205.

[5] Cauchy, A., Mémoire sur quelques propositions fondamentales du calcul des résidus, et sur la théorie des intégrales singulières, Comptes Rendus 19 (1844), 1337-1344, and Oeuvres, Ser. I, Vol. 8, 366-375.

[6] Cheng, S.-Y., and S. T. Yau, Differential equations on Riemannian manifolds and their geometric applications, Comm. Pure Appl. Math. 28 (1975), 333-354.

[7] Cheng, S.-Y., Liouville theorem for harmonic maps, Preprint.

[8] Chern, S.-S., and S. I. Goldberg, On the volume decreasing property of a class of real harmonic mappings, American J. Math. 97 (1975), 133-147.

[9] Chern, S.-S., and R. Osserman, Complete minimal surfaces in euclidean n-space, J. Analyse Math. 19 (1967), 15-34.

[10] De Giorgi, E., Una estensione del teorema di Bernstein, Ann. Scuola Norm. Sup. Pisa 19 (1965), 79-85.

[11] Do Carmo, M., Stability of minimal submanifolds, Lecture notes, Berlin 1979, Preprint.

[12] Do Carmo, M., and C. K. Peng, Stable complete minimal surfaces in R^3 are planes, Preprint.

[13] ———, Stable complete minimal hypersurfaces, Preprint.

[14] Eells, J., and L. Lemaire, A report on harmonic maps, Bull. London Math. Soc. 10 (1978), 1-68.

[15] Eells, J., and J. H. Sampson, Harmonic mappings of Riemannian manifolds, American J. Math. 86 (1964), 109-160.

[16] Fischer-Colbrie, D., Some rigidity theorems for minimal submanifolds of the sphere, Acta Math. 145 (1980), 29-46.

[17] Fischer-Colbrie, D., and R. Schoen, The structure of complete stable minimal surfaces in 3-manifolds of non-negative scalar curvature, Comm. Pure Appl. Math. 33 (1980), 199-211.

[18] Frehse, J., Essential selfadjointness of singular elliptic operators, Bol. Soc. Bras. Mat. 8 (1977), 87-107.

[19] Garber, W.-D., S. N. M. Ruijsenaars, and E. Seiler, and D. Burns, On finite action solutions of the nonlinear σ-model, Annals of Physics 119 (1979), 305-325.

[20] Giaquinta, M., and J. Nečas, On the regularity of weak solutions to nonlinear elliptic systems via Liouville's type property, Comm. Math. Univ. Carolinae 20 (1979), 111-121.

[21] Gordon, W. B., Convex functions and harmonic maps, Proc. Amer. Math. Soc. 33 (1972), 433-437.

[22] Greene, R. E., and H. Wu, Integrals of subharmonic functions on manifolds of nonnegative curvature, Inventiones Math. 27 (1974), 265-298.

[23] ________, C^∞ approximations of convex, subharmonic, and plurisubharmonic functions, Bull. Amer. Math. Soc. 81 (1975), 101-104.

[24] ________, C^∞ convex functions and manifolds of positive curvature, Acta Math. 137 (1976), 209-245.

[25] ________, Function Theory on Manifolds Which Possess a Pole, Lecture Notes in Math., No. 699, Springer-Verlag, New York, 1979.

[26] Hildebrandt, S., and H. Kaul, Two-dimensional variational problems with obstructions, and Plateau's problem for H-surfaces in a Riemannian manifold, Comm. Pure Appl. Math. 25 (1972), 187-223.

[27] Hildebrandt, S., and K.-O. Widman, Some regularity results for quasilinear elliptic systems of second order, Math. Z. 142 (1975), 67-86.

[28] ________, On the Hölder continuity of weak solutions of quasilinear elliptic systems of second order, Ann. Scuola Norm. Sup. Pisa (IV), 4 (1977), 145-178.

[29] Hildebrandt, S., H. Kaul, and K.-O. Widman, An existence theorem for harmonic mappings of Riemannian manifolds, Acta Math. 138 (1977), 1-16.

[30] Hildebrandt, S., Entire solutions of nonlinear elliptic systems, and an approach to Bernstein theorems. Lecture at the Collège de France, March 1979, to appear.

[31] Hildebrandt, S., and K.-O. Widman, Sätze vom Liouvilleschen Typ für quasilineare elliptische Gleichungen und Systeme. Nachr. Akad. Wiss. Göttingen, II. Math.-Phys. Klasse, Jahrgang 1979, Nr. 4, 41-59.

[32] Hildebrandt, S., J. Jost, and K.-O. Widman, Harmonic mappings and minimal submanifolds, Inventiones Math. 62 (1980), 269-298.

[33] Hoffman, D. A., and R. Osserman, The geometry of the generalized Gauss map, Preprint.

[34] ________, The area of the generalized Gaussian image and the stability of minimal surfaces in S^n and R^n, Preprint.

[35] John, F., On quasi-isometric mappings, I and II, Comm. Pure Appl. Math. 21 (1968), 77-110, and 22 (1969), 265-278.

[37] Jones, P. W., A complete bounded complex submanifold of $\mathbb{C}^3$, Proc. Amer. Math. Soc. 76 (1979), 305-306.

[38] Jorge, L. P. de M., and F. Xavier, A complete minimal surface in $\mathbb{R}^3$ between two parallel planes, Preprint.

[39] Jost, J., Eineindeutigkeit harmonischer Abbildungen, Diplomarbeit, Bonn, 1979.

[40] Karp, L., Subharmonic functions on real and complex manifolds, Preprint.

[41] ———, Differential inequalities on Riemannian manifolds: Applications to isometric immersions and harmonic mappings, Preprint.

[42] Kawohl, B., On Liouville theorems, continuity and Hölder continuity of weak solutions to some quasilinear elliptic systems, Darmstadt, Dec. 1979, Preprint No. 515, Fachbereich Mathematik.

[43] Lawson, H. B., and R. Osserman, Non-existence, non-uniqueness and irregularity of solutions to the minimal surface system, Acta Math. 139 (1977), 1-17.

[44] Liouville, J., Remarques de M. Liouville, Comptes Rendus 19 (1844), 1261-1263.

[45] ———, Remarques de M. Liouville, Comptes Rendus 32 (1851), 450-452.

[46] Meier, M., Liouville theorems for nonlinear elliptic equations and systems, Manuscripta Math. 29 (1979), 207-228.

[47] ———, Liouville theorems for nondiagonal elliptic systems in arbitrary dimensions, Math. Z. 176 (1981), 123-133.

[48] Milnor, J., A note on curvature and the fundamental group, J. Diff. Geometry 2 (1968), 1-7.

[49] ———, On deciding whether a surface is parabolic or hyperbolic, Amer. Math. Monthly 84 (1977), 43-46.

[50] Morrey, C. B., Second order elliptic systems of differential equations, in Contributions to the Theory of Partial Differential Equations, Annals of Math. Studies No. 33, Princeton University Press, Princeton, 1954, pp. 101-160.

[51] ———, Multiple integrals in the calculus of variations, Springer-Verlag, Heidelberg-New York, 1966.

[52] Moser, J., On Harnack's theorem for elliptic differential equations, Comm. Pure Appl. Math. 14 (1961), 577-591.

[53] Neuenschwander, E., The Casorati-Weierstrass theorem, Historia Math. 5 (1978), 139-166.

[54] Nitsche, J. C. C., Vorlesungen über Minimalflächen, Springer-Verlag, Berlin-Heidelberg-New York, 1975.

[55] Osserman, R., A Survey of Minimal Surfaces, Van Nostrand, Cincinnati-Toronto-London-Melbourne, 1969.

[56] ———, Minimal surfaces, Gauss maps, total curvature, eigenvalue estimates and stability, Preprint, 1979.

[57] ———, Minimal varieties, Bull. Amer. Math. Soc. 75 (1969), 1092-1120.

[58] Pfluger, A., Theorie der Riemannschen Flächen, Springer-Verlag, Berlin-Göttingen-Heidelberg, 1957.

[59] Reilly, R., Extrinsic rigidity theorems for compact submanifolds of the sphere, J. Diff. Geometry 4 (1970), 487-497.

[60] Ruh, E. A., Asymptotic behaviour of non-parametric minimal hypersurfaces, J. Diff. Geometry 4 (1970), 509-513.

[61] Ruh, E. A., and J. Vilms, The tension field of the Gauss map, Trans. Amer. Math. Soc. 149 (1970), 569-573.

[62] Schoen, R., and S.-T. Yau, Harmonic maps and the topology of stable hypersurfaces and manifolds with nonnegative Ricci curvature, Comm. Math. Helv. 39 (1976), 333-341.

[63] Simons, J., Minimal varieties in Riemannian manifolds, Annals of Math. 88 (1968), 62-105.

[64] Uhlenbeck, K., Regularity for a class of non-linear elliptic systems, Acta Math. 138 (1977), 219-240.

[65] Wiegner, M., A-priori Schranken für Lösungen gewisser elliptischer Systeme, Manuscripta Math. 18 (1976), 279-297.

[66] ———, Regularity theorems for nondiagonal elliptic systems, Preprint No. 338, SFB 72, Bonn, 1980.

[67] Wood, J. C., Singularities of harmonic maps and applications of the Gauss-Bonnet formula, Amer. J. Math. 99 (1977), 1329-1344.

[68] Wu, H., An elementary method in the study of nonnegative curvature, Acta Math. 142 (1979), 57-78.

[69] Xavier, F., The Gauss map of a complete non-flat minimal surface cannot omit 11 points of the sphere, Preprint.

[70] Xin, Y. L., Some results on stable harmonic maps, Preprint.

[71] Yau, S.-T., Harmonic functions on complete Riemannian manifolds, Comm. Pure Appl. Math. 28 (1975), 201-228.

[72] ———, Some function theoretic properties of complete Riemannian manifolds and their applications to geometry, Indiana U. Math. J. 25 (1976), 659-670.

SUBHARMONIC FUNCTIONS, HARMONIC MAPPINGS AND ISOMETRIC IMMERSIONS

Leon Karp*

1. *Introduction*

This paper gives a resumé of some recent results concerning the connections between the geometry of a noncompact manifold and the global behavior of its subharmonic functions, harmonic mappings, and isometric immersions ([19], [20]).

M^n will denote a connected, complete, n-dimensional Riemannian manifold; Δ will denote its Laplacian (with sign conventions such that $\Delta = \frac{d^2}{dx^2}$ on R^1), and a function $u: M^n \to R$ is *subharmonic* (resp. harmonic) if $\Delta u \geq 0$ (resp. $\Delta u = 0$). For simplicity we consider only $u \in C^2(M^n)$.

Note that every divergence-form second-order elliptic operator $L = \frac{\partial}{\partial x^i} a^{ij} \frac{\partial}{\partial x^j}$ on R^n, $n \geq 3$, is (up to a positive factor) the Laplacian of some metric on R^n. Thus we are, in passing, also considering solutions of $Lu \geq 0$ on R^n.

We will be interested in Riemannian analogues of Liouville's theorem: If $\Delta u = 0$ on R^n (or $\Delta u \geq 0$ on R^2) then $\sup u < \infty \Longrightarrow u \equiv \text{constant}$.

*Supported by an NSF Postdoctoral Fellowship

Seminar on Differential Geometry
0-691-08268-5/82/000133-10$00.50/0 (cloth)
0-691-08296-0/82/000133-10$00.50/0 (paperback)

It is of special interest to obtain such results without introducing hypotheses concerning the curvature of M^n.

2. *Integrals of subharmonic and harmonic functions*

Although every noncompact Riemannian manifold admits many harmonic functions (see [11], [12]) and hence also nonnegative subharmonic functions (since $\Delta u^2 \geq 0$ if $\Delta u = 0$), one has the following

THEOREM (Greene-Wu [13], cf. [11]). *Let* M^n *be a complete, noncompact manifold with positive sectional curvature outside some compact subset. If* $u \not\equiv 0$ *is a nonnegative subharmonic function and* $p \geq 1$, *then* $\exists C > 0$ *such that for* $\forall x_0 \in M$

$$\int_{B(x_0,r)} u^p d\mathrm{vol} \geq Cr$$

for all sufficiently large r. *Here* $B(x_0, r)$ *is the geodesic ball of radius* r *centered at* x_0.

Greene and Wu obtained a similar result for manifolds of nonnegative sectional curvature. Moreover, both of these results are valid for continuous functions (see [11] for precise definitions).

Examples show that the hypothesis $u \geq 0$ is necessary. Since we may take $u \equiv$ constant and M^n to be a manifold of positive curvature that is asymptotic to a cylinder, it is clear that the linear growth estimate given above cannot in general be improved. Our first result shows that some improvement is possible (when $p > 1$ and $r \to \infty$) for nonconstant subharmonic functions. To formulate this theorem it is convenient to introduce the following notation:

$$\mathcal{F} \underset{\text{def}}{=} \{F : (0, \infty) \to (0, \infty) | F \text{ is increasing on } (0, \infty) \text{ and } \int_1^\infty \frac{dr}{rF(r)} = +\infty\}.$$

Thus $\log(1+r) \in \mathcal{F}$, $\log(1+r) \cdot \log[1 + \log(1+r)] \in \mathcal{F}$, while $[\log(1+r)]^{1+\varepsilon} \notin \mathcal{F}$ if $\varepsilon > 0$.

The improved growth rate for integrals of nonconstant subharmonic functions needs no hypothesis concerning the curvature of M^n and is given in

THEOREM A ([19]). *If* u *is a nonconstant solution of* $u\Delta u \geq 0$ *on* M^n (*e.g., if* $\Delta u = 0$ *or if* $u \geq 0$ *and* $\Delta u \geq 0$), *then for* $\forall p > 1$, $\forall F \in \mathfrak{F}$ *and* $\forall x_0 \in M^n$, *we have*

$$\text{(a)}\quad \limsup_{r\to\infty} \frac{1}{r^2 F(r)} \int_{B(x_0,r)} |u|^p dvol = +\infty$$

$$\text{(b)}\quad \liminf_{r\to\infty} \frac{1}{r^2} \int_{B(x_0,r)} |u|^p dvol = +\infty .$$

COROLLARY A (Yau [28]). *There are no nontrivial solutions of* $u\Delta u \geq 0$ *in* $L^p(M^n, dvol)$ *if* $p > 1$.

REMARKS:

1) Taking $M^n = \mathbf{R}^n$ with the standard metric and u as in (2.1), it is easy to see that the statement of the theorem is false for $p = 1$ and/or $\frac{1}{r^\alpha F}$, $\alpha > 2$, in place of $\frac{1}{r^2 F}$. Moreover, the conclusion is false if F increases so fast that $F \notin \mathfrak{F}$. In fact, the manifold $M^n = S^{n-1}_\omega \times \mathbf{R}^1_t$ with the metric $ds^2 = dt^2 + [\sqrt{1+t^2}\cdot F(\sqrt{1+t^2})\,d\omega]^2$, where $d\omega^2$ denotes the standard metric on S^{n-1}, admits the bounded harmonic function $h(w,t) = \int_0^t \frac{dr}{\sqrt{1+r^2}F(\sqrt{1+r^2})}$ if $F \notin \mathfrak{F}$. It follows that the conclusion of the theorem is, in general, false for any $F \notin \mathfrak{F}$.

2) O. L. Chung has recently informed us of his construction in 1974-75 of a nontrivial L^1 harmonic function on a complete Riemannian manifold of infinite volume (see [7]). Thus even Corollary A fails for $p = 1$.

These results have analogues for harmonic mappings (see [9], and [8] for definitions). In fact, various authors have recently proved Liouville-type theorems for harmonic mappings $f: M^n \to N^k$ under appropriate hypotheses (see Cheng [5], Garber-Ruisenjaars-Seiler-Burns [10], Hildebrandt-Jost-Widman [15], and Schoen-Yau [24]). All of these results require (besides completeness of the metric) restrictive assumptions concerning the manifold structure of M^n, and/or the behavior of the components of its metric tensor in some preferred coordinate system and/or pointwise conditions on its curvature.

The following corollary (of the proof of Theorem A) gives a sharp estimate of the rate of growth of a nonconstant harmonic mapping without making any assumptions concerning M^n (other than completeness). A Liouville-type theorem follows easily (cf. below, Section 3).

COROLLARY A.2 ([20]). *Suppose that* $f: M^n \to N^k$ *is harmonic and that* N^k *is simply-connected and of nonpositive curvature. Then either* $f \equiv$ *constant or*

$$\limsup_{r\to\infty} \frac{1}{r^2 F(r)} \int_{B(x_0,r)} [\mathrm{dist}_N(f(x), f(x_0))]^p dvol = +\infty$$

for every $x_0 \in M$, $F \in \mathcal{F}$, *and* $p > 1$. Here dist_N denotes Riemannian distance in N^k.

REMARKS:

(1) Corollary A.2 generalizes part of Theorem A, and the examples mentioned above show that the various restrictions imposed are necessary. However, the assumption that N is simply-connected and of nonpositive curvature is not essential, and it can be replaced by the assumption that the range of f lies in a convex ball whose radius is determined (in a specific manner) by the upper bound of the sectional curvature of N. The requisite changes in the arguments of [20] would use, for instance, the comparison lemmata of Hildebrandt-Jost-Kaul-Widman [15], [16].

(2) Professor J. Eells has observed that some other Liouville-type results can be obtained from the Sampson Maximum Principle [23].

3. *Parabolicity and volume growth*

Borrowing some terminology from the theory of Riemann surfaces, let us say that a Riemannian manifold M^n is *strongly parabolic* if it admits no nonconstant negative subharmonic function. Thus R^2 is strongly parabolic while R^n, $n \geq 3$, is not. It turns out that, actually, dimension plays no role here, and the decisive criterion is the rate of volume growth. To formalize this let us say that a Riemannian manifold has *moderate volume growth* if $\exists F \in \mathcal{F}$ such that $\limsup_{r\to\infty} \frac{1}{r^2F(r)} \operatorname{vol} B(x_0, r) < \infty$ for some (and hence all) $x_0 \in M^n$. The following result is a consequence of Theorem A:

THEOREM B ([19]). *M^n is strongly parabolic if it has moderate volume growth, and this condition is sharp.*

In fact, if M has moderate volume growth and $\exists v$ such that $\Delta v \geq 0$ and $v < 0$, then $u = \exp v$ violates the conclusions of Theorem A. The manifolds $S^{n-1} \times R^1$ with the metrics constructed above (see Remark 1 after Theorem A) show that any faster growth rate does not, in general, imply strong parabolicity.

REMARK. A weaker version of Theorem B was obtained by Cheng and Yau [6].

4. *Parabolicity and curvature*

Recall that the curvature of a Riemannian manifold has an effect on the rate of volume growth (cf. Bishop and Crittenden [3]). This relation enables us to obtain the following consequence of Theorem B.

COROLLARY B (cf. [19]). *Let M^2 be a complete 2-dimensional Riemannian manifold. If $\exists x_0 \in M^2$ and $r_0 > 0$ such that the Gaussian curvature $K(x)$ satisfies*

$$K(x) \geq \frac{-1}{r^2(x) \log r(x)}, \quad \forall r(x) = \operatorname{dist}(x, x_0) \geq r_0$$

then M^2 is parabolic.

In fact, it can be shown that $K \geq \frac{-1}{r^2 \log r}$ implies that M^2 has moderate volume growth.

REMARKS:

(1) For the special case of simply-connected manifolds, this result is due to Greene-Wu [14] and follows also from a criterion of Ahlfors [1]. The condition on the curvature cannot be essentially weakened, as shown by Greene-Wu [14] and Milnor [22]. A simple proof for the radially symmetric case may be found in [22].

(2) For related results, valid even when $\dim M > 2$ but under more restrictive hypotheses on the curvature, see Greene-Wu [14], and Siu-Yau [25].

(3) The strong parabolicity of M^2 under the condition $K(x) \geq 0$ was obtained first by Blanc-Fiala-Huber (cf. [17]), and even this condition is not sufficient if $\dim M \geq 3$ (e.g., R^n). However, Yau [27] has shown that the analogous result for harmonic functions obtains in all dimensions under the condition that the Ricci-curvature be nonnegative.

5. *Isometric and minimal immersions*

We have seen above that a noncompact manifold admits no nontrivial bounded subharmonic functions if it has moderate volume growth. A different situation obtains for solutions of $\inf_M (\Delta u) > 0$, which we may call *strongly subharmonic*. Let us say that a Riemannian manifold M^n has *volume growth of exponential type* λ if $\exists x_0 \in M^n$ such that

$$\limsup_{r\to\infty} \frac{\log \operatorname{vol} B(x_0, r)}{r} = \lambda < \infty .$$

Volume growth of exponential type is of particular interest as a result of

THEOREM C ([20]). *If* M^n *has volume growth of exponential type* $\lambda < \infty$, *then every solution of* $\inf(\Delta u) > 0$ *satisfies* $\sup u = +\infty$.

This theorem not only complements Theorem B but also leads to an estimate for the extrinsic radius of a complete n-manifold isometrically immersed in R^{n+k}. This is given in

THEOREM D ([20]). *If a complete Riemannian* n-*manifold* M^n *with volume growth of exponential type is isometrically immersed in a ball of radius* R *in* R^{n+k}, $k > 0$, *then*

$$R \geq \frac{1}{H_0}$$

where $H_0 = \sup \|\mathfrak{N}\|$ *and* $\mathfrak{N}$ = *mean curvature vector of the immersion.*

COROLLARY D ([20]). *If* M^n *is a complete minimal submanifold of* R^{n+k}, $k > 0$, *and has volume growth of exponential type (esp. if* $\inf_M \text{Ricci} > -\infty$*), then* M^n *is unbounded.*

Theorem D follows from Theorem C. In fact, if $\phi : M^n \to R^{n+k}$ is the isometric immersion of Theorem D and, for some $y_0 \in R^{n+k}$, $\|\phi(x)-y_0\|^2_{R^{n+k}} < \frac{1}{H_0}$, then it can be shown that $u(x) = \|\phi(x)-y_0\|^2_{R^{n+k}}$ violates the conclusion of Theorem C.

REMARKS:

(1) The conclusions of Theorems C and D and Corollary D are false if no assumptions are made other than completeness. In fact, P. Jones recently gave an example [18] of a complete bounded two-dimensional minimal submanifold of R^n, $n \geq 6$. The proof of Theorem D, sketched above, thus shows that some assumptions beyond completeness are needed in Theorem C. The result given here gives the weakest known conditions that are sufficient to imply the unboundedness of a minimal submanifold of arbitrary codimension $k > 0$ (however, see Yang [26] for complex hypersurfaces in C^N).

(2) Under the more restrictive assumption that $\inf_M \text{Ricci} > -\infty$, Theorem D and its corollary were obtained by Aminov [2] for $n = 2$. The

corollary itself has been generalized by de M. Jorge and Xavier [21], again under the assumptions of bounded curvature and $n = 2$. For $n \geq 3$ Aminov and Burago [4] obtained weaker estimates of the form $R \geq [H_0 + c(n, K)]^{-1}$, where $c(n, K)$ is positive and depends on the dimension n and the lower bound $-K$ for the Ricci curvature.

(3) The estimate $R \geq \frac{1}{H_0}$ is the best possible, as one sees from the example of a hypersphere of radius R in $\mathbf{R}^{n+1}$.

(4) In Theorem D (and its corollary) $\mathbf{R}^{n+k}$ may be replaced with any simply-connected manifold of nonpositive curvature and dimension $> n$. The proof is essentially the same.

(5) A more precise form of Theorem C [relating the precise values of $\limsup_{r\to\infty} \left(\frac{u(x)}{r(x)}\right)$, $\inf(\Delta u)$, and the exponential type λ] is given in [20].

SCHOOL OF MATHEMATICS
INSTITUTE FOR ADVANCED STUDY
PRINCETON, NEW JERSEY 08540

DEPARTMENT OF MATHEMATICS
PRINCETON UNIVERSITY
PRINCETON, NEW JERSEY 08540

BIBLIOGRAPHY

[1] L. Ahlfors, Sur le type d'une surface de Riemann, *C.R. Acad. Sc. Paris*, 201 (1935), 30-32.

[2] J. Aminov, The exterior diameter of an immersed Riemannian manifold, *Math. USSR Sbornik*, 21 (1973), 449-454.

[3] R. Bishop and R. Crittenden, *Geometry of Manifolds*, Academic Press, New York, 1964.

[4] Ju. Burago, On a theorem of Ju. A. Aminov (Russian); in: *Problems in Global Geometry*, Zap. Naucn. Sem. Leningrad, Otdel. Math. Inst. Steklov, 45 (1974), 53-55.

[5] S. Y. Cheng, Liouville theorem for harmonic maps; in: *Geometry of the Laplace Operator*, Proc. Symp. Pure Math., 36, Amer. Math. Soc., Providence, R. I., 1980.

[6] S. Y. Cheng and S. T. Yau, Differential equations on Riemannian manifolds and their geometric applications, *Comm. Pure Appl. Math.*, 28 (1975), 333-354.

[7] O. L. Chung, Existence of harmonic L^1 functions in complete Riemannian manifolds (preprint).

[8] J. Eells, Jr., and L. Lemaire, A report on harmonic maps, *Bull. London Math. Soc.*, 10 (1978), 1-68.

[9] J. Eells, Jr., and J. H. Sampson, Harmonic mappings of Riemannian manifolds, *Amer. J. Math.*, 86 (1964), 109-160.

[10] W. D. Garber, S. Ruisenaars, E. Seiler, and D. Burns, On finite action solutions of the nonlinear σ-model, *Ann. of Physics*, 119 (1979), 305-325.

[11] R. E. Greene and H. Wu, Integrals of subharmonic functions on manifolds of nonnegative curvature, *Invent. Math.*, 27 (1974), 265-298.

[12] ————, Embedding of open Riemannian manifolds by harmonic functions, *Ann. l'Inst. Fourier*, 25 (1975), 215-235.

[13] ————, C^∞ convex functions and manifolds of positive curvature, *Acta Math.*, 137 (1976), 209-245.

[14] ————, *Function Theory on Manifolds which Possess a Pole*, Lecture Notes in Math. 699, Springer-Verlag, N. Y., 1979.

[15] S. Hildebrandt, J. Jost and K. O. Widman, Harmonic mappings and minimal submanifolds (preprint).

[16] S. Hildebrandt, H. Kaul, and K. O. Widman, An existence theorem for harmonic mappings of Riemannian manifolds, *Acta Math.*, 138 (1977), 1-16.

[17] A. Huber, On subharmonic functions and differential geometry in the large, *Comment. Math. Helv.*, 32 (1957), 13-72.

[18] P. Jones, A complete bounded complex submanifold of C^3, *Proc. Amer. Math. Soc.*, 76 (1979), 305-306.

[19] L. Karp, Subharmonic functions on real and complex manifolds, *Math. Zeit.* (to appear).

[20] ————, Differential inequalities on Riemannian manifolds: Applications to isometric immersions and harmonic mappings (preprint).

[21] L. P. de M. Jorge and X. Xavier, On the existence of complete bounded minimal surfaces in R^n, *Bol. Soc. Bras. Mat.* 10 (1979), 191-174.

[22] J. Milnor, On deciding whether a surface is parabolic or hyperbolic, *Amer. Math. Monthly*, 84 (1977), 43-46.

[23] J. H. Sampson, Some properties and applications of harmonic mappings, *Ann. École Norm. Sup.*, 11 (1978), 211-228.

[24] R. Schoen and S. T. Yau, Harmonic maps and the topology of stable hypersurfaces and manifolds of nonnegative Ricci curvature, *Comm. Math. Helv.*, 51 (1976), 333-341.

[25] Y. T. Siu and S. T. Yau, Complete Kähler manifolds with nonpositive curvature of faster than quadratic decay, *Ann. of Math.*, 105 (1977), 225-264.

[26] P. Yang, Curvatures of complex submanifolds of C^n, *J. Diff. Geom.*, 12 (1977), 499-511.

[27] S. T. Yau, Harmonic functions on complete Riemannian manifolds, *Comm. Pure Appl. Math.*, 28 (1975), 201-228.

[28] ________, Some function theoretic properties of complete Riemannian manifolds and their applications to geometry, *Indiana U. Math. J.*, 25 (1976), 659-670.

Added in Proof. The paper "An inequality between exterior diameter and the mean curvature of bounded immersions" Math. Z. 178 (1981), 77-82, by L. P. de M. Jorge and F. Xavier has just appeared. This paper contains a proof of the conclusion of our Theorem D under the assumption that the scalar curvature of M is bounded from below and also notes that under the assumption "inf Ricci $> -\infty$" the result was also proved by Th. Hasanis and D. Koutroufiotis in "Immersions of bounded mean curvature" Arch. Math. 33 (1979), 170-171, using a theorem of Omori. The former paper also treats the case where the ambient space R^{n+k} is replaced by a more general space with sectional curvature bounded above. The analogous results are true also in our framework–and with the proof as above, as the informed reader will realize. Our choice of R^{n+k} in this note was motivated by its simplicity (cf. Remark 1 after Corollary A.2 and Remark 4 after Corollary D). The various results described here were obtained in and/or before 1979 and were described in the I.A.S. geometry seminar in December of that year.

AN ISOPERIMETRIC INEQUALITY AND WIEDERSEHEN MANIFOLDS

Jerry L. Kazdan[1]

In memory of Rufus Bowen, who was more than a friend.

The concept of a *WIEDERSEHENFLÄCHE* was introduced by Blaschke around 1920. Intuitively, it is a surface M in $\mathbf{R}^3$ such that, starting at any point x on M, if one moves a distance L along any geodesic, then all of the geodesics meet again at some point x′. The obvious example for M is the canonical sphere of radius L/π in $\mathbf{R}^3$. A more precise definition is that M is a complete Riemannian manifold such that for any point x of M, its cut locus is precisely one point, x′. This definition applies equally in higher dimensions, in which case we refer to M as a *Wiedersehen Manifold.*

Blaschke conjectured that the canonical sphere is the only Wiedersehenfläche. This was finally proved by L. W. Green [G] in 1962. The higher dimensional case remained open until 1978, when a proof was found using results of M. Berger [B2], J. Kazdan [K], A. Weinstein [W], and C. T. Yang [Y].

THEOREM A. *The only Wiedersehen Manifold is the standard sphere.*

[1]Supported in part by NSF Grant MCS79-01780.

Seminar on Differential Geometry
0-691-08268-5/82/000143-15$00.75/0 (cloth)
0-691-08296-0/82/000143-15$00.75/0 (paperback)
For copying information, see copyright page.

There are two steps in proving this. The first is an isoperimetric inequality, which we state in the sharp and elegant form recently found by Marcel Berger [B5].

Isoperimetric Inequality (Berger [B2, B5], Kazdan [K]). *If* (M^n, g) *is compact with injectivity radius at least* π, *then*

$$\text{Vol}(M^n, g) \geq \text{Vol}(S^n, \text{can}) , \tag{1}$$

with equality if and only if (M^n, g) *is isometric to* (S^n, can).

Here, and in what follows, (S^n, can) is the canonical sphere S^n of radius 1 in R^{n+1}. By scaling the metric in (1) one obviously obtains the following.

COROLLARY. *Let* (M^n, g) *be compact with injectivity radius* $\geq \rho$. *Then*

$$\text{Vol}(M^n, g) \geq (\rho/\pi)^n \text{Vol}(S^n, \text{can}) \tag{1$'$}$$

with equality if and only if (M^n, g) *is isometric to the standard sphere* S^n *of radius* ρ/π *in* R^{n+1}.

Prior to Berger's recent work, we only knew that the sharp Isoperimetric Inequality (1) held for Wiedersehen manifolds; for general manifolds there was annoying extra factor of $1/2$ on the right side of (1), see [B-K, equation (20)].

The second step uses a topological argument (see also [Besse]).

THEOREM B (A. Weinstein [W], C. T. Yang [Y]). *If* (M^n, g) *is a Wiedersehen manifold* (*normalized to injectivity radius* π) *then*

$$\text{Vol}(M^n, g) = \text{Vol}(S^n, \text{can}) .$$

Theorem B follows from a more general result concerning the volume of a manifold all of whose geodesics are closed and have length 2π (early in this century, Zoll surprised many by showing that there are lots of smooth metrics on S^2 with this property; see [Besse], Chapter 4). Weinstein proved that for these manifolds

$$\mathrm{Vol}(M^n, g) = j\ \mathrm{Vol}(S^n, \mathrm{can}) ,$$

where j is an integer. That $j=1$ was proved for even n by Weinstein, and for odd n (for certain manifolds, including Wiedersehen manifolds) by C.T. Yang.

Theorem A is an immediate consequence of these.

In this lecture we will give a complete proof of the Isoperimetric Inequality including Berger's improvement. Moreover, by reworking a few other details, the proof is shorter and hopefully clearer. This will collect all of the pieces in one place.

The geometric part of the proof is in Section 1, the analytic part in Section 2. Section 3 contains related results and some open questions.

NOTATION. dx is the volume element of (M, g), UM and UM_x are the unit tangent bundle and its fibre over $x \in M$, respectively. If $u \in UM$, then $x = p(u) \in M$ is the usual projection, $u \to \zeta^s(u)$ is the geodesic flow a distance s from x in the direction of u, and du is the element of volume on UM. $B_r(x)$ is the open geodesic bali centered at x with radius r. α_k is the volume of (S^k, can).

1. *Geometric part*

Assume (M, g) is compact with injectivity radius at least π. Since the two points $p(u)$ and $p(\zeta^\pi(u))$ on M are a distance π apart, the balls $B_r(p(u))$ and $B_{\pi-r}(p(\zeta^\pi(u)))$ do not intersect for $0 \leq r \leq \pi/2$.

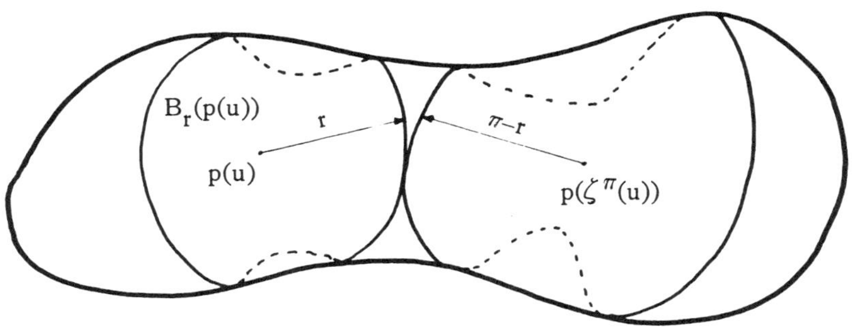

Therefore

$$\text{(2)} \qquad \mathrm{Vol}(M,g) \geq \mathrm{Vol}[B_r(p(u))] + \mathrm{Vol}[B_{\pi-r}(p(\zeta^{\pi}(u)))] .$$

The idea is to average this over all points x and all possible directions (i.e., average over UM), and then average for all $0 \leq r \leq \pi/2$.

Thus, we first integrate (2) over UM. For the second term in (2) use Liouville's theorem concerning invariance under geodesic flow [Besse, 1.125] to find

$$\int_{UM} \mathrm{Vol}[B_{\pi-r}(p(\zeta^{\pi}(u)))]\,du = \int_{UM} \mathrm{Vol}[B_{\pi-r}(p(u))]\,du$$

$$= a_{n-1} \int_M \mathrm{Vol}[B_{\pi-r}(x)]\,dx .$$

Since $\mathrm{Vol}(UM) = a_{n-1}\mathrm{Vol}(M,g)$ this gives

$$\text{(3)} \qquad \mathrm{Vol}(M,g)^2 \geq \int_M (\mathrm{Vol}[B_r(x)] + \mathrm{Vol}[B_{\pi-r}(x)])\,dx .$$

Before going on, it is convenient to introduce polar coordinates and write $dx = f(u,\rho)\,d\sigma_x d\rho$, where $d\rho$ is the radial part and $d\sigma_x$ the angular part, while $f(u,\rho)$ is the usual Jacobian factor (for example, on flat $\mathbf{R}^n$, $f(u,\rho) = \rho^{n-1}$, while on (S^n, can), $f(u,\rho) = \sin^{n-1}\rho$). In this notation

$$\mathrm{Vol}(B_r(x)) = \int_{UM_x}\int_0^r f(u,\rho)\,d\rho\,d\sigma_x$$

so (3) reads

$$\mathrm{Vol}(M,g)^2 \geq \int_{UM}\left(\int_0^r + \int_0^{\pi-r}\right) f(u,\rho)\,d\rho\,du .$$

We again use Liouville's theorem to write this as

$$(4) \qquad = \int_{UM} \left[\int_0^r f(\zeta^{\pi-r}(u), \rho)\, d\rho + \int_0^{\pi-r} f(\zeta^r(u), \rho)\, d\rho \right] du .$$

While somewhat mysterious, this step (4) seems to play a computationally crucial role.

Now average (4) with respect to r for $0 \le r \le \pi/2$; then make the change of variable from $\pi - r$ to r in the first integral to obtain

$$\mathrm{Vol}(M, g)^2 \ge \frac{2}{\pi} \int_{UM} \left[\left(\int_{\pi/2}^{\pi} \int_0^{\pi-r} + \int_0^{\pi/2} \int_0^{\pi-r} \right) f(\zeta^r(u), \rho)\, d\rho\, dr \right] du .$$

Upon combining the integrals, we obtain the following.

LEMMA 1. *If* (M, g) *is a compact manifold with injectivity radius at least* π, *then*

$$(5) \qquad \mathrm{Vol}(M, g)^2 \ge \frac{2}{\pi} \int_{UM} \left[\int_0^{\pi} \int_0^{\pi-r} f(\zeta^r(u), \rho)\, d\rho\, dr \right] du .$$

To go further, we look more closely at the Jacobian factor $f(u, \rho)$. Let $e_1, e_2, \cdots, e_n$ be orthonormal vectors in UM_x with $e_1 = u$ and let y_j be the solution of the Jacobi equation

$$(6) \qquad Y_j'' + RY_j = 0, \text{ with } Y_j(0) = 0, \ Y_j'(0) = e_j ,$$

where R is the curvature of (M, g) along the geodesic starting at u. Then (see [B-C, p. 256])

$$f(u, r) = \det [Y_2(r), \cdots, Y_n(r)] .$$

In the special cases $R = 0$ and $R = I$, (6) can be solved explicitly and gives the expected results $f(u, r) = r^{n-1}$ and $f(u, r) = \sin^{n-1} r$ on flat $\mathbf{R}^n$ and (S^n, can), respectively.

More compactly, let the $(n-1)\times(n-1)$ matrix $A(r;s)$ denote the solution of

$$A''(r;s) + R(r)A(r;s) = 0, \quad \text{with } A(s;s) = 0, \quad A'(s;s) = I, \tag{7}$$

where $' = d/dr$. Then

$$f(\zeta^r(u), \rho) = \det A(r+\rho; r)$$

and (5) reads

$$\mathrm{Vol}(M,g)^2 \geq \frac{2}{\pi} \int_{UM} \left[\int_0^{\pi} \int_0^{\pi - r} \det A(r+\rho; r)\, d\rho\, dr \right] du. \tag{5$'$}$$

2. *Analytic part*

From here on, we could ignore the geometric origins of the problem and just consider solutions of the Sturm-Liouville equation (7), assuming only that $R(r)$ is a self-adjoint matrix and interpreting the geometric assumption that the injectivity radius of (M, g) is at least π as asserting that $A(r, s)$ is invertible, i.e., has no conjugate points, for $s < r < s + \pi$ and all real s. For convenience, let $N = n-1$.

MAIN INEQUALITY. *Let* $R = R^*$ *and let* $A(r;s)$ *be a solution of (7). If* $A(r;s)$ *has no conjugate points for* $s < r < s + \pi$, *then for any* $0 \leq s \leq \pi/2$,

$$\int_0^{\pi - s} \det A(r+s; r)\, dr + \int_0^{s} \det A(r+\pi-s; r)\, dr \geq \pi \sin^N s, \tag{8}$$

with equality if and only if $R = I$, *i.e.,* $A(r;s) = \sin(r-s)I$.

COROLLARY [B-K]. *Under the above assumptions, if* $m(s) \geq 0\ (\not\equiv 0)$ *is any continuous function satisfying* $m(\pi - s) = m(s)$, *then*

$$\int_0^{\pi} \int_s^{\pi} [\det A(r;s)]\, m(r-s)\, dr\, ds \geq \pi \int_0^{\pi/2} \sin^N s\, m(s)\, ds \tag{9}$$

with equality if and only if $R = I$, *i.e.,* $A(r;s) = \sin(r-s)I$.

Proof of Corollary. Multiply (8) by $m(s)$ and integrate over $0 \le s \le \pi/2$.

REMARK. In fact the corollary is equivalent to the Main Inequality. Just let $m(s) = \delta_t(s) + \delta_{\pi-t}(s)$, where $\delta_t(s)$ is the delta function concentrated at $s = t$.

Before proving this inequality, we will apply it to prove the Isoperimetric Inequality.

Proof of Isoperimetric Inequality. Make the change of variables $\rho = t-r$ in (5)′ to find

$$\mathrm{Vol}(M,g)^2 \ge \frac{2}{\pi} \int_{UM} \left[\int_0^{\pi} \int_r^{\pi} \det A(t,r)\, dt\, dr \right] du .$$

Then by the corollary with $m \equiv 1$ we conclude that (recall $N = n-1$)

$$\mathrm{Vol}(M,g)^2 \ge 2 \int_{UM} \left(\int_0^{\pi/2} \sin^{n-1} s\, ds \right) du$$

$$= \alpha_n \mathrm{Vol}(M,g) ,$$

in other words

$$\mathrm{Vol}(M,g) \ge \mathrm{Vol}(S^n, \mathrm{can})$$

with equality if and only if $R = I$. But the only compact manifold with curvature $R = I$ is covered by (S^n, can). By (1) it must be (S^n, can) itself. Q.E.D.

It is time to prove the Main Inequality. Let $A(t) = A(t;0)$; we shall often write $A(t)$ simply as A. To begin with, we express $A(t;s)$ in terms of $A(t;0)$.

LEMMA 2. a) $A^*(t)A'(t) = A'^*(t)A(t)$
and

(9) b) $$A(t;s) = A(t) \int_s^t [A^*(r)A(r)]^{-1} dr\, A^*(s) .$$

Proof. a) From the differential equation, $(A^*A' - A'^*A)' = 0$. But $A(0) = 0$.

b) Since $A(t)$ and $A(t;s)$ are solutions of the same differential equation, only with different initial conditions, we use "reduction of order." Thus, seek $A(t;s)$ in the form $A(t;s) = A(t)C(t;s)$. Substitute this into (7) and use part (a) to obtain $(A^*AC')' = 0$, which can be solved for C and yields (b). Q.E.D.

Since the Main Inequality involves $\det A(t;s)$, we will need the next result.

LEMMA 3. *Let* $\phi(t) = [\det A(t)]^{1/N}$, *so* $\phi(0) = 0$ *and* $\phi'(0) = 1$. *Then*

$$[\det A(t;s)]^{1/N} \geq \phi(t)\phi(s) \int_s^t \phi^{-2}(r)\,dr \tag{10}$$

with equality if and only if $A(t) = \phi(t)I$.

Proof. By a computation of the hessian, one sees that the function $F(B) = (\det B)^{-1}$ is strictly convex in the convex set of positive definite matrices B. Therefore, by Jensen's inequality for convex functions, for any positive measure μ,

$$\det\left[\mu(\Omega)^{-1} \int_\Omega B(r)\,d\mu(r)\right]^{-1} \leq \mu(\Omega)^{-1} \int_\Omega \det[B(r)]^{-1}\,d\mu(r) \tag{11}$$

with equality if and only if $B(r)$ is a constant matrix. [Proof: By Taylor's theorem, $F(B) \geq F(B_0) + F'(B_0)(B - B_0)$, with equality if and only if $B = B_0$. Let $B_0 = \mu(\Omega)^{-1} \int_\Omega B(r)\,d\mu$ and $B = B(r)$. Then integrate the Taylor inequality over Ω.]

Let $B(r) = [A^*(r)A(r)]^{-1}\phi^2(r)$, $d\mu = \phi^{-2}(r)\,dr$, and $\Omega = \{s \leq r \leq t\}$. Then (9) and (11) yield (10) with equality only if $B(r) =$ constant matrix. But by Lemma 2(a), this is equivalent to $0 = (B^{-1})' = 2(A^*/\phi)(A/\phi)'$. Thus, equality occurs only if $A(r)/\phi(r) =$ constant matrix. Evaluating this at $r = 0$ shows that $A(r) = \phi(r)I$. Q.E.D.

Let P_s denote the first integral in (8). Then inequality (10) shows that

$$(12)\quad P_s = \int_0^{\pi-s} \det A(r+s;r)\,dr \geq \int_0^{\pi-s} \left[\phi(r+s)\phi(r)\int_r^{r+s} \phi^{-2}(t)\,dt\right]^N dr$$

with equality if and only if $A(t) = \phi(t)I$. Now $A(t)$ is invertible for $0 < t < \pi$ and $A'(0) = I$ so $\phi(t) > 0$ for $0 < t < \pi$. Since the expected case of equality is $\phi(t) = \sin t$, we are led to make the substitution

$$\phi(t) = \sin t\, e^{u(t)},$$

where $u(\pi)$ may possibly be singular. Note that $\phi'(0) = 1$ implies $u(0) = 0$. Then (12) reads

$$(13)\qquad P_s \geq Q_s(u),$$

where Q_s is the functional

$$Q_s(u) = \int_0^{\pi-s} \left\{\int_r^{r+s} \exp[u(r+s)+u(r)-2u(t)]\,\frac{\sin(r+s)\sin r\,dt}{\sin^2 t}\right\}^N dr .$$

The desired inequality (8) asserts that

$$P_s + P_{\pi-s} \geq \pi \sin^N s .$$

But from (13),

$$P_s + P_{\pi-s} \geq Q_s(u) + Q_{\pi-s}(u)$$

with equality if and only if $A(t) = \phi(t)I = \sin t\, e^{u(t)}I$. Since $u(0) = 0$, the proof of the Main Inequality (8) will be completed as soon as the following inequality is proved.

LEMMA 4. *For any continuous function* u,

$$Q_s(u) + Q_{\pi-s}(u) \geq Q_s(0) + Q_{\pi-s}(0) = \pi \sin^N s ,$$

with equality if and only if $u \equiv \text{constant}$.

Proof. Fix a function u and let $f(\lambda) = Q_s(\lambda u) + Q_{\pi-s}(\lambda u)$, for $0 \le \lambda \le 1$. We wish to show that $f(1) \ge f(0)$, with equality only when $u \equiv$ constant. This will follow, by Taylor's theorem, if we show that

$$f'(0) = 0 \quad \text{and} \quad f''(\lambda) > 0 \quad \text{if} \quad u \not\equiv \text{constant} .$$

Let

$$v = u(r+s) + u(r) - 2u(t) \quad \text{and} \quad h = \frac{\sin(r+s)\sin r}{\sin^2 t} .$$

Then

$$Q_s(\lambda u) = \int_0^{\pi-s} \left\{ \int_r^{r+s} e^{\lambda v} h \, dt \right\}^N dr$$

so

$$\text{(14)} \qquad \frac{d}{d\lambda} Q_s(\lambda u) = N \int_0^{\pi-s} \left\{ \int_r^{r+s} e^{\lambda v} h \, dt \right\}^{N-1} \left(\int_r^{r+s} e^{\lambda v} v h \, dt \right) dr$$

and

$$\frac{d^2}{d\lambda^2} Q_s(\lambda u) = N(N-1) \int_0^{\pi-s} \{ \quad \}^{N-2} \left(\int_r^{r+s} e^{\lambda v} v h \, dt \right)^2 dr$$

$$+ N \int_0^{\pi-s} \{ \quad \}^{N-1} \left(\int_r^{r+s} e^{\lambda v} v^2 h \, dt \right) dr .$$

This, with a similar expression for $d^2 Q_{\pi-s}(\lambda u)/d\lambda^2$, clearly shows that $f''(\lambda) > 0$ unless $v \equiv 0$, which happens only when $u \equiv$ constant.

It remains to show that $f'(0) = 0$, which is elementary but a bit tricky to demonstrate. First

$$\int_r^{r+s} h \, dt = \sin(r+s)\sin r \int_r^{r+s} \sin^{-2} t \, dt = \sin s$$

so that

$$\frac{d}{d\lambda}Q_s(\lambda u)|_{\lambda=0} = N\sin^{N-1}s\int_0^{\pi-s}\left(\int_r^{r+s}[u(r+s)+u(r)-2u(t)]\frac{\sin(r+s)\sin r}{\sin^2 t}dt\right)dr$$

$$(15)\qquad = N\sin^{N-1}s\left\{\sin s\int_0^{\pi-s}[u(r+s)+u(r)]\,dr\right.$$

$$\left.-2\int_0^{\pi-s}\sin(r+s)\sin r[\psi(r)-\psi(r+s)]\,dr\right\},$$

where

$$\psi(r) = \int_r^{\pi/2}\frac{u(t)}{\sin^2 t}\,dt\,, \quad \text{so} \quad \int_r^{r+s} = \psi(r) - \psi(r+s)\,.$$

Make the change of variable $x = r+s$ in the first and last of the four integrals in (15). Then expand $\sin(r+s) = \sin r\cos s + \cdots$ to find that

$$\frac{d}{d\lambda}Q_s(\lambda u)|_{\lambda=0} = N\sin^{N-1}s\left\{\sin s\left(\int_s^{\pi}+\int_0^{\pi-s}\right)u(r)\,dr\right.$$

$$+\,2\cos s\left(\int_s^{\pi}-\int_0^{\pi-s}\right)\sin^2 r\,\psi(r)\,dr$$

$$\left.-\,2\sin s\left(\int_s^{\pi}+\int_0^{\pi-s}\right)\cos r\sin r\,\psi(r)\,dr\right\}.$$

Replacing s by $\pi-s$ we obtain

$$f'(0) = \frac{d}{d\lambda}[Q_s(\lambda u)+Q_{\pi-s}(\lambda u)]_{\lambda=0} = N \sin^{N-1}s\left\{\sin s\left(\int_s^{\pi}+\int_0^{\pi-s}+\int_{\pi-s}^{\pi}+\int_0^{s}\right)u(r)\,dr\right.$$

$$+ 2\cos s\left(\int_s^{\pi}-\int_0^{\pi-s}-\int_{\pi-s}^{\pi}+\int_0^{s}\right)\sin^2 r\psi(r)\,dr$$

$$\left. - 2\sin s\left(\int_s^{\pi}+\int_0^{\pi-s}+\int_{\pi-s}^{\pi}+\int_0^{s}\right)\cos r\sin r\psi(r)\,dr\right\}$$

$$= 2\sin^N s\left\{\int_0^{\pi}u(r)\,dr - 2\int_0^{\pi}\cos r\sin r\psi(r)\,dr\right\}.$$

However, from the definition of ψ and an interchange of the order of integration

$$2\int_0^{\pi}\cos r\sin r\psi(r)\,dr = 2\int_0^{\pi}\cos r\sin r\left(\int_r^{\pi/2}\frac{u(t)}{\sin^2 t}\,dt\right)dr = \int_0^{\pi}u(r)\,dr\,.$$

Consequently $f'(0) = 0$. Q.E.D.

REMARK. It is interesting to note that the proof of the Main Inequality (8) is simply a combination of two convexity inequalities, that of $(\det B)^{-1}$ for B positive definite in Lemma 3, and convexity of $f(\lambda)$ in Lemma 4. Motivated by this, one should probably look for a more direct and basic convexity assertion than (8) concerning the dependence of the solution $A(r,s)$ of the Jacobi equation on the matrix R.

3. *Related results and open questions*

There are three related issues: understand the proof better, apply the method to other questions, and generalize the results.

Perhaps the simplest application of the method is the following, which asserts that if the injectivity radius is at least π, then the volume of the average (n–1 dimensional) geodesic sphere of radius $\pi/2$ is at least that of the corresponding equatorial sphere (S^{n-1}, can) on (S^n, can). Notation: let $S_r(x)$ be the geodesic sphere centered at x and having radius r and let its average volume be

$$\overline{\mathrm{Vol}}\ S_r = V^{-1} \int_M \mathrm{Vol}\,[S_r(x)]\,dx\ ,$$

where $V = \mathrm{Vol}\,(M, g)$.

PROPOSITION (Berger [B5]). *Let* (M^n, g) *be a compact Riemannian manifold with injectivity radius at least* r. *Then for all* $\rho \leq r$

$$\frac{(r-\rho)}{r}\,\overline{\mathrm{Vol}}\ S_\rho + \frac{\rho}{r}\,\overline{\mathrm{Vol}}\ S_{r-\rho} \geq (r/\pi)^{n-1}\,\mathrm{Vol}\,(S^{n-1}, \mathrm{can}) \tag{16}$$

with equality if and only if (M^n, g) *is isometric to* (S^n, can). *In particular, if* $r = \pi$ *and* $\rho = \pi/2$ *then*

$$\mathrm{Vol}\ S_{\pi/2} \geq \mathrm{Vol}\,(S^{n-1}, \mathrm{can})\ . \tag{17}$$

Proof. Scaling (M^n, g), we may assume the injectivity radius is at least π and let $r = \pi$. Then, as just after equation (3),

$$\mathrm{Vol}\ S_\rho(x) = \int_{UM_x} f(u, \rho)\,d\sigma_x$$

so by Liouville's theorem, for any t

$$\overline{\mathrm{Vol}}\ S_\rho = V^{-1} \int_{UM} f(\zeta^t(u), \rho)\,du = V^{-1} \int_{UM} \det A(t+\rho, t)\,du\ .$$

Therefore, integrating over $0 \leq t \leq \pi-\rho$, we find that

$$(\pi-\rho)\,\overline{\mathrm{Vol}}\ S_\rho = V^{-1} \int_{UM} \int_0^{\pi-\rho} \det A(t+\rho, t)\,dt\,du$$

with an analogous formula for $\rho\,\overline{\mathrm{Vol}}\,S_{\pi-\rho}$ (just replace ρ by $\pi-\rho$). Adding these two formulas, we obtain the desired inequality as an immediate consequence of the Main Inequality (8). Q.E.D.

It is reasonable to seek a similar theorem for the volume of the average geodesic ball of radius π (assuming the injectivity radius is at least π). So far, the best result is this direction [B-K, (20)] has an extra factor of $1/2$. One would also like an inequality for the volume of *every* geodesic ball and geodesic sphere. Croke [C] has recently used these results to prove such an inequality; however, his interesting inequalities are not sharp. One result in this direction would be to prove that if (M^n, g) is complete with injectivity radius ρ, then for every x in M

$$\mathrm{Vol}\,(S_{\rho/2}(x)) \geq \left(\frac{\rho}{\pi}\right)^{n-1} \mathrm{Vol}\,(S^{n-1}, \mathrm{can}) ,$$

and a similar inequality for $\mathrm{Vol}\,B_\rho(x)$. Berger [B1] has proved a preliminary version of this for $n \leq 3$.

In a more technical direction, it would be useful to have a direct geometric interpretation of the Main Inequality (8). (See the Remark at the end of Section 2 above.)

These methods have been used to obtain an upper bound for the first eigenvalue of the Laplacian for the Dirichlet problem for a disc in (M, g) in terms of the injectivity radius, (Berger [B3]).

Another application of the method is to obtain information on the general Blaschke Conjecture. (M, g) is a *Blaschke manifold* if for each point x in M, the cut point in any direction occurs at the distance π from x; the conjecture is that the only possibilities are the rank one symmetric spaces with their canonical metrics (see [Besse], p. 143). Berger [B4] has proved the analogue of our Isoperimetric Inequality (1), comparing the volume of a Blaschke manifold M with that of a corresponding rank one symmetric space; however he assumes a totally geodesic condition on M.

REFERENCES

[B1] Berger, M., "Volume et rayon d'injectivité dans les varietes riemanniannes de dimension 3," *Osaka J. Math.*

[B2] ———, "Blaschke's Conjecture for Spheres," Appendix D in the book below by A. Besse.

[B3] ———, "Une inégalité universelle pour la première valeur propre du laplacien," *Bull. Soc. Math. France*, 107 (1974), pp. 3-9.

[B4] ———, "Sur certaines varietes a geodesiques toutes fermées," *Boletin da Socedade Brasileira de Matematica*, 9 (1978).

[B5] ———, "Une borne inferieure pour le volume d'une variété riemannienne en function du rayon d'injectivité," *Ann l'Inst. Fourier, T.30* (1980), pp. 259-265.

[B-K] Berger, M. and Kazdan, J. L., "A Sturm-Liouville Inequality with Applications to an Isoperimetric Inequality for Volume in terms of Injectivity Radius, and to Wiedersehen manifolds," in *General Inequalities 2* (Proceedings of the Second International Conference on General Inequalities, 1978), E. F. Beckenbach (ed.), ISNM47, Birkhauser Verlag, Basel, 1980, pp. 367-377.

[Besse] Besse, A., *Manifolds all of whose Geodesics are Closed*, Ergebnisse der Mathematik No. 93, Springer-Verlag, Berlin, 1978.

[B-C] Bishop, R. and Crittenden, R., *Geometry of Manifolds*, Academic Press, New York, 1964.

[B] Blaschke, W., *Vorlesungen über Differentialgeometrie*, 3rd Edition, Springer-Verlag, Berlin, 1930.

[C] Croke, C., "Some Isoperimetric Inequalities and Eigenvalue Estimates," *Ann. Sci. Ecole Norm. Sup.* (to appear).

[G] Green, L., "Auf Wiedersehenflächen," *Annals of Math.*, Vol. 78 (1963), pp. 289-299.

[K] Kazdan, J. L., "An Inequality Arising in Geometry," Appendix E in the book above by A. Besse.

[W] Weinstein, A., "On the volume of manifolds all of whose geodesic are closed," *J. Diff. Geom.*, Vol. 9 (1974), pp. 513-517.

[Y] Yang, C. T., "Odd dimensional Wiedersehen Manifolds are Spheres," *J. Diff. Geom.* (to appear).

ON THE BLASCHKE CONJECTURE

C. T. Yang*

The following is a lecture on the Blaschke conjecture given at the Institute for Advanced Study.

By a *Blaschke manifold,* we mean a connected closed Riemannian manifold M of dimension > 1 having the following property: There is a number $\ell > 0$ such that, whenever $m \epsilon M$, if $T_m M$ denotes the tangent space of M at m, $\exp_m$ denotes the exponential map of $T_m M$ into M and D_m is the closed disk in $T_m M$ of center 0 and radius ℓ, then $\exp_m | \operatorname{int} D_m$ is a smooth imbedding and $\exp_m | \partial D_m$ is a smooth r-sphere fibration for some integer $r \geqq 0$. It is easily seen that the manifolds listed below together with their usual Riemannian metric are Blaschke manifolds. These Blaschke manifolds will be referred to as *canonical Blaschke manifolds*.

1. The (unit) n-sphere S^n (in the Euclidean (n+1)-space), $n > 1$.
2. The real projective n-space $RP^n (= S^n/Z_2)$, $n > 1$.
3. The complex projective n-space $CP^n (= S^{2n+1}/S^1)$, $n > 0$.
4. The quaternionic projective n-space $HP^n (= S^{4n+3}/S^3)$, $n > 0$.
5. The Cayley projective plane CaP^2.

We note that for the first case, $\ell = \pi$ and $r = n-1$ and for the other cases, $\ell = \pi/2$ and $r = 0, 1, 3, 7$ respectively.

---

*The author is supported in part by the National Science Foundation.

Seminar on Differential Geometry
0-691-08268-5/82/000159-13 $00.65/0 (cloth)
0-691-08296-0/82/000159-13 $00.65/0 (paperback)
For copying information, see copyright page.

THE BLASCHKE CONJECTURE. *Up to a constant factor, every Blaschke manifold is isometric to a canonical Blaschke manifold.*

The conjecture was first given for the 2-sphere only. For an account, see the lecture by Kazdan [3]. Even for that special case, it took mathematicians 30 years to confirm it. Since then, more progress has been made as reported in [3]. But the conjecture in general is still far from being settled.

The concise book by Besse [2] gives a thorough account of results, problems and references related to the Blaschke conjecture up to the date of its publication and the reader is advised to consult the book for further information.

As a start, we state two basic results on Blaschke manifolds, which can be found in [2].

(1) All geodesics in a Blaschke manifold are smoothly simply closed and are of the same length. If ℓ is the number as seen in the definition of a Blaschke manifold M, then closed geodesics in M are of length 2ℓ.

(2) Bott-Samelson theorem. Every Blaschke manifold has the integral cohomology ring of a canonical Blaschke manifold. Moreover, any Blaschke manifold which has the integral cohomology ring of S^n or $\mathbf{R}P^n$ is homeomorphic to S^n or $\mathbf{R}P^n$ and any Blaschke manifold which has the integral cohomology ring of $\mathbf{C}P^n$ has the homotopy type of $\mathbf{C}P^n$.

Let M be a Blaschke manifold of dimension d, let UM be the smooth $(2d-1)$-manifold consisting of unit tangent vectors of M, and let CM be the smooth $(2d-2)$-manifold consisting of oriented closed geodesics in M. Then there is a natural smooth S^{d-1}-fibration

$$\tau : UM \to M$$

and a natural smooth S^1-fibration

$$\pi : UM \to CM$$

such that for any $u \in UM$, u is the unit tangent vector of πu at τu.

Making use of (2) and the Gysin sequences of $\tau : UM \to M$ and $\pi : UM \to CM$, we can compute the integral cohomology groups of UM and CM. For example, we have

(3) If M is homeomorphic to S^n, $n > 1$, then for even n,

$$H^k(UM) = \begin{cases} Z & \text{for } k = 0,\ 2n-1, \\ Z_2 & \text{for } k = n, \\ 0 & \text{otherwise,} \end{cases}$$

$$H^k(CM) = \begin{cases} Z & \text{for } k = 2i,\ i = 0, \cdots, n-1, \\ 0 & \text{otherwise,} \end{cases}$$

and for odd n,

$$H^k(UM) = \begin{cases} Z & \text{for } k = 0, n-1, n, 2n-1, \\ 0 & \text{otherwise,} \end{cases}$$

$$H^k(CM) = \begin{cases} Z & \text{for } k = 2i,\ i = 0, \cdots, n-1 \text{ but } \neq (n-1)/2, \\ Z \oplus Z & \text{for } k = n-1, \\ 0 & \text{otherwise.} \end{cases}$$

Here Z denotes the group of integers and Z_2 denotes the group of integers modulo 2.

(4) If M has the homotopy type of CP^n, $n > 0$, then

$$H^k(UM) = \begin{cases} Z & \text{for } k = 2i \text{ or } 4n-1-2i,\ i = 0, \cdots, n-1, \\ Z_{n+1} & \text{for } k = 2n, \\ 0 & \text{otherwise,} \end{cases}$$

$$H^k(CM) = \begin{cases} (i+1)Z & \text{for } k = 2i \text{ or } 4n-2-2i,\ i = 0, \cdots, n-1, \\ 0 & \text{otherwise.} \end{cases}$$

Here Z_{n+1} denotes the group of integers modulo $n+1$ and $(i+1)Z$ denotes the direct sum of $i+1$ copies of Z.

The Blaschke conjecture for spheres is a consequence of the following two statements. Since any Blaschke manifold homeomorphic to RP^n is covered by a Blaschke manifold homeomorphic to S^n, the Blaschke conjecture for real projective spaces also follows.

(5) Berger [1] and Kazdan [4]. If M is a Blaschke manifold homeomorphic to S^n and if closed geodesics in M are of the same length as those in S^n, then

$$\text{vol } M \geqq \text{vol } S^n$$

and the equality holds iff M is isometric to S^n.

(6) Weinstein [5] and Yang [6]. If M is as in (5), then $\text{vol } M = \text{vol } S^n$.

For a proof of (5), see [3]. We shall sketch the proof of (6) which is based on the following

(7) Weinstein's theorem [5]. Let M be a Blaschke manifold of dimension d in which closed geodesics are of length 2ℓ, and let e be the Euler class of the S^1-fibration $\pi : UM \to CM$. Then

$$i(M) = \frac{1}{2} e^{d-1} \cap [CM]$$

is a positive integer, called the *Weinstein integer* of M, and

$$\text{vol } M = (\ell/\pi)^d \, i(M) \, \text{vol } S^d .$$

(For any oriented closed manifold X, [X] denotes the fundamental homology class of X. Here we let UM and CM be naturally oriented.)

Because of (7), we may reformulate (6) as follows.

(6′) If M is a Blaschke manifold homeomorphic to S^n, then $i(M) = 1$.

A sketch of the proof of (6′). If n is even, say $n = 2m$, we can see from the Gysin sequence of $\pi : UM \to CM$ that

$$e \cup : H^{2i-2}(CM) \to H^{2i}(CM)$$

is an isomorphism for $i = 1, \cdots, 2m-1$ but $\neq m$ and

$$eU : H^{2m-2}(CM) \to H^{2m}(CM)$$

is a monomorphism of cokernel Z_2. Therefore $e^{n-1} \cap [CM] = 2$ and hence $i(M) = 1$.

If n is odd, say $n = 2m+1$, then there are short exact sequences

$$0 \to H^{2m-2}(CM) \to H^{2m}(CM) \to H^{2m}(UM) \to 0 ,$$

$$0 \leftarrow H^{2m+2}(CM) \leftarrow H^{2m}(CM) \leftarrow H^{2m+1}(UM) \leftarrow 0 ,$$

which are parts of the Gysin sequence of $\pi : UM \to CM$ and are dual to each other under Poincaré duality.

Let a be a generator of the image of $H^{2m+1}(UM) \to H^{2m}(CM)$. Then $a^2 = 2g$ with g being a generator of $H^{4m}(CM)$. This can be seen as follows. If x and y are two distinct points of M not far from each other, then $\pi\tau^{-1}x$ and $\pi\tau^{-1}y$ are two 2m-spheres in CM which intersect at exactly two points and transversally. In fact, the points of intersection are oriented closed geodesics in M passing through x and y. This indicates that, if a^* is a generator of $H_{2m}(UM) \to H_{2m}(CM)$, then the intersection number of a^* with itself is equal to 2. Hence our claim follows.

From the short exact sequences we had earlier, it can be seen that there is a basis $\{e^m, b\}$ of $H^{2m}(CM)$ with $ab = g$. Let $a = \beta e^m + \gamma b$, where β and γ are integers. Since $ab = g$, $a^2 = 2g$ and $ae = 0$, it follows that $\gamma = 2$ and β is odd. Therefore b can be so chosen that

$$a = -e^m + 2b .$$

Let $b^2 = rg$, where r is an integer. Then an easy computation yields

$$e^m b = (2r-1)g , \qquad e^{2m} = (4r-2)g .$$

Therefore, by Poincaré duality, the determinant of

$$\begin{pmatrix} 4r-2 & 2r-1 \\ 2r-1 & r \end{pmatrix}$$

is equal to 1 or -1 so that $r = 1$ or 0. Hence we again have $i(M) = 1$.

(6) indicates the importance of the determination of the volume of a Blaschke manifold. Therefore we may formulate the following

THE WEAK BLASCHKE CONJECTURE. The volume of a Blaschke manifold M depends only on the integral cohomology ring $H^*(M)$ of M and the length of closed geodesics in M.

Because of (2) and (7), the weak Blaschke conjecture is actually topological in nature. After reformulating it as follows, one can try to resolve it by studying the ring structure of $H^*(CM)$.

TOPOLOGICAL WEAK BLASCHKE CONJECTURE. If M is a Blaschke manifold, then

$$i(M) = 1 \quad \text{or} \quad 2^{n-1} \quad \text{or} \quad \binom{2n-1}{n-1} \quad \text{or} \quad \frac{1}{2n+1}\binom{4n-1}{2n-1} \quad \text{or} \quad 39$$

when

$$H^*(M) = H^*(S^n) \text{ or } H^*(RP^n) \text{ or } H^*(CP^n) \text{ or } H^*(HP^n) \text{ or } H^*(CaP^2).$$

Of course, the first two cases have been confirmed. Therefore we are going to examine the case that M has the homotopy type of CP^n.

Throughout the rest of this lecture, M denotes a Blaschke manifold having the homotopy type of CP^n. Since CP^1 is a 2-sphere for which the Blaschke conjecture has been confirmed, we assume below that $n > 1$.

Let a' be a generator of $H^2(M)$ and let M be oriented so that $a'^n \cap [M] = 1$. The integral cohomology groups of UM and CM have been given in (4) and they can be described as follows. Let a be an element of $H^2(CM)$ such that $\pi^* a = \tau^* a'$.

(8) For any $i = 1, \cdots, n$, $H^{2i}(UM)$ is generated by $(\pi^* a)^i = (\tau^* a')^i$. For any $i = 1, \cdots, n-1$, $\{a^i, a^{i-1}e, \cdots, ae^{i-1}, e^i\}$ is a basis of $H^{2i}(CM)$. Moreover, the integral cohomology ring $H^*(CM)$ is generated by $\{a, e\}$.

Let

$$\lambda : UM \to UM$$

be the involution defined by $\lambda(u) = -u$. Then λ is orientation-preserving and it induces an orientation-reversing involution

$$\lambda : CM \to CM .$$

(9) The element $a \in H^2(CM)$ can be so chosen that

$$e = a - b \quad \text{with} \quad b = \lambda^* a .$$

For technical reasons, this choice of a is very important. In fact, we suspect that

$$a^{n+1} = 0 .$$

If this can be shown, then

$$\begin{aligned} e^{n+1} &= (-1)^{n-1}\binom{2n-1}{n-1} a^n b^{n-1} + (-1)^n \binom{2n-1}{n} a^{n-1} b^n \\ &= (-1)^{n-1}\, 2\binom{2n-1}{n-1} a^n b^{n-1} \end{aligned}$$

so that

$$i(M) = \frac{1}{2}\, e^{2n-1} \cap [CM] = \binom{2n-1}{n-1} .$$

We note that $a^{n+1} = 0$ holds when $M = CP^n$ and will be shown later for $n = 2$.

(9) can be seen as follows. Let γ be an oriented closed geodesic in M and let p and q be points of γ which divide γ into two arcs of equal length. Then the closed geodesics in M passing through p and q generate a smooth 2-sphere K which can be so oriented that $a' \cap [K] = 1$. Let D and D′ be closed 2-disks in K such that $D \cup D' = K$ and $D \cap D' = \gamma$, and let D and D′ have the same orientation as K. Let $f : K \to UM$ be a map such that for any $x \in K$, $\tau f(x) = x$ and for any $x \in \gamma$, $f(x)$ is tangent to γ. Then πf, $\pi(f|D)$ and $\pi(f|D')$ represent three

elements of $H_2(CM)$, say $\bar{e}$, $\bar{a}$ and $\bar{b}$. It can be shown that

$$\bar{e} = \bar{a} + \bar{b}, \qquad \bar{b} = \lambda_* \bar{a}$$

and that

$$e \cap \bar{e} = 0, \qquad a \cap \bar{e} = 1, \qquad e \cap \bar{a} = 1.$$

Replacing a by $a + re$ for some integer r if necessary, we may assume that

$$a \cap \bar{a} = 1.$$

Then it can be shown that $\{a, b\}$ is a basis of $H^2(CM)$ which is dual to the basis $\{\bar{a}, \bar{b}\}$ of $H_2(CM)$. Hence $b = \lambda^* a$ and $e = a - b$.

Let

$$\tau' : W_1 \to M, \qquad \pi' : W_2 \to CM$$

be the oriented closed disk bundles associated with the oriented sphere bundles

$$\tau : UM \to M, \qquad \pi : UM \to CM$$

respectively. Then W_1 and W_2 are compact smooth 4n-manifolds of boundary UM. Therefore we have a closed smooth 4n-manifold W obtained by pasting together W_1 and W_2 along their common boundary via the identity map. Let W_1 be naturally oriented and let W have the same orientation as W_1.

The involution $\lambda : UM \to UM$ can be naturally extended to an involution

$$\lambda : W \to W$$

having M as its fixed point set.

Since the inclusion map of CM into W induces an isomorphism of $H^2(W)$ onto $H^2(CM)$, we may set

$$H^2(W) = H^2(CM).$$

Then (4), (8) and the Mayer-Vietoris sequence of $(W; W_1, W_2)$ yield

$$(10) \qquad H^k(W) = \begin{cases} (i+1)Z & \text{for } k = 2i \text{ or } 4n-2i,\ i = 0,\cdots,n, \\ 0 & \text{otherwise.} \end{cases}$$

Moreover, for any $i = 1,\cdots,n$, $\{a^i, a^{i-1}e, \cdots, ae^{i-1}, e^i\}$ is a basis of $H^{2i}(W)$ and so is $\{a^i, a^{i-1}b, \cdots, ab^{i-1}, b^i\}$, where

$$e = a - b, \qquad b = \lambda^* a .$$

Furthermore, the integral cohomology ring $H^*(W)$ is generated by $\{a, b\}$.

(11) There is an oriented closed smooth 2n-manifold N of W which is diffeomorphic to CP^n and has the property that $b \cap [N] = 0$, $a^n \cap [N] = e^n \cap [N] = 1$ and $[M] \cap [N] = 1$.

In the proof of (9), we have constructed a smooth imbedding $h : CP^1 \to CM \subset W$ representing $\bar{a}$. Since $\pi_k(W) = 0$ for $k = 3, \cdots, 2n-1$, h can be extended to a smooth imbedding $h : CP^n \to W$. Let

$$N = h(CP^n)$$

and let N be oriented so that $h : CP^n \to N$ is orientation-preserving. Then

$$b \cap [N] = 0, \qquad a^n \cap [N] = e^n \cap [N] = 1 .$$

The imbedding h can be so chosen that $N \cap CM = h(CP^{n-1})$ and $N \cap W_2 = \pi'^{-1} h(CP^{n-1})$. Then $N \cap UM$ is a (2n–1)-sphere. It can be shown that $N \cap UM$ represents a generator of $\pi_{2n-1}(UM)$. Therefore we may assume that $N \cap W_1$ is a closed 2n-disk which intersects M at exactly one point and transversally. Hence $[M] \cap [N] = 1$.

By Poincaré duality, each element of $H_{2n}(W)$ can be uniquely written as

$$c \cap [W]$$

with $c \in H^{2n}(W)$, and by (10), each element of $H^{2n}(W)$ can be uniquely written as a homogeneous polynomial of a and b of degree n with integral coefficients. Let p(a, b) and q(a, b) be the polynomials such that

$$[M] = p(a,b) \cap [W], \qquad [N] = q(a,b) \cap [W].$$

(12) (i) $p(a,b) = p(b,a)$, $p(1,0) = 1$ and $p(1,1) = n+1$.

(ii) $q(1,1) = 1$.

(iii) $aq(b,a) - bq(a,b) = q(0,1)(a-b)\,p(a,b)$.

That $p(a,b) = p(b,a)$ is a consequence of the fact that M is the fixed point set of the involution $\lambda : W \to W$.

Let

$$p(a,b) = \alpha_0 a^n + \alpha_1 a^{n-1} b + \cdots + \alpha_{n-1} a b^{n-1} + \alpha_n b^n .$$

Then

$$1 = [M] \cap [N] = p(a,b) \cap [N] = \alpha_0 a^n \cap [N] = \alpha_0 .$$

Since $a^i b^{n-i} \cap [M] = 1$, $i = 0, \cdots, n$, we infer that

$$\begin{aligned} p(1,1) &= p(a,b) \cap [M] = [M] \cap [M] \\ &= \text{Euler characteristic of } M = n+1 . \end{aligned}$$

Let

$$q(a,b) = \beta_0 a^n + \beta_1 a^{n-1} b + \cdots + \beta_{n-1} a b^{n-1} + \beta_n b^n .$$

Then $q(a,b) = q(1,1)a^n + r(a,b)e$ for some polynomial $r(a,b)$. Therefore

$$1 = [M] \cap [N] = q(a,b) \cap [M] = q(1,1)a^n \cap [M] = q(1,1) .$$

Let $P_k(a,b)$ be the group of homogeneous polynomials in a and b of degree k with integral coefficients. Then

$$P_n(a,b) = H^{2n}(W)$$

and there is a natural homomorphism

$$h : P_{n+1}(a,b) \to H^{2n+2}(W) .$$

Clearly h is surjective and its kernel is free and contains

$$aq(b,a), \quad bq(a,b), \quad (a-b)\,p(a,b) .$$

Since $aP_n(a,b)$ is a direct summand of $P_{n+1}(a,b)$ and $P_{n+1}(a,b)/aP_n(a,b) = Z$, there is an $r(a,b) \in P_{n+1}(a,b)$ such that

$$r(1,0) \geqq 0$$

and that $\{aq(b,a), r(a,b)\}$ is a basis of $\ker h$. Replacing $r(a,b)$ by $r(a,b) - r(1,1)aq(b,a)$ if necessary, we may assume that

$$r(1,1) = 0 .$$

Let s and t be integers such that

$$(a-b)\,p(a,b) = saq(b,a) + tr(a,b) .$$

Then $s = sq(1,1) = 0$ and $tr(1,0) = p(1,0) = 1$. Therefore $s = 0$, $t = 1$ and

$$r(1,0) = 1 .$$

Hence

$$r(a,b) = (a-b)\,p(a,b) .$$

Let s and t be integers such that

$$bq(a,b) = saq(b,a) + tr(a,b) .$$

Then $s = 1$ and $t = -q(0,1)$. Hence

$$aq(b,a) - bq(a,b) = q(0,1)\,r(a,b) = q(0,1)\,(a-b)\,p(a,b) .$$

We conclude the lecture by proving

THEOREM. *Let* M *be a Blaschke manifold which has the homotopy type of the complex projective plane* CP^2 *and in which closed geodesics are of the same length as those in* CP^2. *Then*

$$\mathrm{vol}\, M = \mathrm{vol}\, CP^2 .$$

Proof. As we remarked earlier, it is sufficient to show that the element a of $H^2(CM) = H^2(W)$ has the property that $a^3 = 0$.

Let

$$p(a,b) = \alpha_0 a^2 + \alpha_1 ab + \alpha_2 b^2 , \qquad q(a,b) = \beta_0 a^2 + \beta_1 ab + \beta_2 b^2 .$$

By (12), (i),

$$\alpha_0 = \alpha_1 = \alpha_2 = 1 .$$

By (12), (iii),

$$\beta_0 = \beta_1 = \beta \text{ say} .$$

Then, by (12), (ii),

$$\beta_2 = 1-2\beta .$$

Since $\{a^2, ab, b^2\}$ is a basis of $H^4(W)$ and has

$$\{q(a,b), p(a,b) - q(a,b) - q(b,a), q(b,a)\}$$

as its dual basis, it follows that the coefficients of

$$q(a,b) , \quad p(a,b) , \quad q(b,a)$$

form a 3×3 matrix whose determinant is equal to ± 1. Therefore

$$\pm 1 = \begin{vmatrix} \beta & \beta & 1-2\beta \\ 1 & 1 & 1 \\ 1-2\beta & \beta & \beta \end{vmatrix} = -(1-3\beta)^2 .$$

Hence

$$\beta_0 = \beta_1 = \beta = 0 , \qquad \beta_2 = 1 .$$

From this result, we infer that

$$a^3 = aq(b,a) = 0 .$$

ADDENDUM. The author may have succeeded to prove the weak Blaschke conjecture for CP^n by proving $a^{n+1} = 0$ in $H^*(W)$, even though details of the proof still need to be checked. The idea of the proof is to construct a smooth imbedding f of W into $CP^N \times CP^N$, N large, such that the image of $f^*: H^*(CP^N \times CP^N) \to H^*(W)$ is isomorphic to $H^*(CP^n \times CP^n)$. In order to apply induction on n, we deal with a more general 4n-manifold X instead of W, which is allowed to have a larger homology group in the middle dimension. Then a smooth imbedding f of X into $CP^N \times CP^N$ is constructed such that f is transverse regular at $CP^{N-1} \times CP^{N-1}$ and $X' = f^{-1}(CP^{N-1} \times CP^{N-1})$ has the same property as X so that the induction hypothesis can be applied.

UNIVERSITY OF PENNSYLVANIA
PHILADELPHIA, PENNSYLVANIA 19104

REFERENCES

[1] Berger, M., Blaschke's conjecture for spheres, Appendix D in the book below by A. Besse.

[2] Besse, A., Manifolds All of Whose Geodesics are Closed, Ergebnisse der Mathematik, No. 93, Springer-Verlag, Berlin, 1978.

[3] Kazdan, J. L., An isoperimetric inequality and Wiedersehen manifolds, this volume, pp. 143-157.

[4] ________, An inequality arising in geometry, Appendix E in the book above by A. Besse.

[5] Weinstein, A., On the volume of manifolds all of whose geodesics are closed, J. Diff. Geom., vol. 9 (1974), pp. 513-517.

[6] Yang, C. T., Odd-dimensional Wiedersehen manifolds are spheres, J. Diff. Geom., 15 (1980), pp. 91-96.

BEST CONSTANTS IN THE SOBOLEV IMBEDDING THEOREM
THE YAMABE PROBLEM

Thierry Aubin

I. The Best Constants

1. *Introduction*

Before discussing Sobolev's spaces, let us introduce the notion of best constants in general. Consider three Banach spaces B_1, B_2 and B_3. The well-known result of Lions concerns the case in which $B_1 \subset B_2 \subset B_3$ with the imbedding $B_1 \subset B_2$ compact and that of $B_2 \subset B_3$ continuous. Under these hypotheses, for any $\varepsilon > 0$ there is some $A(\varepsilon)$ such that all $\phi \in B_1$ satisfy

$$\|\phi\|_{B_2} \leq \varepsilon \|\phi\|_{B_1} + A(\varepsilon)\, \|\phi\|_{B_3} .$$

Now let $B_1 \subset B_2$ be continuous but non-compact and let $B_1 \subset B_3$ be compact. Obviously there exist constants A and C such that all $X \in B_1$ satisfy

$$\|X\|_{B_2} \leq C\|X\|_{B_1} + A\|X\|_{B_3} .$$

But if we set $K = \inf C$ such that $A(C)$ exists, we have $K > 0$ (instead of $K = 0$ in the preceding case).

Proof. There is at least some sequence $\{X_i\}_{i \in N}$ in B_1 which satisfies $X_i \to 0$ in B_3, $\|X_i\|_{B_1} \leq 1$ and $\|X_i\|_{B_2} \geq \alpha > 0$ for some α. Letting $i \to \infty$ in the preceding inequality with $X = X_i$ we obtain $C \geq \alpha$.

Seminar on Differential Geometry
0-691-08268-5/82/000173-12 $00.60/0 (cloth)
0-691-08296-0/82/000173-12 $00.60/0 (paperback)

2. *The Sobolev spaces*

Let (M_n, g) be a C^∞ Riemannian manifold. Consider $\mathcal{C}$ the space of the C^∞ functions on M_n which along with their gradients belong to $L_q (q \geq 1)$. On the vector space $\mathcal{C}$ we define the following norm

$$\|\phi\|_{H_1^q} = \|\phi\|_q + \|\nabla\phi\|_q .$$

The Sobolev space $H_1^q(M)$ is defined as the completion of $\mathcal{C}$ with respect to the preceding norm.

The Sobolev imbedding theorem asserts that $H_1^q \subset L_r$ when $q \leq r \leq p = nq/(n-q)$ if $1 \leq q < n$ and the imbedding is continuous. Sobolev's theorem holds for Riemannian manifolds with bounded curvature and global injectivity radius, and for compact Riemannian manifolds with Lipschitzian boundary, Aubin [4].

Recall now Kondrakov's theorem: The imbedding $H_1^q \subset L_r$ is compact when $r < p$. This theorem holds for compact Riemannian manifolds with Lipschitzian boundary or without boundary. On the other hand the imbedding $H_1^q \subset L_p$ is not compact.

Thus, applying the result of the first paragraph with $B_1 = H_1^q(M)$, $B_2 = L_p(M)$ and $B_3 = L_q(M)$ we find that if all $\phi \in H_1^q(M)$ satisfy

$$\|\phi\|_p \leq C\|\nabla\phi\|_q + A\|\phi\|_q$$

then $C \geq K > 0$. For the present the best constant K depends on q and on the manifold. But it is possible to prove the following.

3. *Theorem*

Let M_n *be a Riemannian manifold of dimension* n *with bounded curvature and global injectivity radius and let* $1 \leq q < n$. *Then all* $\phi \in H_1^q(M_n)$ *satisfy:*

(1) $$\|\phi\|_p \leq C\|\nabla\phi\|_q + A\|\phi\|_q$$

for C *and* A *some constants and* $p = nq/(n-q)$.

Moreover $K = \{\inf C$ *such that* A *exists*$\}$, *depends only on* n *and* q, $K = K(n,q)$. *If the manifold is compact with differentiable boundary the same result holds but the best constant* $K = 2^{1/n} K(n,q)$.

For the proof see Aubin [3] and for a survey of Sobolev's spaces on Riemannian manifold Aubin [4].

4. *What is the use of best constants?*

When we study non-linear problems in Riemannian Geometry and try to solve them by analysis, limiting cases often arise and best constants must be introduced. Suppose we use the variational method and we are able to solve the problem if $\mu < f(K)$, where μ is the inf of the functional considered over a set of functions and f depends on the best constant K. If K also depends on the manifold, both sides of the inequality $\mu < f(K)$ depend on the manifold and we get nothing really powerful.

But if it is a question of best constants in Sobolev inequalities we know that these constants do not depend on the manifold and we get a theorem. If $\mu < f[K(n,q)]$ the problem is solved. For instance in Section II we will see that Yamabe's problem is solved if $\mu < 4\frac{n-1}{n-2}[K(n,2)]^{-2}$ (here $n > 2$). For the problem of Berger [8] we obtain as condition for solving that the Euler-Poincaré caracteristic of the manifold of dimension 2 must satisfy $\chi < f(K)$.

To go further, the preceding examples show that we must know the value of the considered best constant. In fact in Berger's problem we have $f > 0$. So the problem is solved if $\chi \leq 0$, but one can do better. If we compute $f(K)$, we find $f(K) = 2$. Therefore the problem cannot be solved if we are on the sphere S_2.

5. *The values of the best constants*

Since the best constants in Sobolev inequalities do not depend on the manifold we have only to consider the case of $\mathbf{R}^n$. All $\phi \in H_1^q(R^n)$ satisfy $\|\phi\|_p \leq K(n,q)\|\nabla\phi\|_q$. The best constant is equal to the norm of the imbedding $H_1^q(R^n) \subset L_p(R^n)$. To compute it we reduce the problem to a one-dimensional one making use of symmetrization. Thus it is sufficient to find the norm of the imbedding for radially symmetric functions. Now this variational problem was solved by Bliss [9]. Thus $K(n, q)$ is known.

The reader can find the formula for $K(n, q)$ in Aubin [3] and [4]. Talenti [11] also found the value of the norm of the imbedding $H_1^q(R^n) \subset L_p(R^n)$. For the symmetrization he used the Haussdorf measure. I myself used only Lebesgue measure because before symmetrization I approximate the C^∞ function with compact support $\overline{\Omega}$ uniformly in C^1 on $\overline{\Omega}$ by C^∞ functions whose critical points are non-degenerate.

6. *Improvement of the best constants*

For some problems, such as that of Nirenberg, see Aubin [7], one has $\mu = f(K)$, so the best constant must be lowered. This is possible if the functions satisfy some natural orthogonality conditions. In Aubin [7] the following result is proved:

THEOREM. *Let $\{f_i\}_{i \in J}$ be a family of C^1 functions whose gradient is uniformly bounded and such that the $|f_i|^q$ $(i \in J)$ is a partition of unity. Then the functions $\phi \in H_1^q(M_n)$ which verify $\int |\phi|^p f_i |f_i|^{p-1} dV = 0$ for all $i \in J$ satisfy inequality (1) with pair $[C, A(C)]$, where C can be chosen equals to $2^{-1/n} K(n, q) + \varepsilon$ for any $\varepsilon > 0$. Moreover if ϕ satisfies more conditions we find a sequence of best constants $\{m^{-1/n} K(n, q)\}_{m \in N}$, as for an eigenvalue sequence, except here all terms are known when we know the first.*

7. *Open questions*

a) The values of best constants for spaces of higher derivatives than one are unknown. If $\phi \in H_k^q$ with $k > 1$ or if $\Delta\phi \in L_q$ for instance. We know the values of best constants for H_1^q because we have Bliss' result. For $k > 1$ we need similar results.

b) Are the best constants attained? (i.e. does $A(K)$ exist?). There are only partial results. The best constants are attained for manifolds of dimension two and for those with constant curvature. This question is related to the following isoperimetric problem: Let $\bar{M}$ be a compact ball of dimension n endowed with a Riemannian metric g whose sectional curvature is bounded above by k. Given a measurable set $\Omega \subset M$ whose measure is $\mu_g(\Omega)$ we consider on the sphere S_n whose curvature is equal to k, a ball B with the same measure $\mu_{S_n}(B) = \mu_g(\Omega)$. The question is: Is the area of ∂B smaller than $\mathcal{A}_g(\partial\Omega)$ the area of $\partial\Omega$? The result would be local, M is small enough so that B exists and if $\partial\Omega$ is not rectifiable, $\mathcal{A}_g(\partial\Omega) = +\infty$. It is possible to prove this result if M has constant curvature or if Ω is convex. Therefore the result is true if $n = 2$.

c) Find the best possible inequality. As results we have: For $\mathbf{R}^n$ or H_n the hyperbolic space, inequality (1) holds with $C = K(n,q)$ and $A = 0$. For the sphere S_n of volume one (Aubin [5]) all $\phi \in H_1^2(S_n)$ satisfy:

$$\|\phi\|^2_{2n/(n-2)} \leq K^2(n,2)\,\|\nabla\phi\|_2^2 + \|\phi\|_2^2 \,.$$

In that case obviously $A \geq 1$ and inequality (1) holds with $A = 1$. The value of $K(n,2)$ is $2[n(n-2)]^{-1/2}\omega_n^{-1/n}$ with ω_n the volume of the sphere $S_n(1)$ of radius 1.

II. The Yamabe Theorem

Yamabe's problem cannot be solved without considering the best constants. In this paragraph we will solve the easy case ($\mu \leq 0$ see below) for which this notion is not useful.

Let (M_n, g) be a C^∞ compact Riemannian manifold of dimension $n \geq 3$, R its scalar curvature. Consider a conformal metric $g' = \phi^{4/(n-2)}g$ where ϕ is a strictly positive C^∞ function. It is easy to verify that the scalar curvature R' related to g' satisfies the equation:

(2) $$4\frac{n-1}{n-2}\Delta\phi + R\phi = R'\phi^{(n+2)/(n-2)}, \quad \text{where } \Delta\phi = -\nabla^{\nu}\nabla_{\nu}\phi .$$

Yamabe's problem is: does there exist a conformal metric g' whose scalar curvature R' is constant. As we saw previously this problem is equivalent to proving the existence of a strictly positive C^{∞} solution ϕ of (2) with $R' = \text{Const}$. For this define

$$I_q(\phi) = \left[4\frac{n-1}{n-2}\int \nabla^{\nu}\phi\, \nabla_{\nu}\phi\, dv + \int R\phi^2\, dv\right]\|\phi\|_q^{-2}$$

with $2 \le q \le N = 2n/(n-2)$ and $\mu_q = \inf I_q(\phi)$ for all $\phi \not\equiv 0$ belonging to H_1^2. Yamabe considered this functional because its Euler equation is equation (3) which is, for $q = N$, equation (2) with $R' = \text{Const}$. For simplicity set $\mu = \mu_N$ and $I(\phi) = I_N(\phi)$.

First of all Yamabe [13] proved:

THEOREM 1. *For* $2 < q < N$, *there exists a* C^{∞} *positive function* ϕ_q *satisfying* $\|\phi_q\|_q = 1$ *and*

(3) $$4\frac{n-1}{n-2}\Delta\phi_q + R\phi_q = \mu_q\phi_q^{q-1}. \quad \text{So } I_q(\phi_q) = \mu_q .$$

Then he claimed that the functions ϕ_q, $q \in]2, N[$ are uniformly bounded. Now that is wrong generally. For instance on the sphere the functions $\phi(r)$ of Theorem 7 are not uniformly bounded. Yamabe's proof contains a mistake. We can overcome this gap only when $\mu \le 0$. And thus we obtain the following result which may be called *Yamabe's*

THEOREM 2 (Aubin [1, p. 386], Trüdinger [12]). *Let* M_n *be a* C^{∞} *compact Riemannian manifold* $(n \ge 3)$. *There exists a conformal metric whose scalar curvature is either a non-positive constant or is everywhere positive.*

Proof. Pick $p \in]2, N[$ and set $g' = \phi_p^{4/(n-2)} g$ where ϕ_p is the function from Theorem 1. Equations (2) and (3) together lead to $R' = \mu_p \phi_p^{p-N}$. In the positive case $(\mu > 0)$, μ_p is positive, so it is possible to make the scalar curvature everywhere positive. In the null case $(\mu = 0)$, $R' \equiv 0$. Thus it remains to prove Proposition 2 below. We have distinguished three cases according to the sign of μ. Indeed it is possible to prove the following:

PROPOSITION 1. *μ_q is a continuous function of q for $q \in [2, N]$. This function is either everywhere positive, or everywhere zero, or everywhere negative.*

PROPOSITION 2. *In the negative case $(\mu < 0)$, the functions ϕ_q, $q \in]2, N[$ are uniformly bounded.*

Proof. Choose $G(P, Q)$ the Green's function of the Laplacian non-negative (see Aubin [2]). As $\phi_q > 0$ satisfies (3) and as μ_q is negative, $\Delta\phi_q \leq \frac{-(n-2)}{4(n-1)} R \phi_q$ and we get

$$\phi_q(P) \leq \|\phi\|_1 \left[\int dv\right]^{-1} + \frac{n-2}{4(n-1)} \sup |R| \int G(P, Q)\phi_q(Q) dv(Q) .$$

According to the properties of $G(P, Q)$, iterating r times, $r > n/2$, the preceding inequality yields $\phi_q \leq$ Const. because $\|\phi_q\|_1 \leq$ Const. since $\|\phi_q\|_q = 1$. Thus Proposition 2 is established and it follows that equation (2) has a strictly positive C^∞ solution with $R' = \mu$ in the negative case. Indeed as the functions $\{\phi_q\}$ solution of (3) are uniformly bounded in C^0, they are uniformly bounded in C^1. Ascoli's theorem then implies that a subsequence converges uniformly to a function ϕ_N which satisfies equation (2) with $R' = \mu$. Obviously ϕ_N is Lipschitzian and according to the usual regularity theorem $\phi_N \in C^2$. On the other hand $\phi_N \geq 0$ since $\phi_q > 0$. Applying the maximum principle leads to either

$\phi_N > 0$ or $\phi_N \equiv 0$. Now the former case is impossible since $\|\phi_N\|_N = 1$. Lastly $\phi_N \in C^\infty$ by induction using the regularity theorem.

The preceding results show that the sign of μ is important. It is easy to verify the following.

PROPOSITION 3. *μ is a conformal invariant.*

But if we consider other change of metric than a conformal one it is always possible to exhibit a metric for which $\mu < 0$, Aubin [1, p. 388]. Thus it is proved:

THEOREM 3. *A compact Riemannian manifold M_n of dimension $n \geq 3$, always carries a metric whose scalar curvature is a negative constant.*

III. The Positive Case

This case is hardest. We will use inequality (1) with $q = 2$, the value of the best constant $K(n, 2)$ is $2(\omega_n)^{-1/n}[n(n-2)]^{-1/2}$, ω_n being the volume of the sphere $S_n(1)$ of radius 1. Without loss of generality we suppose henceforth that the volume is equal to one. Following Yamabe, let us consider the set $\{\phi_q\}$ $(q \in]2, N[)$ of functions from Theorem 1. It is bounded in H_1^2: $\|\phi_q\|_2 \leq \|\phi_q\|_q = 1$ and

$$4\,\frac{n-1}{n-2}\,\|\nabla\phi_q\|_2^2 \leq \mu_q + \sup|R| \leq \int R\,dv + \sup|R| \,.$$

Thus there exist $\tilde{\phi} \in H_1^2$ and a sequence $q_i \to N$ such that ϕ_{q_i} converges to $\tilde{\phi}$ weakly in H_1^2 (the unit ball in H_1^2 is weakly compact), strongly in L_2 (Kondrakov's theorem) and almost everywhere. From (3), for all $\psi \in H_1^2$:

$$4\,\frac{n-1}{n-2}\int \nabla^\nu\psi\,\nabla_\nu\phi_{q_i}\,dv + \int R\psi\,\phi_{q_i}\,dv = \mu_{q_i}\int \psi\phi_{q_i}^{q_i-1}\,dv \,.$$

Passing to the limit leads to:

$$(4) \qquad 4\,\frac{n-1}{n-2}\int \nabla^{\nu}\psi\,\nabla_{\nu}\tilde{\phi}\,dv + \int R\psi\,\tilde{\phi}\,dv = \mu\int \psi\,\tilde{\phi}^{N-1}\,dv\,.$$

Indeed $\phi_{q_i}^{q_i-1}$ converges weakly to $\tilde{\phi}^{N-1}$ in $L_{N/(N-1)}$ as we can verify. Thus $\tilde{\phi}$ is a weak solution of equation (2) with $R' = \mu$. According to a theorem of Trüdinger [12], $\tilde{\phi} \in C^{\infty}$. Now it remains to prove that $\tilde{\phi}$ is strictly positive. For the present we know that $\tilde{\phi} \geq 0$ as ϕ_{q_i}, and by the maximum principle, $\tilde{\phi}$ cannot be zero somewhere without to be constant. In order to exclude the case $\tilde{\phi} \equiv 0$ we are going to establish that $\|\tilde{\phi}\|_2 \neq 0$. Since $\phi_{q_i} \to \tilde{\phi}$ strongly in L_2, it is sufficient to prove that $\|\phi_{q_i}\|_2$ is bounded away from zero. From (1) and (3) together we get:

$$1 = \|\phi_q\|_q^2 \leq \|\phi_q\|_N^2 \leq [K^2(n,2)+\varepsilon]\,\frac{n-2}{4(n-1)}\left[\mu_q - \int R\phi_q^2\,dv\right] + A(\varepsilon)\,\|\phi_q\|_2^2$$

for any $\varepsilon > 0$ and all ϕ_q. If $\mu < n(n-1)\omega_n^{2/n}$, using the value of $K(n,2)$ (see the end of I), we can pick ε small enough so that the preceding inequality implies $\liminf_{q\to N} \|\phi_q\|_2 \geq \text{Const.} > 0$. So we have proved:

THEOREM 4 (Aubin [5]). *If* $\mu < n(n-1)\omega_n^{2/n}$, *there exists* $\tilde{\phi} \in C^{\infty}$ *a strictly positive* C^{∞} *solution of (2) with* $R' = \mu$ *and* $\|\tilde{\phi}\|_N = 1$.

Likewise we can prove:

THEOREM 5. *The equation*

$$(5) \qquad 4\,\frac{n-1}{n-2}\,\Delta\phi + h\phi = \eta f\,\phi^{(n+2)/(n-2)}$$

with $h \in C^{\infty}$, $f \in C^{\infty}$ *given* $(f > 0)$ *and* $\eta(=0, 1$ *or* $-1)$ *to determine, has a* C^{∞} *strictly positive solution if* λ, *the* inf. *of*

$$J(\phi) = \left[4\,\frac{n-1}{n-2}\int \nabla^{\nu}\phi\nabla_{\nu}\phi\,dv + \int h\phi^2\,dv\right]\left[\int f\phi^N\,dv\right]^{-2/N} \quad \textit{for all } \phi \in H_1^2\ \phi \neq 0\,,$$

satisfies : $\lambda < n(n-1)\omega_n^{2/n}\,[\sup f]^{-2/N}$.

Computing an asymptotic expansion of $J(\phi_k)$ for a special sequence of functions ϕ_k yields the following for $n \geq 4$.

PROPOSITION 4. *If, at a point* P *where* f *is maximum* $h(P) - R(P) + \frac{n-4}{2}\frac{\Delta f(P)}{f(P)} < 0$, *then equation (5) has a* C^∞ *strictly positive solution.*

IV. Geometrical Applications

Consider again Yamabe's problem, which is equivalent to solving equation (5) with $f(P) = 1$ and $h(P) = R(P)$. The assumption of Proposition 4 is not satisfied. So we see that Yamabe's equation is twice limiting case. For one thing with the exponent of the non-linear part, and secondly with the function R in the linear part.

Computing one more term in the asymptotic expansion of $I(\phi_k)$, we get a term whose sign is $-W_{ijkl}W^{ijkl}$ when $n \geq 6$, W_{ijkl} being the Weyl tensor at P an arbitrary point of M. Thus we obtain:

THEOREM 6. *If* $M_n (n \geq 6)$ *is a compact non-locally conformally flat Riemannian manifold, then* $\mu < n(n-1)\omega_n^{2/n}$. *So in this case the minimum* μ *is attained and equation (2) has a* C^∞ *strictly positive solution with* $R' = \mu$.

If M_n is locally conformally flat with finite fundamental group we are brought back to the case of the sphere where Yamabe's problem is solved. On the other hand it is solved if $\int R\,dv \leq n(n-1)\omega_n^{2/n}$. Thus the problem is open when $\int R\,dv > n(n-1)\omega_n^{2/n}$ if $n \leq 5$ or if the manifold is locally conformally flat with infinite fundamental group. For instance the product of a circle with the sphere is of this kind. But this manifold has constant scalar curvature. The question is if μ is not equal to $n(n-1)\omega_n^{2/n}$. Lastly let us consider the case of the sphere.

THEOREM 7. *For the sphere* S_n, $(n \geq 3)$, $\mu = n(n-1)\omega_n^{2/n}$ *and equation (2) with* $R' = R$ *has an infinity of solutions. The functions* $\phi(r) = (\beta - \cos \alpha r)^{1-n/2}$, *with* $1 < \beta$ *a real number and* $\alpha^2 = R/n(n-1)$, *are solutions of (2) with* $R' = R(\beta^2-1)$. *Moreover on the sphere* S_n *with* $\int dV = 1$, *all* $\phi \in H_1$ *satisfy* $\|\phi\|_N^2 \leq K^2(n,2)\|\nabla\phi\|_2^2 + \|\phi\|_2^2$.

Proof. According to Kazdan and Warner [10] equation (2) has no solution on the sphere when $R' = R + \varepsilon \cos \alpha r$, $\varepsilon \neq 0$. If $\mu < n(n-1)\omega_n^{2/n}$, we can choose ε small enough in order to apply Theorem 5 to equation (5) with $h = R$ and $f = R + \varepsilon \cos \alpha r$. So we get a contradiction.

When μ is attained, computing the second variation of $I(\phi)$ yields:

PROPOSITION 5. *When* μ *is attained, let* $I(\phi_0) = \mu$. *In the corresponding metric* g_0 *whose scalar curvature* R_0 *is constant, the first non-zero eigenvalue of the Laplacian* $\lambda_1 \geq R_0/(n-1)$.

UNIVERSITE PIERRE ET MARIE CURIE
4 PLACE JUSSIEU
75230 PARIS CEDEX 05

REFERENCES

[1] Aubin, T., Métriques riemanniennes et courbure, J. Diff. Geo. 4 (1970), 383-424.

[2] ________, Fonction de Green et valeurs propres du Laplacien, J. Math pures et appl., T. 53 (1974), 347-371.

[3] ________, Problèmes isopérimétriques et espaces de Sobolev, J. Diff. Geo. 11 (1976), 573-598. Resume dans C.R. Acad. Sc., Paris - Série A 280 (1975), 279.

[4] ________, Espaces de Sobolev sur les varietes Riemanniennes, Bull. Sci. Math. 100 (1976), 149-173.

[5] ________, Equations différentielles non-linéaires et problème de Yamabe concernant la courbure scalaire, J. Math. Pures Appl. 55 (1976), 269-296.

[6] ________, The scalar curvature. Differential Geometry and Relativity, Cahen and Flato, editors, Reider Publishing Company (1976).

[7] ________, Meilleurs constantes dans le theoreme d'inclusion de Sobolev et un theoreme de Fredholm non-lineaire pour la transformation conforme de la courbure scalaire Journal of functional Analysis vol. 32 nº 2 (1979), 148-174.

[8] Berger, M.S., On Riemannian structures of prescribed Gaussian curvature for compact 2-manifolds, J. Diff. Geo. 5 (1971), 325-332.

[9] Bliss, G.A., An integral inequality, J. London Math. Soc. 5 (1930), 40-46.

[10] Kazdan, J.L. and Warner, F.W., Scalar curvature and conformal deformation of Riemannian structure, J. Diff. Geo. vol. 10 (1975), 113-134.

[11] Talenti, G., Best constant in Sobolev inequality, Ann. Mat. Pura, Appl. 110 (1976), 353-372.

[12] Trüdinger, N.S., Remarks concerning the conformal deformation of Riemannian structures on compact manifolds, Ann. Scuola Norm. Sup. Pisa, (3), vol. 22 (1968), 265-274.

[13] Yamabe, Y., On a deformation of Riemannian structures on compact manifolds, Osaka Math. J., vol. 12 (1960), 21-37.

GAUSSIAN AND SCALAR CURVATURE, AN UPDATE

Jerry L. Kazdan*

In this lecture, we will report on recent progress on prescribing Gaussian and scalar curvature on compact and complete Riemannian manifolds (M, g) of dimension n. To avoid needless repetition, this report will be a continuation of our earlier (1973) survey [K-W, 5], and assume the reader is acquainted with that survey (or is willing to refer to it). Here emphasis will be placed on more recent results and open problems.

1. *Gaussian curvature* ($\dim M = 2$)

Let k and Δ be the Gaussian curvature and Laplacian, respectively, of (M, g). We use the sign convention that $\Delta u = u_{xx} + u_{yy}$ on flat $\mathbb{R}^2$. Given a function K on M, we wish to know when the candidate K is actually the Gaussian curvature of a metric g_1 that is pointwise conformal to g, so $g_1 = e^{2u}g$ for some function u. This leads directly to solving the nonlinear elliptic partial differential equation

$$\Delta u = k - Ke^{2u} . \tag{1.1}$$

On compact M, by a change of variable [$u = v - z$ with $\Delta z = k - \bar{k}$, and $k = (\int k)/(\text{area } M)$], this equation reduces to

*Supported in part by NSF Grant MCS79-01780.

Seminar on Differential Geometry
0-691-08268-5/82/000185-7 $00.50/0 (cloth)
0-691-08296-0/82/000185-7 $00.50/0 (paperback)
For copying information, see copyright page.

$$(1.1)' \qquad \Delta u = c - Ke^{2u}$$

where $c = \bar{k}$ and the function K is obviously modified. We discuss only the case where M is compact. Integrating (1.1) over M we find, by Gauss-Bonnet, that

$$\int_M Ke^{2u}dA_g = \int_M kdA_g = 2\pi\chi(M) ,$$

which imposes an obvious necessary sign condition on the candidate K. Is this sign condition sufficient to solve (1.1)? Note that as a pure problem in differential equations, this question makes sense on any compact M (of any dimension).

Case i) $c < 0$. If $K < 0$ everywhere, one can always solve (1.1), while if K is not too positive too often, then one can solve (1.1) —on any compact M of any dimension. This situation is fairly well understood [K-W, 1].

Case ii) $c = 0$. If $\dim M = 2$ there are necessary and sufficient conditions [K-W, 1]. No sufficient conditions are known if $\dim M \geq 3$.

Case iii) $c > 0$. Very little is known except on S^2 with the canonical metric, so $k = 1$. Moser [M] proved that if $K(x) = K(-x)$ and K is positive somewhere, then one can solve (1.1), i.e., one can solve (1.1) on P^2 if the obvious sign condition is satisfied. In fact, he showed that one can solve (1.1)′ on (S^2, can) if $0 < c < 1$ and $K > 0$ somewhere.

However, Kazdan-Warner [K-W, 1] showed that on (S^2, can) with $k = 1$, i.e., $c = 1$, if u is a solution of (1.1), then

$$(1.2) \qquad \int_M (\nabla K \cdot \nabla F) e^{2u} dA = 0$$

for all first order spherical harmonics $F \, (-\Delta F = 2F)$. In particular, if $K = F + \text{const}$, then one *cannot* solve (1.1).

One is led to find (1.2) by the following procedure (see [K] for a similar derivation of the standard Bianchi identities). Since the Möbius transformations ought to be useful on S^2, let $\phi_\lambda : S^2 \to S^2$ be the conformal maps of S^2 induced under stereographic projection by the Möbius transformation $z \to \lambda z$ of $\mathbb{C}$. Compose both sides of (1.1) with ϕ_λ and take the derivative of the resulting equation with respect to λ, evaluated at $\lambda = 1$. Since $(d\phi_\lambda/d\lambda)|_{\lambda=1} = \nabla F$ for some first order spherical harmonic, after a somewhat tricky calculation – using $(\Delta u)\nabla u \cdot \nabla F$ = divergence of something – one finds that

$$\text{div}\,(\text{something}) = \nabla K \cdot \nabla F\, e^{2u} ,$$

so (1.2) follows by integration. No similar identity for (1.1) is known on any other manifold of any dimension.

Except for Moser's result, there are few existence theorems. We do know that if K is positive somewhere on (S^2, can), then one can solve (1.1): (a) with K replaced by $K \circ \psi$ for some diffeomorphism ψ [K-W, 4], and also (b) with K replaced by $K + F$ for some (unknown) first order spherical harmonic F [A, 3].

Of course, the natural conjecture is that if K satisfies (1.2), then on (S^2, can) a solution must exist. There is no guess as to what happens on other manifolds, even in dimension two. Note that in higher dimensions, the equation (1.1) arises for Hermitian manifolds (see [B, §III]); there too there is no information except for the easy case $\int k < 0$, where we already know the main facts in all dimensions. This concludes our discussion of case iii).

Equation (1.1) can also be viewed as the one (complex) dimensional version of the equation

$$(1.3) \qquad \det(g_{i\bar{j}} + u_{i\bar{j}}) = \exp(f - cu) \det g_{i\bar{j}}$$

concerning Kähler-Einstein manifolds [Yau, A-4]. If $c \leq 0$, there is a solution. Although geometric proofs of non-existence are known for the case $c > 0$ (see [SP, pp. 135-147]), one suspects that there should also be an identity like our (1.2) which would also prove non-existence.

In a slightly different direction, one can ask if every function $K(x,y)$ is the Gaussian curvature of some surface $z = u(x,y)$ in $\mathbb{R}^3$, at least in some neighborhood of the origin. Thus, one must solve the equation

$$u_{xx}u_{yy} - u_{xy}^2 = K(x,y)(1+|\nabla u|^2)^2 \tag{1.4}$$

locally. If $K(0) > 0$ (< 0), this equation is elliptic (hyperbolic) so standard results give local existence. However, if $K \in C^\infty$, has $K(0) = 0$, it is unknown if a solution $u \in C^2$ exists. One suspects that, in fact, for certain functions K there is no solution — which makes the problem all the more enticing.

2. *Scalar curvature* (dim $M \geq 3$)

Let k be the scalar curvature of g. The first question is, what are the topological obstructions to scalar curvature? Aubin [A, 1] proved that *every* compact M has a metric of negative scalar curvature. However, Lichnerowicz [L] found that there are topological obstructions to positive scalar curvature. These were subsequently improved by Hitchin [H], who showed that certain exotic spheres do not admit metrics with positive scalar curvature (and hence no positive sectional curvature). Kazdan-Warner [K-W, 3] found obstructions to zero scalar curvature and asked if tori T^n $(n \geq 3)$ have metrics of positive scalar curvature. Schoen-Yau [S-Y, 1,2] showed that for many manifolds, including T^n, $n \leq 7$, there are no positive scalar curvature metrics. By an extension of this technique, they subsequently resolved the "positive mass and action conjectures" in general relativity. Schoen-Yau also obtained a fairly complete classification of all 3-manifolds having positive scalar curvature metrics. Slightly later, by generalizing Lichnerowicz's approach, Gromov-Lawson found new topological obstructions in all dimensions [G-L, 1,2]. In particular, they show that T^n has no positive scalar curvature metric for any n.

Again turning to conformal deformations, given (M, g) and a function K, is there a metric g_1 conformal to g (so $g_1 = u^{4/(n-2)}g$ with $u > 0$) having scalar curvature K? Then $u > 0$ must satisfy

$$L_g u \equiv -\alpha\Delta u + ku = Ku^{(n+2)/(n-2)}, \tag{2.1}$$

where $\alpha = 4(n-1)/(n-2)$. One key observation in [K-W, 3,5] is that the sign of the first eigenvalue, $\lambda_1(g)$ of L_g is a conformal invariant, and that a manifold admits a metric of positive (respectively negative, zero) scalar curvature if and only if it admits a metric with $\lambda_1(g) > 0$ $(< 0, = 0)$. These and other properties of $\lambda_1(g)$ have made it a valuable tool in subsequent work on scalar curvature. Its primary virtue is that the functional $\lambda_1(g)$ is a number, not a function. (It would be nice to find a similar functional for Ricci curvature.)

The well-known Yamabe conjecture ([Y], he wanted to use analysis to resolve the Poincaré conjecture) is to solve (2.1) with K a constant. This was resolved in some cases, including $\lambda_1(g) \leq 0$, by Trüdinger [T], while for $\lambda_1(g) > 0$ with $n \geq 6$ the metrics which are not conformally flat near some point were treated by Aubin [A, 2]. The problem has turned out to be surprisingly recalcitrant.

Solvability of (2.1) given not necessarily constant functions K was discussed by Kazdan-Warner in [K-W, 3] by techniques similar to those used on (1.1). As usual, the negative curvature case, $\lambda_1(g) < 0$, is easier, while on (S^n, can), $n \geq 3$, an obstruction to solvability was found (this obstruction is similar to (1.2), and may be proved by the same method). One question which arises from [K-W, 4, Theorem 6.4] is, "if M admits a metric of positive scalar curvature, does it admit a metric (not necessarily conformal) of positive *constant* scalar curvature?" If the Yamabe conjecture is true, then the answer is "yes." In any case, by Aubin's work one knows the answer is "yes" if $n \geq 6$ (since he showed that the answer is "yes" for a dense set of metrics). The answer to this fairly primitive question is open if $3 \leq n \leq 5$. It would show that there are no pinching theorems for scalar curvature.

So far we have only discussed compact manifolds. Little is known for complete manifolds (M, g). Schoen-Yau [S-Y, 3] have some topological obstructions to positive scalar curvature if $\dim M = 3$, while Ni [N] has some results on solving (2.1).

BIBLIOGRAPHY

[A, 1] Aubin, T. Métriques riemanniennes et courbure, *J. Differential Geometry* *4* (1970), 383-424.

[A, 2] ———, "Equations différentielles non linéaires et problème de Yamabe concernant la courbure scalaire," *J. Math. Pures. Appl.* *55* (1976), 269-296.

[A, 3] ———, "Meilleures constantes dans le théorème d'inclusion de Sobolev," *J. Funct. Anal,* *32* (1979), 148-174.

[A, 4] ———, "Equations du type Monge-Ampère sur les variétés Kahlériennes compactes," *Bull. Sc. Math,* 2e serie, *102* (1978), 63-95.

[B] Berger, Melvyn S., "Constant Scalar Curvature Metrics for Complex Manifolds," *Proc. Symp. Pure Math,* Vol. *27* (1975), Amer. Math. Soc., 153-170.

[G-L, 1] Gromov, M., and Lawson, H. B., "Spin and scalar curvature in the presence of a fundamental group, I," *Ann. of Math.* *111* (1980), 209-230.

[G-L, 2] ———, "The classification of simply-connected manifolds of positive scalar curvature," *Ann. of Math.* *111* (1980), 423-434.

[H] Hitchin, N., "Harmonic Spinors," *Adv. in Math.* *14* (1974), 1-55.

[K] Kazdan, Jerry L., "Another Proof of Bianchi's Identity in Riemannian Geometry," *Proc. Amer. Math. Soc.* (to appear in 1981).

[K-W, 1] Kazdan, Jerry L., and Warner, F. W., "Curvature functions for compact 2-manifolds," *Ann. of Math.* (2) *99* (1974), 14-47.

[K-W, 2] ———, "Curvature functions for open 2-manifolds," *Ann. of Math.* *99* (1974), 203-219.

[K-W, 3] ———, "Scalar curvature and conformal deformation of Riemannian structure," *J. Differential Geometry* *10* (1975), 113-134.

[K-W, 4] ———, "Existence and conformal deformation of metrics with prescribed Gaussian and scalar curvature," *Ann. of Math.* *101* (1975), 317-331.

[K-W, 5] ———, "Prescribing Curvatures," *Proc. Symp. Pure Math.,* Vol. *27* (1975), Amer. Math. Soc., 309-319.

[K-W, 6] ———, "A Direct Approach to the Determination of Gaussian and Scalar Curvature Functions," *Inventiones Math,* Vol. *28* (1975), 227-230.

[L] Lichnerowicz, A., "Spineurs harmoniques," *C. R. Acad. Sci. Paris* *257*(1963), 7-9. MR27 #6218.

[M] Moser, J., "On a nonlinear problem in differential geometry," *Dynamical Systems* (M. Peixoto, Editor), Academic Press, New York, 1973.

[N] Ni, Wei-Ming, "Conformal Metrics with Zero Scalar Curvature and a Symmetrization Process via Maximum Principle," this same volume.

[SP] Seminaire Palaiseau, "Première classe de Chern et courbure de Ricci: preuve de la conjecture de Calabi," in *Astérisque, 58* (1978).

[S-Y, 1] Schoen, R., and Yau, S. T., "Existence of incompressible minimal surfaces and the topology of three dimensional manifolds with non-negative scalar curvature," *Ann. of Math. 110*(1979), 127-142.

[S-Y, 2] ———, "The structure of manifolds with positive scalar curvature," *Manuscripta Math. 28*(1979), 159-183.

[S-Y, 3] ———, to appear.

[T] Trüdinger, N. S., Remarks concerning the conformal deformation of Riemannian structures on compact manifolds, *Ann. Scuola Norm. Sup. Pisa* (3) *22*(1968), 265-274, MR39 #2093.

[Y] Yamabe, H., On a deformation of Riemannian structures on compact manifolds, *Osaka Math. J. 12*(1960), 21-37, MR23 #A2847.

[Yau] Yau, S. T., "On the Ricci Curvature of a Compact Kähler Manifold and the Complex Monge-Ampere Equation, I," *Comm. Pure. Appl. Math.*, Vol. *31* (1978), 339-411.

CONFORMAL METRICS WITH ZERO SCALAR CURVATURE AND A SYMMETRIZATION PROCESS VIA MAXIMUM PRINCIPLE

Wei-Ming Ni*

§1. *Introduction*

This expository paper consists of two parts. In the first part, we describe some of the results in [4]. As a special case, we consider the positive solutions of the equation

$$\Delta u + u^{\frac{n+2}{n-2}} = 0\,, \qquad \left(\Delta = \sum_{i=1}^{n} \frac{\partial^2}{\partial x_i^2}\right) \tag{1}$$

in $\mathbf{R}^n$, which is a special model for the Yang-Mills equation when $n=4$, and also arises in geometry:

Let (M, g) be a Riemannian manifold of dimension ≥ 3, and K be a given function on (M, g). One may ask: can we find a new metric g_1 on M, such that K is the scalar curvature of g_1 and g_1 is *conformal* to g (i.e. $g_1 = u^{\frac{4}{n-2}} g$, for some $u > 0$)? This problem is equivalent to the problem of finding the positive solutions of the equation

$$\frac{4(n-1)}{n-2}\Delta u - ku + Ku^{\frac{n+2}{n-2}} = 0 \tag{2}$$

*Supported in part by NSF Grants No. MCS 77-18723(02) and MCS 79-01780.

Seminar on Differential Geometry
0-691-08268-5/82/000193-10$00.50/0 (cloth)
0-691-08296-0/82/000193-10$00.50/0 (paperback)

where Δ, k are the Laplacian and scalar curvature in the g metric, resp. In case $M = \mathbf{R}^n$, and g is the standard metric in $\mathbf{R}^n$, then $k \equiv 0$ and (2) reduces to

$$\frac{4(n-1)}{n-2} \Delta u + K u^{\frac{n+2}{n-2}} = 0 . \tag{3}$$

As a special case of a result in [4], we have the following theorem:

THEOREM 1. *Let* $u > 0$ *be a solution of (1) and* $u = O\left(\frac{1}{|x|^{n-2}}\right)$ *at* ∞. *Then*

$$u(x) = \frac{[n(n-2)\lambda^2]^{\frac{n-2}{4}}}{(\lambda^2 + |x-x_0|^2)^{\frac{n}{2}-1}} \tag{4}$$

for some $\lambda > 0$ *and* $x_0 \in \mathbf{R}^n$.

The key step in the proof is to show u is radially symmetric about some point in $\mathbf{R}^n$. It was R. Schoen who pointed out to us that *Theorem 1* is equivalent to the following theorem of M. Obata [10]:

A Riemannian metric on S^n (n-sphere), which is conformal to the usual metric and having the same constant scalar curvature, is, in fact, the pullback of the usual metric under a conformal map of S^n onto itself.

The key step in Obata's proof is to show, in fact, this given metric has constant sectional curvatures. The method he used is entirely different from ours.

Our method of showing rotational symmetry of u in Theorem 1 is using the moving plane technique (which goes back to A.D. Alexandroff) and various extensions of E. Hopf's boundary point lemma; in fact, we can prove more general results (cf. [4]). In particular, we have obtained the following result which is useful in analysis:

THEOREM 2. *Let* $u > 0$ *be a solution in* $C^2(\bar{D})$ *of* $\Delta u + f(u) = 0$ *with* $u|\partial D = 0$ *where* f *is locally Lipschitz and* D *is a ball. Then,* u *is radially symmetric and* $\frac{\partial u}{\partial r} < 0$ *for* $r > 0$.

We shall sketch the proof of this theorem in Section 2, to illustrate the idea of using "moving plane" technique.

In the second part of this paper, we wish to describe some work in progress (cf. [7]) concerning the existence of positive solutions of (3) when $K \geq 0$. In case of $K \leq 0$, there have been much work concerning the non-existence of positive solutions of (3). For example, [1], [2], [5], [6] and [11].

A special case of one of our main results is the following

THEOREM 3. *Let* $K(x)$ *be a Hölder continuous function in* R^n *with* $K(x) \geq 0$ *for all* $x \in R^n$ *and* $K(x) \leq \frac{C}{|x|^\ell}$ *near infinity, for some constants* $C > 0$ *and* $\ell > 2$. *Then (3) has infinitely many positive solutions in* R^n *such that each of them tends to a positive constant at infinity.*

REMARK. Not all positive solutions of (3) (with K in Theorem 3) are bounded away from zero; and the power in Theorem 3 is sharp! In fact, the decay of K is needed only in a 3-dimensional subspace and the sign condition of K may be dropped (cf. [7]).

From this theorem, we obtain immediately that $g_1 = u^{\frac{4}{n-2}} g$ (where g is the standard metric in R^n) has K as its scalar curvature; moreover, g_1 is a *complete metric* (since $u \to \alpha > 0$ at infinity).

In case K depends only on $|x|$, better results are obtained. Such as:

THEOREM 4. *Let* $K \geq 0$ *be a smooth function of* $r = |x| \geq 0$. *Suppose* K *satisfies (5) or (6):*

(5) $$K(r) \leq \frac{C}{r^2} \text{ at infinity, for some constant } C > 0,$$

(6) $$K' \leq 0 \text{ for all } r > 0.$$

Then, (3) has infinitely many positive solutions.

It should be remarked that for some $K(r)$ in Theorem 4, all positive solutions do tend to 0 at infinity. We also have some a priori estimates and existence theorems of the solutions of (3) in various cases (including $K \leq 0$ and K changes signs), however, we shall not mention them here. We also remark that (6) can be considerably improved by allowing K to oscillate "mildly," but we shall not spell out the technical condition here!

Let us also look at a non-existence result in the case $K \geq 0$.

THEOREM 5. (i) *The equation (3) does not possess any positive radially symmetric solutions in case* $K(x) = |x|^{\ell}$, *for some* $\ell > 0$.

(ii) *The equation (3) does not possess any positive solutions if* $\tilde{K}(r) \geq C \cdot r^{\ell}$ *at* $r = \infty$ *for some* $\ell \geq 2$, $C > 0$, *where*

$$\tilde{K}(r) \equiv \left(\frac{1}{\omega_n r^{n-1}} \int_{|x|=r} \frac{dS_x}{K(x)^{\frac{n-2}{4}}}\right)^{-\frac{4}{n-2}}$$ *and* ω_n = *the surface area of unit sphere in* $\mathbb{R}^n$.

The proofs of Theorems 3 and 5 (i) will be sketched in Section 3, and the proof of Theorem 4 will not be discussed here (although it is entirely different from that of Theorem 3). The details of the proofs together with other theorems and estimates will appear elsewhere [7] and [13].

Acknowledgement

I thank J. Kazdan for suggesting the existence problem concerning (3); R. Schoen, for several conversations on the geometrical significance of the results presented here; S. T. Yau, for his interest in this work; and L. Nirenberg, for his continuous encouragement and support, much of the work in the second part of this paper is influenced by him while we have worked together in the past three years.

§2. *Sketch of the proofs of Theorems 1 and 2*

We start with the proof of Theorem 2. For simplicity, we shall only prove it in the case $0 \leq f \in C^1$. The general case can be proved in a similar manner, but requires more work. First, let us recall maximum principle and Hopf boundary point lemma:

Let $u \leq 0$ in Ω (bounded domain in $\mathbf{R}^n$), and satisfies

$$Lu \equiv a_{ij}(x)u_{x_ix_j} + b_i(x)u_{x_i} + c(x)u \geq 0$$

where L is uniformly elliptic.

MAXIMUM PRINCIPLE. *If* $u = 0$ *at some point in* Ω, *then* $u \equiv 0$ *in* Ω.

HOPF LEMMA. *Suppose* $u \in C^2(\Omega)$ *and* $u(P) = 0$ *for some point* $P \in \partial\Omega$. *If* $u \not\equiv 0$ *and there exists a ball* $B \subseteq \Omega$ *with* $P \in \partial B$, *then,* $\frac{\partial u}{\partial \nu}(P) > 0$ *where* ν *is any outward direction.*

(It should be remarked that no sign condition is imposed on $c(x)$ above.)

Now, we begin to prove Theorem 2 in the case $0 \leq f \in C^1$. Define $\Sigma_\lambda = \{x = (x_1, \cdots, x_n) \in D \mid x_1 > \lambda\}$. T_λ = the hyperplane which is perpendicular to x_1-axis at $x_1 = \lambda$, and Σ'_λ = the reflection of Σ_λ w.r.t. T_λ. Let $(*)_\lambda$ denote the following statement:

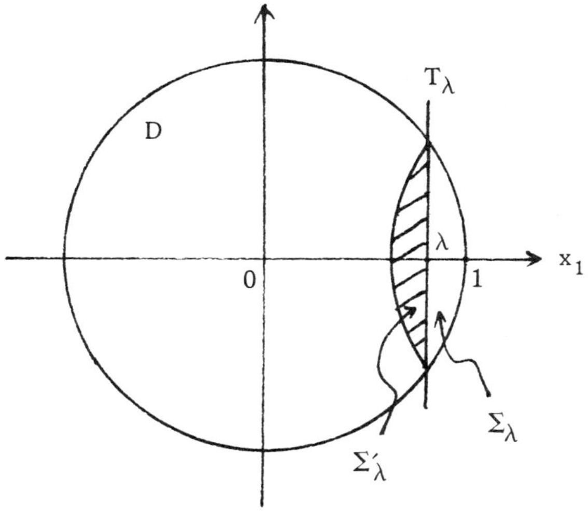

$(*)_\lambda$ $\quad u(x) < u(x^\lambda)$, *for all* $x \in \Sigma_\lambda$ *and* $u_{x_1}(x) < 0$ *on* $D \cap T_\lambda$

where x^λ is the reflection of x w.r.t. T_λ.

And we claim the set $\Lambda = \{\lambda \in (0,1) \mid (*)_\lambda \text{ holds}\}$ is non-empty, open, and closed in $(0,1)$, thus $\Lambda = (0,1)$. So, $u(x) \leq u(x^0)$. The fact that

$\lambda \in \Lambda$ for λ sufficiently close to 1 is easy to see from the hypothesis $u > 0$ in D, $u|_{\partial D} = 0$ and Hopf Lemma. The fact that Λ is both open and closed follows from the following

LEMMA. *Assume for some* $\lambda > 0$, *we have*: $u(x) \leq u(x^{\lambda})$ *for every* $x \in \Sigma_{\lambda}$ *and* $u_{x_1} \leq 0$ *in* Σ_{λ}. *If* $u(x) \not\equiv u(x^{\lambda})$, *then*, $(*)_{\lambda}$ *holds*.

Proof. In Σ'_{λ}, define $v(x) = u(x^{\lambda})$, then, v also satisfies $\Delta v + f(v) = 0$ in Σ'_{λ}. Consider $w = u - v$, we have, $w \geq 0$ in Σ'_{λ} and $w > 0$ in $\partial\Sigma'_{\lambda} \setminus T_{\lambda}$, so, $w \not\equiv 0$. Moreover, $0 = \Delta(u-v) + f(u) - f(v) = \Delta w + c(x)w$ by the mean-value theorem.

Then apply Hopf Lemma to conclude $(*)_{\lambda}$ holds. q.e.d.

Now, reverse x_1-axis. We get $u(x) = u(x^0)$. Since x_1-axis may be any direction, Theorem 2 is proved. We now turn to the proof of Theorem 1. The major difficulty now is how to get started, i.e. to conclude Λ contains sufficiently large λ's. This can be achieved by a suitable choice of the origin and a tedious calculation, we shall not reproduce it here.

We conclude this section by the following

REMARKS. (i) Both Theorems 1 and 2 can be extended to *fully nonlinear* elliptic and parabolic equations as well as various domains. In order to do this, we have to extend Hopf's boundary lemma. For the case P in the intersection of two transversally intersect hypersurfaces, cf. [4; Appendix]. If P is in the intersection of more than two transversally intersect hypersurfaces or, if P is the vertex of an arbitrary cone in $\mathbf{R}^n$, cf. [8].

(ii) Theorem 2 has some useful applications in analysis. We only mention [3] for a priori estimates of positive solutions of some elliptic boundary value problem, and [9] for global vortex flow in fluid dynamics.

(iii) This ingenious device of moving parallel planes used in the proof above is due to A.D. Alexandrov, and was applied elegantly to a problem in fluid dynamics by J. Serrin [12].

§3. *Sketch of the proofs of Theorems 3 and 5 (i)*

We shall only sketch a proof of Theorem 3 under a stronger hypothesis on K, namely, $0 \le K(x) \le \frac{c}{|x|^{n+2}}$. A basic step in the proof is

THEOREM 6. *Let* $v_a(r)$ *denote the solution of*

$$\begin{cases} v'' + \frac{n-1}{r} v' + v^{\frac{n+2}{n-2}} = 0 \\ v'(0) = 0, \quad v(0) = a > 0 \end{cases} \tag{7}$$

Suppose $0 \le K(x) \le v_1^{\frac{n+2}{n-2}}(r)$ *and* K *is radially symmetric, then, the* IVP

$$\begin{cases} u'' + \frac{n-1}{r} u' + K(r) u^{\frac{n+2}{n-2}} = 0 \\ u'(0) = 0, \quad u(0) = a > 0 \end{cases} \tag{8}$$

has a positive solution in R^+ *for each* $a \le 1$ *and* $u \ge v_a$.

Proof. Calculation shows $K(r) u^{\frac{n+2}{n-2}} \le v_a^{\frac{n+2}{n-2}}$ for $a \le 1$. Set $w = u - v_a$, then,

$$\Delta w + [K(r) u^{\frac{n+2}{n-2}} - v_a^{\frac{n+2}{n-2}}] = 0$$

implies $\Delta w \ge 0$. Thus, maximum principle implies $w(r)$ is increasing in $r > 0$. Since $w(0) = 0$, then $w(r) > 0$, for $r > 0$. On the other hand, u is decreasing in r when $u > 0$. This concludes the proof. q.e.d.

From the above proof, it is not hard to obtain

THEOREM 7. *If we assume, in addition,* $K(x) = K(|x|)$ *then, Theorem 3 holds.*

Now we proceed to sketch the proof of Theorem 3. At first, choose $K_1(r) \ge 0$ such that $K_1(r) \ge K(x)\,(r = |x|)$, for all x and $K_1(r) = \frac{c}{r^{n+2}}$ at infinity for some $c > 0$. By Theorem 7, let v_1 be a positive solution of $v_1'' + \frac{n-1}{r} v_1' + K_1(r) v_1^{\frac{n+2}{n-2}} = 0$ with $v_1'(0) = 0$ $v_1(0) > 0$, then, v_1 decreases to $\alpha > 0$ at infinity. Choose $K_2(r)$ = nonnegative radial function which is less than or equal to $K(x)$, and let v_2 be a solution of $v_2'' + \frac{n-1}{r} v_2' + K_2(r) v_2^{\frac{n+2}{n-2}} = 0$ with $v_2'(0) = 0$, $v_2(0) > 0$, and $v_2(0) \le \alpha$. Then it is easy to check

(9) $$\Delta v_1 + K(x) v_1^{\frac{n+2}{n-2}} \le 0 \text{ and } v_1(r) \ge \alpha, \qquad \forall r > 0$$

(10) $$\Delta v_2 + K(x) v_2^{\frac{n+2}{n-2}} \ge 0 \text{ and } v_2(r) \le \alpha, \qquad \forall r > 0 .$$

Now, consider the boundary value problem

(11) $$\begin{cases} \Delta u + K(x) u^{\frac{n+2}{n-2}} = 0 & \text{in ball } D_R \text{ of radius } R \\ u|_{\partial D_R} = \alpha \end{cases}$$

(9), (10) show that v_1, v_2 are super-solution and sub-solution of (11), respectively. Since $v_1 \ge v_2$, we conclude (11) has a solution u_R such that $v_1 \ge u_R \ge v_2$. Now elliptic interior estimates show there exists a sequence $R_n \uparrow \infty$ such that $u_{R_n} \to u$ in $C^2(\Omega)$ on every compact set $\Omega \subseteq \mathbf{R}^n$. Thus, u is a solution we required. If we take $K_2 \equiv 0$, $v_2 \equiv \alpha$, we have $u(x) \to \alpha$ as $|x| \to \infty$. This finishes the proof of Theorem 3.

Now we turn to the proof of part (i) of Theorem 5. At first, we prove

THEOREM 8. *The elliptic boundary value problem*

(12) $$\begin{cases} \Delta u + |x|^{\ell} u^{\frac{n+2}{n-2}} = 0 & \text{in } D = \text{a ball in } \mathbf{R}^n \\ u|_{\partial D} = 0 \end{cases}$$

where $\ell > 0$, *a real number, possesses a radially symmetric solution which is positive in* D.

The proof of this theorem makes use of the well-known "Mountain Pass Lemma" in variational approach. Theorem 8 can be generalized to the form

$$(13) \qquad \begin{cases} \Delta u + b(|x|)\,p(u) = 0 & \text{in } D \\ u|_{\partial D} = 0 \end{cases}$$

with appropriate conditions on b, p (cf. [13]). We shall not get into this.

Let u be a solution in Theorem 8. Then the different degrees of homogeneity of the two terms Δu and $|x|^{\ell} u^{\frac{n+2}{n-2}}$ together with the following uniqueness theorem constitute the proof of Theorem 5 (i).

THEOREM 9. *Let* u_1, u_2 *be two solutions of*

$$(14) \qquad \begin{cases} u'' + \frac{n-1}{r}\, u' + c(r) f(u) = 0 \\ u'(0) = 0 \qquad u(0) = a > 0 \end{cases}$$

where $f'(a) > 0$ *and* $c(r) \geq 0$ *in* $(0, \delta)$, *for some* $\delta > 0$. *Then* $u_1 \equiv u_2$.

The proof depends essentially on the following *comparison Lemma*: *Let* w *be a solution of*

$$\begin{cases} w'' + \frac{n-1}{r}\, w' + h(r)\, w \geq 0 \\ w(0) = w'(0) = 0 \end{cases}$$

where $h(r) \geq 0$ *for all* $r \in (0, \delta)$, *for some* $\delta > 0$. *Then,* $w \geq 0$ *in* $(0, \varepsilon)$ *for some* $\varepsilon > 0$.

We leave these proofs to the readers.

SCHOOL OF MATHEMATICS
INSTITUTE FOR ADVANCED STUDY
PRINCETON, NEW JERSEY 08540

SCHOOL OF MATHEMATICS
UNIVERSITY OF MINNESOTA
MINNEAPOLIS, MINNESOTA 55455

REFERENCES

[1] Calabi, E. An extension of E. Hopf maximum principle with an application to Riemannian geometry, Duke Math. J. 25 (1958), 45-56.

[2] Cheng, S.-Y. and Yau, S.-T. Differential equations on Riemannian manifolds and their geometric applications, Comm. Pure Applied Math. 38 (1975), 333-354.

[3] de Figueiredo, D. G., Lions, P.-L. and Nussbaum, R. D. Estimations a priori pour les solutions positives de problèmes elliptiques semi-linéaires, C. R. Acad. Sci. Paris, Ser. A, 290 (1980), 217-220.

[4] Gidas, B., Ni, W.-M. and Nirenberg, L. Symmetry and related properties via the maximum principle, Comm. Math. Phys. 68 (1979), 209-243.

[5] Kazdan, J. and Warner, F. Scalar curvature and conformal deformation of Riemannian structure, J. Diff. Geom. 10 (1975), 113-134.

[6] Loewner, C. and Nirenberg, L. Partial differential equations invariant under conformal and projective transformations, Contributions to Analysis, 245-272, Academic Press (1974).

[7] Ni, W.-M. On the elliptic equation $\Delta u + K(x) u^{\frac{n+2}{n-2}} = 0$, Indiana Univ. Math. J., (to appear).

[8] ———. Hopf boundary lemma on singular domains and its applications (in preparation).

[9] ———. On the existence of global vortex flow, J. d'Analyse Math. 37 (1980), 208-247.

[10] Obata, M. The conjectures on conformal transformations of Riemannian manifolds, J. Diff. Geom. 6 (1971), 247-258.

[11] Osserman, R. On the inequality $\Delta u \geq f(u)$, Pacific J. Math. 7 (1957), 1641-1647.

[12] Serrin, J. A symmetry problem in potential theory, Arch. Rat. Mech. Analy. 43 (1971), 304-318.

Added in proof:

[13] Ni, W.-M. A nonlinear Dirichlet problem on the unit ball and its applications, Indiana Univ. Math. J., (to appear).

For the related question of finding conformal metrics with prescribed Gaussian curvatures on $\mathbf{R}^2$, see

[14] Ni, W.-M. On the elliptic equation $\Delta u + K(x) e^{2u} = 0$ and conformal metrics with prescribed Gaussian curvatures (in preparation).

RIGIDITY OF POSITIVELY CURVED MANIFOLDS WITH LARGE DIAMETER

Detlef Gromoll and Karsten Grove

Let M be a complete Riemannian manifold of dimension n with positive sectional curvature K bounded away from zero. We normalize the metric so that $1 \leq K$. By Myers' theorem the diameter of M satisfies $d \leq \pi$, and M is compact. Recall that Toponogov [T] proved that $d = \pi$ iff M is isometric to the standard sphere S^n of constant curvature $K = 1$. A strong extension of this rigidity result is the Diameter Sphere Theorem of Grove and Shiohama [GS], a homotopy version of which had been obtained earlier by Berger: If $d > \pi/2$, then M is homeomorphic to S^n. This also generalizes the classical Sphere Theorem of Rauch, Berger, and Klingenberg, which arrives at the same conclusion under the assumptions M is simply connected and $1 \leq K < 4$. Those conditions imply $d > \pi/2$ via Klingenberg's lemma on the injectivity radius. The simply connected symmetric spaces of rank 1 endowed with their standard homogeneous metric have curvature $1 \leq K \leq 4$ and diameter $d = \pi/2$, so either one of the above sphere theorems is optimal. In [B], Berger proved the celebrated rigidity theorem for 1/4-pinched manifolds: If M is simply connected and $1 \leq K \leq 4$, then M is either homeomorphic to a sphere or isometric to a rank 1 symmetric space.

In this note we outline how to generalize Berger's result and extend the Diameter Sphere Theorem. We can completely analyze the structure of

Seminar on Differential Geometry
0-691-08268-5/82/000203-5 $00.50/0 (cloth)
0-691-08296-0/82/000203-5 $00.50/0 (paperback)
For copying information, see copyright page.

M when $1 \leq K$ and $d = \pi/2$. Since it is fairly simple to classify the non-simply connected manifolds in this case if the universal covering is a standard symmetric space of rank 1 [SS], [S], the essential part of our result is contained in the following

THEOREM 1. *Let* M *be a complete Riemannian manifold with* $1 \leq K$ *and* $d = \pi/2$. *Then* M *is either homeomorphic to a sphere or the universal covering* $\tilde{M}$ *of* M *is isometric to a rank* 1 *symmetric space.*

Some partial results in low dimensions had been announced earlier. Berger's Rigidity Theorem can be derived from Theorem 1, though non-trivially, using that under the above hypotheses the injectivity radius and thus the diameter is at least $\pi/2$. Actually, a complete proof of this fact (which was also essential in Berger's original arguments) has been given only recently [CG]. We shall now describe the main steps in the proof of Theorem 1.

Convexity. We say that a subset $S \subset M$ is totally a-convex, $a > 0$, if for any $p_1, p_2 \in S$ and any normal geodesic $c : [0, \ell] \to M$ with $c(0) = p_1$, $c(\ell) = p_2$ and $\ell < a$ we have $c[0, \ell] \subset S$.

Now choose $p \in M$ so that

$$A = \{x \in M | d(x, p) = \pi/2\}$$

is non-empty, and let

$$A' = \{x \in M | d(x, A) = \pi/2\} .$$

From the generalized triangle comparison theorem of Toponogov [CE] it follows that

PROPOSITION 2. A *and* A′ *are totally* π*-convex.*

In particular, A and A′ are topological manifolds with possibly empty boundary whose interior is smooth and totally geodesic. Using the Synge argument one then gets from the second variation formula

PROPOSITION 2. $\dim A + \dim A' \leq \dim M - 1$.

Topological duality and non-rigidity. Consider now an arbitrary point $x \in M \setminus (A \cup A')$ and let c (resp. c') be any minimal geodesic from x to A (resp. A'). Then again from the triangle comparison theorem of Toponogov it follows that $\measuredangle(\dot{c}(0), \dot{c}'(0)) > \pi/2$. As in [GS] this leads to the construction of a non-singular smooth gradient field on the complement of suitable tubular ε-neighborhoods of A and A' whose integral curves are transversal to the boundaries of these neighborhoods. In particular, we get the following duality of A and A'.

PROPOSITION 4. A *(resp.* A'*) is a strong deformation retract of* $M \setminus A'$ *(resp.* $M \setminus A$ *).*

THEOREM 5. *If* A *and* A' *have non-empty boundaries then* M *is homeomorphic to a sphere.*

The latter fact follows since for sufficiently small ε, the ε-neighborhoods of A and A' are discs of class C^1. This in turn is a consequence of results on positively curved manifolds with non-trivial convex boundary and Proposition 2.

Riemannian fibrations of Euclidean spheres. As a consequence of Theorem 5 we may assume that say A has no boundary, i.e. A is a closed totally geodesic submanifold of M.

Now fix an arbitrary point $p' \in A'$ and a minimal normal geodesic $c: \left[0, \frac{\pi}{2}\right] \to M$ from p' to A. Using Toponogov's Triangle Theorem we conclude that the parallel translate of $-\dot{c}\left(\frac{\pi}{2}\right)$ along any geodesic in A defines a minimal geodesic from A to p'. This leads to the construction of a fiber bundle over A whose fiber is the orbit of $-\dot{c}\left(\frac{\pi}{2}\right)$ under the action of the normal holonomy group of A. From the constructions in connection with Proposition 4 and Theorem 5 it follows that this bundle is a subbundle of a bundle over A with total space homeomorphic to a sphere (the normal sphere to A' at p' if $\partial A' = \emptyset$, and the unit normal bundle to

A if $\partial A' \neq \emptyset$). An application of the homotopy lifting property for fibrations then provides the following important information about A, A′ and the geodesics from A′ to A.

PROPOSITION 6. *A′ is either a point or it has no boundary. Furthermore, any unit normal vector at A′ (resp. A) defines a minimal geodesic to A (resp. A′), and the maps thus defined from each unit normal sphere of A′ (resp. A) to A (resp. A′) are Riemannian submersions.*

The latter statement of the above proposition is yet another consequence of Toponogov's Triangle Comparison Theorem.

In [GG_1] we give a complete classification of Riemannian fibrations of Euclidean spheres, which was only known in the already locally fairly rigid case that all fibers are totally geodesic [E]. The general situation is considerably more difficult. From this classification and Proposition 6 above we obtain

PROPOSITION 7. *There are only the following possibilities for A and A′: A point, a simple closed geodesic, or a finite quotient of a rank 1 symmetric space with standard metric.*

Rigidity. First suppose M is not simply connected, and let $\tilde{M}$ be the universal covering of M. Obviously, $d(\tilde{M}) \geq \pi/2$, and as a consequence of all the previous propositions, the Maximal Diameter Theorem of Toponogov [T], and the main result of [GS] one gets

PROPOSITION 8. *If $d(\tilde{M}) > \pi/2$, then $d(\tilde{M}) = \pi$, and $\tilde{M}$ is isometric to a sphere of constant curvature 1.*

Now we assume M is simply connected and not homeomorphic to the sphere. If in addition $\dim M \geq 3$ ($\dim M = 2$ is simple), an argument combining Morse Theory and Proposition 4 yields

PROPOSITION 9. *A and A′ are simply connected.*

If A (resp. A′) is not a point it has to be isometric to a standard complex, quaternionic, or Cayley projective space (plane). By the definitions of A and A′ we may then assume that A′ is a point. Now Proposition 4 together with Proposition 6 imply that one has the same picture from every point in M. This leads to the conclusion that M is a symmetric space of rank 1.

The complete proof [GG_2] will appear elsewhere.

REFERENCES

[B] M. Berger, Les variétés riemanniennes 1/4-pincées, Ann. Scuola Norm. Sup. Pisa 14 (1960), 161-170.

[CE] J. Cheeger and D. Ebin, Comparison Theorems in Riemannian Geometry (1975), North Holland Publ.

[CG] J. Cheeger and D. Gromoll, On the lower bound for the injectivity radius of 1/4-pinched manifolds, J. Diff. Geom. 15 (1980), 437-442.

[E] R. Escobales, Riemannian submersions with totally geodesic fibers, J. Diff. Geom. 10 (1975), 253-276.

[GG_1] D. Gromoll and K. Grove, Riemannian fibrations of euclidean spheres.

[GG_2] ———, A generalization of Berger's Rigidity Theorem for positively curved manifolds.

[GS] K. Grove and K. Shiohama, A generalized Sphere Theorem, Ann. of Math. 106 (1977), 201-211.

[SS] T. Sakai and K. Shiohama, On the Structure of Positively Curved Manifolds with certain diameter, Math. Zeitschrift 127 (1972), 75-82.

[S] T. Sakai, On the diameter of some riemannian manifolds, Archiv der Mathematik 30 (1978), 427-434.

[T] V. Toponogov, Riemannian spaces having their curvature bounded below by a positive number, Amer. Math. Soc. Transl. 37 (1969), 291-336.

COMPLETE THREE DIMENSIONAL MANIFOLDS WITH POSITIVE RICCI CURVATURE AND SCALAR CURVATURE

Richard Schoen and Shing-Tung Yau*

Let M be a complete manifold with nonnegative Ricci curvature or nonnegative scalar curvature. Then it is a fundamental question in geometry to determine the topology of M. In this paper, we shall restrict our attention to noncompact manifolds.

Let us begin by surveying some results on Ricci curvature. Based on a beautiful theorem of Cheeger-Gromoll [11] on complete noncompact manifolds with nonnegative curvature, one is tempted to believe that any complete noncompact manifold with nonnegative Ricci curvature can be deformed to a finite simplicial complex.

While this question remains to be answered, we consider an analogue of a special case of the theorem of Cheeger-Gromoll which was first due to Gromoll-Meyer [12]. They assert that any complete noncompact manifold with positive curvature is diffeomorphic to the Euclidean space. The naive analogue of this statement turns out to be false. One can simply take the product of a parabola with the two-dimensional sphere. Less nontrivially, Nabonnand recently constructed examples of complete noncompact manifolds with positive Ricci curvature whose fundamental

*Research was partially supported by the National Science Foundation.

Seminar on Differential Geometry
0-691-08268-5/82/000209-20$01.00/0 (cloth)
0-691-08296-0/82/000209-20$01.00/0 (paperback)

group is infinite. (This example destroys the belief that the fundamental group of such manifolds should be finite.) One may explain the failure of the analogue of the Gromoll-Meyer theorem as follows. If we consider positivity of the curvature as a system of differential inequalities with the metric tensor as unknown, then we have an overdetermined system. However, if the curvature tensor is replaced by the Ricci tensor, then we have a determined system. (Recall that the study of the Ricci tensor is equivalent to the study of the Einstein field equation in the category of positive definite metrics.) Naturally we expect that an overdetermined system should impose more stringent conditions on the manifold. From a geometric point of view, the sectional curvature of a manifold measures the curvature of the manifold in every direction and hence controls the geodesic behavior very nicely. The Ricci curvature is obtained by averaging the sectional curvature. It cannot be used effectively to deal with geodesics and hence we gain less information about the topology of the manifold.

In this paper, we use the theory of minimal surfaces to prove that every complete noncompact three-dimensional manifold with positive Ricci curvature is diffeomorphic to Euclidean space. Therefore the exact analogue of the Gromoll-Meyer theorem holds for three-dimensional manifolds. We can also classify complete three-dimensional manifolds with nonnegative Ricci curvature. For simplicity of presentation, we postpone it for a later publication.

The essential feature of the method is quite similar to our previous paper [9]. It is based on the study of stable minimal surfaces. However, much stronger results on minimal surfaces are needed for the purposes of this paper. It turns out that since the appearance of our previous paper, new results on stable minimal surfaces were obtained by Fisher-Colbrie and Schoen [5] on the one hand and Meeks-Simon-Yau [7] on the other. The first one is needed to deal with the homotopic problems and the second one is needed to deal with diffeomorphic properties of the three-dimensional manifold. In fact, the second one was already used in [7] to prove that three-dimensional homotopy spheres which are not prime do not admit metrics with positive Ricci curvature. The generalization of these arguments to the present situation is nontrivial because of the complications caused by the behavior of the metric at infinity.

In the second part of the paper, we generalize the results of [9] to complete noncompact manifolds. We prove that if M is a complete three-dimensional manifold with positive scalar curvature, then $\pi_1(M)$ does not contain a subgroup which is isomorphic to the fundamental group of a compact surface with genus ≥ 1. Apparently, this is the first known topological constraint for complete noncompact manifolds with positive scalar curvature whose dimension is greater than two. The arguments here can be generalized to high dimension. One can prove, for example, that $T^3 \times R$ cannot admit a complete metric with positive scalar curvature. This is related to our previous work on the positive action conjecture in general relativity.

Finally, it should be mentioned that based on the works here, it is reasonable to conjecture that for a complete manifold with positive Ricci curvature, the fundamental group is an almost polycyclic group which contains no abelian subgroup with rank $\geq n-3$.

1. *Preliminary results*

We state the theorems we will use in the subsequent sections, and extend the results of [5] on the classification of complete stable minimal surfaces in three-dimensional manifolds of nonnegative scalar curvature. The first theorem follows from the splitting theorem of Cheeger and Gromoll [2].

THEOREM 1. *Let* M *be a complete manifold with nonnegative Ricci curvature. If* M *has more than one end at infinity, then* M *is isometric to a Riemannian product of the real line with a lower dimensional manifold.*

The second theorem we will use was done by Fischer-Colbrie and Schoen [5] except in one case which we prove below. The case when M is Euclidean was also proved by do Carmo-Peng [3]. The compact case was in Schoen-Yau [9].

THEOREM 2. *Let* M *be a three-dimensional manifold with nonnegative scalar curvature. Let* Σ *be a complete minimal surface in* M *whose area is stable with respect to compactly supported deformations. If* Σ *is compact, then* Σ *is either a sphere or a flat totally geodesic torus. If* Σ *is not compact, then* Σ *is conformally diffeomorphic to the complex plane or the punctured plane. If the latter case occurs then* Σ *is totally geodesic and the scalar curvature of* M *is zero along* Σ.

Moreover, if M *has nonnegative Ricci curvature, then* Σ *is totally geodesic, and the Ricci curvature of* M *normal to* Σ *vanishes at all points of* Σ.

The results of Theorem 2 are proved in [5, Theorem 3] with the exception of the assertion that if M has strictly positive scalar curvature than Σ cannot be a punctured plane. (This follows from [5, Theorem 3] only if Σ is assumed to have finite absolute total curvature.) We give here a proof of this result.

Assume on the contrary that Σ is a complete stable surface in M conformally equivalent to the punctured plane such that Σ is not totally geodesic or R does not vanish along Σ. The stability condition is equivalent to the assumption that the first eigenvalue of the linear operator

$$L = \Delta + \left(R - K + \frac{1}{2}|A|^2\right)$$

is positive on each compact domain of Σ (see [9]) where Δ is the Laplace operator for the induced metric on Σ, K is the Gauss curvature of Σ, and $|A|^2$ is the square length of the second fundamental form of Σ in M.

Applying [5, Theorem 1] we conclude that there is a positive function f on Σ satisfying $Lf = 0$. If ds^2 denotes the metric on Σ, we define a metric on $\tilde{M} = \Sigma \times S^1$ (see [5, Theorem 4]) by $\widetilde{ds}^2 = ds^2 + f^2 d\theta^2$, where θ is a coordinate on S^1 of period one. Clearly $\tilde{M}$ is complete, and by an easy calculation the scalar curvature $\tilde{R}$ of $\tilde{M}$ is

$$\tilde{R} = f^{-1}(Kf - \Delta f) = R + \frac{1}{2}|A|^2 \tag{1.1}$$

where we have used $Lf = 0$. Thus $\tilde{M}$ has strictly positive scalar curvature at some point. The reason that $\tilde{M}$ is useful is because the metric of $\tilde{M}$ is of a simple form. We next replace Σ by a new surface $\tilde{\Sigma}$ whose metric is also of a special form. Specifically, we note that $\tilde{M}$ has a circle group of isometries obtained by rotation about the S^1 factor. We seek a new stable surface $\tilde{\Sigma}$ in $\tilde{M}$ which is invariant under this group. To construct such a $\tilde{\Sigma}$, we consider for any curve γ in Σ the functional

$$I[\gamma] = \int_\gamma f\,dx$$

where x is an arclength parameter along γ. Observe that $I[\gamma]$ is the area of the two-dimensional surface $\gamma \times S^1 \subseteq \tilde{M}$ with metric $dx^2 + f^2\,d\theta^2$ obtained by taking the orbit of γ under the circle group. Since Σ is topologically $S^1 \times \mathbf{R}$, we can find a properly imbedded curve Γ in Σ which minimizes the functional $I[\cdot]$ on each compact subinterval of Γ. This curve Γ may be constructed by choosing two sequences of points tending to infinity in opposite directions on Γ, minimizing $I[\cdot]$ for each pair of points, and taking the limiting curve. We then let $\tilde{\Sigma} = \Gamma \times S^1$ with metric $dx^2 + f^2\,d\theta^2$ where x is an arclength parameter along Γ. It is not difficult to check from the minimizing property of Γ that $\tilde{\Sigma}$ is a stable, minimal surface in $\tilde{M}$. In particular, the operator

$$\tilde{L} = \tilde{\Delta} + \left(\tilde{R} - \tilde{K} + \frac{1}{2}|\tilde{A}|^2\right)$$

has positive first eigenvalue on each compact domain of $\tilde{\Sigma}$. We now prove the following lemma.

LEMMA 1. *Let* $(x, \theta) \in \mathbf{R} \times S^1$ *be coordinates with* $-\infty < x < \infty$, θ *of period one. Let* $\mu(x)$ *be a positive function on* $\mathbf{R}$, *and let* Σ *be* $\mathbf{R} \times S^1$ *endowed with the (complete) metric* $ds^2 = dx^2 + \mu^2(x)\,d\theta^2$. *If the first eigenvalue of the operator* $\Delta - K$ *(with respect to* ds^2*) is positive on each compact domain of* Σ, *then* μ *is a constant function and* ds^2 *is flat.*

Proof. We first note that by direct calculation we have $K = -\mu^{-1}\dfrac{d^2\mu}{dx^2}$. The variational inequality corresponding to $\Delta - K$ is

$$\int_\Sigma \frac{d^2\mu}{dx^2}\zeta^2\,dx\,d\theta \le \int_\Sigma \left[\left(\frac{d\zeta}{dx}\right)^2 + \mu^{-2}\left(\frac{d\zeta}{d\theta}\right)^2\right]\mu\,dx\,d\theta$$

for any function ζ with compact support on Σ. If we require ζ to be a function of x alone, then we get the following inequality

$$\int_{-\infty}^{\infty} \frac{d^2\mu}{dx^2}\zeta^2\,dx \le \int_{-\infty}^{\infty}\left(\frac{d\zeta}{dx}\right)^2\mu\,dx \tag{1.2}$$

where ζ is any function of compact support on $\mathbf{R}$. We will show that inequality (1.2) implies that μ is a constant function. We note first that by choosing $\zeta = \mu^{-1/2}\phi$ where ϕ has compact support we get

$$\int_{-\infty}^{\infty} \mu^{-1}\frac{d^2\mu}{dx^2}\phi^2\,dx \le 2\int_{-\infty}^{\infty}\left(\frac{d\zeta}{dx}\right)^2\mu\,dx$$

(1.3)

$$= \int_{-\infty}^{\infty}\left[2\left(\frac{d\phi}{dx}\right)^2 - 2\phi\,\mu^{-1}\frac{d\mu}{dx}\frac{d\phi}{dx} + \frac{1}{2}\phi^2\mu^{-2}\left(\frac{d\mu}{dx}\right)^2\right]dx$$

where we have multiplied the right-hand side of (1.2) by two, and have expanded by the product rule. The middle term on the right can be written

$$-\int_{-\infty}^{\infty}\frac{d(\phi^2)}{dx}\frac{d\mu}{dx}\mu^{-1}\,dx = \int_{-\infty}^{\infty}\phi^2\mu^{-1}\frac{d^2\mu}{dx^2}\,dx$$

$$-\int_{-\infty}^{\infty}\phi^2\mu^{-2}\left(\frac{d\mu}{dx}\right)^2dx$$

after integration by parts. Putting this into (1.3) and cancelling the second derivative terms we get

$$\int_{-\infty}^{\infty} \phi^2 \mu^{-2}\left(\frac{d\mu}{dx}\right)^2 dx \leq 4\int_{-\infty}^{\infty} \left(\frac{d\phi}{dx}\right)^2 dx$$

for any ϕ with compact support on $\mathbf{R}$. Choosing for any $\sigma > 1$, the function ϕ to be one on $[-\sigma, \sigma]$, zero outside $[-2\sigma, 2\sigma]$, and with $\left|\frac{d\phi}{dx}\right| \leq \sigma^{-1}$ on $\mathbf{R}$, we get

$$\int_{-\sigma}^{\sigma} \mu^{-2}\left(\frac{d\mu}{dx}\right)^2 dx \leq 16\sigma^{-1} .$$

Letting σ tend to infinity we conclude that μ is a constant function, and hence Σ is flat. This concludes the proof of Lemma 1.

We now continue the proof of Theorem 2. By Lemma 1, we have $\tilde{K} \equiv 0$ and hence $\tilde{\Sigma}$ is the flat cylinder. The variational inequality for $\tilde{L}$ now implies

$$\int_{\tilde{\Sigma}} \left(\tilde{R} + \frac{1}{2}|\tilde{A}|^2\right)\zeta^2 dv \leq \int_{\tilde{\Sigma}} |\nabla\zeta|^2 dv \tag{1.4}$$

for any function ζ with compact support on $\tilde{\Sigma}$ where dv is the volume element of $\tilde{\Sigma}$ and $\nabla\zeta$ is the metric gradient of ζ. Let ρ denote the geodesic distance to a given point of $\tilde{\Sigma}$, and choose ζ to be a function of ρ which is one for $\rho \leq \sigma$, vanishes for $\rho > 2\sigma$, and satisfies $|\nabla\zeta| \leq 2\sigma^{-1}$. If B_σ denotes the ball $\{\rho < \sigma\}$, then by (1.4)

$$\int_{B_\sigma} \left(\tilde{R} + \frac{1}{2}|\tilde{A}|^2\right) dv \leq 4\sigma^{-2} \int_{B_{2\sigma}} dv .$$

Since the area of $B_{2\sigma}$ is at most a constant times σ, we can let σ tend to infinity to show $\tilde{R} \equiv 0$ on $\tilde{\Sigma}$. This contradicts (1.1) and the assumption that either R or $|A|^2$ is strictly positive somewhere on Σ. Thus such Σ cannot exist, and hence no such Σ can exist satisfying our assumption. This completes the proof of Theorem 2.

2. *Complete manifolds with positive Ricci curvature*

We will use the results of Section 1 to prove that a complete noncompact three-dimensional manifold with positive Ricci curvature is diffeomorphic to $\mathbf{R}^3$.

LEMMA 2. *Let* M *be a three dimensional complete manifold with nonnegative Ricci curvature. Then* $\pi_2(M) = 0$ *unless* M *is isometrically covered by the product of the real line with a two-dimensional surface of nonnegative curvature.*

Proof. If $\pi_2(M) \neq 0$, then $\pi_2(\tilde{M}) \neq 0$ where $\tilde{M}$ is the universal cover of M (with the pullback metric). The sphere theorem [6] implies that there is an embedded S^2 in $\tilde{M}$ which is not homotopically trivial. If this S^2 does not divide $\tilde{M}$ into more than one component, we can find a closed Jordan curve in M which intersects the S^2 at only one point, contradicting the fact that $\pi_1(\tilde{M}) = 1$. Thus S^2 divides $\tilde{M}$ into two components. If one of these components is compact, then this component would be simply connected and hence homotopically a three-dimensional ball, contradicting the fact that S^2 is not homotopically trivial in $\tilde{M}$. Therefore, S^2 divides $\tilde{M}$ into two noncompact components, so we may apply Theorem 1 to assert that $\tilde{M}$ is isometric to the product of the real line with a nonnegatively curved surface. Since $\pi_2(\tilde{M}) \neq 0$, this surface must be diffeomorphic to S^2. Thus $\tilde{M}$ has a unique parallel vector field, and the group of isometries of $\tilde{M}$ must preserve this vector field. This completes the proof of Lemma 2.

LEMMA 3. *If* M *is a complete noncompact three-dimensional manifold with positive Ricci curvature, then* M *is contractible.*

Proof. Since $\pi_2(M) = 0$ and $\dim M = 3$, it follows that M is a $K(\pi, 1)$, i.e., all homotopy groups of the universal cover $\tilde{M}$ vanish. Since infinitely many cohomology groups of a finite cyclic group are nonzero, it is a well-known fact that a finite group cannot act freely on $\tilde{M}$ and hence $\pi_1(M)$ is torsion free.

By passing to a covering space of M, we may assume that M is orientable and $\pi_1(M) = \mathbb{Z}$. We may write M as an increasing union of compact subdomains M_i so that for each i, ∂M_i is a disjoint union of smooth closed two-dimensional surfaces. Let σ be a closed Jordan curve in M which represents the generator of $\pi_1(M)$. Then no nontrivial multiple of σ is homologous to zero in M. We may suppose without loss of generality that $\sigma \subseteq M_i$ for all i.

By Poincaré duality, we can find for each i a compact orientable surface Σ_i so that $\partial\Sigma_i \subseteq \partial M_i$ and the oriented intersection number of Σ_i with σ is nonzero. Perturbing the metric of M near ∂M_i so that ∂M_i has positive mean curvature with respect to the outward normal, we can minimize area (for the perturbed metric) among surfaces in M_i which are homologous to Σ_i and which have the same boundary $\partial\Sigma_i$. (See [4, Chapter 5] for the existence.) We denote the minimal surface also by Σ_i, and observe that Σ_i must intersect σ.

Since Σ_i minimizes area in homology, it follows by comparison that the area of Σ_i inside any compact domain Ω of M has a uniform bound independent of i. The compactness and regularity theorems for minimal surfaces then guarantee that a subsequence of $\Sigma_i \cap \Omega$ converges smoothly to a (possibly empty) limiting surface. Since $\Sigma_i \cap \sigma \neq \phi$ for each i, we see that a subsequence of Σ_i converges smoothly on compact subsets of M to a nontrivial properly embedded limiting surface Σ in M with $\Sigma \cap \sigma \neq \phi$. Since each Σ_i is area minimizing, the surface Σ is stable on each compact subset, and hence we have a contradiction to Theorem 2. Thus $\pi_1(M) = 1$, and M is contractible.

THEOREM 3. *Let* M *be a complete noncompact three-dimensional manifold with positive Ricci curvature. Then* M *is diffeomorphic to* $\mathbb{R}^3$.

Proof. By a theorem of Stallings [10], a contractible three-dimensional manifold is diffeomorphic to $\mathbf{R}^3$ if and only if it is irreducible and simply connected at infinity.

We first show that M is simply connected at infinity. Otherwise, we would have a compact set $K \subseteq M$ and a sequence of Jordan curves $\{\sigma_i\}$ tending uniformly to infinity with the property that any sequence of disks $\{D_i\}$ with $\partial D_i = \sigma_i$ has the property that $D_i \cap K \neq \phi$ for each i. Perturbing the metric near infinity as we indicated in the proof of Lemma 3, we can span σ_i by a disk D_i which minimizes area with respect to the perturbed metric and has $\partial D_i = \sigma_i$. Choose a point $x_i \in D_i \cap K$, and note that $\lim_{i\to\infty} \operatorname{dist}(x_i, \sigma_i) = \infty$ since K is a compact set. Since we are assuming that the Ricci curvature of M is everywhere positive, it has a positive lower bound near K, and by [8, Theorem 2] the maximum radius of a stable surface about a point of K can be estimated. This contradicts the fact that $\operatorname{dist}(x_i, \sigma_i)$ is arbitrarily large. This shows that M is simply connected at infinity.

It remains to prove that M is irreducible, i.e., any embedded two-sphere in M bounds a standard three-ball. The idea of this proof is quite simple, but some technical problems arise because of the nonsmoothness of the Buseman function. The idea is to use the results of Meeks-Simon-Yau [7] to assert the existence of an embedded minimal surface Σ which is a limit of spheres surrounding a given fake cell. Since M need not be homogeneously regular at infinity, one has to allow Σ to be noncompact. More importantly, to guarantee that such a Σ exists, we use the level sets of the Buseman function, which essentially have positive mean curvature, to show that the sequence of spheres does not go everywhere to infinity.

Let γ be a geodesic ray in M parametrized by arclength on $[0, \infty)$. Associated to γ we can define the Buseman function B_γ (see [2]) by

$$B_\gamma(x) = \lim_{t\to\infty} (t - d(x, \gamma(t)))$$

where $d(\cdot,\cdot)$ denotes geodesic distance in M. The function B_γ is in general a Lipschitz function whose gradient has length one at almost every point of M. If the Ricci curvature of M is nonnegative, then it can be shown that B_γ is subharmonic (see [2, Theorem 1]). We will need the following elementary property of B_γ.

LEMMA 4. *Suppose* M *is a complete* n*-dimensional manifold with strictly positive Ricci curvature. Suppose* Σ *is a smooth compact minimal hypersurface with boundary* $\partial\Sigma$. *Suppose* $a \in R$ *is such that* $\partial\Sigma \subseteq \{B_\gamma \leq a\}$. *Then it follows that* $(\Sigma \sim \partial\Sigma) \subseteq \{B_\gamma < a\}$.

Proof. If the conclusion were not true, we could find an interior point of Σ in the set $\{B_\gamma \geq a\}$, so Σ may be perturbed to a new hypersurface $\hat{\Sigma}$ satisfying for any preassigned number $\varepsilon > 0$

(i) the mean curvature $\hat{H}$ of $\hat{\Sigma}$ satisfies $|H| < \varepsilon$,

(ii) $\partial\hat{\Sigma} = \partial\Sigma$,

(iii) $\max\{B_\gamma(x) : x \in \hat{\Sigma}\} > a$.

It follows that there is a number $\hat{t} = \hat{t}(\varepsilon)$ such that for $t \geq \hat{t}$ there is $x_t \in \hat{\Sigma} \sim \partial\hat{\Sigma}$ satisfying

$$d(x_t, \gamma(t)) = \max_{x \in \hat{\Sigma}} d(x, \gamma(t)) < t - a .$$

It is clear that any minimizing geodesic γ_t from $\gamma(t)$ to x_t meets $\hat{\Sigma}$ orthogonally. Suppose γ_t is parametrized on $[0,\ell]$, $\ell = d(x_t, \gamma(t))$ with $\gamma_t(0) = \gamma(t)$, $\gamma_t(\ell) = x_t$. We now note that by perturbing $\hat{\Sigma}$ if necessary we may assume that x_t is not conjugate to $\gamma(t)$ along γ_t. For if x_t and $\gamma(t)$ are conjugate, we can define a new hypersurface by $x \mapsto \exp_x(\phi(x)\nu(x))$ for $x \in \hat{\Sigma}$ where $\nu(x)$ is a unit normal of $\hat{\Sigma}$ chosen so that $\nu(x_t) = -\dot{\gamma}_t(\ell)$, and ϕ is a nonnegative function vanishing near $\partial\hat{\Sigma}$ and achieving its maximum at x_t. The triangle inequality then implies that the point $\exp_{x_t}(\phi(x_t)\nu(x_t))$ is a nearest point on the perturbed hypersurface to $\gamma(t)$, and we may take as a minimizing geodesic from

$\gamma(t)$ to $\exp_{x_t}(\phi(x_t)\nu(x_t))$ a strict subarc of γ_t. These points then are not conjugate to each other. We denote this perturbed surface also by $\hat{\Sigma}$.

Since γ_t contains no pairs of conjugate points, we can invert the exponential map $\exp_{\gamma(t)}: M_{\gamma(t)} \to M$ locally near γ_t and define a smooth function $\hat{d}$ near x_t by $\hat{d}(x) = |\exp^{-1}_{\gamma(t)}(x)|$ where $|\cdot|$ is Euclidean length in the tangent space $M_{\gamma(t)}$. Clearly $\hat{d}$ has the property that $\hat{d}(x_t) = d(x_t, \gamma(t))$ and $\hat{d}(x) \geq d(x, \gamma(t))$ for x near x_t. Since the function $d(x, \gamma(t))$ has a local minimum at x_t, it follows that $\hat{d}(x)$ also has a local minimum at x_t. On the other hand, it is easy to estimate $\Delta\hat{d}(x_t)$ where Δ is the Laplace operator in the induced metric on $\hat{\Sigma}$. This can be done as follows. Let $e_1, \cdots, e_{n-1}$ be an orthonormal basis for $\hat{\Sigma}$ at x_t, and extend the e_i by parallel translation along radial geodesics of $\hat{\Sigma}$ to a neighborhood of x_t in $\hat{\Sigma}$. Thus $D_{e_i}e_j$ is normal to $\hat{\Sigma}$ at x_t, where D is differentiation in M. We then have

$$\Delta\hat{d}(x_t) = \sum_{i=1}^{n-1} e_i e_i \hat{d}(x_0) .$$

If $\sigma_i(s)$ denotes the radial geodesic in $\hat{\Sigma}$ tangent to e_i, we can consider a one-parameter family of geodesics from $\gamma(t)$ to $\sigma_i(s)$ whose lengths realize $\hat{d}(\sigma_i(s))$. If V_i denotes the resulting Jacobi field along $\gamma_t(\tau)$, $0 \leq \tau \leq \ell = d(x_t, \gamma(t))$, we have by the second variation formula (see e.g. [1, page 20])

$$\begin{aligned}\Delta\hat{d}(x_t) &= \sum_{i=1}^{n-1} \langle D_{e_i}e_i(x_t), \dot{\gamma}_t(\ell)\rangle + \sum_{i=1}^{n-1}\int_0^{\ell} (\|\dot{V}_i\|^2 - \langle R(V_i, \dot{\gamma}_t)\dot{\gamma}_t, V_i\rangle)\, d\tau \\ &\leq \hat{H}(x_0) + \frac{n-1}{\ell} - \int_0^{\ell} \frac{\tau^2}{\ell^2}\, \mathrm{Ric}\,(\dot{\gamma}_t(\tau), \dot{\gamma}_t(\tau))\, d\tau\end{aligned}$$

where we have used the definition of $\hat{H}$ and standard inequalities (see [2, Lemma 2]). Since we are assuming the Ricci curvature of M is strictly positive, there is a continuous positive function $k(x)$ on M so that $\mathrm{Ric}(v,v) \geq k(x)$ for any unit vector $v \in M_x$. Thus we have

$$\Delta\hat{d}(x_t) \leq \varepsilon + \frac{n-1}{d(x_t,\gamma(t))} - \int_0^{\ell} \frac{\tau^2}{\ell^2}\, k(\gamma_t(\tau))\, d\tau .$$

If ε is chosen sufficiently small and t is chosen sufficiently large we have $\Delta\hat{d}(x_t) < 0$ contradicting the fact that $\hat{d}$ has a local minimum at x_t. This finishes the proof of Lemma 4.

We use Lemma 4 to approximate the level set $S_a = \{x : B_\gamma(x) = a\}$ for any $a \in \mathbf{R}$ by smooth embedded hypersurfaces of positive mean curvature. We first give a piecewise smooth approximation. Let $\varepsilon > 0$ be any given number. By [4, 4.5.9(12)] almost every level set of B_γ defines a locally rectifiable $(n-1)$-current in M, so we choose $a' \in (a-\varepsilon/2, a]$ so that $S_{a'}$ represents such a current. Given a point $x \in S_{a'}$, and a ball $B_r(x)$ with r chosen so small that the ball is convex, we may take the $(n-2)$-current $\Gamma_{r,x} = \partial(S_{a'} \llcorner B_r(x))$ as a boundary, where "$\llcorner$" indicates restriction of a current. We can then solve the least area problem for all $(n-1)$-currents with support in $B_r(x)$ and with boundary $\Gamma_{r,x}$ to get an absolutely area minimizing rectifiable current $\Sigma_{r,x}$ (see [4, 5.1.6] for existence). For $n \leq 7$, the regularity theorem (see [4, 5.4] for an account) implies that the interior of $\Sigma_{r,x}$ denoted $\mathring{\Sigma}_{r,x} = (\text{support } \Sigma_{r,x}) \sim \Gamma_{r,x}$ is a smooth embedded oriented minimal hypersurface in $B_r(x)$. Now note that by Lemma 4 we have

(2.1) $$\mathring{\Sigma}_{r,x} \subseteq \{x : B_\gamma(x) < a'\} .$$

We let $\Omega_{r,x} \subseteq \{x : B_\gamma(x) < a'\}$ be the open set determined by $\Sigma_{r,x}$ satisfying

$$\partial\Omega_{r,x} = (S_{a'} \sim B_r(x)) \cup \Sigma_{r,x} .$$

We now choose a sequence of convex balls $\{B_{r_i}(x_i)\}$ with $r_i < \varepsilon/2$ which forms a locally finite covering of $S_{a'}$, and let $\Omega_i = \Omega_{r_i,x_i}$ be the open set defined as above. We then set $\Omega = \bigcap_i \Omega_i$. Since the covering was locally finite, Ω is an open set, and $\bar{\Omega} \subseteq \{B_\gamma < a'\}$ by (2.1). A point of $\partial\Omega$ must lie on a finite number of the hypersurfaces $\mathring{\Sigma}_{r_i,x_i}$, so that $\partial\Omega$ is piecewise smooth. It is clear that

$$\partial\Omega \subseteq \{a - \varepsilon < B_\gamma < a\} .$$

We now describe a method of smoothing $\partial\Omega$ so that the resulting hypersurface is smooth and has positive mean curvature with respect to the outer normal. We first note that $(\partial\Omega) \cap \mathring{\Sigma}_{r_i,x_i}$ is a compact subset of $\mathring{\Sigma}_{r_i,x_i}$ for any i. Since $\Sigma_i = \Sigma_{r_i,x_i}$ minimizes area, it follows that any compact subdomain of $\mathring{\Sigma}_i$ is strictly stable, so by the implicit function theorem it is possible to perturb Σ_i slightly so that the mean curvature of Σ_i is strictly positive in a neighborhood of $(\partial\Omega) \cap \mathring{\Sigma}_i$. Thus we may assume without loss of generality that the mean curvature of $\mathring{\Sigma}_i$ is strictly positive for each i. By transversality arguments, we may assume that all intersections $\mathring{\Sigma}_i \cap \mathring{\Sigma}_j$ are transverse. At a point of intersection of $(\partial\Omega) \cap \mathring{\Sigma}_i$ and $(\partial\Omega) \cap \mathring{\Sigma}_j$ the hypersurfaces make an angle of less than 180 degrees, so it is possible to round off the corner preserving the positivity of the mean curvature. In this way $\mathring{\Sigma}_i \cup \mathring{\Sigma}_j$ may be replaced by a single smooth hypersurface of positive mean curvature which can be perturbed to be transverse to the remaining faces. By induction, we can smooth $\partial\Omega$ preserving positive mean curvature as well as the conditions $\Omega \subseteq \{B_\gamma < a\}$ and $\partial\Omega \subseteq \{a - \varepsilon < B_\gamma < a\}$. We have established the following result.

LEMMA 5. *Let* M *be a complete manifold of positive Ricci curvature with* $n = \dim M \leq 7$. *For any* $a \in \mathbb{R}$, $\varepsilon > 0$, *there is a domain* $\Omega \subseteq \{B_\gamma < a\}$ *such that* $\partial\Omega$ *is a smooth properly embedded hypersurface of positive mean curvature with respect to the outward normal vector, and* $\partial\Omega \subseteq \{a - \varepsilon < B_\gamma < a\}$.

We are now ready to prove that a complete noncompact three-manifold M of positive Ricci curvature is irreducible. Suppose on the contrary that M contains a 2-sphere bounding a fake 3-ball. We want to find a stable minimal surface in M. Let M_i be a sequence of bounded smooth domains exhausting M. We assume that for each i, M_i contains the fake cell. Let a be a number large enough so that $\{B_\gamma < a-1\}$ contains the fake cell. By Lemma 5, let $\Omega \subseteq \{B_\gamma < a+1\}$ be a domain so that $\partial\Omega \subseteq \{a < B_\gamma < a+1\}$, and $\Sigma = \partial\Omega$ is smooth with positive mean curvature. Let $\rho(x) = \operatorname{dist}(x, \Sigma)$, and let $\Sigma_t \subseteq \Omega$ denote the parallel hypersurface $\Sigma_t = \{x \in \Omega : \rho(x) = t\}$ for $t > 0$. By perturbing M_i, we may assume that for some $\varepsilon_i > 0$, ∂M_i intersects Σ_t orthogonally for $t \leq \varepsilon_i$ and that $\Sigma_t \cap M_i$ is a smooth embedded hypersurface of positive mean curvature for $t \in [0, \varepsilon_i]$. Let $D_i = M_i \cap \Omega$, and note that D_i contains the fake cell. We now find a smooth function r_i on $\overline{M}_i$ and a number $\delta_i > 0$ with the following properties:

(i) $0 < r_i(x) \leq 1$ for each $x \in M_i$, $r_i(x) = 1$ if $\operatorname{dist}(x, \partial M_i) \geq 2\delta_i$, and $r_i(x) = 0$ for $x \in \partial M_i$.

(ii) $\frac{1}{2} \leq \|Dr_i(x)\| \leq 1$ for x with $\operatorname{dist}(x, \partial M_i) \leq \delta_i$.

(iii) $\langle Dr_i(x), D\rho(x)\rangle = 0$ for $x \in D_i$ with $\rho(x) \leq \varepsilon_i$.

Properties (i), (ii) insure that r_i behaves like the distance function to ∂M_i for points near ∂M_i. Property (iii) says that the level sets of r_i intersect the Σ_t orthogonally at points near Σ. To construct the function r_i, we introduce the following notation: for $x \in D_i$ with $\rho(x) \leq \varepsilon_i$, let x' denote the closest point of Σ to x, and we can express the point x as a pair $x = (x', \rho)$. Let $d(x)$ denote the distance function to ∂M_i, and for a point $x \in D_i$ with $\rho(x) \leq \varepsilon_i$, $d(x) \leq 2\delta_i$ we set

$$\overline{r}_i(x) = d(x) - \int_0^\rho \langle Dd(x', t), D\rho(x', t)\rangle dt$$

where the integration is taken over the shortest geodesic from Σ to x.

We then have $<D\bar{r}_i, D\rho> = 0$ by construction, and since ∂M_i is orthogonal to each Σ_t for $t \leq \varepsilon_i$, we also have $\bar{r}_i(x) = 0$ if $x \in \partial M_i$ and if δ_i is chosen sufficiently small we have property (ii) satisfied as well as the inequality $\bar{r}_i(x) > \frac{3}{2}\delta_i$ for $d(x) = 2\delta_i$, $\rho(x) \leq \varepsilon_i$. It is then a simple matter to extend $\bar{r}_i$ to the set $\{x \in M_i : d(x) \leq 2\delta_i\}$ so that $\bar{r}_i$ satisfies $0 < \bar{r}_i \leq 2\delta_i$, $\frac{1}{2} \leq \|D\bar{r}_i\| \leq 1$, and for $x \in \partial M_i$ we can take $\bar{r}_i(x) = 0$. We may also require that $\bar{r}_i(x) > \frac{3}{2}\delta_i$ for $x \in M_i$ with $d(x) = 2\delta_i$. Let $\zeta(s)$ be a smooth nondecreasing function of s satisfying

$$\zeta(s) = \begin{cases} s & \text{for} \quad s \leq \delta_i \\ 1 & \text{for} \quad s \geq \frac{3}{2}\delta_i \end{cases},$$

and define $r_i(x) = \zeta(\bar{r}_i(x))$ for $x \in M_i$ (with $r_i \equiv 1$ on $\{d > 2\zeta\}$). The function r_i then satisfies (i), (ii), (iii).

We will use the functions r_i, ρ to blow up the metric near ∂D_i: Let $k(s)$ be a smooth nonincreasing real valued function satisfying

$$k(s) = \begin{cases} \frac{1}{s} & \text{if} \quad 0 < s \leq \varepsilon_i \\ 1 & \text{if} \quad s \geq 2\varepsilon_i \end{cases}.$$

If ds^2 denotes the given metric on M, we define a new metric $\overline{ds}^2$ on D_i by

$$\overline{ds}^2 = r_i^{-2} k(\rho)^2 ds^2 .$$

It is easy to check that $\overline{ds}^2$ is a complete, homogeneously regular metric on D_i. Moreover, it is easy to calculate the mean curvature $\bar{H}_t$ of Σ_t with respect to $\overline{ds}^2$ as

$$\bar{H}_t = r_i k(\rho_i)^{-1}[H_t - 2<D\rho, D \log(r_i^{-1} k(\rho))>]$$

where H_t is the mean curvature of Σ_t in the metric ds^2. By condition (iii) we thus have

$$\bar{H}_t = r_i k(\rho)^{-1}[H_t - 2k'(\rho)] .$$

In particular, $\bar{H}_t > 0$ on Σ_t for $t \in (0, \varepsilon_i]$.

By the theorem of Meeks-Simon-Yau [7], there is an embedded stable minimal two-sphere S_i in D_i which bounds a fake three-cell. Since the level sets of ρ have positive mean curvature $\bar{H}_t$, the sphere S_i must lie entirely in the set $\{x \in D_i : \rho(x) > \varepsilon_i\}$, for otherwise it would touch one of the Σ_t tangentially from the inside, a situation which is impossible. In particular, the surface $S_i \cap \{x \in D_i : d(x) > 2\delta_i\}$ is a stable minimal surface with respect to the original metric ds^2.

We now assert that such an S_i cannot exist for i sufficiently large. To see this, let K be a compact, connected domain containing the fake cell and such that $K \cap \{x : B_\gamma(x) > a+1\} \neq \emptyset$. Then $K \cup \{B_\gamma > a+1\}$ is a connected unbounded region. It follows that for each i, $S_i \cap K \neq \emptyset$. Since the Ricci curvature of M has a positive lower bound on K, by [8, Theorem 2] any stable surface passing through a point of K must have a boundary point inside a fixed bounded set. But since $\{M_i\}$ exhausts M, we have ∂M_i tending to infinity. This contradiction shows that M is irreducible. This finishes the proof of Theorem 3.

3. *Complete three-manifolds with positive scalar curvature*

In this section we generalize our previous work on positive scalar curvature to include complete noncompact manifolds. We have the following theorem.

THEOREM 4. *Let* M *be a three-dimensional manifold. Suppose* $\pi_1(M)$ *contains a subgroup which is isomorphic to the fundamental group of a compact surface* Σ *of positive genus. Then* M *cannot carry a complete metric of positive scalar curvature.*

Proof. Suppose on the contrary that M has such a metric. Then by taking a covering of M we may assume that both M and Σ are orientable and $\pi_1(M) = \pi_1(\Sigma)$. We assert that M has at least two ends M_1 and M_2, and there is a sequence of compact surfaces Σ_i in M so that $\partial\Sigma_i$ is a disjoint union of two collections of Jordan curves C_i^1 and C_i^2 with $C_i^1 \subseteq M_1$, $C_i^2 \subseteq M_2$. Furthermore, we require that both $[C_i^1]$ and $[C_i^2]$ represent nonzero elements of $H_1(M)$, and that both C_i^1 and C_i^2 tend uniformly to infinity as i goes to infinity.

To prove these assertions, we proceed as follows. Let Q be a compact subdomain of M with smooth boundary so that the inclusion map induces an isomorphism of $\pi_1(Q)$ with $\pi_1(M)$. (Such a Q exists by [6, Theorem 8.6] which is a result of G. P. Scott.) Let $\mathcal{P}(Q)$ be the unique compact three-dimensional manifold such that $\mathcal{P}(Q)$ contains no fake three-cells and $\hat{Q} = \mathcal{P}(Q) \# F$ where $\hat{Q}$ is obtained from Q by capping off the two spheres in ∂Q with three balls, and F is a family of homotopy three spheres. By Theorem 10.6 of [6], $\mathcal{P}(Q)$ is an I-bundle over a closed surface with fundamental group $\pi_1(\Sigma)$. Since both M and Σ are orientable, $\mathcal{P}(Q)$ is homeomorphic to $\Sigma \times I$. Therefore, Q is homeomorphic to a manifold obtained from $\Sigma \times I$ by removing some three-cells, and taking connected sum with some homotopy spheres. In particular, ∂Q is the union of 2-spheres together with $\Sigma \times \{0\}$ and $\Sigma \times \{1\}$. Now $\Sigma \times \{0\}$ and $\Sigma \times \{1\}$ lie on the boundary of components M_1 and M_2 of $M \sim Q$ respectively. We claim that $M_1 \neq M_2$. Otherwise we could find a closed curve σ in M which intersects $\Sigma \times \{0\}$ transversally at one point. This curve σ represents a nontrivial element of $H_1(M, Z)$, and Σ represents a nontrivial element of $H_2(M, Z)$ whose intersection number with $[\sigma]$ is not zero. Since all elements of $H_1(\Sigma, Z) \cong H_1(M, Z)$ can be moved away from $\Sigma \times \{0\}$, they have zero intersection number with $[\Sigma]$. This contradiction shows that $M_1 \neq M_2$. The same argument shows that $\partial M_1 = \Sigma \times \{0\}$ and $\partial M_2 = \Sigma \times \{1\}$.

Next observe that both M_1 and M_2 are noncompact because their boundaries are surfaces of positive genus. (See [6, Lemma 6.8] which shows that if M_i were compact some nonzero element of $H_1(\partial M_i)$ would bound in M_i.)

For $j = 1, 2$ we can write M_j as an increasing union of compact smooth domains Ω_i^j so that for each i, $\partial\Omega_i^1 \supseteq \Sigma \times \{0\}$ and $\partial\Omega_i^2 \supseteq \Sigma \times \{1\}$. Let σ be a closed curve in $\Sigma \times \{0\}$ which represents a nonzero element in $H_1(\Sigma, Z)$. Then $[\sigma] \neq 0$ in $H_1(\Omega_i^1 \cup Q \cup \Omega_i^2)$ for each i, so σ gives rise to smooth map $f: \Omega_i^1 \cup Q \cup \Omega_i^2 \to S^1$ so that $f_*[\sigma] = [S^1]$. Choose a point $y \in S^1$ so that a component of $f^{-1}(y)$ is a properly embedded surface Σ_i whose algebraic intersection number with σ is one. Hence $\Sigma_i \cap (\Sigma \times \{0\})$ is a nonzero element of $H_1(M, Z)$, and hence both $\partial\Sigma_i \cap ((\partial\Omega_i^1) \sim (\Sigma \times \{0\}))$ and $\partial\Sigma_i \cap ((\partial\Omega_i^2) \sim (\Sigma \times \{1\}))$ are nonzero in $H_1(M)$. These Σ_i satisfy the properties which we required.

We now choose a smooth exhaustion of M by domains U_i with the property that $\Sigma_i \subseteq U_i$ for each i. Perturbing the metric of M on the region $U_{i+1} \sim \bar{U}_i$ so that ∂U_{i+1} has positive mean curvature with respect to the outward normal, we can minimize area in the homology class of Σ_i relative to $\partial\Sigma_i$. The minimizing surface we get has a connected component S_i so that both $(\partial S_i) \cap M_1$ and $(\partial S_i) \cap M_2$ are nontrivial in $H_1(M)$. Then it follows that $S_i \cap Q \neq \emptyset$ for each i, and hence we can prove that the S_i converge smoothly on compact subsets of M to a properly embedded stable limiting surface S.

We assert that some connected component of S has at least two ends, one in M_1 and the other in M_2. In fact, let σ_i be a shortest geodesic on S_i joining $(\partial S_i) \cap M_1$ to $(\partial S_i) \cap M_2$. Then a subsequence of the σ_i converges to a finite union of geodesic lines in S. Clearly one of these lines passes from infinity in M_1 to infinity in M_2. The component of S containing this line then has at least two ends, and is not homeomorphic to R^2. Since the scalar curvature of M is assumed to be strictly positive, this contradicts Theorem 2.

DEPARTMENT OF MATHEMATICS
UNIVERSITY OF CALIFORNIA, BERKELEY
BERKELEY, CALIFORNIA 94720

THE INSTITUTE FOR ADVANCED STUDY
PRINCETON UNIVERSITY
PRINCETON, NEW JERSEY 08540

DEPARTMENT OF MATHEMATICS
UNIVERSITY OF CALIFORNIA, SAN DIEGO
LA JOLLA, CALIFORNIA 92093

REFERENCES

[1] J. Cheeger, D. Ebin, *Comparison Theorems in Riemannian Geometry*, North Holland, 1975.

[2] J. Cheeger, D. Gromoll, The splitting theorem for manifolds of non-negative Ricci curvature, J. Diff. Geom. *6* (1) 1971.

[3] M. do Carmo, C. K. Peng, Stable complete minimal surfaces in R^3 are planes, Bull. AMS *1* (1979), 903-905.

[4] H. Federer, *Geometric Measure Theory*, Springer-Verlag, 1969.

[5] D. Fischer-Colbrie, R. Schoen, The structure of complete stable minimal surfaces in 3-manifolds of nonnegative scalar curvature, CPAM *33* (1980), 199-211.

[6] J. Hempel, *Three-manifolds*, Ann. Math. Studies, Princeton Univ. Press, *86* (1976).

[7] W. Meeks, L. Simon, S. T. Yau, to appear.

[8] R. Schoen, Estimates for stable minimal surfaces in three dimensional manifolds, to appear.

[9] R. Schoen, S. T. Yau, Existence of incompressible minimal surfaces and the topology of three dimensional manifolds with nonnegative scalar curvature, Ann. of Math. *110* (1979), 127-142.

[10] J. Stallings, *Group Theory and Three Dimensional Manifolds*, Yale Univ. Press, 1971.

[11] J. Cheeger, D. Gromoll, On the structure of complete manifolds of nonnegative curvature, Ann. of Math. *96* (3), 1972, 413-443.

[12] D. Gromoll, W. Meyer, On Complete open manifolds of positive curvature, Ann. of Math. *90* (1), 1969.

ENTIRE SPACELIKE HYPERSURFACES OF CONSTANT MEAN CURVATURE IN MINKOWSKI SPACE

Andrejs E. Treibergs

Complete spacelike hypersurfaces of constant mean curvature are becoming increasingly interesting in general relativity. Serving as initial surfaces for the Cauchy problem for the field equations, their understanding might be useful in the study of the propogation of gravitational waves and the formation of singularities. We give a classification scheme for these surfaces in the prototypical spacetime, Minkowski space. We can refine the method to construct constant mean curvature hypersurfaces asymptotic to arbitrary C^2 perturbations of the light cone in Minkowski space of any dimension.

The mathematical question treated here is analogous to the uniqueness and existence questions for hypersurfaces of constant mean curvature in Euclidean space. In contrast to the Euclidean case, where the existence of an entire zero mean curvature graph implies that the graph is a plane only for dimensions $n < 8$, [see, e.g. 1], in the Minkowski case, the conclusion is true for all dimensions [2]. However, unlike the Euclidean case where a maximally defined graph is unique up to translation [3], in the Minkowski case, for the entire constant mean curvature hypersurfaces, there is still some indeterminacy of solutions beyond translation and isometry. We determine the extent of this indeterminacy. Details of the proofs sketched here may be found in [4, 10].

Seminar on Differential Geometry
0-691-08268-5/82/000229-10$00.50/0 (cloth)
0-691-08296-0/82/000229-10$00.50/0 (paperback)

Minkowski space, M^{n+1}, is R^{n+1} endowed with the Lorentzian metric $ds^2 = \sum_1^n dx_i^2 - dx_{n+1}^2$. Let N^n be a hypersurface such that ds^2 restricts to a positive definite metric on N^n. Such N are called *spacelike* and can be represented as graphs of functions $x_{n+1} = z(x)$, $x \in R^n$, with gradient $|Dz| < 1$. The entire hypersurfaces of constant mean curvature are solutions of

$$(1) \quad Lu = (1-|Du|^2) \sum_{i=1}^n D_{ii}u + \sum_{i,j=1}^n D_i u D_j u D_{ij} u - nH(1-|Du|^2)^{3/2} = 0$$

$$|Du| < 1 \quad \text{on} \quad R^n .$$

Calabi [5] proposed the investigation of Bernstein properties of the solutions of (1) with zero H. He showed that all entire zero mean curvature solutions of (1) must be planes for $n \leq 4$. Cheng and Yau [2] showed that this was true for all n. Choquet-Bruhat [6] showed that constant mean curvature H solutions which are within $o(|x|^{-1})$ of the hyperboloid of mean curvature H must be the hyperboloid. Goddard [7] found that constant mean curvature infinitesimal perturbations of the three-dimensional hyperboloid could only be translations and conjectured that the only entire hypersurfaces of constant mean curvature are hyperboloids. We construct many counterexamples in the process of classifying these hypersurfaces.

The prototypical examples of spacelike constant mean curvature H hypersurfaces are the hyperboloids. For $k = 1 \cdots n$ these are given by

$$z(x) = \left(k^2 n^{-2} H^{-2} + \sum_{i=1}^k x_i^2\right)^{1/2} .$$

Another example is given by the semitrough [8]:

$$R^+ \times R \ni (r,t) \mapsto (2r - \coth r, \coth r \sinh t, \coth r \cosh t) .$$

This is an example of a surface of rotation about the x_1 axis. Another class of examples was recently constructed by Stumbles [9]. Although her results apply to more general spacetimes, for M^4 they say that for sufficiently small $d > 0$ and arbitrary $f \in C^5(S^2)$ such that $\|f\|_{C^5} < d$, one can construct an entire spacelike hypersurface of constant mean curvature, u, such that $u \to |x| + f(x/|x|)$ as $|x| \to \infty$. We shall show that these examples are representative of the general situation.

We are motivated by the observation that entire spacelike hypersurfaces of constant mean curvature $H > 0$ are convex. Thus we define the *projective boundary values* of a convex spacelike function f to be the positively homogeneous function of degree one, V_f, gotten by blowing down:

$$V_f(x) = \lim_{r\to\infty} \frac{f(rx)}{r} .$$

We call V_f the *boundary cone* for short.

The classification we obtain is the identification of the set of boundary cones associated to entire constant mean curvature spacelike hypersurfaces. It is the purpose of Theorems 1 and 2 to show that this set coincides with the set Q of positively homogeneous of degree one convex functions with gradient of length one. Theorem 3 extends Stumbles' result without the restriction on the smallness of d or to dimension three: Among solutions with projective boundary $|x|$, there exist those asymptotic to arbitrary $C^2(S^{n-1})$ perturbations of $|x|$.

THEOREM 1. *Let* u *be an entire spacelike hypersurface of positive constant mean curvature. Then* $V_u \in Q$.

Proof. Our argument uses the maximum principle. V_u is convex since u is, and so differentiable at almost every point. Suppose for contradiction that at some point $y \neq 0$,

$$|DV_u(y)| < c < 1 .$$

Hence there exists $d > 0$ so that for every x such that $|x-y| = d$,

$$V_u(y) + cd > V_u(x) .$$

For R sufficiently large,

$$V_u(y) + cd \geq \frac{u(Rx)}{R} .$$

Thus by the comparison principle for equations of mean curvature RH we may relate two solutions within the circle $|x-y| \leq d$ by

$$V_u(y) + cd + (R^{-2}H^{-2} + |x-y|^2)^{1/2} - (R^{-2}H^{-2} + d^2)^{1/2} \geq \frac{u(Rx)}{R} .$$

In particular this holds at $x = y$. By letting $R \to \infty$ we obtain

$$V_u(y) - (1-c)d \geq V_u(y)$$

which is a contradiction.

THEOREM 2. *Let* W *be a positively homogeneous convex function whose gradient has length one wherever defined. Then there exists an entire spacelike hypersurface of positive constant mean curvature* u *such that* $V_u = W$.

Sketch of proof. The proof breaks down into five steps.

Step 1. Without loss of generality we may assume that $W(x) > 0$ for $x \neq 0$. We observe that solutions and boundary cones are invariant under the ambient isometries of M^{n+1}. A suitable isometry may be found that moves the cone so that it is everywhere nonnegative. This results in W which is either positive away from zero or which is the product of a subspace and a lower dimensional cone W', positive away from the origin. Now the product of a lower dimensional solution u' corresponding to W' and the subspace yields the desired solution.

Step 2. Assuming that W is positive away from zero we construct global sub- and supersolutions z_1 and z_2 that guarantee the right boundary values of a sandwiched solution. The structure of Q is crucial for this step. In fact we may give formulas for these barriers by exploiting the availability of explicit solutions. The subsolution is defined to be

$$z_1(x) = \sup\{z_{a,b}(x) : a = DW(p) \neq b = DW(q) \text{ some } p,q \in R^n\}$$

where $z_{a,b}$ is the trough solution obtained by taking an isometric image of $t(x) = (n^{-2}H^{-2} + x_1{}^2)^{1/2}$ so that $DVz_{a,b}(R^n) = \{a,b\}$. The supersolution is defined to be

$$z_2(x) = \inf\{W(y) + (H^{-2} + |x-y|^2)^{1/2} : y \in R^n\}.$$

One checks that these blow down correctly. In fact,

$$W(x) < z_1(x) < z_2(x) < W(x) + \text{constant}.$$

Step 3. An approximating sequence of solutions $\{u_m\}$ is constructed by solving the Dirichlet problem for equation (1) in an increasing sequence of domains $\{G_m\}$. Let G_m be a convex $C^{2,\alpha}$ domain closely approximating $z_1((-\infty, m])$ from within. Then we can solve the following Dirichlet problem for the u_m; L is defined by equation (1):

$$\text{(DP)} \qquad \begin{aligned} L\,u_m &= 0 \quad \text{in} \quad G_m \\ u_m &= m \quad \text{on} \quad \partial G_m . \end{aligned}$$

The key to solving (DP) is the establishment of an apriori global gradient bound away from one. This follows from the fact that the gradient of spacelike solutions of $Lu = 0$ satisfies the maximum principle and that at the boundary there exist sharp barriers. Indeed, for any $y \in \partial G_m$, by rotating and translating we may arrange that $y = (R, 0, \cdots, 0)$ and that G_m lies in the slab $|x_1| \leq R$ where $2R$ is the diameter of G_m. Then for $x \in G_m$, u_m lies between the sub- and supersolutions

$$m + (n^{-2}H^{-2} + x_1{}^2)^{1/2} - (n^{-2}H^{-2} + R^2)^{1/2} \leq u_m(x) \leq m.$$

Hence we may estimate the gradient by the normal derivatives

$$\text{(2)} \qquad \sup_{G_m} |Du_m| \leq R(n^{-2}H^{-2} + R^2)^{-1/2}.$$

Now we can conclude that there exists $u_m \in C^{2,\alpha}(\overline{G_m})$ solving (DP) from Schauder Theory by taking proper account of (2).

Step 4. We wish to show that on a compact subset $I \subset R^n$ a subsequence of the $\{u_m\}$ can be extracted that converges to a solution of (1). This will follow from bounds on the third derivatives which hold for each I and are independent of M. These in turn follow from curvature estimates and a size assumption that is verified in step 5. We describe the curvature estimates now.

Let $\{e_1,\cdots,e_{n+1}\}$ be a local orthonormal frame in TM so that $e_1,\cdots,e_n$ are tangent to N and e_{n+1} a timelike normal. Let the dual frame of one-forms be denoted w^s; we have $w^s(e_t) = \delta_t^s$. The connection one-forms $w_s^{\ t}$ are defined by the equations

$$\begin{aligned} dw^k &= \sum w^j \wedge w_j^{\ k} - w^{n+1} \wedge w_{n+1}^{\ \ k}, \qquad k = 1,\cdots,n, \\ dw^{n+1} &= \sum w^j \wedge w_j^{\ n+1}, \\ 0 &= w_s^{\ t} + w_t^{\ s}. \end{aligned}$$

On N, $w^{n+1} = 0$, so by exterior differentiating,

$$0 = dw^{n+1} = \sum w^j \wedge w_j^{\ n+1}.$$

By Cartan's lemma there exist functions $h_{ij} = h_{ji}$ such that

$$w_k^{\ n+1} = \sum h_{kj} w^j.$$

The second fundamental form is $\sum h_{ij} w^i \otimes w^j$. Covariant differentiation of the second fundamental form is given by

$$\sum h_{ijk} w^k = dh_{ij} - \sum h_{ik} w_j^{\ k} - \sum h_{kj} w_i^{\ k}.$$

The mean curvature, the length of the second fundamental form, and the length of the covariant derivatives of the second fundamental form are defined by

$$nH = \sum {}_i h_{ii}$$
$$II = \sum {}_{i,j} h_{ij}{}^2$$
$$III = \sum {}_{i,j,k} h_{ijk}{}^2 .$$

We have the following curvature estimates:

PROPOSITION 1 [2, Theorem 2]. *Suppose* N *is a spacelike hypersurface of* M^{n+1} *such that for some point* $p \in N$ *the geodesic ball of radius* a *and center* p *is compact. Then if the mean curvature* H *of* N *is constant,*

$$II(x) \leq \frac{n^2H^2a^4 + ca^3(1+H^2)}{(a^2-r^2)^2}$$

where c *is a constant depending only on* n, *and* r *is the intrinsic distance function from* p.

PROPOSITION 2 [4]. *If in addition to the hypotheses of Proposition 1 we assume that there is a constant* B *such that* $II(x) \leq B$ *for all* x *in the geodesic ball of radius* a *about* p, *then*

$$III^{\frac{1}{2}}(x) \leq \frac{(c_1B + c_2H^2)a^4 + c_3a^3(1+H^2)}{(a^2-r^2)^2}$$

where the constants c_1, c_2, *and* c_3 *depend only on* n. *Moreover there is a constant* $k(n, H, a, B) < 1$ *so that if* N *is the graph of* u,

(3) $$\sup_{r \leq a} |Du| \leq k .$$

In view of (3) applied to the approximating solutions, it is possible to estimate the second and third derivatives of u_m in terms of the bounds on II and III. Thus we obtain

$$\|u_m\|_{C^3(I)} \leq c(n, H, r_1, r_2)$$

provided that the following size condition holds independently of m.

Size condition. There exist $0 < r_1 < r_2$ such that for m sufficiently large the following holds:

$$r_m(x) < r_1 \quad \text{for all} \quad x \in I \tag{4}$$

$$r_m(x) > r_2 \quad \text{for all} \quad x \in \partial G_m \tag{5}$$

where $r_m(x)$ is the intrinsic distance from $u_m(0)$ to $u_m(x)$ in u_m.

Step 5. The size condition is verified for the approximating sequence. (4) follows from the fact that $r_m(x) \le |x|$. Since we may approximate z_1 arbitrarily closely by G_m we may suppose that the G_m are given by a spacelike defining function $z_b(x)$ that satisfies

$$\begin{aligned} G_m(x) &= z_b^{-1}((-\infty, m)) , \\ z_1(x) \le z_b(x) &< z_2(x) , \\ z_b(0) &= z_1(0) . \end{aligned}$$

Define the functions q_m and q by

$$\begin{aligned} q(x) &= |x|^2 - (z_b(x) - z_b(0))^2 , \\ q_m(x) &= |x|^2 - (u_m(x) - u_m(0))^2 . \end{aligned}$$

From [2, Proposition 1] we know that q and q_m are proper on the spacelike surfaces z_b and u_m. Thus it follows that

$$J_m = \inf_{x \in \partial G_m} q_m(x) \ge \inf_{x \in \partial G_m} q_b(x) \to \infty \quad \text{as} \quad m \to \infty .$$

Moreover, $q_m(x)$ is related to $r_m(x)$. By applying [2, equation 3.24] we derive

$$\inf_{q_m = \frac{1}{2} J_m} r_m(x) \ge c(n, H) \log\left(\tfrac{1}{2} J_m + 1\right) \to \infty \quad \text{as} \quad m \to \infty .$$

Hence (5) is verified and the proof completed.

THEOREM 3 [10]. *Suppose* $f \in C^2(S^{n-1})$. *Then there exists an entire spacelike constant mean curvature hypersurface,* u, *such that*

$$u(x) \to |x| + f(x/|x|) \quad \textit{as} \quad x \to \infty .$$

Proof. All steps from the previous proof work here except the second where a refinement is needed in constructing barriers which take on the asymptotic boundary values. Extend f to $R^n - \{0\}$ by $f(x) = f(x/|x|)$. Since $f \in C^2$ we may find a constant M so that

$$|f(x) - f(y) - Df(y)(x-y)| \leq 2M|x-y|^2 = -2My \cdot (x-y)$$

for $x, y \in S^{n-1}$. Defining $p_i(y) = Df(y) + 2(-1)^i My$ for $i = 1, 2$ gives these linear bounds on f:

$$p_1(y) \cdot (x-y) \leq f(x) - f(y) \leq p_2(y) \cdot (x-y) .$$

Sub- and supersolutions given by

$$z_1(x) = \sup\{f(y) - p_1(y) \cdot y + z(x + p_1(y)) : y \in S^{n-1}\}$$

$$z_2(x) = \inf\{f(y) - p_2(y) \cdot y + z(x + p_2(y)) : y \in S^{n-1}\}$$

where $z(x) = (H^{-2} + |x|^2)^{1/2}$ satisfy

$$z_1(x) < z_2(x)$$

for $x \in R^n$ and

$$\lim_{r \to \infty} (z_i(ry) - r) = f(y)$$

for $y \in S^{n-1}$. Hence z_1 and z_2 may serve as the improved barriers. This completes the proof.

DEPARTMENT OF MATHEMATICS
UNIVERSITY OF CALIFORNIA, BERKELEY
BERKELEY, CALIFORNIA 94720

REFERENCES

[1] Bombieri, E., Theory of Minimal Surfaces and a Counter-Example to the Bernstein Conjecture in High Dimensions, Courant Institute Lecture Notes, Spring 1970.

[2] Cheng, S. Y. and S. T. Yau, Maximal Spacelike Hypersurfaces in the Lorentz-Minkowski spaces, Annals of Math., 104 (1976), 407-419.

[3] Giusti, E., On the equation of surfaces of prescribed mean curvature. Existence and uniqueness without boundary cond., Inv. Mat. 46 (1978), 2.111-13.

[4] Treibergs, A., Entire spacelike hypersurfaces of constant mean curvature in Minkowski space, Stanford Thesis 1980.

[5] Calabi, E., Examples of Bernstein problems for some nonlinear equations, Proc. Symp. Pure & Appl. Math. 15, 1968.

[6] Choquet-Bruhat, Y., Maximal submanifolds and submanifolds of constant extrinsic curvature, Ann. Scou. Norm. Pisa 3 (1976), 361-376.

[7] Goddard, A., Some remarks on the existence of spacelike hypersurfaces of constant mean curvature, Math. Proc. Camb. Phil. Soc., 82 (1977), 489-495.

[8] Calabi, E., private communication 1979.

[9] Stumbles, S., Hypersurfaces of constant mean extrinsic curvature, preprint.

[10] Treibergs, A., Entire spacelike hypersurfaces of constant mean curvature in Minkowski space, preprint.

APPLICATIONS OF THE MONGE-AMPERE OPERATORS TO THE DIRICHLET PROBLEM FOR QUASILINEAR ELLIPTIC EQUATIONS

I. Bakelman

This lecture is devoted to the applications of the Monge-Ampere elliptic operators to the Dirichlet problem for quasilinear elliptic equations. The Monge-Ampere elliptic operators considered below represent in their integral form the so-called R-curvature of a convex hypersurface. The definition and the main properties of the R-curvature are given in §1. In the same paragraph there are given the estimates of the height of convex hypersurfaces depending on the properties of their R-curvature in connection with the Dirichlet problem. In §2 on the basis of these estimates there are given two theorems of upper and lower estimates of the solutions of the Dirichlet problem for general quasilinear elliptic equations.

In §3 I consider briefly the Dirichlet problem for hypersurfaces with the given mean curvature in (n+1)-dimensional Euclidean space.

In §§1 and 2 both the results of the papers ([1]-[4], [6]) and new results of the author obtained in spring 1980 are included. It is being published for the first time.

The Dirichlet problem for hypersurfaces with the given mean curvature in Euclidean, Lobachevskii spaces was solved in the papers of I. Bakelman [4], [5], [6] and J. Serrin [7], [8].

I wish to express my gratitude to S. T. Yau and J. Serrin for useful discussions.

Seminar on Differential Geometry
0-691-08268-5/82/000239-20$01.00/0 (cloth)
0-691-08296-0/82/000239-20$01.00/0 (paperback)

§1. *The estimates of convex functions in the terms of their* R*-curvature*

1.1. R*-curvature of convex functions*

Let $x_1, x_2, \cdots, x_n$, $x_{n+1} = z$ be Cartesian coordinates in $(n+1)$-dimensional Euclidean space E^{n+1}. E^n is the hyperplane $x_{n+1} = 0$ in E^{n+1} and G is an open bounded domain in E^n. $x = (x_1, x_2, \cdots, x_n)$ is a point of E^n, $(x, z) = (x_1, x_2, \cdots, x_n, z)$ is a point of E^{n+1}; S_z is the graph of the function $z : G \to R$; $W^+(G)$ is the set of all convex functions in G, $W^-(G)$ is the set of all concave functions in G and $W(G) = W^+(G) \cup W^-(G)$. If $z(x) \in W(G)$ then S_z is called a convex (concave) hypersurface.

We fix the arbitrary function $z(x) \in W(G)$; let α be an arbitrary supporting hyperplane to S_z and $p_0 = (p_1^0, p_2^0, \cdots, p_n^0)$ is the system of the angle coefficients of α. We consider the n-dimensional Euclidean space $\tilde{R}^n = \{(p_1, p_2, \cdots, p_n)\}$. The point $p_0 = (p_1^0, p_2^0, \cdots, p_n^0)$ is called the normal image of the supporting hyperplane α and is denoted by $\chi_z(\alpha)$. We construct the set

$$\chi_z(x_0) = \bigcup_{\alpha} \chi_z(\alpha)$$

for each point $x_0 \in G$ where α is an arbitrary supporting hyperplane to S_z in the point $(x_0, z(x_0)) \in S_z$. The set $\chi_z(x_0)$ is called the normal image of the point x_0 (relative to the function $z(x)$). The set $\chi_z(x_0)$ is a closed convex set in $\tilde{R}^n$. The set

$$\chi_z(e) = \bigcup_{x_0 \in e} \chi_z(x_0)$$

is called the normal image of the subset $e \subset G$ (relative to the function $z(x)$).

We may consider χ_z as a mapping which transforms each subset $e \subset G$ into some subset $\chi_z(e) \subset \tilde{R}^n$. We called this mapping *the normal mapping* of the function $z(x)$. If $z(x) \in W(G) \cap C^2(G)$ then the normal mapping can be considered as a mapping of points. In this case the normal mapping is the tangential mapping.

Let $R(p) > 0$ be a locally summable function on $\tilde{R}^n$. We consider the function of sets

$$\omega(R, z, e) = \int_{\chi_z(e)} R(p)\,dp, \qquad e \subset G \tag{1.1}$$

for each $z(x) \in W(G)$. The function $\omega(R, z, e)$ is nonnegative and completely additive on the ring of Borel subsets of the convex domain G for each function $z(x) \in W(G)$. This function is called *the* R-*curvature of the convex function* $z(x)$. We put

$$A(R) = \int_{\tilde{R}^n} R(p)\,dp\,. \tag{1.2}$$

It is clear that $A(R) > 0$ the case $A(R) = +\infty$ is not excluded. Since

$$\chi_z(G) \subset \tilde{R}^n \tag{1.3}$$

for each $z(x) \in W(G)$ we have

$$\omega(R, z, G) = \int_{\chi_z(G)} R(p)\,dp \leq A(R)\,. \tag{1.4}$$

Moreover if $z(x) \in W(G) \cap C^2(G)$ then

$$\omega(R, z, e) = \int_e |\det \|z_{ij}\|\,|\,R(Dz)\,dx\,. \tag{1.5}$$

The formula (1.5) is also correct if $z(x) \in W(G) \cap W_n^{(2)}(G)$.

The concepts of normal mapping and of R-curvature are introduced by I. Bakelman in [1]. Proofs of the above assertions can be found in [1], [2], [3].

1.2. *Geometric lemmas*

Let G be a convex bounded open domain in E^n and let $u(x) \in C(\overline{G})$ be an arbitrary convex function satisfying the condition

$$u|_{\partial G} = 0 . \tag{1.6}$$

We consider the convex cone K with the vertex in the point $(x_0, u(x_0))$ and the basis ∂G, where x_0 is an inner point of G.

LEMMA 1. *Let* $R(p) > 0$ *be an arbitrary locally summable function in* $R^n = \{p = (p_1, \cdots, p_n)\}$. *Then*

$$\omega(R, u, G) \geq \omega(R, K, G) \geq \int\limits_{|p| \leq \rho} R(p)\, dp \tag{1.7}$$

where $\rho = \dfrac{|u(x_0)|}{d(G)}$ *and* $d(G)$ *is the diameter of* G. *(The function* $u(x)$ *and the cone* K *are defined above.)*

Proof. Since

$$\chi_u(G) \supset \chi_K(G)$$

then

$$\omega(R, K, G) = \int\limits_{\chi_K(G)} R(p)\, dp \leq \int\limits_{\chi_u(G)} R(p)\, dp = \omega(R, u, G) . \tag{1.8}$$

We consider the n-ball $S \subset E^n$ with the center in the point x_0 and the radius $d(G)$. Let K_0 be the convex cone of revolution with the vertex in the point $(x_0, u(x_0))$ and with the basis ∂S.

It is clear that

$$\chi_{K_0}(S) \subset \chi_K(G) .$$

Therefore

$$\omega(R, K_0, S) \leq \omega(R, K, G) . \tag{1.9}$$

The set $x_{K_0}(S)$ is the n-ball in $\tilde{R}^n$, which has the center $(0,0,\cdots,0)$ and the radius

$$\rho = \frac{|u(x_0)|}{d(G)} .$$

From (1.7), (1.8) and (1.9) we obtain

$$\omega(R,u,G) \geq \omega(R,K,G) \geq \int_{|p|\leq\rho} R(p)\,dp .$$

The Lemma 1 is proved.

REMARK. If we consider the condition

$$u|_{\partial G} = h = \text{const} \tag{1.10}$$

instead of the condition (1.6) then the inequalities (1.7) take the following form

$$\omega(R,u,G) \geq \omega(R,K,G) \geq \int_{|p|\leq\rho_h} R(p)\,dp \tag{1.11}$$

where

$$\rho_h = \frac{|h-u(x_0)|}{d(G)} . \tag{1.12}$$

Let $R(p) > 0$ be a locally summable function in $\tilde{R}^n$. We consider the function

$$g_R(\rho) = \int_{|p|\leq\rho} R(p)\,dp \tag{1.13}$$

for $\rho \in [0,+\infty)$. It is evident that $g_R(\rho)$ is strictly increasing and continuous and

$$g_R(0) = 0\,, \quad g_R(\infty) = A(R) . \tag{1.14}$$

We denote by $T_R : [0, A(R)) \to [0,+\infty)$ the inverse function for the function $g_R(\rho)$.

LEMMA 2. *Let* $u(x)$ *be a convex function in* G *which satisfies two conditions:*

a) $u|_{\partial G} = h = \text{const}$,

b) $\omega(R, u, G) < A(R)$.

Then

$$h \geq u(x) \geq h - T_R(\omega_u) \cdot d(G) \tag{1.15}$$

where $\omega_u = \omega(R, u, G)$.

If $u(x)$ *is a concave function in* G *then the inequalities*

$$h \leq u(x) \leq h + T_R(\omega_u) \cdot d(G) \tag{1.16}$$

hold everywhere in G.

Proof. Let K_0 be the cone of revolution which we consider in Lemma 1. Then

$$g_R(\rho_h) = \int_{|p| \leq \rho_h} R(p)\, dp = \omega(R, K_0, S)$$

where $\rho_h = \dfrac{|h - u(x_0)|}{d(G)}$.

We can take $x_0 \in G$ so that the equality

$$\sup_G |h - u(x)| = |h - u(x_0)|$$

could hold.

Therefore

$$\rho_h = T_R(\omega(R, K_0, S)) \leq T_R(\omega(R, u, G)) = R_R(\omega_u)$$

because T_R is an increasing function. Therefore

$$|h - u(x_0)| \leq T_R(\omega_u) \cdot d(G)\,. \tag{1.17}$$

From (1.17) it follows that

$$h \geq u(x) \geq h - T_R(\omega_u) \cdot d(G) \tag{1.18}$$

everywhere in G if u(x) is a convex function and

$$h \leq u(x) \leq h + T_R(\omega_u) \cdot d(G) \tag{1.19}$$

everywhere in G if u(x) is a concave function.

Lemma 2 is proved.

1.3. *Estimates of convex and concave functions in terms of their R-curvature*

THEOREM 1. *Let* G *be a convex bounded domain in* E^n *and let* $V(\omega_0) = \{z(x)\}$ *be the set of all convex and concave functions belonging to* W(G) *and satisfying the conditions*

$$1) \quad -\infty < m \leq z|_{\partial G} \leq M < +\infty , \tag{1.20}$$

$$2) \quad \omega(R, z, G) \leq \omega_0 < A(R) . \tag{1.21}$$

Then the inequalities

$$m - T_R(\omega_0) d(G) \leq z(x) \leq M \tag{1.22}$$

if $z(x) \in W^+(G)$ *and*

$$m \leq z(x) \leq M + T_R(\omega_0) d(G) \tag{1.23}$$

if $z(x) \in W^-(G)$ *hold in* G.

The proof of this theorem follows from Lemmas 1 and 2. (See [2], [3], [4].)

§2. *Estimates of solutions of the Dirichlet problem for quasilinear equations*

2.1. *The first theorem of estimates of solutions*

We consider the Dirichlet problem

$$\sum_{i,k=1}^{n} a_{ik}(x, u, Du) u_{ik} = b(x, u, Du) \tag{2.1}$$

$$u|_{\partial\Omega} = h(x) \tag{2.2}$$

where Ω is a bounded open domain in E^n, $\partial\Omega \in C$ and $Du = \{u_1, u_2, \cdots, u_n\}$, $u_i = \frac{\partial u}{\partial x_i}$, $u_{ik} = \frac{\partial^2 u}{\partial x_i \partial x_k}$ $(i,k = 1,2,\cdots,h)$.

We assume that:

1. The functions $a_{ik}(x,u,p)$, $b(x,u,p)$ are defined in $\overline{\Omega} \times R \times R^n$ and in each point $(x,u,p) \in \overline{\Omega} \times R \times \tilde{R}^n$ the form

$$\sum_{i,k=1}^{n} a_{ik}(x,u,p)\,\xi_i \xi_k$$

is positive and $\det \|a_{ik}(x,u,p)\| > 0$.

2. In each point $(x,u,p) \in \overline{\Omega} \times R \times \tilde{R}^n$ the inequality

$$\frac{b^n(x,u,p)}{n^n \det \|a_{ij}(x,u,p)\|} \le \frac{\phi_1(x)}{R_1(p)} \tag{2.3}$$

holds if $b(x,u,p) \ge 0$, and the inequality

$$\frac{|b^n(x,u,p)|}{n^n \det \|a_{ij}(x,u,p)\|} \le \frac{\phi_2(x)}{R_2(p)} \tag{2.4}$$

holds if $b(x,u,p) \le 0$.

In the inequalities (2.3) and (2.4) the functions $\phi_1(x)$, $\phi_2(x)$ are nonnegative, $\phi_1(x)$, $\phi_2(x) \in L(\Omega)$ and the functions $R_1(p)$, $R_2(p)$ are positive and locally summable in $\tilde{R}^n$.

3. $h(x) \in C(\partial\Omega)$ and $h = \inf_{\partial\Omega} h(x)$, $H = \sup_{\partial\Omega} h(x)$.

THEOREM 2. *Let* $u(x) \in W_n^{(2)}(\Omega) \cap C(\overline{\Omega})$ *be a solution of the Dirichlet problem. We denote by* ω_1 *and* ω_2 *the numbers*:

$$\omega_1 = \int_\Omega \phi_1(x)\,dx, \quad \omega_2 = \int_\Omega \phi_2(x)\,dx. \tag{2.5}$$

If the inequalities

$$\omega_1 < A(R_1) \quad \text{and} \quad \omega_2 < A(R_2) \tag{2.6}$$

are fulfilled then the following estimates are valid for $u(x)$ *in* G

$$h - T_{R_1}(\omega_1)d(\Omega) \leq u(x) \leq H + T_{R_2}(\omega_2)d(\Omega) \tag{2.7}$$

where the numbers $A(R_i)$ $(i=1,2)$ *are defined in* §1.

The proof of this theorem can be found in [2], [3], [4]. The short scheme of proof of the lower estimate in (2.7) is the following: We construct the set $\overline{G} = \overline{Co}(\overline{\Omega})$. It is clear that $d(\overline{G}) = d(\Omega)$. We consider the set

$$\overline{G}_h = \{(x,z),\ x \in \overline{G},\ z = h\}.$$

Let S_u be the graph of the function $u(x)$ and $S_{u,h}$ be the part of S_u lies under the hyperplane $z = h$. We construct the hypersurface

$$S = \partial\{\overline{Co}\,[\overline{G}_h \cup S_{u,h}]\} \setminus \overline{G}_h .$$

The hypersurface S is the graph of certain function $f(x) \in W^+(G)$. I call $f(x)$ the convex function spanning $u(x)$ from below. It is clear that

$$f|_{\partial G} = h \tag{2.8}$$

$$f(x) \leq u(x). \tag{2.9}$$

Without loss of generality we may suppose that $\inf\limits_{\overline{\Omega}} u(x) < h$ for otherwise the estimate

$$h - T_{R_1}(\omega_1)d(\Omega) \leq u(x)$$

is trivial. Let $\varepsilon > 0$ be any number less than $h - \inf\limits_{\overline{\Omega}} u(x)$. We denote by S^ε the part of S which lies under the hyperplane $z = h-\varepsilon$. Let M^ε be the projection of the set $S^\varepsilon \cap S_u$ on the hyperplane $z = 0$. It is clear that M^ε is closed set, moreover $M^\varepsilon \subset \Omega$ and $\operatorname{dist}(M^\varepsilon, \partial\Omega) > 0$. We denote by $\overline{G}^{\varepsilon}$ the projection of $\overline{S}^{\varepsilon}$ on E^n and by G^{ε} the set $\operatorname{int} \overline{G}^{\varepsilon}$. It is clear that $G \supset \overline{G}^{\varepsilon} \supset G^{\varepsilon}$.

From the theory of convex hypersurfaces (see [3]) we have

$$\omega(R_1, f, G) \leq \lim_{\varepsilon \to 0} \omega(R_1, f, G^{\varepsilon}) . \tag{2.10}$$

In [2] and [4] the following assertions were proved:

$$\omega(R_1, f, G^{\varepsilon}) \leq \omega(R_1, f, \overline{G}^{\varepsilon}) = \omega(R_1, f, M^{\varepsilon})$$

and

$$\omega(R_1, f, M^{\varepsilon}) = \int_{M^{\varepsilon}} R_1(Du) \cdot \det \|u_{jk}\| \, dx . \tag{2.11}$$

All the points $(x, u(x))$ of the graph of $u(x)$ will be convex points, if $x \in M^{\varepsilon}$. Therefore the form d^2u is nonnegative for each $x \in M^{\varepsilon}$.

We consider the equation

$$\sum_{i,k=1}^{n} a_{ik}(x, u, Du) u_{ik} = b(x, u, Du)$$

only on the set M^{ε}. Then

$$b(x, u, Du) = \sum_{i,k=1}^{n} a_{ik}(x, u, Du) u_{ik} \geq n \cdot (\det \|u_{ik}\|)^{\frac{1}{n}} (\det \|a_{ik}\|)^{\frac{1}{n}}$$

since the forms $a_{ik} \xi_i \xi_k$ and d^2u are correspondingly positive and nonnegative. Therefore the inequality

$$0 \leq \det \|u_{jk}\| \leq \frac{b^n(x, u, Du)}{n^n \det \|a_{ik}(x, u, Du)\|}$$

holds in any points $x \in M^{\varepsilon}$. Therefore from (1.6) it follows that

$$\omega(R_1, f, M^\varepsilon) = \int_{M^\varepsilon} R_1(Du)\det\|u_{jk}\|\,dx \le \int_{M^\varepsilon} \phi_1(x)\,dx$$

$$\le \int_\Omega \phi_1(x)\,dx = \omega_1$$

because

$$R_1(Du)\det\|u_{jk}\| \le \phi_1(x)$$

holds in each point $x \in M^\varepsilon$ and $M^\varepsilon \subset \Omega$. From (1.5) we obtain

$$\omega(R_1, f, G) \le \lim_{\varepsilon\to 0} \omega(R_1, f, M^\varepsilon) \le \int_\Omega \phi_1(x)\,dx = \omega_1 . \tag{2.12}$$

Now we can apply Theorem 1 to obtain the lower estimate of the functions $f(x)$ and $u(x)$. First, from (2.12) and the conditions of the Theorem 2 we have

$$\omega(R_1, f, G) \le \omega_1 < A(R_1) . \tag{2.13}$$

Then from Theorem 1 it follows that

$$h - T_{R_1}(\omega_1)\,d(\Omega) \le f(x) \le u(x) .$$

2.2. *The second theorem of estimates of solutions*

We consider the Dirichlet problem

$$\sum_{i,k=1}^{n} a_{ik}(x, u, Du)\,u_{ik} = b(x, u, Du) , \tag{2.14}$$

$$u|_{\partial\Omega} = h(x) \tag{2.15}$$

where Ω is a bounded open domain in E_n, $\partial\Omega \in C$.

We assume that:

1. The functions $a_{ik}(x, u, p)$, $b(x, u, p)$ are defined in $\bar{\Omega} \times R \times \tilde{R}^n$ and the form

$$\sum_{i,k=1}^{n} a_{ik}(x, u, p) \xi_i \xi_k$$

is positive and $\det \|a_{ik}(x, u, p)\| > 0$ in each point $\bar{\Omega} \times R \times \tilde{R}^n$.

2. The function $\dfrac{c(x, u, p)}{[\det \|A_{ik}(x, u, p)\|]^{\frac{1}{n}}}$ increases with u if $x \in \bar{\Omega}$ and $p \in \tilde{R}^n$ are fixed.

3. The inequalities

$$\begin{aligned} -d_M^-(x) - C_M^-(x)\,|p| \le \frac{b(x, M, p)}{(\det \|A_{ik}(x, x, p)\|)^{\frac{1}{n}}} \le \\ \le d_M^+(x) + C_M^+(x)\,|p| \end{aligned} \tag{2.16}$$

hold for each number M, where $C_M^+(x)$, $C_M^-(x)$, $d_M^+(x)$, $d_M^-(x)$ are non-negative functions in Ω, which belong to $L_n(\Omega)$.

We'll use the following notation

$$H = \sup_{\partial\Omega} h(x), \qquad h = \inf_{\partial\Omega} h(x).$$

THEOREM 3. *Let* $u(x) \in W_n^{(2)}(\Omega) \cap C(\bar{\Omega})$ *be a solution of the Dirichlet problem (2.14-15). Then the inequalities*

$$\begin{aligned} h - q_n d(\Omega) \|d_h^+(x)\|_{L_n} \exp(\tilde{q}_n \|C_H^-(x)\|_{L_n}^n) \le \\ \le u(x) \le H + q_n d(\Omega) \|d_H^-(x)\|_{L_n} \exp(\tilde{q}_n \|C_H^-(x)\|_{L_n}^n) \end{aligned} \tag{2.17}$$

hold everywhere in Ω, *where*

$$q_n = \frac{2}{n} \cdot \left(\frac{e}{\mu_n}\right)^{\frac{1}{n}}, \qquad \tilde{q}_n = \frac{2^n}{n^{n+1} \mu_n},$$

μ_n is the volume of the unit n-ball, if the equation (2.14) satisfies the conditions 1, 2, 3 of §2.2.

This is a new theorem I have just proved. It is being published for the first time.

Proof. Let $f(x)$ be the convex function spanning $u(x)$ from below (see the proof of Theorem 2). We consider the functions $u(x)$ and $f(x)$ only on the set M^ε. For this case we have proved the inequality

$$0 \le (\det \|u_{ik}\|)^{\frac{1}{n}} \le \frac{b(x, u, Du)}{n[\det \|A_{ik}(x, u, Du)\|]^{\frac{1}{n}}} \tag{2.18}$$

everywhere on M^ε for each sufficiently small $\varepsilon > 0$. (See the proof of Theorem 2.)

It is clear that

$$u(x) \le h(x) - \varepsilon \le h(x)$$

everywhere on M^ε. From the conditions of Theorem 3 we obtain the inequality

$$0 \le (\det \|u_{ik}\|)^{\frac{1}{n}} \le \frac{1}{n}[C_h^+(x)|Du| + d_h^+(x)] \tag{2.19}$$

in each point $x \in M^\varepsilon$, where $|Du| = (u_1^2 + \cdots + u_n^2)^{1/2}$.

Now from (2.19) it follows that

$$0 \le (\det \|u_{ik}\|)^{\frac{1}{n}} \le \frac{1}{n}[[C_h^+(x)]^n + \delta^n[d_h^+(x)]^n]^{\frac{1}{n}} \times$$
$$\times \left[|Du|^{\frac{n}{n-1}} + \delta^{-\frac{n}{n-1}}\right]^{\frac{n-1}{n}}$$

for each $x \in M^\varepsilon$, where $\delta > 0$ is an arbitrary number.

Let $p = (p_1, \cdots, p_n)$ be an arbitrary vector in $\tilde{R}^n$. We consider the function

$$R(p) = [|p|^n + \delta^{-n}]^{-1}$$

in $\tilde{R}^n$. Then

$$\inf_{\tilde{R}^n} \frac{\left[|p|^{\frac{n}{n-1}} + \delta^{-\frac{n}{n-1}}\right]^{-(n-1)}}{R(p)} = \inf_{[0,+\infty)} \frac{\xi^n + \delta^{-n}}{\left(\xi^{\frac{n}{n-1}} + \delta^{-\frac{n}{n-1}}\right)^{n-1}} \geq 2^{-n} .$$

Therefore we obtain the inequality

$$R(Du) \det \|u_{ik}\| \leq \left(\frac{2}{n}\right)^n [[C_h^+(x)]^n + \delta^n [d_h^+(x)]^n]$$

everywhere on M^ε for each sufficiently small $\varepsilon > 0$.

From the last inequality it follows that

$$\omega(R, f, G) \leq \left(\frac{2}{n}\right)^n \|C_h^+(x)\|_{L_n}^n + \delta^n \left(\frac{2}{n}\right)^n \|d_h^+(x)\|_{L_n}^n \tag{2.20}$$

where $f(x)$ is the convex function spanning $u(x)$ from below. From the method of construction of the function $f(x)$ it follows that

$$f|_{\partial G} = h = \inf_{\partial\Omega} h(x)$$

(see the proof of theorem 2). Only the case when $f(x_0) = \inf_{\Omega} f(x) < \inf_{\partial\Omega} f(x) = h$, $x_0 \in \Omega$ is interesting. It is evident that

$$f(x_0) = u(x_0)$$

and

$$h - u(x_0) > 0 .$$

From Lemma 1 (see §1.2) we obtain

$$\omega(R, f, G) \geq \int_{|p| \leq \rho_h} R(p)\, dp$$

where $\rho_h = \dfrac{h - f(x_0)}{d(G)} = \dfrac{h - u(x_0)}{d(\Omega)}$. Therefore

$$\omega(R, f, G) \geq \int_{|p| \leq \rho_h} \frac{dp}{|p|^n + \delta^{-n}} =$$

$$= \sigma_{n-1} \int_0^{\frac{h-u(x_0)}{d(\Omega)}} \frac{p^{n-1} dp}{p^n + \delta^{-n}} = \mu_n \ell n \left(1 + \left[\frac{h - u(x_0)}{d(\Omega)}\right]^n \delta^n\right), \tag{2.21}$$

where μ_n is the volume of the unit n-ball and σ_{n-1} is the area of the unit (n−1)-sphere.

From the inequalities (2.20) and (2.21) we obtain the lower estimate of u(x) (see (2.17). The upper estimate of u(x) in (2.17) is proved similarly. This completes the proof of Theorem 3.

§3. *Existence theorems for hypersurfaces with given mean curvature*

3.1. *Introduction*

Here we consider the applications of the results to the Dirichlet problem of the reconstruction of the smooth hypersurface with given mean curvature in Euclidean space E^{n+1}. We consider only the simplest case of this problem.

Let G be a bounded open domain in E^n and $\partial G \in C^{m+2,\lambda}$ ($m \geq 1$, $0 < \lambda < 1$). We consider the function $H(x) \in C^{m,\lambda}(\overline{G})$ and the function $h(x) \in C^{m+2,\lambda}(\partial G)$.

Then our geometric problem is reduced to the following Dirichlet problem

$$\left(1 + \sum_{i=1}^n z_i^2\right)\Delta z - \sum_{i,k=1}^n z_i z_k z_{ik} = nH(x)\left(1 + \sum_{i=1}^n z_i^2\right)^{3/2} \tag{3.1}$$

$$z|_{\partial G} = h(x) . \tag{3.2}$$

We consider the family of the Dirichlet problems with the parameter $t \in [0, 1]$ which are obtained by means of change of the functions $H(x)$ and $h(x)$ correspondingly for the functions $tH(x)$ and $th(x)$ in (3.1) and (3.2). We denote by $z_t(x)$ the solutions of these Dirichlet problems. If we can prove the uniform estimates

$$\|z_t(x)\|_{C^1(\bar{G})} \leq Z_0 = \text{const} < +\infty \tag{3.3}$$

then the Dirichlet problem (3.1-2) will have a solution $z(x) \in C^{m+2,\lambda'}(\bar{G})$ where $0 < \lambda' \leq \lambda < 1$.

The process of obtaining the estimates (3.3) consists of two parts:

a) the proof of the existence of uniform estimates $\|z_t(x)\|_{C(\bar{G})}$,

b) the proof of the existence of uniform estimates of $\|\text{grad } z_t(x)\|_{C(\bar{G})}$.

The results of §§1, 2 are applied to the obtaining of both types of these a priori estimates in Theorem 4 and of the a priori estimate of $\|z_t(x)\|_{C(\bar{G})}$ in Theorem 5 (see §2 and §3.3). The corresponding sufficient conditions for these estimates are (3.4) and (3.5).

3.2. *Existence theorems*

We consider two existence theorems. In the first one G is a convex bounded domain.

THEOREM 4. *Let all the principle normal curvatures of* G *in each point of* ∂G *be not less than* $\kappa_0 = \text{const} > 0$ *and let*

$$H_1 = \sup_{\bar{G}} H_+(x) < \kappa_0, \quad H_2 = \sup_{\bar{G}} H_-(x) < \kappa_0 . \tag{3.4}$$

($H_+(x)$ *and* $H_-(x)$ *are correspondingly the positive and negative parts of the function* $H(x)$) *then there exists one and only one hypersurface* $F : z = z(x) \in C^{m+2,\lambda'}(\bar{G})$ $(0 < \lambda' \leq \lambda < 1)$ *such that* $z|_{\partial G} = 0$ *and the mean curvature of* F *coincides with the function* $H(x)$.

If G *is the* n-*ball of the radius* $\frac{1}{\kappa_0}$ *and* $H(x) = H_0 = \text{const} > 0$ *then the condition (3.4) is necessary for the existence of hypersurface* $F : z = z(x) \in C^2(\bar{G})$ *with given mean curvature* $H(x)$.

Theorem 4 can be generalized for the boundary condition

$$z|_{\partial G} = h(x) \in C^{m+2,\lambda}(\partial G)$$

(see [4], [6]).

Now we consider a nonconvex domain G and a boundary condition $z|_{\partial G} = h(x)$ where $h(x)$ is an arbitrary function in $C^{m+2,\lambda}(\partial G)$.

THEOREM 5. *Let* G *be an open bounded domain in* E^n, *whose boundary is of class* $C^{m+2,\lambda}$ *and let* $H(x)$ *and* $h(x)$ *be correspondingly the functions of classes* $C^{m,\lambda}(\overline{G})$ *and* $C^{m+2,\lambda}(\partial G)$. *We suppose that*

1) $\|H_{\pm}(x)\|_{L_n(\overline{G})} < \mu_n^{1/n}$; (3.5)

2) $\tilde{H}(x) \geq \frac{n}{n-1} |H(x)|$ (3.6)

everywhere on ∂G, *where* $\tilde{H}(x)$ *is the mean curvature of* ∂G *in the direction of the intrinsic normal. Then there exists one and only one hypersurface*

$$F : z = z(x) \in C^{m+2,\lambda'}(\overline{G}) \quad (0 < \lambda' \leq \lambda < 1)$$

such that $z|_{\partial G} = h(x)$ *and the mean curvature of* F *coincides with the function* $z(x)$.

The sufficient condition (3.5) is interlocked with the necessary condition

$$\|H_{\pm}(x)\|_{L_n(\overline{G})} \leq \mu_n^{1/n} \tag{3.7}$$

in the case $H(x) = H_0 = \text{const} > 0$ and G is the n-ball. The condition (3.6) is necessary if we consider arbitrary functions $h(x) \in C^{m+2,\lambda}(\partial G)$ in the boundary condition $z|_{\partial G} = h(x)$. If $n \geq 3$ then G can be a nonconvex domain. For example there are toruses with a positive mean curvature in E^3.

The condition (3.6) was obtained by Serrin [8] and by Bakelman [5],[6] independently by means of different methods.

The Theorems 4 and 5 have been proved by Bakelman [4], [5], [6]. The Theorems 4 and 5 are non-overlapping, because in the special case when G is the unit n-ball of the radius R and $H(x) \equiv H_0 = \text{const} > 0$ the inequality (3.6) takes the following form

$$H_0 < \frac{n-1}{n} \frac{1}{R} .$$

The inequalities (3.4) (see Theorem 4) have the form

$$H_0 < \frac{1}{R} .$$

J. Serrin [8] has found series of interesting conditions for the estimates of $\|z(x)\|_{C(\bar{G})}$ by means of other geometric constructions. If we consider Serrin's conditions instead of the condition (3.5) we will have Serrin's existence and uniqueness theorems (see [8]) for hypersurfaces with given mean curvature.

3.3. *Estimates of* $\|z(x)\|_{C(\bar{G})}$ *for the hypersurfaces in terms of its mean curvature*

Let S be the hypersurface defined by $z(x) \in C^2(\Omega) \cap C(\bar{\Omega})$ and let $H(x)$ be its mean curvature. Since $z(x)$ satisfies throughout Ω the equation

$$\Big(1+\sum_{i=1}^{n} z_i^2\Big)\Delta z - \sum_{i,j=1}^{n} z_i z_j z_{ij} = nH(x)\Big(1+\sum_{i=1}^{n} z_i^2\Big)^{3/2} \tag{3.8}$$

we can on the basis of Theorem 2 obtain a two-sided estimate for $z(x)$ depending on properties of $H(x)$. We use the following notation

$$\omega\pm = \int_{\Omega} [H\pm(x)]^n dx , \qquad h = \inf_{\partial\Omega} z(x) , \qquad M = \sup_{\partial\Omega} z(x) ,$$

where $H_+(x)$ and $H_-(x)$ are, respectively, the positive and negative parts of $H(x)$.

In this case the function

$$\frac{b(x,z,p)}{(\det \|a_{ij}(x,z,p)\|)^{1/n}}$$

takes the form $nH(x)(1+|p|^2)^{\frac{n+2}{2n}}$. Hence we have the estimates

$$-nH_-(x)\frac{1}{(1+|p|^2)^{-\frac{n+2}{2n}}} \leq \frac{b}{(\det \|a_{ij}\|)^{1/n}} \leq nH_+(x)\frac{1}{(1+|p|^2)^{-\frac{n+2}{2n}}} .$$

If we put $R(p) = (1+|p|^2)^{-\frac{n+2}{2}}$, then $A(R) = \mu_n$, where μ_n is the volume of the unit n-ball. On the Theorem 2 we see that when the inequalities

$$\omega_+ < \mu_n, \quad \omega_- < \mu_n \tag{3.9}$$

hold, we obtain the following estimates for $z(x)$ in $\overline{\Omega}$

$$h - T_R(\omega_+)d(\Omega) \leq z(x) \leq H + T_R(\omega_-)d(\Omega) . \tag{3.10}$$

In our case $T_R(\tau)$ has the form

$$T_R(\tau) = \left(\frac{\tau^{2/n}}{\mu_n^{2/n} - \tau^{2/n}}\right)^{\frac{1}{2}} .$$

Thus, finally the required estimates for the height of the hypersurface in terms of integrals of the n-th powers of the positive and negative parts of its mean curvature are as follows:

$$h - \left(\frac{\omega_+^{2/n}}{\mu_n^{2/n} - \omega_+^{2/n}}\right)^{\frac{1}{2}} d(\Omega) \leq z(x) \leq H + \left(\frac{\omega_-^{2/n}}{\mu_n^{2/n} - \omega_-^{2/n}}\right)^{\frac{1}{2}} d(\Omega) . \tag{3.11}$$

In the case where $H(x) = H_1 = \text{const} > 0$ and Ω is a ball of radius R, the conditions (3.9), under which (3.11) was obtained, take the form

$$H_1 < \frac{1}{R}, \quad 0 < \frac{1}{R} .$$

This shows that the conditions (3.9), which are sufficient for the estimates (3.11), interlock with the necessary conditions for the solvability of the geometric problem of reconstructing a hypersurface from its mean curvature.

The results of §3.2 and §3.3 can be generalized for the Lobachevskii space and wide classes of the Riemannian spaces (see [5], [6]). In these works I. Bakelman has proved existence and uniqueness theorems for hypersurfaces in the Lobachevskii and the Riemannian spaces.

UNIVERSITY OF MINNESOTA

REFERENCES

[1] I. Bakelman. Generalized solutions of the Monge-Ampere equations. Dokl. Akad. Nauk SSSR, 114 (1957), 1143-1145.

[2] ———. On the theory of quasilinear elliptic equations. Sibirsk. Mat. Zh. (1961), 179-186.

[3] ———. Geometric methods of solving elliptic equations. Izdat. Nauka, Moscow (1965).

[4] ———. Hypersurfaces with given mean curvature and quasilinear elliptic equations with strong non-linearity. Math. Sborniik, 75 (1968), 604-638; Math. USSR Sb. 4 (1968), 561-596.

[5] ———. Mean curvature and quasilinear elliptic equations. Sibirsk. Math. Zh. 9 (1968), 1014-1040; Siberian Math J. 9 (1968), 752-771.

[6] ———. Geometric problems in quasilinear elliptic equations. Uspekhi Math. Nauk., 25, 3, 1970, 49-112; Russian Math Surveys (1970), 25, no. 3, 45-109.

[7] J. Serrin. The Dirichlet problem for the minimal surface equation in higher dimensions, J. reine Angew. Math. 223 (1968), 170-187.

[8] ———. The problem of Dirichlet for quasilinear elliptic differential equations with many independent variables. Philos. Trans. Roy. Soc. (1969), 413-496.

EXTREMAL KÄHLER METRICS

Eugenio Calabi*

Introduction

The gauge field problem in its most general formulation seeks a connection (vector potential) in a given vector bundle with orthogonal group structure (the gauge group), which minimizes the L_2 norm (action), defined by integrating a natural pointwise squared absolute value of the curvature (gauge field). In recent years the problem has attracted the attention of geometers in a setting that goes beyond the scope that made it interesting to physicists. The most natural geometric problem that falls in this classification deals with the special case of the tangent bundle of a compact differentiable manifold, where one may choose an arbitrary Riemannian metric, which fixes a unique torsion-free connection (the Levi-Civita connection), subject to some reasonable constraints (for instance, fixing the total volume, or within a pre-assigned conformal class).

A special case of the Riemannian problem was proposed several years ago by the author [3, 4]+. It deals with the set of all Kähler metrics that can be introduced in a given, compact complex manifold M. A natural, additional restriction (involving at most a finite number of parameters)

*Supported in part by N.S.F., Contract No. MCF 78-02285.

+Arabic numerals in brackets refer to the bibliography of research papers; Roman numerals to textbooks and monographs.

Seminar on Differential Geometry
0-691-08268-5/82/000259-32 $01.60/0 (cloth)
0-691-08296-0/82/000259-32 $01.60/0 (paperback)

that is appropriate to impose is to fix the cohomology class of the Kähler metrics. This amount to using an initially given Kähler metric, expressed by

$$ds^2 = 2g_{\alpha\overline{\beta}}(z,\overline{z})\,dz^\alpha\, d\overline{z}^{\beta}$$

in terms of a local system of holomorphic parameters $z = (z^1, \cdots, z^n)$ and their complex conjugates $\overline{z}$; a new metric tensor $g'_{\alpha\overline{\beta}}(z,\overline{z})$ is cohomologous to $(g_{\alpha\overline{\beta}})$, if and only if there exists a real valued scalar function u, globally defined on the manifold M, such that, at each point

(1) $$g'_{\alpha\overline{\beta}} = g_{\alpha\overline{\beta}} + \frac{\partial^2 u}{\partial z^\alpha \partial \overline{z}^\beta}\,.$$

On the other hand, given $g_{\alpha\overline{\beta}}$, the function u in (1), which may be called a *metric distortion potential*, may be chosen arbitrarily, subject only to a condition of differentiability (for the present purposes, differentiability of class C^6 is sufficient), and with appropriate bounds on the second derivatives to ensure that the resulting Kählerian form $g'_{\alpha\overline{\beta}}\,dz^\alpha\, d\overline{z}^{\beta}$ be positive definite everywhere.

We present here some preliminary results on the variational problem of minimizing the functional on the function space of all Kähler metrics $(g_{\alpha\overline{\beta}})$, which assigns to each metric the L_2-norm of the corresponding curvature tensor (in terms of the metric itself), integrated over M with respect to the volume element of the metric. The first section is devoted to setting up the notations and conventions followed in the sequel, and to reducing the variational problem of minimizing the L_2-norm of the Riemann curvature to the corresponding, equivalent problems of minimizing the L_2-norm of either the Ricci curvature or of the scalar curvature; since the formal computations with the scalar curvature are more tractable than the corresponding ones for the Ricci or the Riemannian tensors, the second and third sections mention the problem only in terms of the scalar curvature. In the second section we derive the Euler equation for the variational problem considered and prove that the second variation is essentially positive definite. The third section describes some special types of compact, complex manifolds, for which the extremal Kähler metrics for our variational problem are explicitly constructible; what makes the examples interesting is the fact that, for these manifolds, the extremal Kähler metric does not (and cannot) have constant scalar curvature.

1. *Preliminary results*

All manifolds in this paper are understood to be connected. Let M be a compact, complex manifold with n (complex) dimensions, admitting at least one Kähler metric, expressed locally, in terms of a holomorphic system of parameters $z = (z^1, \cdots, z^n)$ and their conjugates $\overline{z} = (\overline{z^1}, \cdots, \overline{z^n})$, by

(1.1)
$$ds^2 = 2g_{\alpha\overline{\beta}}\, dz^\alpha \, d\overline{z^\beta} .$$

We use Greek indices, ranging in values from 1 to n, in relation with the local parameters $(z^1, \cdots, z^n)$, barred Greek indices referring to the conjugate parameters $\overline{z}$, and Latin indices ranging from 1 to 2n referring to both z and $\overline{z}$, where formally $\overline{z^\alpha}$ is identified with $z^{n+\alpha}$, and $n+\alpha$ is denoted by $\overline{\alpha}$. The Kähler condition on the metric is expressed by the existence of a locally defined, real valued, smooth function $\Phi = \Phi(z, \overline{z})$, independent of the local coordinates $(z, \overline{z})$ but not unique, such that

(1.2)
$$g_{\alpha\overline{\beta}}(z, \overline{z}) = \frac{\partial^2 \Phi}{\partial z^\alpha \partial \overline{z^\beta}} .$$

The complex analytic structure of M has an intrinsic invariant, the function space of all Kähler metrics (we assume this space to be nonempty); to each such metric, the associated, real valued, exterior form of bidegree (1.1) in dz, $d\overline{z}$,

$$\omega = \sqrt{-1}\, g_{\alpha\overline{\beta}}\, dz^\alpha \wedge d\overline{z^\beta}$$

is closed, because of (1.2), and therefore determines a 2-dimensional deRham cohomology class $[\omega]$, called the *principal cohomology class* of the metric. The set of all deRham classes admissible as principal cohomology classes for some Kähler metric is an open, convex cone in the Hodge-Dolbeault cohomology group $H^{1,1}(M, \mathbf{R})$ (deRham classes representable by closed (1,1)-forms), of a shape that is unknown in general, but conjectured to be determined by a finite number of real analytic inequalities.

In this paper we are interested only in the space of Kähler metrics in a fixed principal class $[\omega]$, determined for instance by one given Kähler metric (1.1). Any other Kähler metric $(g'_{\alpha\bar{\beta}})$ in this class is then obtained from $(g_{\alpha\bar{\beta}})$ be means of a globally defined, suitable, real valued function u on M by means of the local equations

$$g'_{\alpha\bar{\beta}} = g_{\alpha\bar{\beta}} + \frac{\partial^2 u}{\partial z^\alpha \partial \overline{z^\beta}} .$$

For further details on the elementary properties of Kähler manifolds, we refer to any one of the available texts on the subject, such as [I] through [VIII].* The notations that we use in this paper represent a compromise between the different ones followed by the various authors to whom we refer. The point of departure of this paper lies in a preliminary abstract [4] concerning the existence and properties of a Kähler metric that is extremal with respect to certain functionals involving the Ricci tensor. In order to fix the notations, we recall that the local components of the metric connection (Γ^i_{jk}), of the Kahler metric (1,1), that are not identically zero are

$$\Gamma^\lambda_{\alpha\beta} = g^{\lambda\bar{\mu}} \frac{\partial g_{\alpha\bar{\mu}}}{\partial z^\beta} = g^{\lambda\bar{\mu}} \frac{\partial^3 \Phi}{\partial z^\alpha \partial z^\beta \partial \overline{z^\mu}} \qquad ((g^{\lambda\bar{\mu}}) = {}^t(g_{\lambda\bar{\mu}})^{-1})$$

---

*Roman numerals in brackets refer to the general textbook references; Arabic numerals, to the bibliography on related research papers.

and

$$\Gamma^{\overline{\lambda}}_{\overline{\alpha}\overline{\beta}} = (\overline{\Gamma^{\alpha}_{\alpha\beta}}) .$$

The curvature tensor $(R_{ijk\ell})$ is skew-hermitian in each of the pairs (i, j) and (k, ℓ) of indices; in other words $\Omega_{ij} = R_{ijk\ell}\, dz^k \wedge dz^\ell = R_{k\ell ij} dz^k \wedge dz^\ell$ is a tensorial form of bidegree $(1, 1)$, and the tensor $(R_{ijk\ell})$ is completely determined by the components

$$R_{\overline{\mu}\alpha\overline{\beta}\lambda} = -R_{\alpha\overline{\mu}\overline{\beta}\lambda} = g_{\gamma\overline{\mu}} \frac{\partial \Gamma^{\gamma}_{\alpha\lambda}}{\partial z^{\overline{\beta}}} ; \tag{1.3}$$

the Ricci tensor $(R_{ij}) = (g^{\ell m} R_{i\ell jm})$ is hermitian and satisfies a Kähler condition as follows

$$R_{\alpha\overline{\beta}} = g^{\lambda\overline{\mu}} R_{\alpha\overline{\mu}\overline{\beta}\lambda} = -\frac{\partial^2 \log \det (g_{\lambda\overline{\mu}})}{\partial z^\alpha \partial z^{\overline{\beta}}} , \tag{1.4}$$

and the scalar curvature (which we renormalize to be one half of the usual value found in Riemannian geometry texts) is expressed by the equation

$$R = \frac{1}{2} g^{ij} R_{ij} = g^{\alpha\overline{\beta}} R_{\alpha\overline{\beta}} .$$

To any Kähler metric (1.1), that we can symbolize by (g), we associate the following invariants:

a) the volume element

$$dV(g) = \det (g_{\alpha\overline{\beta}}) \bigwedge_{\lambda=1}^{n} (\sqrt{-1}\, dz^\lambda \wedge \overline{dz^\lambda}) = \omega^{[n]} ,$$

where $\omega^{[0]} = 1$, $\omega^{[k]} = \frac{1}{k}\, \omega \vee \omega^{[k-1]}$, denotes the divided exterior powers of the form ω, and

b) the following real numbers:

i) The total volume

$$V(g) = \int_M dV(g)$$

ii) the curvature integrals

$$\overline{R}(g) = \int_M R\,dV(g), \qquad S(g) = \int_M R^2 dV(g)\,,$$

$$Q(g) = \frac{1}{2}\int_M R_{ij}R^{ij}\,dV(g) = \int_M g^{\alpha\overline{\mu}}g^{\lambda\overline{\beta}}R_{\alpha\overline{\beta}}R_{\lambda\overline{\mu}}\,dV(g)\,,$$

and

$$K(g) = \frac{1}{4}\int_M R^{ijk\ell}R_{ijk\ell}\,dV(g) = \int_M g^{\alpha\overline{\alpha'}}g^{\beta'\overline{\beta}}g^{\lambda\overline{\lambda'}}g^{\mu'\overline{\mu}}R_{\alpha\overline{\mu}\overline{\beta}\lambda}R_{\mu'\overline{\alpha'}\overline{\lambda'}\beta'}\,dV(g)\,.$$

Thus $V(g)$, $\overline{R}(g)$, $S(g)$, $Q(g)$ and $K(g)$ are continuous functionals on the space of all Kähler metrics on M, where this function space may be viewed with any topology T at least as strong as the uniform C^2-topology. We shall restrict our attention to metrics (g) in a fixed principal class $[\omega]$, with a topology T that is at least as strong as the C^4-topology, denoting it by $\mathfrak{g}(\omega, T)$ or more briefly by $\mathfrak{g}$.

PROPOSITION 1.1. *The following functionals are constant over* $\mathfrak{g}$: $V(g)$, $\overline{R}(g)$, $Q(g) - S(g)$, $K(g) - Q(g)$.

Proof. The volume $V(g)$ of M with respect to the metric g is given, as we have mentioned earlier, by

$$V(g) = \int_M \omega^{[n]}\,,$$

where $\omega^{[n]}$ denotes the n'th divided exterior power of ω (the n-fold exterior product of ω with itself divided by $n!$). Since ω is a closed 2-form in a fixed de Rham cohomology class, the independence of $V(g)$ from g for $g \in \mathfrak{g}$ is verified.

In order to prove the invariance of $\overline{R}(g)$, we recall that the Ricci form $\Sigma_1 = \sqrt{-1}\, R_{\alpha\overline{\beta}}\, dz^\alpha \wedge dz^{\overline{\beta}}$ is closed, because of (1.4); furthermore its deRham class is fixed, since for any two Kähler metrics (g) and (g'), the difference between the corresponding Ricci forms Σ_1 and Σ'_1 is

$$\Sigma'_1 - \Sigma_1 = -\sqrt{-1}\, \frac{\partial^2}{\partial z^\alpha \partial \overline{z}^\beta}\left(\log \frac{\det(g'_{\lambda\overline{\mu}})}{\det(g_{\lambda\overline{\mu}})}\right) dz^\alpha \wedge dz^{\overline{\beta}}$$

and $\log \dfrac{\det(g'_{\lambda\mu})}{\det(g_{\lambda\overline{\mu}})}$ is the local representation of a globally defined, real valued scalar (indeed, the deRham class of $\frac{1}{2\pi}\Sigma_1$ is the first Chern class of M).

Next, we recall the algebraic identity

$$\Sigma_1 \wedge \omega^{[n-1]} = (\Lambda \Sigma_1)\, \omega^{[n]} ,$$

where, for any p-form $\phi = \phi_{i_1,\cdots,i_p}\, dz^{i_1} \wedge \cdots \wedge dz^{i_p}$ with $p \geq 2$, $\Lambda\phi$ is the (p–2)-form

$$\Lambda\phi = -\sqrt{-1}\, g^{\alpha\overline{\beta}}\, \phi_{\alpha\overline{\beta} i_3,\cdots,i_p}\, dz^{i_3} \wedge \cdots \wedge dz^{i_p} ,$$

so that $\Lambda\Sigma_1 = R$. Therefore

$$\overline{R}(g) = \int_M \Lambda\Sigma_1 \omega^{[n]} = \int_M \Sigma_1 \wedge \omega^{[n-1]} ,$$

and the deRham class of $\Sigma_1 \wedge \omega^{[n-1]}$ is fixed. For $n = 1$ (closed Riemann surfaces) the proof of the proposition is now complete, since in that case the integrands in the definitions of $S(g)$, $Q(g)$ and $K(g)$ are identical.

We next prove the constancy of $S(g) - Q(g)$ as (g) varies in $\mathfrak{g}$. This follows from the fact that the deRham class of $\Sigma_1^{[2]} \wedge \omega^{[n-2]}$ is fixed and from the two identities

(1.5) $$\Sigma_1^{[2]} \wedge \omega^{[n-2]} = \frac{1}{2}(\Lambda^2 \Sigma_1^{[2]})\omega^{[n]}$$

and

(1.6) $$\Lambda^2\Sigma_1^{[2]} = (\Lambda\Sigma_1)^2 - g^{\alpha\bar{\mu}}g^{\lambda\bar{\beta}}R_{\alpha\bar{\beta}}R_{\lambda\bar{\mu}} = R^2 - R_{\alpha\bar{\beta}}R^{\alpha\bar{\beta}} .$$

Therefore

$$S(g) - Q(g) = 2\int_M \Sigma_1^{[2]} \wedge \omega^{[n-2]} ,$$

where the last expression is clearly constant as (g) varies in $\mathfrak{g}$.

Finally, in order to prove that $Q(g) - K(g)$ is constant, we introduce the closed, exterior (2,2)-form Σ_2, defined from the curvature form, such that $(2\pi)^{-2}\Sigma_2$ represents the second Chern class

$$\Sigma_2 = (R^{\lambda}{}_{\mu\alpha\bar{\beta}}R^{\mu}{}_{\lambda\gamma\bar{\delta}} - R^{\lambda}{}_{\lambda\alpha\bar{\beta}}R^{\mu}{}_{\mu\gamma\bar{\delta}})\, dz^{\alpha} \wedge dz^{\bar{\beta}} \wedge dz^{\gamma} \wedge dz^{\bar{\delta}}$$

and apply the algebraic identities similar to (1.5) and (1.6)

$$\Sigma_2 \wedge \omega^{[n-2]} = \frac{1}{2}(\Lambda^2\Sigma_2)\omega^{[n]}$$

and

$$\Lambda^2\Sigma_2 = 2R_{\alpha\bar{\beta}}R^{\alpha\bar{\beta}} - R_{\bar{\mu}\alpha\lambda\bar{\beta}}R^{\alpha\bar{\mu}\beta\lambda} - R^2 .$$

Since the integral of $\Sigma_2 \wedge \omega^{[n-2]}$ over M is obviously constant as (g) varies in $\mathfrak{g}$, it follows that $2Q(g) - K(g) - S(g)$ is constant, and since $S(g) - Q(g)$ was just proved to be constant, so is $Q(g) - K(g)$. This completes the proof of the proposition.

COROLLARY. *Let* V *and* $\bar{R}$ *denote respectively the constant values of* V(g) *and* $\bar{R}(g) = \int_M R\,dV(g)$ *as* (g) *varies in the function space* $\mathfrak{g}$, *and let* S_0, Q_0 *and* K_0 *denote respectively the greatest lower bounds of the nonnegative functionals* S(g), Q(g) *and* K(g). *Then we have the following estimates for* S_0, Q_0 *and* K_0:

a) $S_0 \geq \frac{\bar{R}^2}{V}$,

b) $Q_0 \geq S_0 - \left(1 - \frac{1}{n}\right) \frac{\bar{R}^2}{V}$,

c) $K_0 \geq \frac{2(2n+1)}{(n+1)(n+2)} S_0 - \frac{4(n-1)}{n(n+2)} \frac{\bar{R}^2}{V}$.

Proof. The first inequality is an immediate application of the Schwarz inequality. In fact, for any $g \in \mathfrak{g}$,

$$S(g)V = \int_M R^2 dV(g) \int_M dV(g) \geq \left(\int_M R dV(g)\right)^2 = \bar{R}^2 .$$

The inequality b) is seen by looking at the Hodge orthogonal decomposition of the closed (1,1)-form Σ_1 ,

$$\Sigma_1 = H\Sigma_1 + (\Sigma_1 - H\Sigma_1) ,$$

where $H\phi$ denotes the Hodge orthogonal projection of any given exterior form ϕ into the space of harmonic forms. Recalling that, in a compact Kähler manifold, H commutes with the contraction operator Λ and substituting $\Sigma_1 - H\Sigma_1$ for Σ_1 in (1.5) and (1.6), we have the following inner product identity, since $\Sigma_1 - H\Sigma_1$ is an exact (1,1)-form

$$(\Sigma_1 - H\Sigma_1 , \Sigma_1 - H\Sigma_1) = (\Lambda(\Sigma - H\Sigma_1), \Lambda(\Sigma_1 - H\Sigma_1)) =$$

$$= (R, R) - (HR, HR) \geq S_0 - \frac{\bar{R}^2}{V} .$$

At the same time, we note that, for the harmonic 2-form $H\Sigma_1$ we have a further orthogonal decomposition

$$H\Sigma_1 = \left(H\Sigma_1 - \frac{1}{n} \Lambda(H\Sigma_1)\omega\right) + \frac{1}{n} (\Lambda H\Sigma_1)\omega .$$

Therefore we have the inequality, since $\Lambda H\Sigma_1$ is equal to the constant $V^{-1}\bar{R}$,

$$Q(g) = (\Sigma_1, \Sigma_1) \geq \left(\tfrac{1}{n}(\Lambda H\Sigma_1)\omega,\ \tfrac{1}{n}(\Lambda H\Sigma_1)\omega\right) + (\Sigma_1 - H\Sigma_1, \Sigma_1 - H\Sigma_1)$$

$$\geq \frac{\overline{R}^2}{nV} + \left(S_0 - \frac{\overline{R}^2}{V}\right)$$

which is the estimate asserted.

We observe, incidentally, that the estimate a) is sharp and attainable, if and only if there exists a Kähler metric $(g) \in \mathfrak{g}$ with constant scalar curvature R. This condition is equivalent to the Ricci form Σ_1 being harmonic. In fact, if R is constant,

$$-*d*\Sigma_1 = -\sqrt{-1}\, g^{\lambda\bar{\mu}}(\nabla_{\bar{\mu}} R_{\lambda\bar{\beta}}\, dz^{\bar{\beta}} - \nabla_\lambda R_{\alpha\bar{\mu}}\, dz^\alpha)$$

$$= \sqrt{-1}\left(\frac{\partial R}{\partial z^\alpha} dz^\alpha - \frac{\partial R}{\partial z^{\bar{\beta}}} dz^{\bar{\beta}}\right) = 0\ ;$$

conversely, if Σ_1 is harmonic, then so is $\Lambda\Sigma_1 = R$, and therefore R is constant. Inequality b) is sharp, only if Σ_1 is cohomologous to a real multiple of ω but has been proved attainable, under that hypothesis, [10], by S. T. Yau, only when Σ_1 is cohomologous to a nonpositive multiple of ω.

Finally, we prove the estimate c) as follows. We introduce the Kählerian complex projective curvature tensor [VII]

$$C_{\alpha\bar{\mu}\bar{\beta}\gamma} = R_{\alpha\bar{\mu}\bar{\beta}\gamma} - \frac{1}{n+2}(R_{\alpha\bar{\beta}} g_{\gamma\bar{\mu}} + R_{\alpha\bar{\mu}} g_{\gamma\bar{\beta}} + R_{\gamma\bar{\beta}} g_{\alpha\bar{\mu}} + R_{\gamma\bar{\mu}} g_{\alpha\bar{\beta}}) +$$

$$+ \frac{1}{(n+1)(n+2)} R(g_{\alpha\bar{\beta}} g_{\gamma\bar{\mu}} + g_{\gamma\bar{\mu}} g_{\gamma\bar{\beta}})$$

and the trace-free part of the Ricci curvature tensor

$$P_{\alpha\bar{\beta}} = R_{\alpha\bar{\beta}} - \frac{1}{n} R g_{\alpha\bar{\beta}}\ .$$

From these two tensors we obtain a pointwise orthogonal, 3-way decomposition of the full curvature tensor

$$R_{\alpha\bar{\mu}\beta\bar{\gamma}} = C_{\alpha\bar{\mu}\beta\bar{\gamma}} + \frac{1}{n+2}(P_{\alpha\bar{\mu}}g_{\gamma\bar{\beta}} + P_{\alpha\bar{\beta}}g_{\gamma\bar{\mu}} + P_{\gamma\bar{\beta}}g_{\alpha\bar{\mu}} + P_{\gamma\bar{\mu}}g_{\alpha\bar{\beta}})$$
$$+ \frac{R}{n(n+1)}(g_{\alpha\bar{\mu}}g_{\gamma\bar{\beta}} + g_{\alpha\bar{\beta}}g_{\gamma\bar{\mu}}) = I + II + III .$$

We take the pointwise squared absolute value of $(R_{\alpha\bar{\mu}\beta\bar{\gamma}})$ thus decomposed and, neglecting the first term, integrate it over M with respect to $dV(g)$, with the following inequality as a result

$$K(g) \geq \frac{4}{n+2}\int_M P_{\alpha\bar{\mu}}P^{\alpha\bar{\mu}}\,dV(g) + \frac{2}{n(n+1)}\,S(g) . \tag{1.7}$$

Since $P_{\alpha\bar{\mu}}P^{\alpha\bar{\mu}} = R_{\alpha\bar{\mu}}R^{\alpha\bar{\mu}} - \frac{1}{n}R^2$ and because of the estimate b) for Q_0, we reduce (1.7) to the following

$$K(g) \geq \frac{4}{n+2}\left(Q(g) - \frac{1}{n}\,S(g)\right) + \frac{2}{n(n+1)}\,S(g)$$
$$\geq \frac{4(n-1)}{n(n+2)}\left(S_0 - \frac{\bar{R}^2}{V}\right) + \frac{2}{n(n+1)}\,S_0 =$$
$$= \frac{2(2n+1)}{(n+1)(n+2)}\,S_0 - \frac{4(n-1)}{n(n+2)}\,\frac{\bar{R}^2}{V} ,$$

which is the estimate asserted in c). This last estimate is probably not sharp, because there are other inequalities satisfied by the curvature tensor in any Kähler manifold, which are not applied here. The corollary is now proved.

A consequence of Proposition 1.1 is that the variational problems of minimizing the three functionals $K(g)$, $Q(q)$ and $S(g)$ for $(g) \in \mathfrak{g}$ are pairwise equivalent. For this reason we shall limit our consideration in the sequel to the formally simpler problem of minimizing $S(g)$. We will conclude this section by showing that the estimate $\frac{\bar{R}^2}{V}$ for a lower bound S_0 of $S(g)$, that is estimate a) of the corollary just proved, is not sharp in the case of some special compact Kähler manifolds. This fact is a sort of a converse of a theorem proved first, in a weaker form, by Y. Matsushima [8] and shortly after, in the form needed here, by A. Lichnérowicz [7; IV].

THEOREM 1.2. *Let* M *be a compact,* n*-dimensional complex manifold admitting Kahler metrics, and assume that* M *admits a nontrivial, connected Lie group of holomorphic transformations. If the maximal connected group* G *of holomorphic transformations of* M *is not reductive, then the lower bound* $\frac{\overline{R}^2}{V}$ *cannot be achieved by the functional* S(g) *for any Kähler metric in any cohomology class.*

Proof. We recall the theorem of Bochner and Montgomery [1, 2] to the effect that the maximal connected group of holomorphic automorphisms of any compact, complex manifold is a complex Lie group, which we denote by G. The group G is reductive, if it is the smallest complex Lie group containing its maximal compact subgroup (as a real Lie subgroup). If the lower bound $\overline{R}^2/V$ for S(g) is achieved by some Kähler metric $(g) \in \mathfrak{g}$, this means that the scalar curvature R with respect to the metric (g) is constant. The theorem of Matsushima-Lichnérowicz [7], [IV, pp. 153-158] asserts that, if a compact Kähler manifold M has constant scalar curvature, then the group G of holomorphic transformations of M is reductive, contradicting our assumption. The information that this theorem does *not* provide, under the stated assumptions, is a positive lower bound for $S_0 - (\overline{R}^2/V)$, although very probably such a bound could be found in terms of the structure of the Lie algebra of G (as a Lie algebra of holomorphic tangent vector fields in M) and of the cohomology class $[\omega]$ of the Kähler metrics under consideration.

2. *Variational formulas*

Let M be a compact, n-dimensional complex manifold with a principal class $[\omega]$ determined by the principal (1,1)-form of a given Kähler metric (1.1), and let $\mathfrak{g}$ be the function space of all Kähler metrics belonging to the same class, with a sufficiently strong function space topology (for instance, the uniform C^4 topology). A one-parameter family of metrics in $\mathfrak{g}$, parametrized by a real variable $t\ (t_0<t<t_1,\ t_0<0<t_1)$, is representable, locally, as a one-parameter family of positive definite hermitian forms (g_t) with coefficient matrices

$$g_{\alpha\overline{\beta}}(t;z,\overline{z}) = g_{\alpha\overline{\beta}}(z,\overline{z}) + \frac{\partial^2 u(t;z,\overline{z})}{\partial z^\alpha \partial z^{\overline{\beta}}}$$

where $u(t;z,\overline{z})$ is a smooth, real-valued function, globally defined in $(t_0,t_1)\times M$. Without loss of generality we may set $g_{\alpha\overline{\beta}}(z,\overline{z}) = g_{\alpha\overline{\beta}}(0;z,\overline{z})$ and $u(0;z,\overline{z})=0$. The infinitesimal variational formulas represent the derivatives with respect to t, at $t=0$, of various geometric objects (connections, curvature, etc.) attached to (g_t), expressed as the result of some linear differential operators on the infinitesimal distortion potential $u(z,\overline{z}) = \left.\frac{\partial u(t;z,\overline{z})}{\partial t}\right|_{t=0}$.

Thus, denoting the infinitesimal variations by the usual prefix δ, we have

$$\delta g_{\alpha\overline{\beta}} = u_{,\alpha\overline{\beta}}$$

$$\delta(\det g_{\alpha\overline{\beta}}) = \det(g_{\alpha\overline{\beta}})\, g^{\lambda\overline{\mu}} u_{,\lambda\overline{\mu}} ;$$

therefore the variation of the Ricci curvature is given by

$$\delta R_{\alpha\overline{\beta}} = \frac{\partial^2}{\partial z^\alpha \partial z^{\overline{\beta}}}\, \Delta u ,$$

where Δ denotes the Laplace operator $-g^{\lambda\overline{\mu}}\nabla_\lambda \nabla_{\overline{\mu}}$, and

$$\begin{aligned}
\delta R &= g^{\alpha\overline{\beta}}\delta R_{\alpha\overline{\beta}} - g^{\lambda\overline{\beta}} g^{\alpha\overline{\mu}}(\delta g_{\lambda\overline{\mu}}) R_{\alpha\overline{\beta}} \\
&= -\Delta^2 u - R_\mu{}^\lambda u_{,\lambda}{}^\mu = -u_{,\alpha}{}^\alpha{}_\beta{}^\beta - u_{,\lambda}{}^\mu R_\mu{}^\lambda \\
&= (-u_{,\alpha\beta}{}^\alpha + u_{,\lambda} R^\lambda{}_{\alpha\beta}{}^\alpha)^{,\beta} - u_{,\lambda}{}^\mu R_\mu{}^\lambda \\
&= -u_{,\alpha\beta}{}^{\alpha\beta} + g^{\lambda\overline{\mu}} u_{,\lambda} R_{,\overline{\mu}} .
\end{aligned}$$

From the last expression we obtain the variational formula for $S(g) = \int_M R^2 dV(g)$

$$\delta S(g) = \int_M \delta(R^2 \det(g_{\alpha\overline{\beta}}))(-1)^{\frac{n}{2}} dz^1 \wedge d\overline{z}^1 \wedge \cdots \wedge dz^n \wedge d\overline{z}^n ,$$

where

$$\delta(R^2 \det(g_{\alpha\overline{\beta}})) = \det(g_{\alpha\overline{\beta}})(-R^2\Delta u + 2R(-u_{,\alpha\beta}{}^{\alpha\beta} + g^{\lambda\overline{\mu}}u_{,\lambda}R_{,\overline{\mu}}))$$

$$= \det(g_{\alpha\overline{\beta}})(-2Ru_{,\alpha\beta}{}^{\alpha\beta} + g^{\lambda\overline{\mu}}(R^2u_{,\lambda})_{,\overline{\mu}}) .$$

Since the integral of the second term on M vanishes, we have the formula, expressed in terms of functional inner products

(2.2) $$\delta S(g) = \delta(R, R) = -2(R, u_{,\alpha\beta}{}^{\alpha\beta}) - 2\int_M Ru_{,\alpha\beta}{}^{\alpha\beta} dV(g) ;$$

we obtain thus the following result

THEOREM 2.1. *The Euler-Lagrange equation for a Kähler metric* $g \in \mathfrak{g}$ *for which* $S(g) = (R, R)$ *is stationary is the following condition on the scalar curvature*

(2.3) $$R_{,\overline{\alpha}\overline{\beta}} = 0 \quad (\text{or equivalently } R_{,\alpha\beta} = 0).$$

In other words, the Euler equation expresses the condition that the real vector field on M

$$X = g^{\lambda\mu}\left(R_{,\lambda}\frac{\partial}{\partial\overline{z}^{\mu}} + R_{,\mu}\frac{\partial}{\partial z^{\lambda}}\right)$$

generates a holomorphic flow (possibly trivial, if R *is constant).*

Proof. We start with the variational formula (2.2) for $S(g)$. Equating it to zero for all real valued functions u, and applying Gauss' theorem four times, we obtain the equation

$$(R^{,\alpha\beta}{}_{\alpha\beta}, u) = 0$$

which, if satisfied for all functions u, means

$$R^{,\alpha\beta}{}_{\alpha\beta} = 0 . \tag{2.4}$$

Next we prove the following lemma

LEMMA. *If a (real or complex valued) function* ϕ *satisfies the equation* $\phi^{,\alpha\beta}{}_{\alpha\beta} = 0$ *in a compact Kähler manifold, then* $\phi_{,\overline{\alpha}\overline{\beta}} \equiv 0$.

Proof. If $\phi^{,\alpha\beta}{}_{\alpha\beta} \equiv 0$ identically, then

$$\begin{aligned} 0 \leq g^{\alpha\overline{\lambda}} g^{\beta\overline{\mu}} \phi_{,\overline{\lambda}\overline{\mu}} (\phi_{,\overline{\alpha}\overline{\beta}}) &= (\phi^{,\alpha\beta} \overline{\phi}_{,\alpha} - \phi^{,\alpha\beta}{}_{\alpha} \overline{\phi})_{,\beta} + \phi^{,\alpha\beta}{}_{\alpha\beta} \overline{\phi} \\ &= *d*(g_{\beta\overline{\mu}} (\phi^{,\alpha\beta}{}_{\alpha} \overline{\phi} - \phi^{,\alpha\beta} \overline{\phi}_{,\alpha}) dz^{\overline{\mu}}) ; \end{aligned}$$

were the last expression is written as a divergence; thus, because of Gauss' theorem, $\phi_{,\overline{\lambda}\overline{\mu}} = 0$, proving the lemma.

We apply the lemma to the scalar curvature R, obtaining from (2.4), the asserted equation

$$R_{,\overline{\alpha}\overline{\beta}} = 0 ,$$

which is equivalent to the statement that the gradient vector field of R, grad $R = g^{\alpha\overline{\beta}} \left(\frac{\partial R}{\partial z^{\overline{\beta}}} \frac{\partial}{\partial z^{\alpha}} + \frac{\partial R}{\partial z^{\alpha}} \frac{\partial}{\partial z^{\overline{\beta}}} \right)$, generates a one-parameter group of holomorphic transformations of M. This completes the proof of the theorem.

In the next section we shall exhibit an example of a class of Kähler manifolds with metrics satisfying this Euler equation, but where the scalar curvature is not constant.

COROLLARY. *If a Kähler metric satisfies the Euler equation (2.3), then the nonnegative functionals* Q(g) *and* K(g) *are stationary, as well as* S(g), *under infinitesimal deformations of the Kähler metric preserving the principal class.*

This is an immediate consequence of Proposition 1.1.

There is strong intuitive reason to believe that every Kähler metric satisfying (2.3) achieves a minimum value for S(g) under changes of the metric, not only in the "trivial" cases where R is constant. The main indication comes from a computation of the second variation

THEOREM 2.2. *Let* M *be an* n-*dimensional, compact Kähler manifold with a fixed class* $[\omega]$ *of Kähler metrics. Let* $(g) \in [\omega]$ *be any metric on* M *satisfying the Euler equation*

$$R_{,\bar{\alpha}\bar{\beta}} = 0$$

for a critical point of the functional $S(g) = (R, R)$. *Then the second variation of* S(g) *with respect to any infinitesimal deformation with*

$$\delta g_{\alpha\bar{\beta}} = u_{,\alpha\bar{\beta}}$$

is positive semi-definite, and is zero, if and only if $\delta g_{\alpha\bar{\beta}}$ *is induced by a holomorphic flow.*

Proof. From the first variation of (R, R)

$$\delta(R, R) = -2(R_{,\bar{\alpha}\bar{\beta}}{}^{\bar{\alpha}\bar{\beta}}, u)$$

$$= -2 \int_M R_{,\bar{\alpha}\bar{\beta}}{}^{\bar{\alpha}\bar{\beta}} u dV(g) ,$$

one calculates the second variation $\delta^2(R, R)$. This calculation is tedious, since it involves the variation of the connection in the fourth order covariant derivatives. However, the final terms are greatly simplified, owing to the fact that $R_{,\bar{\alpha}\bar{\beta}} = 0$. The final outcome is given by the following expression, after which one may apply Gauss's theorem

$$\delta^2(R,R) = 2(u_{,\lambda\mu}{}^{\lambda\mu\alpha\beta}{}_{\alpha\beta}, u) = 2(u_{,\lambda\mu}{}^{\lambda\mu}, \overline{u_{,\overline{\lambda}\overline{\mu}}{}^{\overline{\lambda}\overline{\mu}}})$$

to show that $\delta^2(R,R) > 0$ except if $u_{,\overline{\lambda}\overline{\mu}}{}^{\overline{\lambda}\overline{\mu}} = 0$. The latter equation is satisfied, according to the lemma of Theorem 2.1, if and only if

$$u_{,\overline{\lambda}\overline{\mu}} = 0$$

or, equivalently, since u is real valued, if the real vector field

$$U = g^{\lambda\overline{\mu}}\left(u_{,\overline{\mu}}\frac{\partial}{\partial z^\lambda} + u_{,\lambda}\frac{\partial}{\partial z^{\overline{\mu}}}\right)$$

generates a holomorphic flow. It is then immediate to prove that

$$\delta(g_{\alpha\overline{\beta}}) = \frac{1}{2}\mathcal{L}_U(g_{\alpha\overline{\beta}}),$$

the last expression denoting the Lie derivative of the metric with respect to the gradient field U of u. This proves the theorem.

3. *Some examples*

The theorem of Lichnérowicz-Matsushima (quoted here as Theorem 1.2) states that any compact Kählerian manifold M whose maximal connected group G of automorphisms is not reductive does not admit any Kähler metric with constant scalar curvature. Therefore, if the variational problem of minimizing (R, R) over the space of all metrics with a given principal class has a solution, it must be one for which the vector field $X = g^{\alpha\overline{\beta}}\left(R_{,\overline{\beta}}\frac{\partial}{\partial z^\alpha} + R_{,\alpha}\frac{\partial}{\partial z^{\overline{\beta}}}\right)$ (generating a one-parameter group of holomorphic transformations) is nontrivial. It is interesting to note that, in the case of one complex dimension (compact Riemann surfaces), the variational problem of minimizing (R, R) (here, fixing the principal class is equivalent to fixing the total area of M) has R = constant as the only stationary metric. A lemma equivalent to this fact was proved, independently, also by J. Kazdan and F. Warner [6].

THEOREM 3.1. *The only solution of the Euler equation for metrics satisfying* $R_{,\overline{\alpha\beta}} = 0$ *on a compact Riemann surface are metrics of constant scalar curvature* R.

Proof. In the case of genus ≥ 2, there are no holomorphic tangent vector fields, so that the condition $R_{,\overline{\alpha\beta}} = 0$, equivalent to the vector field $g^{\alpha\overline{\beta}} R_{,\overline{\beta}} \frac{\partial}{\partial z^\alpha}$ being holomorphic, implies $R_{,\overline{\beta}} = 0$ and hence $R = \text{constant}$. In the case of genus one, the nontrivial holomorphic tangent vector fields are nowhere vanishing and hence $g^{\alpha\overline{\beta}} R_{,\overline{\beta}} \frac{\partial}{\partial z^\alpha}$, which vanishes at the critical points of R, cannot be holomorphic unless it vanishes identically.

Finally, in the case of the sphere (genus zero), if there existed a metric with $g^{\alpha\overline{\beta}} R_{,\overline{\beta}} \frac{\partial}{\partial z^\alpha}$ holomorphic and not identically zero, then this vector field should have at least two zeros (critical points of R), but then also at most two zeros (fixed points of a one-parameter group of conformal transformations). Furthermore the rotated vector field

$$JX = g^{\alpha\overline{\beta}} \left(\sqrt{-1} \frac{\partial R}{\partial \overline{z^\beta}} \frac{\partial}{\partial z^\alpha} - \sqrt{-1} \frac{\partial R}{\partial z^\alpha} \frac{\partial}{\partial \overline{z^\beta}} \right)$$

generates a compact group of isometries, or rotations of the sphere. A 2-sphere with a one-parameter group of isometries can be parameterized (as a Riemannian manifold) by two real parameters (u, v), where $0 \leq u \leq a$ for some constant $a > 0$, and $v \in \mathbf{R}/2\pi\mathbf{Z}$, with each of the boundary circles, represented respectively by $u = 0$ and $u = a$, collapsed to a corresponding point. The isometry group is represented by the translations of the parameter v and the metric is expressed in terms of these parameters by

$$ds^2 = du^2 + (\phi(u))^2 \, dv^2 ,$$

where $\phi(u)$ is a smooth, odd, periodic function of $u \in \mathbf{R}$ with period 2a, satisfying the following conditions in the closed interval $[0, a]$:

a) $\phi(u) > 0$ for $0 < u < a$

b) $\phi(0) = \phi(a) = 0$, $\phi'(0) = -\phi'(a) = 1$.

The Gaussian (scalar) curvature R is a function depending only on u, namely $R = -\frac{\phi''(u)}{\phi(u)}$ for $0 < u < a$, with R bounded at the end points. One shows that, under these conditions, R must have more than two critical points on the sphere. In fact, the function

$$\psi(u) = (\phi'(u))^2 + R(\phi(u))^2$$

has, by assumption, the same value (unity) for $u = 0$ or a, and hence $\psi'(u)$ must vanish at least at one point $u_0 (0 < u_0 < a)$. But

$$\begin{aligned}\psi'(u_0) &= 2\phi'(u_0)(\phi''(u_0) + R(u_0)\phi(u_0)) + R'(u_0)(\phi(u_0))^2 \\ &= R'(u_0)(\phi(u_0))^2\end{aligned}$$

and $\phi(u_0) > 0$, so that $R'(u_0) = 0$, verifying that R, as a function on the sphere, must have critical points not only at each of the two "poles," $u = 0$, $u = a$, of the sphere, but also at each point of some circle $\{u = u_0,\ 0 \le v < 2\pi\}$. Thus the gradient field of R cannot generate a one-parameter group of conformal transformations of M. This completes the proof of the theorem.

An analogous statement for extremal Kähler metrics in higher dimensions is easy to prove in the case of Kähler manifolds of a general type (complex manifolds M with no holomorphic tangent vector fields) or in the case of products of a manifold of this general type with a complex torus, since each one-parameter group of holomorphic transformations in this case has no fixed point. However, in the case of manifolds with transformation groups admitting fixed points, there are considerable difficulties in proving the existence or any local compactness property of the space of Kähler metrics with $R_{,\overline{\alpha}\overline{\beta}} = 0$. For instance, it is not known, in the case of compact, higher dimensional, *homogeneous*, simply connected, complex manifolds admitting Kähler metrics, whether the equation $R_{,\bar{\alpha}\bar{\beta}} = 0$ implies that R = constant.

We shall conclude this paper with a construction of compact, complex manifolds M each admitting a Kähler metric g satisfying $R_{,\bar{\alpha}\bar{\beta}} = 0$ with R not constant; the idea of this construction is to assume M to be sufficiently "symmetric" to reduce the equation to an ordinary differential equation.

The manifolds M, with n complex dimensions, are in all cases complex projective line bundles of a certain type over a compact, symmetric, $(n-1)$-dimensional Kähler manifold; for the sake of simplicity we shall limit the description to bundles over the $(n-1)$-dimensional, complex projective $(n-1)$-space $P^{n-1}_{\mathbf{C}}$. The latter is then represented as the quotient of the homogeneous coordinate space $\mathbf{C}^n \setminus \{0\}$ (with numerical coordinates $(x_1, \cdots, x_n)$) by the action of the group of nonzero scalar multiplications. The base space $P^{n-1}_{\mathbf{C}}$ is covered by n coordinate domains $U_\lambda (1 \leq \lambda \leq n)$; each U_λ is characterized by the condition $x_\lambda \neq 0$ and covered by the (nonhomogeneous) holomorphic coordinate system $({}_\lambda z^\alpha)$ $(1 \leq \alpha \leq n;\ \alpha \neq \lambda)$, ranging over $\mathbf{C}^{n-1}$, where ${}_\lambda z^\alpha = \frac{x_\alpha}{x_\lambda}$ (we use this equation also, with $\alpha = \lambda$ so that, formally, ${}_\lambda z^\lambda = 1$).

We describe a bundle $M \xrightarrow{\pi} P^{n-1}_{\mathbf{C}}$ by introducing a projective, holomorphic fibre coordinate $y_\lambda \in \mathbf{C} \cap \{\infty\}$ in $\pi^{-1}(U_\lambda)$ with the transition relation in $\pi^{-1}(U_\lambda \cap U_\mu)$

$$y_\mu = \left(\frac{x_\mu}{x_\lambda}\right)^k y_\lambda = ({}_\lambda z^\mu)^k y_\lambda, \tag{3.1}$$

for any fixed, integral exponent k; without loss of generality we set $k \geq 0$ (otherwise one may replace each y_λ by y_λ^{-1} and k by $-k$). The resulting manifold will be denoted by $M_k (k = 0, 1, 2, \cdots)$. The manifold M_0 is the trivial bundle $P^{n-1}_{\mathbf{C}} \times P^1_{\mathbf{C}}$ and will be excluded from our consideration; for $k \geq 1$ the bundle $M_k \xrightarrow{\pi} P^{n-1}_{\mathbf{C}}$ has two distinguished cross sections $s_0, s_\infty : \mathbf{P}^{n-1}_{\mathbf{C}} \to M_k$, whose images in M_k, denoted by S_0 and S_∞, are described respectively by the equations $y_\lambda = 0$ and $y_\lambda = \infty$ for each λ. The complement $M' = M_k \setminus (S_0 \cup S_\infty)$ of these two cross sections can be globally parametrized by the homogeneous coordinate space $C^n \setminus \{0\}$ under the k-to-one map that assigns to any point $(x_1, \cdots, x_n)$ with, let us say, $x_\lambda \neq 0$, the point in $M' \cap \pi^{-1}(U_\lambda)$ with coordinates

$$(3.2) \qquad ((_\lambda z^\alpha);\, y_\lambda) = \left(\left(\frac{x_\alpha}{x_\lambda}\right);\, (x_\lambda)^k\right) \quad (1 \le \alpha \le n;\ \alpha \neq \lambda)\,.$$

We shall describe now the maximal connected group G of holomorphic transformations of M_k. This group is a complex Lie group [1, 2]; since it is connected, its action preserves the fibration $\pi : M_k \to \mathbf{P}^{n-1}_{\mathbf{C}}$, and hence has a homomorphism $\overline{\pi}$ into the complex projective group $PGL(n, \mathbf{C})$ acting on $\mathbf{P}^{n-1}_{\mathbf{C}}$. On the other hand G contains a compact subgroup G_K, represented by the action of the unitary group $U(n)$ on the complex parameters $(x_1, \cdots, x_n)$ occurring in (3.2), where the action of $U(n)$ on M_k is effective, modulo the central subgroup Z_k consisting of scalar multiplications by the k'th roots of unity. Hence G contains the complexification of G_K, which is isomorphic to $GL(n, \mathbf{C})/Z_k$, and, since $\overline{\pi}$ obviously maps the latter group onto $PGL(n, \mathbf{C})$, the homomorphism $\overline{\pi} : G \to PGL(n, \mathbf{C})$ is surjective.

The kernel $B \subset G$ of $\overline{\pi}$ is the subgroup of G mapping each fibre $\pi^{-1}(z) \subset M_k$ $(z \in \mathbf{P}^{n-1}_{\mathbf{C}})$ onto itself; each element can be described by extending (by continuity) its action in any coordinate domain $\pi^{-1}(U_\lambda)$, in terms of the coordinates $((_\lambda z^\alpha);\, y_\lambda)$ of (3.2), mapping the point $(_\lambda z^\alpha), y_\lambda)$ into the point $(_\lambda z^\alpha), y'_\lambda)$ with

$$(y'_\lambda)^{-1} = c(y_\lambda)^{-1} + P_k(_\lambda z^\alpha)\,,$$

where c is any complex constant $\neq 0$ and P_k any polynomial of degree $\le k$ in the $n-1$ variables $(_\lambda z^\alpha)$ $(\alpha \neq \lambda)$. Thus B is a complex, solvable

Lie group with $\binom{n+k-1}{k}+1$ complex dimensions, G is $\left(n^2+\binom{n+k-1}{n}\right)$-dimensional, has trivial center and has the n^2-(real)-dimensional subgroup G_K as its maximal compact subgroup. Therefore G is not reductive.

We shall now construct a Kähler metric in M_k satisfying the Euler equation $R_{,\alpha\overline{\beta}} = 0$, by a method similar to one developed in [5] for other purposes, and where the construction is described in greater detail. We begin by assuming that such a metric is, in addition, invariant under the action of a maximal compact subgroup G_K ($G_K \cong U(n)/Z_k$) of transformations of M_k. It is unknown (but plausible to speculate) whether this restriction is necessary for the solution of the Euler equation in general for M_k. The calculations are carried out most conveniently in terms of the coordinate covering $(x_1, \cdots, x_n)$ of $M_k \setminus (S_0 \cup S_\infty)$, since the action of G_K is the easiest to describe in terms of these coordinates. The orbits under the action of G_K are parametrized by the real parameter $t = \log \sum_{\alpha=1}^{n} |x_\alpha|^2$; any Kähler metric $g_{\alpha\overline{\beta}}(x, \overline{x})$ in $\mathbb{C}^n \setminus \{0\}$ invariant under the action of U(n) can be generated by a Kähler potential $\Phi(x, \overline{x})$ $\quad g_{\alpha\overline{\beta}} = \dfrac{\partial^2 \Phi}{\partial x_\alpha \partial \overline{x}_\beta}$ such that

$$\Phi(x, \overline{x}) = u(t) \quad \left(t = \log\left(\sum_{\alpha=1}^{n} |x_\alpha|^2\right)\right)$$

for some function u(t) which is differential for all real t. Conversely, any function u(t) generates a positive definite Kähler metric in $\mathbb{C}^n \setminus \{0\}$ (obviously invariant under U(n)), and hence in $M' = M_k \; (S_0 \cup S_\infty)$, if and only if it satisfies the differential inequations

$$u'(t) > 0 , \quad u''(t) > 0 . \tag{3.3}$$

In fact, for any smooth function u(t), for $t = \log \sum_{\alpha=1}^{n} |x_\alpha|^2$,

(3.4) $$g_{\alpha\bar{\beta}} = \frac{\partial}{\partial x_\alpha \partial \bar{x}_\beta} u\left(\log \sum_\lambda |x_\lambda|^2\right) = e^{-t}u'(t)\,\delta_{\alpha\beta} + e^{-2t}\bar{x}_\alpha x_\beta (u''(t) - u'(t))$$

and

(3.5) $$\det(g_{\alpha\bar{\beta}}) = e^{-nt}(u'(t))^{n-1}u''(t) ,$$

so that (3.3) is necessary and sufficient in order that $g_{\alpha\bar{\beta}}$ be positive definite. The property that any Kähler metric (3.4) in $C^n \setminus \{0\}$, pulled into $M' = M_k \setminus (S_0 \cup S_\infty)$, be extendable by continuity to a positive definite, smooth metric in all of M_k can be translated into the following asymptotic properties of $u(t)$ as $t \to \pm\infty$:

a) There exists a real constant $a > 0$ such that the function $\bar{u}_0(r)$, defined for all $r > 0$ by the equation

$$u_0(e^{kt}) = u(t) - at ,$$

is extendable by continuity to a smooth function at $r = 0$, satisfying the additional condition

(3.6) $$u'_0(0) > 0 .$$

b) There exists a real constant $b > 0$ such that the function $\overline{u_\infty}(\bar{r})$, defined for all positive $\bar{r}$ by the equation

$$\overline{u_\infty}(e^{-kt}) = u(t) - bt ,$$

is extendable by continuity to a smooth function at $\bar{r} = 0$, satisfying the additional condition

(3.7) $$u'_\infty(0) > 0 .$$

It is easy to verify the necessity and sufficiency of the above conditions by calculating the metric in a neighborhood of S_0 and S_∞ respectively.

We note that the positive constants a, b describe the cohomology class of the resulting Kähler metric. In fact, the second homology group of M_k with real coefficients is generated by the two 2-cycles represented by two complex projective lines lying one in each of the two cross sections S_0 and S_∞; the restriction of the Kähler metric to each of these cross sections is a Fubini-Study metric with scalar curvature respectively $\frac{n(n-1)}{a}$ and $\frac{n(n-1)}{b}$. Therefore the integral of $\omega = \sqrt{-1}\, g_{\alpha\overline{\beta}}\, dz^\alpha \wedge dz^{\overline{\beta}}$ on each of these two projective lines is respectively $2\pi a$ and $2\pi b$. Furthermore $b > a$, since $a = \lim_{t\to-\infty} u'(t)$, $b = \lim_{t\to\infty} u'(t)$, and $u''(t) > 0$ for all real t.

For any function $u(t)$ satisfying (3.3) and the asymptotic conditions (3.6), (3.7) in terms of preassigned constants a, b $(0<a<b)$, define the function (cf. (3.5)).

$$v(t) = -\log\det(g_{\alpha\overline{\beta}}(x,\overline{x})) = nt-(n-1)\log u'(t) - \log u''(t)\;; \tag{3.8}$$

from this one calculates the Ricci tensor

$$R_{\alpha\overline{\beta}}(x,\overline{x}) = \frac{\partial^2}{\partial x_\alpha \partial \overline{x}_\beta}\, v\left(\log\sum_{\lambda=1}^{n}|x_\lambda|^2\right) = e^{-t}v'(t)\delta_{\alpha\beta} + e^{-2t}\overline{x}_\alpha x_\beta(v''(t)-v'(t))\,.$$

Next, one computes the contravariant components of the metric from (3.4) and (3.5),

$$g^{\alpha\overline{\beta}}(x,\overline{x}) = e^t(u'(r))^{-1}\delta^{\alpha\beta} + x_\alpha\overline{x}_\beta((u''(t))^{-1} - (u'(t))^{-1})\;;$$

these are used in computing the scalar curvature $R = g^{\alpha\overline{\beta}}R_{\alpha\overline{\beta}}$; we get $R = w(t)$ where

$$w(t) = (n-1)\frac{v'(t)}{u'(t)} + \frac{v''(t)}{u''(t)}\,. \tag{3.9}$$

Finally the equation $R_{,\overline{\alpha}\overline{\beta}} = 0$, equivalent to the condition that the components of the tangent vector field $g^{\alpha\overline{\beta}}\frac{\partial R}{\partial \overline{x}_\beta}\frac{\partial}{\partial x_\alpha}$ be holomorphic, is expressed in terms of $w(t) = R$ as follows: since

$$g^{\alpha\overline{\beta}} \frac{\partial R}{\partial \overline{x}_{\beta}} = g^{\alpha\overline{\beta}}(x, \overline{x}) e^{-t} x_{\beta} w'(t) = x_{\alpha} \frac{w'(t)}{u''(t)}$$

and since x_{α} is holomorphic and $\frac{w'(t)}{u''(t)}$ real valued, the Euler equation $R_{,\overline{\alpha}\overline{\beta}} = 0$ is equivalent to the equation

$$\frac{w'(t)}{u''(t)} = \text{constant} ;$$

we set the constant (for typographical convenience) to be $(n+1)(n+2)c_1$; the differential equation

$$w' = (n+1)(n+2)c_1 u''$$

is immediately integrable to

$$R = w'(t) = (n+1)(n+2)c_1 u'(t) + n(n+1)c_2 . \tag{3.10}$$

Replacing $w(t)$ by its expression in terms of $u(t)$, $v(t)$ and their derivatives according to (3.9), we obtain a second order differential relation between u and v that admits $(u'(t))^{n-1}u''(t)$ as an integrating factor. The new integration yields

$$(u'(t))^{n-1}v'(t) = (n+2)c_1(u'(2))^{n+1} + (n+1)c_2(u'(t))^n + c_3 . \tag{3.11}$$

We replace $v'(r)$ by the derivative of the equivalent expression of v in terms of $u(t)$ according to (3.8) and multiply both sides by the integrating factor $u''(t)$, obtaining the equation

$$n(u'(t))^{n-1}u''(t) - (n-1)(u'(t))^{n-2}(u''(t))^2 - (u'(t))^{n-1}u''(t)\,u'''(t) =$$
$$= \{(n+2)c_1(u'(t))^{n+1} + (n+1)c_2(u'(t))^n + c_3\}\, u''(t) .$$

This equation is integrable directly to the first order differential equation with respect to $\psi(t) = u'(t)$

$$(u'(t))^n - (u'(t))^{n-1}u''(t) = c_1(u'(t))^{n+2} + c_2(u'(t))^{n+1} + c_3u'(t) + c_4 ,$$

or

$$\frac{d\psi(t)}{dt} = -c_1\psi^{n+2} - c_2\psi^{n+1} + \psi^n - c_3\psi - c_4 .$$

The separation of the variables t and ψ gives

$$dt = \frac{\psi^{n-1}\,d\psi}{-c_1\psi^{n+2} - c_2\psi^{n+1} + \psi^n - c_3\psi - c_4} .$$

The integration of both sides establishes an implicit functional relation and $\psi(t) = u'(t)$. The solution is a function $u(t)$ satisfying the inequalities (3.3), if and only if the constants c_1, c_2, c_3, c_4 are such that the rational function of ψ,

$$f(\psi) = \frac{\psi^{n-1}}{-c_1\psi^{n+2} - c_2\psi^{n+1} + \psi^n - c_3\psi - c_4} , \tag{3.12}$$

is strictly positive valued in an interval I in the half-line $\psi > 0$. The completeness of the range of values of the "independent" variable t in the whole real line requires that the interval I in which ψ varies be bounded by two poles of the rational function (3.12) of ψ. Finally the asymptotic conditions of $u(t)$ as $t \to -\infty$ and as $t \to +\infty$, expressed respectively by (3.6) and (3.7) in terms of the positive constants a, b $(0 < a < b)$, can be translated into the following properties of (3.12):
i) the rational function (3.12) has simple poles at $\psi = a$ and at $\psi = b$ with residues respectively equal to $\frac{1}{k}$ and $-\frac{1}{k}$; ii) the rational function of ψ,

$$\frac{\psi^{n-1}(b-\psi)(\psi-a)}{-c_1\psi^{n+2} - c_2\psi^{n+1} + \psi^n - c_3\psi - c_4}$$

is analytic and strictly positive valued in the closed interval $[a,b]$. These conditions correspond, for any given constants a, b $(0<a<b)$ to a system of four linear equations in c_1, c_2, c_3, c_4 with a unique set of solutions, which we include for the record:

(3.13)*
$$c_1 = \frac{(k-1)b^{2n}+n(n-k)a^{n-1}b^{n+1}-2(n^2-1)a^n b^n+n(n+k)a^{n+1}b^{n-1}-(k+1)a^{2n}}{b^{2n+2}-(n+1)^2 a^n b^{n+2}+2n(n+2)a^{n+1}b^{n+1}-(n+1)^2 z^{n+2}b^n+a^{2n+2}},$$

$$c_2 = \frac{(-k+2)b^{2n+1}-(n+1)(n-k)a^{n-1}b^{n+2}+(n+2)(n-k-1)a^n b^{n+1}+(n+2)(n+k-1)a^{n+1}b^n-(n+1)(n+k)a^{n+2}b^{n-1}+(k+2)a^{2n+1}}{b^{2n+2}-(n+1)^2 a^n b^{n+2}+2n(n+2)a^{n+1}b^{n+1}-(n+1)^2 a^{n+2}b^n+a^{2n+2}},$$

$$c_3 = \frac{(n-k)a^{n-1}b^{2n+2}+(n+1)(k-2)a^n b^{2n+1}-(n+2)(k-1)a^{n+1}b^{2n}+(n+2)(k+1)a^{2n}b^{n+1}-(n+1)(k+2)a^{2n+1}b^n+(n+k)a^{2n+2}b^{n-1}}{b^{2n+2}-(n+1)^2 a^n b^{n+2}+2n(n+2)a^{n+1}b^{n+1}-(n+1)^2 a^{n+2}b^n+a^{2n+2}},$$

$$c_4 = \frac{-(n-k-1)a^n b^{2n+2}-n(k-2)a^{n+1}b^{2n+1}+(n+1)(k-1)a^{n+2}b^{2n}-(n+1)(k+1)a^{2n}b^{n+2}+n(k+2)a^{2n+1}b^{n+1}-(n+k-1)a^{2n+2}b^n}{b^{2n+2}-(n+1)^2 a^n b^{n+2}+2n(n+2)a^{n+1}b^{n+1}-(n+1)^2 a^{n+2}b^n+a^{2n+2}}.$$

*In each of the following four fractions the numerators are divisible by $(b-a)^3$ and the denominators by $(b-a)^4$; the simplified form, however, is clumsy to express.

The verification of all the inequalities required of the rational function (3.12) of ψ,

$$f(\psi) = \frac{\psi^{n-1}}{-c_1\psi^{n+2} - c_2\psi^{n+1} + \psi^n - c_3\psi - c_4},$$

is an exceedingly tedious task to carry out in its full generality. However, in the case of surfaces $(n=2)$, i.e. when the base space is a projective line, the rational function reduces to

$$\frac{\psi}{-c_1\psi^4 - c_2\psi^3 + \psi^2 - c_3\psi - c_4} =$$

$$= \frac{b-a}{k(\psi-a)(b-\psi)} + \frac{(b-a)((k-1)b+(k+1)a)}{k((k-1)b+(k+1)a)(\psi^2+ab) + k(b^2+2kab-a^2)\psi},$$

which trivially satisfies the inequalities for all choices of integers $k \geq 1$ and $0 < a < b$. While the solution of the equation (3.10) to determine $u(r)$ in closed form is impossible (except perhaps for some special values of a, b, k, none of which have been found), nevertheless one can calculate some of the properties of the resulting manifolds. In particular we are interested in the total volume V of M_k under the metric g determined by the cohomology class represented by the pair (a, b) $(0<a<b)$, and in the integrals

$$\bar{R} = \int_{M_k} R\,dV(g), \quad S(g) = \int_{M_k} R^2\,dV(g).$$

Let σ_n denote the volume of the unit Euclidean ball in $2n$ (real) dimensions, $\sigma_n = \pi^n(n!)^{-1}$; then we refer to (3.5) in expressing the volume element induced by the metric in $C^n\backslash\{0\}$,

$$dV_0 = \det(g_{\lambda\bar{\mu}}(x,\bar{x})) \bigwedge_{\alpha=1}^{n} (\sqrt{-1}\,dx_\alpha \vee d\bar{x}_\alpha) = 2^n n\sigma_n(u'(t))^{n-1}u''(t)\,dt.$$

Recalling that the parametrization map of $C^n\backslash\{0\}$ onto $M_k \ (S_0 \cup S_\infty)$ is k-to-one, we obtain the volume formula for M_k,

$$(3.14) \qquad V = 2^n k^{-1} \sigma_n \int_{-\infty}^{\infty} n(u'(t))^{n-1} u''(t)\, dt = 2^n k^{-1} \sigma_n (b^n - a^n) .$$

The scalar curvature R is expressed conveniently by (3.9), so that

$$(3.15) \quad RdV = 2^n k^{-1} n\sigma_n ((n-1)\, v'(t)\, (u'(t))^{n-2} d(u'(t)) + (u'(t))^{n-1} d(v'(t)) =$$

$$= 2^n k^{-1} n\sigma_n d\{(u'(t))^{n-1} v'(t)\} .$$

The integral of the above expression for $-\infty < t < \infty$ is obtainable from (3.11) and (3.13). Thus

$$(3.16) \quad \bar{R} = \int_{M_k} RdV = 2^n k^{-1} n\sigma_n \{(n+2)c_1 (b^{n+1} - a^{n+1}) + (n+1)c_2 (b^n - a^n)\} =$$

$$= 2^n k^{-1} n\sigma_n ((n+k) b^{n-1} - (n-k) a^{n-1}) .$$

Similarly we can calculate the stationary value of $S(g) = \int_{M_k} R^2 dV(g)$ from the metric just constructed as follows. We use the differential form of (3.14) to express the element of volume $dV(g)$ and (3.10) to express R; this yields

$$(3.17) \quad \int R^2 dV = 2^n \frac{n}{k} \sigma_n \int_{-\infty}^{\infty} ((n+1)(n+2)c_1 u'(t) + n(n+1)c_2)^2 (u'(t))^{n-1} d(u'(t))$$

$$= \frac{2^n n(n+1)}{k} \sigma_n \{(n+1)(n+2)c_1^2 (b^{n+2} - a^{n+2}) + 2n(n+2)c_1 c_2 (b^{n+1} - a^{n+1}) + n(n+1)c_2^2 (b^n - a^n$$

and substitute for the constants c_1, c_2 the values in terms of n, k, a, b given by (3.13).

It is very likely that this last value represents the absolute minimum value of $\int R^2 dV(g)$ for any metric Kähler metric (g) in M_k in the cohomology class represented by (a, b).

4. *Conclusion*

The problem of minimizing $\int_M R^2 dV(g)$ for all Kähler metrics of any fixed cohomology class $[\omega]$ in a given, compact complex manifold M can be viewed as a generalization of the uniformization problem from closed Riemann surfaces to higher dimensions. We believe that several interesting geometrical consequences may be derived, both from its solution (assuming that there is one), as well as from the conceivable methods of obtaining it. Concerning the existence problem of extremal Kähler metrics in our sense in general, the following observations can be made.

1. If the first Chern class of M is negative (i.e. if the canonical bundle is ample) Yau has proved [10] that the extremal metric exists, and it has necessarily constant (negative) scalar curvature, at least when the principal class is chosen to be proportional to the first Chern class.

2. More generally, for closed Kählerian manifolds admitting no holomorphic tangent vector fields other than zero, the parabolic equation with respect to a real parameter $t \geq 0$

$$\frac{\partial g_{\alpha\overline{\beta}}(t; z, \overline{z})}{\partial t} = \frac{\partial^2}{\partial z^\alpha \partial \overline{z}^\beta} R(g(t))$$

with $g_{\alpha\overline{\beta}}(0; z, \overline{z})$ equal to any given Kähler metric is one whose solutions decrease $\int R^2 dV(g)$ as a function of t. In estimating whether the solution exists and converges, as $t \to \infty$, to the one with $R = \text{constant}$, it is important to keep control of the lowest eigenvalue $\lambda(t)$ of the self-adjoint positive definite, fourth order elliptic operator $\square$ of Lichnérowicz, defined, for each Kähler manifold, by

$$\Box u = \Delta^2 u - R^{\alpha\overline{\beta}} u_{,\alpha\overline{\beta}} = u_{,\alpha\beta}{}^{\alpha\beta} - g^{\alpha\overline{\beta}} u_{,\alpha} R_{,\overline{\beta}}$$

$$= u^{,\alpha\beta}{}_{\alpha\beta} - g^{\alpha\overline{\beta}} R_{,\alpha} u_{,\overline{\beta}} \,.$$

The crucial estimate which is still lacking is one showing that, for some suitable Sobolev norm $\| \ \|_g$ defined in terms of any Kähler metric (g), the space of all Kähler metrics (g) in a fixed cohomology class, and for which $\|R(g)\|_g$ is bounded by any given positive constant, is compact with respect to a similar norm on the metrics (g) themselves.

3. For compact Kähler manifolds M admitting holomorphic tangent vector fields $\neq 0$, the analogous problem seems much harder. In this case the space of all Kähler metrics (in a fixed cohomology class) that are equivalent to any given one under the maximal connected, complex Lie group G of all holomorphic automorphisms of M, is not compact; however it is locally compact in view of [1, 2]. In this case one needs to devise a transformation semigroup on the function space $\mathfrak{g}$ of all admissible Kähler metrics, decreasing the functional $S(g) = \int_M R^2(g)\, dV(g)$, and with relatively compact trajectories, and hopefully converging to metrics (g) for which S(g) is stationary.

BIBLIOGRAPHY AND REFERENCES

[I] Chern, S. S., *Complex Manifolds*, Universitade do Recife, Recife (Braz.), *Textos de Matematica*, N. 5, 1959.

[II] ———, *Complex Manifolds without Potential Theory*, Princeton, N. J., Van Nostrand, 1967.

[III] Flaherty, E. J., *Hermitian and Kählerian Geometry in Relativity*, Berlin-New York, Springer-Verlag, *Lecture Notes in Physics*, N. 46, (1976).

[IV] Lichnérowicz, A., *Géométrie des groupes de transformations*, Paris, Dunod, 1958.

[V] Weil, A., *Introduction à l'étude des variétés kählériennes*, Paris, Hermann, *Actualités Scientifiques et Industrielles*, N. 1267, 1971.

[VI] Wells, R. O., *Differential Analysis on Complex Manifolds*, Englewood Cliffs, N. J., Prentice Hall, 1973.

[VII] Yano, K. and Bochner, S., Curvature and Betti Numbers, Princeton, N. J., Princeton University Press, *Annals of Mathematics Studies*, No. 32, 1953.

[VIII] Séminaire Palaiseau, *Première classe de Chern et courbure de Ricci,* Soc. Math. de France, Astérisque, *58,* 1978.

[1] Bochner, S. and Montgomery, D., *Locally Compact Groups of Differentiable Transformations*, Ann. of Math., 47(1946), 639-653.

[2] ———, *Groups on Analytic Manifolds*, Ann. of Math., *48*(1947), 659-669.

[3] Calabi, E., *The Variation of Kähler Metrics,* II *A Minimum Problem,* Amer. Math. Soc., Bull., *60*(1954) Abstract N. 294, p. 168.

[4] ———, *The Space of Kähler Metrics*, Proc. Intern. Congress of Math., Amsterdam, 1954, vol. II, 206-207.

[5] ———, *Métriques kählériennes et fibrés holomorphes*, Ann. Sci. Ec. Norm. Sup. Paris, 4me série, *12*(1979), 269-294.

[6] Kazdan, J. L. and Warner, F. W., *Surfaces of Revolution with Monotonic Increasing Curvature ...*, Amer. Math. Soc., Proc., *32* (1972), 139-141.

[7] Lichnérowicz, A., *Sur les transformations analytiques des variéties kählériennes*, C. R. Acad. Sci. Paris, *244*(1957), 3011-3014.

[8] Matsushima, Y., *Sur la structure du groupe d'homéomorphismes analytiques d'une certaine variété kählérienne,* Nag. Math. J., *11* (1957), 145-150.

[9] Myers, S. B. and Steenrod, N. E., *The Group of Isometries of a Riemannian Manifold,* Ann. Math., *40*(1939), 400-416.

[10] Yau, S. T., *On the Ricci Curvature of a Compact Kähler Manifold and the Complex Monge-Ampère Equation,* Comm. P. & Appl. Math., *31* (1978), 339-411.

L^2 HARMONIC FORMS ON COMPLETE MANIFOLDS

Jozef Dodziuk*

In this paper, all manifolds are complete, oriented and Riemannian unless explicitly stated otherwise. We shall discuss the following question.

What are the relations between the geometry or topology of a manifold and the spaces of L^2 harmonic forms on it?

For other topological applications of L^2 harmonic forms on noncompact manifolds see [Che], [T].

Before going on we remark that harmonic forms are C^∞ and the L^2 condition is a condition on growth at infinity. That some growth conditions have to be imposed was made clear by Greene and Wu [GW1], who proved that on an open manifold M^n (possibly incomplete) there are enough harmonic functions to embed the manifold in R^{2n+1}. We also remark that by a theorem of Andreotti and Vesentini [dR, Theorem 26] the spaces $\mathcal{H}^*(M)$ of L^2 harmonic forms on a manifold M consist of forms which are closed and co-closed, i.e.,

$$\mathcal{H}^*(M) = \left\{\omega \,\middle|\, \int_M \omega \wedge *\omega < \infty,\ \ d\omega = d*\omega = 0\right\}. \tag{1}$$

*Supported in part by NSF Grant MCS 78-02285.

Seminar on Differential Geometry
0-691-08268-5/82/000291-12 $00.60/0 (cloth)
0-691-08296-0/82/000291-12 $00.60/0 (paperback)

Moreover, if $\dim M = 2k$, the Hodge star operator on k-forms depends only on the conformal structure. Hence $\mathcal{H}^k(M)$ is a conformal invariant.

§1. *Examples*

In this section we discuss examples showing that in special cases topology influences the existence of L^2 harmonic forms.

1) Let M be the unit disc in the complex plane with Poincaré metric $ds^2 = \frac{|dz|^2}{(1-|z|^2)^2}$. Then $\mathcal{H}^0(M) \approx \mathcal{H}^2(M) = *\mathcal{H}^0(M) = \{0\}$, and

$$\mathcal{H}^1(M) = \left\{\omega = a\,dx + b\,dy \,|\, b_x - a_y = 0,\ a_x + b_y = 0,\ \iint_M (a^2+b^2)\,dx\,dy < \infty\right\}.$$

Thus $\mathcal{H}^1(M)$ can be identified with the space of holomorphic functions $f(z) = a - ib$ which are square-integrable with respect to the Lebesgue measure.

2) Let $M = S^1 \times \mathbf{R}$ with the product metric. Since the volume of M is infinite $\mathcal{H}^0(M) = \mathcal{H}^2(M) = \{0\}$. $\mathcal{H}^1(M)$ is also trivial.

It appears that the two examples above have nothing to do with topology. With a certain amount of overkill they can be explained on topological grounds (see Example 6 below). These examples illustrate, however, that the naive expectation that L^2 harmonic forms are in some sense intermediate between the cohomology with compact supports and the ordinary cohomology is false.

3) Let N be a compact, Riemannian manifold with boundary ∂N such that the metric is a product near ∂N. Let M be the union of N with $\partial N \times [0, \infty)$ identified along the common boundary.

Such manifolds were used in [APS], where it is shown that $\mathcal{H}^*(M)$ is isomorphic with the image of $H^*(N, \partial N)$ in $H^*(N)$ (i.e., with the image of $H^*_{comp}(M)$ in $H^*(M)$). This follows from the fact (proved by separation of variables) that L^2 harmonic forms decay exponentially on $\partial N \times [0, \infty)$ as the second coordinate tends to infinity. It is interesting to consider what happens with $\mathcal{H}^*(M)$ when the metric on $\partial N \times [0, \infty)$ varies. If the product metric is replaced by $g_1 + f(t)^2 dt^2$ near infinity, where g_1 is the metric on ∂N, then $\mathcal{H}^*(M)$ is unchanged if $f(t)$ has moderate growth (e.g. $f(t) = 0(t^{n-1})$, $n = \dim N$). On the other hand, if $f(t)$ grows exponentially and N is a disc $\mathcal{H}^*(M)$ is infinite dimensional. We see that, in addition to the topology, the size of N near infinity influences existence of L^2 harmonic forms.

4) The following proposition is essentially due to Gaffney [G1]. It may be used to prove existence of nontrivial L^2 harmonic forms.

PROPOSITION. *Suppose that* ω_0 *is smooth, closed and has compact support. Let* $H\omega_0$ *denote the projection of* ω_0 *onto* $\mathcal{H}^*(M)$. *For every cycle* C

$$\int_C \omega_0 = \int_C H\omega_0 .$$

COROLLARY 1. *Suppose* C *is a cycle of codimension one in* M^n *which does not disconnect. Then there exists* $\omega \in \mathcal{H}^{n-1}(M)$ *such that* $\int_C \omega = 1$ *and all other periods are equal to zero. In particular* $\mathcal{H}^{n-1}(M) \approx \mathcal{H}^1(M) \neq \{0\}$.

COROLLARY 2. *Let* N^n *be a compact (not necessarily orientable) manifold with nonzero Euler characteristic* $\chi(N)$. *For every complete metric on* $M = TN$, $\mathcal{H}^n(M) \neq \{0\}$.

Corollary 2 was suggested by F. Connolly.

5) Let M be an open Riemann surface. The space $\mathcal{H}^1(M)$ is a conformal invariant. It breaks up as an orthogonal direct sum

$$\mathcal{H}^1(M) = \mathcal{H}_1 \oplus \mathcal{H}_2 .$$

$\mathcal{H}_1$ is spanned by forms constructed as in Corollary 1, and $\mathcal{H}_2$ is either $\{0\}$ or infinite dimensional depending on whether the ideal boundary of M has zero or positive capacity (cf. [N, Ch. 10, §7]). We see that $\mathcal{H}_1$ is determined by the topology and $\mathcal{H}_2$ by the geometry of M near the ideal boundary (cf. Examples (1) and (2)). Determining whether $\mathcal{H}_2$ is trivial or not for a given Riemann surface is very difficult.

6) This is a special case of the situation considered in [A]. Let $\tilde{M} \to M = \tilde{M}/\Gamma$ be a Galois covering of a compact, oriented Riemannian manifold M. We equip $\tilde{M}$ with the pull-back metric so that Γ acts on M by isometries.

THEOREM [D1]. *The equivalence class of the representation of* Γ *on* $\mathcal{H}^*(M)$ *is a homotopy invariant of* M.

COROLLARY. *The normalized dimensions* (cf. [A]) $\dim_\Gamma \mathcal{H}^p(M) = B^p_\Gamma(M)$ *are homotopy invariants.*

We will make some remarks about the proof of the theorem in §3 below. We remark here that, as a consequence of L^2 index theorem of [A],

$$\Sigma(-1)^p B^p_\Gamma(M) = \chi(M) . \tag{2}$$

This formula can be used to show, by considering the Poincaré disc as a covering of a compact Riemann surface of negative Euler characteristic, that the L^2 harmonic forms on the Poincaré disc form an infinite dimensional space ($\dim_\Gamma \mathcal{H}^1 = -\chi > 0$ which implies that the ordinary dimension of $\mathcal{H}^1$ is infinite). Similarly the cylinder in Example 2 covers a torus whose Euler characteristic is zero. Hence $\dim_\Gamma \mathcal{H}^1 = 0$ i.e., $\mathcal{H}^1 = \{0\}$.

§2. *Vanishing theorems*

It is well known that certain positivity conditions on the curvature tensor imply vanishing of cohomology for compact manifolds. It turns out that the same vanishing theorems hold for L^2 harmonic forms on complete manifolds. This should be expected, as pointed out by E. Calabi, in view of the following

THEOREM [G2]. *Suppose* ω *is a smooth form of degree* $n-1$ *such that* ω *and* $d\omega$ *are* L^1. *Then*

$$\int_M d\omega = 0 .$$

REMARK. Generalizations of this theorem were obtained by Yau [Y] and Karp [K].

The Weitzenbock identity [dR, §26]

$$\Delta = \nabla^*\nabla + \mathcal{R} ,$$

where ∇ is the covariant derivative, ∇^* its formal adjoint and $\mathcal{R}$ is an algebraic operator depending only on the curvature tensor, is combined with the integration by parts to yield

$$(\Delta\omega, \omega) = (\nabla\omega, \nabla\omega) + (\mathcal{R}\omega, \omega)$$

in the compact case. In case M is complete one can integrate by parts, as suggested by the theorem above, to obtain vanishing theorems of Bochner type for L^2 harmonic forms if $\mathcal{R} \geq 0$. The statements and proofs are exactly, as in the compact case [D2]. We remark that the same vanishing theorems (with a possible exception of Matsushima vanishing theorem [D2, Theorem 3]) hold for L^p harmonic forms as shown by Greene and Wu [GW2] and Yau [Y]. Their proofs are not straightforward generalizations of the proofs in the compact case.

§3. *DeRham-Hodge isomorphism*

We describe now some conditions under which the spaces of L^2 harmonic forms are isomorphic with L^2 cohomology defined simplicially.

Let K be a simplicial complex which is uniformly locally finite, i.e., there exists an integer $N > 0$ such that for every vertex $v \in K$ the number of simplexes of K having v as a vertex does not exceed N. For such a complex we define

$$C_2^p(K) = \left\{ f \in C^p(K, \mathbf{R}) \mid \sum_{\sigma^p} f(\sigma)^2 < \infty \right\}.$$

$C_2^p(K)$ is a Hilbert space with the inner product

$$(f, g) = \sum_{\sigma^p} f(\sigma) \cdot g(\sigma) .$$

The uniform local finiteness implies that the simplicial coboundary d_c defines a bounded operator

$$d_c : C_2^p(K) \to C_2^{p+1}(K) .$$

The adjoint $\delta_c = d_c^*$ is the usual boundary operator (if we identify chains with cochains) and, by analogy with the smooth case, we define the combinatorial Laplacian

$$\Delta_c = d_c \delta_c + \delta_c d_c$$

and $\mathcal{H}^p(K) = \{ f \in C_2^p(K) \mid \Delta_c f = 0 \}$.

PROPOSITION. *We have the following orthogonal decomposition*

$$C_2^p(K) = \overline{d_c C_2^{p-1}(K)} \oplus \mathcal{H}^p(K) \oplus \overline{\delta_c C_2^{p+1}(K)} .$$

Moreover $\ker d_c | C_2^p(K) = \overline{d_c C_2^{p-1}(K)} \oplus \mathcal{H}^p(K)$ *and therefore*

$$\mathcal{H}^*(K) \approx \frac{\operatorname{Ker} d_c}{\overline{\operatorname{Im} d_c}} .$$

If K is a triangulation of a surface $\mathcal{H}^1(K)$ has the following interpretation. Think of $K^{(1)}$, the one-dimensional skeleton of K, as an electric network with one ohm resistor on every edge. Harmonic cochains in $\mathcal{H}^1(K)$ represent current flows with finite heat output satisfying both Kirchoff laws.

It turns out that for sufficiently well-behaved metrics on M and sufficiently nice triangulations K of M the spaces $\mathcal{H}^*(M)$ and $\mathcal{H}^*(K)$ are isomorphic via integration [D3]. This isomorphism certainly holds for manifolds which are coverings of compact ones and constitutes the main step in the proof of the theorem quoted in Example 6. It is not clear that integration of an L^2 harmonic p-form over simplexes in K yields a cochain in $C_2^p(K)$ in general.

What one needs is an estimate of the pointwise norm of an L^2 harmonic form in terms of the L^2 norm in a neighborhood of the point in question. Moreover the constants in such estimates should be independent of the point. Once such estimates are available, it is easy to conclude that every cochain obtained by integrating an L^2 harmonic form is square summable, and an argument patterned after Whitney's proof [W] of the de Rham theorem gives the isomorphism of $\mathcal{H}^*(M)$ and $\mathcal{H}^*(K)$ [D1], [D3]. It remains to describe the geometric assumptions under which required estimates can be proved.

We say that the metric of M satisfies the condition C_m, for a non-negative integer m, provided

a) The injectivity radius is bounded from below by a positive constant.

b) $R, \nabla R, \nabla^2 R, \cdots, \nabla^m R$ are uniformly bounded, where R denotes the Riemann curvature tensor and ∇ is the covariant derivative.

Complete metrics satisfying C_m exist on every open manifold by the result of [G].

A triangulation K of M will be called uniform if

a) There exist four positive constants C_1, C_2, C_3, C_4 such that

$$C_1 \leq (\operatorname{diam} \sigma)^n \leq C_2 \operatorname{vol} \sigma \leq C_3 (\operatorname{diam} \sigma)^n \leq C_4$$

for every simplex σ of K of dimension $n = \dim M$.

b) There exists a constant $C > 0$ such that for every vertex v of K the barycentric coordinate function ϕ_v satisfies

$$|\nabla \phi_v| \leq C .$$

The condition a) requires that simplexes of K be roughly of the same size and not too thin. The second condition means that simplexes do not curve too much. According to an unpublished result of Calabi uniform triangulations exist on every manifold satisfying C_1. The precise statement of the theorem alluded to above is the following [D3].

THEOREM. *Suppose* M^n *is a complete, oriented Riemannian manifold satisfying the condition* C_m *for some* $m > \frac{n}{2} - 1$. *If* K *is a uniform triangulation of* M, *then* $\mathcal{H}^*(M)$ *and* $\mathcal{H}^*(K)$ *are isomorphic via integration.*

§4. *Open problems*

Formula (2) suggests several problems (cf. [A]).

1) If M is compact, $\chi(M) \neq 0$ then every Galois covering carries an L^2 harmonic form by (2). Is the same true if the covering is not Galois? Even the answer to the following question is not known.

Suppose $\tilde{S} \to S$ is an infinite covering of a compact Riemann surface S of genus $g \geq 2$. Does $\tilde{S}$ carry a nonzero L^2 harmonic differential?

The answer is positive if $\tilde{S} \to S$ is Galois or if $\tilde{S}$ is not planar, i.e., if $\tilde{S}$ has a nondividing cycle (cf. Example 4, Corollary 1). Thus the case in which the answer is not known is that of planar, non-Galois coverings. Because of de Rham-Hodge isomorphism the problem reduces to a combinatorial one, which has been studied by D. DeBaun in a University of Pennsylvania Ph.D. Thesis. DeBaun was not able to answer the question but he obtained several partial results of which the most interesting is perhaps the following

THEOREM. *Let* $\tilde{S} \to S$ *be as above and assume that* $\tilde{S}$ *is planar. Let* K *be a triangulation of* S *and let* $\tilde{K}$ *be the pull-back triangulation of* $\tilde{S}$. $\mathcal{H}^1(S) \neq \{0\}$ *if and only if the random walk on* $\tilde{K}$ *is transient.*

A random walk is called transient if a particle undergoing it has a nonzero probability of escaping to infinity. The above theorem points again to the connection between the existence of L^2 harmonic forms and the size of the manifold near infinity.

2) The L^2 Betti numbers $B^p_\Gamma(M)$ appearing in (2) are a priori real. No examples are known in which they are irrational.

What are possible values of $B^p_\Gamma(M)$?

This question was asked by Atiyah in [A]. If Γ is abelian $B^p_\Gamma(M)$ are rational as was proved independently by Cohen [Co] and Donnelly [D]. Nothing is known for general Γ.

3) Formula (2), as observed by Singer, opens a possibility of using L^2 harmonic forms to prove the following conjecture of Hopf (cf. [Ch, p. 49]).

If M^{2k} is compact and has negative sectional curvature then $(-1)^k\chi(M) > 0$. In particular one might try to work with L^2 harmonic forms on the universal covering $\tilde{M}$, which is diffeomorphic with R^{2k} by Hadamard-Cartan theorem. The following conjecture (cf. [D4]) holds in all cases in which one can carry out the computations. It also implies the conjecture of Hopf.

Let M be complete, diffeomorphic with R^n and have negative sectional curvature $K \leq -\delta < 0$. Then

$$\mathcal{H}^p(M) = \begin{cases} 0 \text{ if } p \neq \frac{n}{2} \\ \text{infinite dimensional space if } p = \frac{n}{2} . \end{cases}$$

It is possible that it is not so much the curvature as the growth of M at infinity that determines the existence of L^2 harmonic forms. This point of view is supported by the situation in two dimensions (cf. Example 4) and by the following.

THEOREM [D4]. *Suppose* M *is diffeomorphic with* R^n *and has a* C^∞ *metric which in terms of geodesic polar coordinates centered at some point of* M *can be written as*

$$ds^2 = dr^2 + f(r)^2 d\theta^2$$

Then

$$\mathcal{H}^p(M) = \{0\} \quad \textit{if} \quad p \neq 0, \frac{n}{2}, n .$$

$$\mathcal{H}^0(M) \simeq \mathcal{H}^n(M) \simeq \begin{cases} \{0\} & \textit{if} \quad \int_0^\infty f^{n-1}\, dr = \infty \\ R & \textit{if} \quad \int_0^\infty f^{n-1} dr < \infty . \end{cases}$$

If n *is even and* $\int_1^\infty \frac{dr}{f} = \infty$, $\mathcal{H}^{n/2}(M) = \{0\}$.

If n *is even and* $\int_1^\infty \frac{dr}{f} < \infty$, $\mathcal{H}^{n/2}(M)$ *is an infinite dimensional Hilbert space.*

The conjecture about L^2 harmonic forms on a simply connected, negatively curved manifold involves vanishing of $\mathcal{H}^p(M)$ for $p \neq \frac{n}{2}$. We remark that Bochner type vanishing theorems are useless in this context since they require some positivity assumptions about the curvature. A small step toward proving the required vanishing was made following a suggestion of S. T. Yau. Namely if $K \leq -1$ and M is simply connected one can prove that

$$\frac{(n-1)^2}{4} \int_M |\omega|^2 dV + \int_M <\mathcal{R}\omega, \omega> dV \leq 0 .$$

This yields some vanishing results if we assume an appropriate pinching of sectional curvature but is much too weak to prove the conjecture.

4) We conclude with the following general problem. According to Vesentini [V, §7] if M is flat outside a compact subset, the spaces $\mathcal{H}^*(M)$ are finite dimensional. Do they admit a topological interpretation?

REFERENCES

[A] M.F. Atiyah, Elliptic operators, discrete groups and von Neumann algebras, Asterisque 32-33, Societe Mathematiques de France, Paris, 1976, pp. 43-72.

[APS] M.F. Atiyah, V.K. Patodi, and I.M. Singer, Spectral asymmetry and Riemannian geometry I, Math. Proc. Comb. Phil. Soc. 77(1975), 43-69.

[Ch] S.S. Chern, Differential geometry, its past and its future, Actes du Congres. International du Mathematiciens (1970), vol. 1, Gauthier-Villars, Paris, 1971, pp. 41-53.

[Che] J. Cheeger, On the Hodge theory of Riemannian pseudomanifolds, Proc. Sympos. Pure Math. 36, AMS, Providence, 1980, pp. 91-146.

[Co] J. Cohen, Von Neumann dimension and the homology of covering spaces, Quart. J. Math. Oxford 30(1979), 133-142.

[D1] J. Dodziuk, De Rham-Hodge theory for L^2 cohomology of infinite coverings, Topology 16(1977), 157-165.

[D2] ________, Vanishing theorems for L^2 harmonic forms, to appear in the volume in memory of V.K. Patodi, Tata Institute of Fundamental Research.

[D3] ________, Sobolev spaces of differential forms and de Rham-Hodge isomorphism, J. Diff. Geom., to appear.

[D4] ________, L^2 harmonic forms on rotationally symmetric Riemannian manifolds, Proc. Amer. Math. Soc. 77(1979), 395-400.

[D] H. Donnelly, On L^2 Betti numbers of abelian groups, preprint.

[dR] G. de Rham, Varieties differentiables, formes, conrantes, formes harmoniques, 2nd, ed., Actualite Sci. Indust., no. 122a, Hermann, Paris, 1960.

[G1] M. Gaffney, The heat equation method of Milgram and Rosenblum for open Riemannian manifolds, Ann. of Math. 60(1954), 458-466.

[G2] ________, A Special Stokes' theorem for complete Riemannian manifolds, Ann. of Math. 60(1954), 140-145.

[G] R. Greene, Complete metrics of bounded curvature on noncompact manifolds, Archiv der Mathematik, 31(1978), 89-95.

[GW1] R. Greene, H. Wu, Embedding of open Riemannian manifolds by harmonic functions, Ann. L'Inst. Fourier 25(1975), 215-235.

[GW2] ________, Harmonic forms on noncompact Riemannian and Kähler manifolds, preprint.

[H] F. Hirzebruch, New topological methods in algebraic geometry, Springer-Verlag, New York, 1966.

[K] L. Karp, On Stokes' theorem for noncompact manifolds, preprint.

[N] R. Nevanlinna, Uniformisierung, Berlin, 1953.

[T] N. Teleman, Combinatorial Hodge Theory and Signature Theorem, preprint.

[V] E. Vesentini, Lectures on Levi convexity of complex manifolds and cohomology vanishing theorems, Tata Institute of Fundamental Research, Bombay, 1967.

[W] H. Whitney, Geometric Integration Theory, Princeton University Press, Princeton, N. J. 1957.

[Y] S. T. Yau, Some function-theoretic properties of Riemannian manifolds and their applications to geometry, Indiana University Mathematics Journal 25(1976), 659-670.

L^2-COHOMOLOGY AND INTERSECTION HOMOLOGY OF SINGULAR ALGEBRAIC VARIETIES

Jeff Cheeger, Mark Goresky, and Robert MacPherson

§1. *Introduction*

In this largely expository paper we describe some generalizations to the singular case of the special cohomological properties of nonsingular complex projective algebraic varieties which are usually proven using Kähler geometry. One avenue of generalization is to suitably modify the properties so that they remain true of the ordinary cohomology of a singular variety (see 1.3 below). Here we take the point of view that the properties themselves need not be weakened provided that the cohomology theory to which they apply is suitably modified.

1.1. Let M be a compact complex algebraic manifold of dimension n embedded in complex projective space $\mathbb{C}P^N$, and let Ω be the restriction to M of the Kähler form on $\mathbb{C}P^N$. The cohomology of M with complex coefficients satisfies several remarkable theorems which we refer to collectively as the *Kähler package*:

1. Pure Hodge decomposition:

$$(1.1) \qquad H^i(M) \simeq \bigoplus_{p+q=i} H^{p,q} \quad (\text{with } \overline{H^{p,q}} = H^{q,p}) .$$

Seminar on Differential Geometry
0-691-08268-5/82/000303-38$01.90/0 (cloth)
0-691-08296-0/82/000303-38$01.90/0 (paperback)
For copying information, see copyright page.

2. Hard Lefschetz: the map

$$H^{n-i}(M) \xrightarrow{\cup[\Omega]^i} H^{n+i}(M) \tag{1.2}$$

is an isomorphism.

3. Poincaré duality: the pairing

$$H^i(M) \times H^{2n-i}(M) \to \mathbb{C} \tag{1.3}$$

given by cup product and evaluate on the fundamental class is, nonsingular. (Equivalently the pairing

$$H_i(M) \times H_{2n-i}(M) \to \mathbb{C} \tag{1.4}$$

given by intersecting transverse cycles and counting points with their multiplicities, is nonsingular.)

4. The Lefschetz hyperplane theorem.

5. The Hodge signature theorem.

1.2. Aside from the Lefschetz hyperplane theorem, all the theorems of the Kähler package can be deduced using L^2 analytic methods. (Of course Poincaré duality and Hard Lefschetz [D4] can also be proved by other means.)

The first step is De Rham's theorem that the topological invariant $H^*(M)$ has an analytic expression as $H^*_{DR}(M)$, the cohomology of differential forms. Then the results can be viewed as formal consequences of a) the Hodge theorem that every cohomology class contains exactly one harmonic form, b) the fact that the action of the almost complex structure J on differential forms carries harmonic forms to harmonic forms; and c) the fact that the harmonic forms are just the forms which are closed and co-closed.

1.3. There have been several deep studies of how to extend various theorems of the Kähler package to the cohomology of a singular variety. The Zeeman spectral sequence [ZE], [MC], studies Poincaré duality; Ogus'

De Rham depth [O] studies the Lefschetz hyperplane theorem; and Deligne's mixed Hodge theory [D1], [D2], [D3] studies the Hodge (p, q) decomposition. All of these theories proceed by filtering the cohomology (roughly by how "tied to the singularities" a cocycle is). Then they express the "degree of failure" of the theorem as stated in the nonsingular case in terms of this filtration. The picture we present here does not oppose, but rather compliments, that of these theories, (see §7).

1.4. Let X be a complex n dimensional projective algebraic variety. The invariants of X which concern us in this paper are the middle intersection homology groups $IH_*(X)$. Roughly, $IH_*(X)$ is the homology of a certain subcomplex of the homology chain complex of X, defined by certain geometric conditions on how the chains enter the singularities of X. (These will be recalled in §2; see [GM1], [GM2], [GM3] for details.) For cycles of this special type the intersection pairing is well defined and leads to a perfect pairing (1.3). The groups $IH_*(X)$ are topological invariants but not homotopy invariants. There are natural maps

(1.5)
$$\begin{array}{ccc} & IH_i(X) & \\ \nearrow & & \searrow \\ H^{2n-i}(X) & \xrightarrow[\cap[X]]{} & H_i(X)\,. \end{array}$$

If X is nonsingular these are all isomorphisms, but in some cases $IH_*(X)$ is much bigger than either homology or cohomology.

The local calculation of $IH_*(X)$ has an interesting result. For any $x \in X$ let B be the intersection of X with a small open ball of radius ε about x and let S be the intersection of X with a sphere of radius $\varepsilon/2$ about x. Then (using chains with compact support)

(1.6)
$$IH_i(B) = \begin{cases} IH_i(S) & \text{for} \quad i < n \\ 0 & \text{for} \quad i \geq n \end{cases}.$$

In fact, an appropriately stated version of this local calculation characterizes the groups $IH_*(X)$. (See §2.2.)

1.5. The main idea of this paper is the following program: the theorems of the Kähler package should hold without modification in the singular case, provided that intersection homology is used in place of ordinary homology.

We present several conjectures relating to this program and possible L^2 methods of proof. We also give a number of examples and consequences.

The current status of the program is summarized in §6. Briefly, all parts of the Kähler package but those relating directly to the Hodge (p,q) decomposition have been proved in general (by other than L^2 methods, for the most part). The (p,q) decomposition is known in many classes of examples.

1.6. The original motivation for the program was the existence of L^2 methods on certain singular varieties which turned out to be appropriate for the study of intersection homology theory [C1], [C2], [C3].

Let $X \subset \mathbb{C}P^N$ be a singular variety and let Σ be its singularity set. One studies the *smooth incomplete* Riemannian manifold $X-\Sigma$ with the metric Ω induced by the inclusion. The ith L^2 cohomology group $H^i_{(2)}(X)$ of X is just the quotient space of smooth i forms θ on $X-\Sigma$ which are L^2, by the subspace $\{d\psi\}$, where ψ is a smooth $i-1$ form with ψ, $d\psi \in L^2$ (see §3).

If X has conical singularities (see §3 for definitions) then two somewhat surprising results hold [C3], §6.3:

1) $H^i_{(2)}(X)$ is isomorphic to the space of closed and co-closed harmonic forms.

2) The calculation of the local groups is dual to that of (§1.6) and therefore $H^*_{(2)}(X)$ is the cohomology theory dual to $IH_*(X)$.

These results are analogues of the Hodge theorem and the De Rham theorem so they lead one to expect L^2 proofs of the Kähler package theorem for intersection homology theory.

1.7. Although the possible existence of an appropriate L^2 theory on the incomplete Kähler manifold $X-\Sigma$ was the main source of motivation for the above conjectures, of course it is not the only possible method of attack. In fact, at present the most general results have been obtained by other methods. Moreover, as the first author realized after the publication of [C3], the assertion made there that the conjectures are formal consequences of Hodge Theory, is *not* correct in the context of *incomplete* metrics. The reason is that on incomplete manifolds, an L^2 harmonic form is not necessarily closed and co-closed. Thus although a) and b) of §1.3 above are automatic, c) is not. In particular, the assertion of [C3], that the Hodge Theorem proved there implies the "Kähler package" for algebraic varieties for which the induced metric has conical singularities, is still unsubstantiated, except in certain special cases in which it can be checked directly that J preserves the space $\mathcal{H}^i$ closed and co-closed L^2 harmonic forms; (see §6.4 and [C4]). A general proof of this assertion, (which seems extremely delicate), must make use of the assumption that the singularities of the metric are conical in a complex analytic (rather than just piecewise smooth) sense. Otherwise, the topological conclusions can definitely be false.

1.8. The above-mentioned difficulty disappears for the case of *complete* metrics on $X-\Sigma$, since on a complete manifold L^2 harmonic forms are automatically closed and co-closed; (see [DR], [AV] and §3). Thus another possible approach is to attempt to construct a complete metric on $X-\Sigma$ (or on $X-\Sigma'$ for suitable $\Sigma' \supset \Sigma$), for which the space $\mathcal{H}^i$ is dual to IH_i. But in return for the great advantage offered by completeness one pays a certain price in that $\mathcal{H}^i$ becomes more difficult to calculate in general.* In either program formidable difficulties arise at the outset because even in very simple cases, the riemannian geometry of singular algebraic varieties and their complements has been little studied and, it seems fair to say, is very poorly understood.

*See however, [M], [ZU1], and [ZU2].

1.9. In summary, although at present, the results on the "Kähler package" which can be obtained from L^2 methods are quite modest as compared to those which can be obtained by other techniques, we emphasize the L^2 theory here for several reasons. First, it was the original source of motivation for the conjectures and should be basic to an eventual complete understanding. Second, just as in the nonsingular case, these methods are based on the existence of a Kähler metric rather than the more special property of being an algebraic variety. Thus, when successful they work in more generality. Third, the Kähler metrics (complete and incomplete) on the nonsingular parts of algebraic varieties, provided a rich source of as yet unstudied problems in geometry and analysis.

1.10. *Historical note*

The ideas in this paper had three independent sources.

1) The intersection homology groups were defined in piecewise linear topology by Goresky and MacPherson to study intersection theory of cycles on a singular variety.
2) The study of L^2 cohomology on the nonsingular part of a variety with conical singularities was initiated by Cheeger; this was suggested by points which arose in the study of analytic torsion.
3) A procedure for extending to all of M, a variation of Hodge structure on $M-D$, (where M is a nonsingular complex algebraic variety, and D is a divisor with normal crossings), was discovered by Deligne, generalizing work of Zucker, and was used to study the mixed Hodge theory of sheaf cohomology.

All of these procedures led to the same seemingly strange local calculation of (§1.6). This coincidence was observed in 1976 by Sullivan (for 1) and 2)) and by Deligne (for 1) and 3)). Deligne then proposed a sheaf theoretic construction of intersection homology.

The natural hypothesis was that all three approaches give the same group—i.e., that a single invariant has definitions by topological, analytic, and algebraic means. This hypothesis led, in conversations between

MacPherson and Cheeger in 1977, to the conjecture that the theorems of the Kähler package hold for that invariant. Now the hypothesis has been proved ((§6.3) and [C3] for 1) and 2); [GM3] for 1) and 3)) and much of the Kähler package has been established as well.

We would like to thank the I.H.E.S. for its hospitality while this paper was being written. We are grateful to D. Burns, P. Deligne, O. Gabber, W. Fulton, and D. Sullivan for helpful conversations.

§2. *Intersection homology theory*

2.1. In this section we recall the definition and basic properties of the "middle" intersection homology groups $IH_k(X)$ ([GM1], [GM2]). Since there is no canonical piecewise-linear structure on an algebraic variety, it is technically convenient to use subanalytic chains in the definition of $IH_k(X)$. However, we remark that one could just as well choose a P.L. structure on X, use P.L. chains to define $IH_k(X)$; and the result is independent of the P.L. structure. This is the point of view which is adopted at the end of Chapter 3. The reader who is unfamiliar with the subanalytic category can think in terms of P.L. chains in this section as well.

Let X be an n-dimensional complex analytic variety contained in some nonsingular variety. We choose an analytic Whitney stratification ([MA], [T]). This consists of a filtration by (closed) analytic subvarieties

$$X_0 \subset X_1 \subset X_2 \subset \cdots \subset X_n = X$$

such that:

(a) $X_i - X_{i-1}$ is a (possibly empty) complex analytic i-dimensional manifold (the components of which are called the *strata* of complex dimension i).

(b) Whitney's condition B holds with respect to any pair of strata R and S, i.e., suppose $x_i \in S$ is a sequence of points converging to some point $x \in R$ and suppose $y_i \in R$ is a sequence converging to x. Suppose the secant lines $\overline{x_i, y_i}$ converge to some line ℓ and the tangent planes $T_{x_i}S$ converge to some limiting plane τ (with respect to any coordinate system on the ambient nonsingular variety). Then we demand that $\ell \subset \tau$.

The groups $IH_k(X)$ will be constructed using this stratification but the result is independent of the stratification.

Let $C_*(X)$ denote the chain complex of compact (real) subanalytic chains on X with complex coefficients. The homology of this complex coincides with the singular homology groups $H_*(X; \mathbb{C})$ by Hardt [H]. For any $\xi \in IC_i(X)$ we let $|\xi|$ denote the *support* of ξ; it will be a (real) i-dimensional subanalytic subset of X.

Define the subcomplex $IC_*(X)$ of *allowable* chains by the condition $\xi \in IC_i(X)$ if

$$\dim |\xi| \cap X_k \leq i - n + k - 1$$

and

$$\dim |\partial\xi| \cap X_k \leq i - n + k - 2 .$$

DEFINITION. $IH_i(X)$ is defined to be the ith homology group of this chain complex, $IC_*(X)$.

It is possible to define an intersection product $IH_i(X) \times IH_j(X) \to IH_{i+j-2n}(X)$ by following the original method of Lefschetz [L]. If $\xi \in IC_i(X)$ and $\eta \in IC_j(X)$ are transverse, then the intersection $|\xi| \cap |\eta|$ carries the structure of an $i+j-2n$ dimensional subanalytic chain such that

$$\partial(\xi \cap \eta) = (\partial\xi) \cap \eta + (-1)^i \xi \cap \partial\eta .$$

For any chain ξ, almost all chains η are transverse to ξ. Therefore transverse intersection induces a pairing on the intersection homology groups.

THEOREM (Poincaré duality). *If X is compact, the intersection pairing in complementary dimensions is nonsingular, i.e., the composition*

$$IH_i(X) \times IH_{2n-i}(X) \to H_0(X) \to \mathbb{C}$$

induces an isomorphism

$$IH_i(X) \cong \mathrm{Hom}(IH_{2n-1}(X), \mathbb{C}) .$$

Relation with cohomology

The intersection homology groups lie "between" cohomology and homology, in the sense that for compact X we have the following:

THEOREM. *There are canonical homomorphisms*

$$H^i(X) \to IH_{2n-i}(X) \to H_{2n-i}(X)$$

which factor the Poincaré duality map $\cap [X] : H^i(X) \to H_{2n-i}(X)$. *If* X *is nonsingular, these are both isomorphisms. The intersection product extends to a module structure*

$$H^i(X) \times IH_j(X) \to IH_{j-i}(X)$$

which is compatible with cup and cap products.

These maps may be constructed for compact varieties X which are embedded in some nonsingular variety of dimension N as follows: choose a subanalytic neighborhood with boundary $(U, \partial U)$ of X in the ambient space, such that X is a deformation retraction of U, and ∂U is a topological manifold. By Lefschetz duality $H^i(X) \cong H_{2N-i}(U, \partial U)$ so any cohomology class may be represented by a (relative) subanalytic chain ξ on $(U, \partial U)$ which we may take to be transverse to X. The intersection $\xi \cap X$ then satisfies the allowability condition, so it determines a chain in $IC_{2n-i}(X)$, thus inducing $H^i(X) \to IH_{2n-i}(X)$.

Similarly if $\eta \in IC_j(X)$ then ξ may also be chosen transverse to η so $\xi \cap \eta \in IC_{j-i}(X)$, thus inducing the module structure.

Finally the map $IH_i(X) \to H_i(X)$ is induced by the inclusion of chain complexes $IC_*(X) \subset C_*(X)$.

Functorality

DEFINITION. Let $f: Y \to X$ be the inclusion of a subvariety Y. f is said to be a *normally nonsingular inclusion* (of relative complex dimension c) if f is proper and there is an analytic manifold M (of codimension c) in the ambient projective space, such that M is transverse to each stratum of X and $Y = M \cap X$. Such a subvariety inherits a stratification from that of X.

Such a map f determines a pushforward

$$f_* : IH_i(Y) \to IH_i(X)$$

since $f(\xi)$ is an allowable chain in $IC_i(X)$ whenever ξ is an allowable chain in $IC_i(Y)$. Dualizing we obtain a pullback:

$$f^* : IH_k(X) \cong \mathrm{Hom}(IH_{2n-k}(X), \mathbf{C}) \to \mathrm{Hom}(IH_{2n-k}(Y), \mathbf{C}) \cong IH_{k-2c}(Y) .$$

DEFINITION. Let $f: Y \to X$ be a proper smooth map. This will be a topological fibre bundle whose fibres are complex manifolds. We shall assume they all have dimension d. Such a map is called a *normally nonsingular projection* of relative dimension $-d$. The stratification of X pulls back to a stratification of Y.

Such a projection f induces a pullback

$$f^* : IH_i(X) \to IH_{i+2d}(Y)$$

since the pre-image $f^{-1}(\xi)$ satisfies the allowability conditions on Y whenever $\xi \in IC_i(X)$. We obtain by duality a push forward map

$$f_* : IH_k(Y) \to IH_k(X) .$$

DEFINITION. A normally nonsingular map f is any map which can be factored into a normally nonsingular inclusion followed by a normally nonsingular projection. The relative dimension of f is defined to be the sum of the relative dimensions of its factors.

For normally nonsingular maps the homology, cohomology, and intersection homology groups map *both* ways. ([FM], [GO]).

Branched coverings

If $f: Y \to X$ is a finite branched covering then f induces homomorphisms $f_*: IH_i(Y) \to IH_i(X)$ and $f^*: IH_i(X) \to IH_i(Y)$. Simply stratify X and Y compatibly with the map f and observe that for any allowable chain ξ in Y the image $f(\xi)$ is allowable in X. Similarly, for any allowable chain η in X, the pre-image $f^{-1}(\eta)$ is allowable in Y.

2.2. *The local calculation*

We now give an intuitive description of the local calculation (1.6). In this case, any allowable cycle $\xi \in IC_i(B)$ is the boundary of the allowable chain $c(\xi)$ (the *cone* on ξ) provided $i \geq n$. Thus $IH_i(B) = 0$ if $i \geq n$.

For $i < n$, any allowable chain $\xi \in IC_i(B)$ cannot contain the singular point p, so it may be deformed into a chain on S by "pushing along the cone lines." (This corresponds to the homotopy operator of (3.27).) Furthermore, any allowable chain $\eta \in IC_i(S)$ is allowable in B, so we conclude $IH_i(B) \cong IH_i(S)$ for $i < n$. This agrees with the analogous calculation for L^2 cohomology (3.23) since for nonsingular S we have $H_i(S) \cong IH_i(S)$.

Using this local calculation in the Mayer-Vietoris exact sequence, we obtain the following

COROLLARY. *Suppose* X *has a single isolated singularity* x. *Then*

$$IH_i(X) = \begin{cases} H_i(X) & \text{if } i > n \\ \text{Image } (H_i(X-x) \to H_i(X)) & \text{if } i = n \\ H_i(X-x) & \text{if } i < n\,. \end{cases}$$

2.3. *Axiomatic characterization*

A *complex of sheaves* on X is a collection of sheaves $\{\underline{\underline{S}}^p\}$ of $\mathbb{C}$-vector spaces, together with sheaf maps

$$\underline{\underline{S}}^p \xrightarrow{d} \underline{\underline{S}}^{p+1} \xrightarrow{d} \underline{\underline{S}}^{p+2}$$

such that $d \circ d = 0$. If each $\underline{\underline{S}}^p$ is *fine*, we shall use $H^p(X; \underline{\underline{S}}^\cdot)$ to denote the pth cohomology group of the complex

$$\Gamma(X; \underline{\underline{S}}^0) \to \Gamma(X; \underline{\underline{S}}^1) \to \Gamma(X; \underline{\underline{S}}^2) \to \cdots$$

while the *local cohomology sheaf* $\underline{\underline{H}}^p(\underline{\underline{S}}^\cdot)$ denotes the sheaf $(\underline{\underline{\ker}}\, d^p / \underline{\underline{\mathrm{Im}}}\, d^{p-1})$.

Using the same method as that in [GM3], we obtain the following characterization of the intersection homology groups:

THEOREM. *Let* $\underline{\underline{S}}^\cdot$ *be a complex of fine sheaves on* X *such that*:

(1) $\underline{\underline{S}}^k = 0$ *for all* $k < 0$.

(2) *The local cohomology sheaves* $\underline{\underline{H}}^p(\underline{\underline{S}}^\cdot)$ *are locally constant on each stratum* $X_i - X_{i-1}$.

(3) *There is a sheaf map* $\underline{\underline{C}}_{X-X_{n-1}} \to \underline{\underline{S}}^\cdot|(X - X_{n-1})$ *which induces isomorphisms on the local cohomology sheaves, where* $\underline{\underline{C}}_{X-X_{n-1}}$ *denotes the constant sheaf on* $X - X_{n-1}$.

(4) *For each point* $x \in X_i - X_{i-1}$ *there is a neighborhood* $U \subset X - X_{i-1}$ *of* x, *such that*

$$H^k(U; \underline{\underline{S}}^\cdot|U) = 0 \quad \textit{for all} \quad k \geq n-i$$

and $H^k(U; \underline{\underline{S}}^\cdot|U) \cong H^k(U - X_i; \underline{\underline{S}}^\cdot|(U - X_i))$ *for all* $k \leq n-i-1$ *where the isomorphism is induced by restriction.*

Then $H^k(X; \underline{\underline{S}}^\cdot) \cong IH_{2n-k}(X)$.

REMARKS (1) Condition (4) above is a functorial version of the local homology calculation (1.6) and §2.2.

(2) The above theorem remains valid when X is a pseudomanifold with a stratification by strata of even dimension (see §3.4).

§3. L^2-*Cohomology*

In this section we recall the basic definitions and facts concerning L^2-cohomology, and indicate the connection with IH_* in the conical case; (see [C3] for further details).

3.1. Let Y be any (in general incomplete) riemannian manifold with metric g. Assume $\partial Y = \phi$. Let Λ^i, Λ^i_0 denote the spaces of C^∞ i-forms and C^∞ i-forms of compact support, and let L^2 denote the space of square integrable i-forms with measurable coefficients. Let $d_i : \Lambda^i \to \Lambda^{i+1}$ be exterior differentiation. We can define an unbounded operator (in the Hilbert space sense), also to be denoted by d_i, by setting

$$\text{dom } d_i = \{\alpha \epsilon \Lambda^i \cap L^2 | d\alpha \epsilon \Lambda^{i+1} \cap L^2\} . \tag{3.1}$$

As usual,

$$\ker d_i = \{\alpha \epsilon \Lambda^i | d\alpha = 0\} , \tag{3.2}$$

$$\text{range } d_i = \{\eta \epsilon \Lambda^{i+1} | d_i\alpha = \eta , \text{ for some } \alpha \epsilon \text{dom } d_i\} \tag{3.3}$$

The simplest definition of L^2-cohomology is

$$H^i_{(2)}(Y) = \ker d_i / \text{range } d_{i-1} . \tag{3.4}$$

Note that $H^i_{(2)}(Y)$ depends only on the quasi isometry class of g; (g' is quasi isometric to g if for some $k > 0$, $\frac{1}{k} g \leq g' \leq kg$).

Set $d_{i,0} = d_i | \Lambda^i_0$, $\delta_{i,0} = \delta_i | \Lambda^i_0$, where δ_i is the differential operator $(-1)^{n(i+1)+1} {}_* d_*$, and dom δ_i is defined as in (3.1). Let A^* denote the adjoint of an operator A. Then, since dom $\delta_{i+1,0} = \Lambda^{i+1}_0$ is dense, and by Stokes' Theorem,

$$<d\alpha, \beta> = <\alpha, \delta\beta> \tag{3.5}$$

whenever $\alpha \epsilon \text{ dom } d_i$, $\beta \epsilon \text{ dom } \delta_{i+1,0}$, it follows that d_i has a well-defined *weak* closure, $\delta^*_{i+1,0}$. There is also a *strong* closure $\overline{d}_i$, of $d_i \cdot \overline{d}_i \cdot \alpha = \eta$ means that $\alpha \epsilon L^2$ and there exist $\alpha_j \epsilon \text{ dom } d_i$ such that $\alpha_j \to \alpha$, $d\alpha_j \to \eta$. Clearly $\overline{d}_i$, $\delta^*_{i+1,0}$ are closed operators (i.e. they have closed graphs) and dom $\overline{d}_i \subseteq \text{dom } \delta^*_{i+1,0}$, $\delta^*_{i+1,0} | \text{dom } \overline{d}_i = \overline{d}_i$. In fact, one can show that in general, $\overline{d}_i = \delta^*_{i+1,0}$; (see [C3], [GA1]). Then one might also consider

(3.6) $$H^i_{(2),\#}(Y) = \ker \bar{d}_i/\text{range } \bar{d}_i$$

as a possible candidate for L^2-cohomology. However, as shown in [C3], the natural map $H^i_{(2)}(Y) \to H^i_{(2),\#}(Y)$ is always an isomorphism. Thus one can use either definition as convenience dictates. There are natural pseudo norms on $H^i_{(2)}(Y)$, $H^i_{(2),\#}(Y)$, given by

(3.7) $$\|U\| = \inf_{\alpha \epsilon U} \|\alpha\| .$$

These are preserved by the isomorphism above. It follows immediately that the pseudo norm is a norm if and only if the range of $\bar{d}_i$ is a closed subspace; (i.e. if $\bar{d}_i \gamma_j \to \eta$ implies $\eta = \bar{d}_i \psi$ for some ψ). Since $\bar{d}_i$ is a closed operator, it is a standard consequence of the Open Mapping Theorem that range d_{i-1} is closed if $H^i_{(2)}(Y) = H^i_{(2),\#}(Y)$ is finite dimensional.

EXAMPLE 3.1. Let $Y = R$, the real line. If f is a C^∞ function such that $f(x) = \frac{1}{x}$ for $|x| \geq 1$, then $f dx \,\epsilon \ker \bar{d}_1$. Clearly, the most general function α such that $d\alpha = f(x)\,dx$ satisfies $\alpha = \log x + c$ for $x > 1$. But then, $\alpha \notin L^2$ so $f dx \notin$ range d_0. Since $H^1_{(2)} = H^1_{(2),\#}$, also $f dx \notin$ range $\bar{d}_0$. Let ϕ be a smooth function which is supported on $[-2,2]$, such that $\phi|[-1,1] \equiv 1$. Set $\phi_n(x) = \phi(x/n)$. Then easy estimates show $d(\phi_n \alpha) \to f dx$ in L^2. Thus range d_0 is not closed and $H^1_{(2)}(r)$ is infinite dimensional.

Let $\bar{V}$ denote the closure of a subspace V. Define $\mathcal{H}^i$ to be the space of i-forms h, such that $h \,\epsilon\, L^2$, $dh = \delta h = 0$. Kodaira has observed that one always has the *Weak Hodge Theorem*

(3.8) $$L^2 = \overline{\text{range } \delta_{i+1,0}} \oplus \overline{\text{range } d_{i-1,0}} \oplus \mathcal{H}^i$$

where the sum is orthogonal and preserves $\Lambda^i \cap L^2$. This is a consequence of *local* elliptic regularity for the Laplacian $\Delta = d\delta + \delta d$.

Note that there is a natural map, $i_{\mathcal{H}}: \mathcal{H}^i \to H^i_{(2)}(Y)$. We say that the *Strong Hodge Theorem* holds for Y, if $i_{\mathcal{H}}$ is an isomorphism, or equivalently if range $\bar{d}_{i-1} = \overline{\text{range } d_{i-1,0}}$. Clearly, this property depends only on the quasi isometry class of the metric. The *surjectivity* of $i_{\mathcal{H}}$ is equivalent to range $\bar{d}_{i-1} \supseteq \overline{\text{range } d_{i-1,0}}$, and follows in particular if range d_{i-1} is closed.

The *injectivity* of $i_{\mathcal{H}}$ follows if $\bar{d} = \delta^*$, or equivalently, (since $A^{**} = A$ for closed operators), if $\bar{d}^* = \bar{\delta}$. In this case, as usual,

$$\langle h, \bar{d}\beta \rangle = \langle \bar{\delta} h, \beta \rangle = 0 \,. \tag{3.9}$$

As indicated above, in general, $\delta^*_{i+1,0} = \bar{\delta}^*_{i+1,0} = \bar{d}_i$. Thus,

$$\bar{d}_i = \bar{\delta}^*_{i+1} \iff \bar{d}^*_i = \bar{\delta}_{i+1} \iff \tag{3.10}$$
$$\bar{d}_i = \bar{d}_{i,0} \iff \bar{\delta}_{i+1} = \bar{\delta}_{i+1,0} \,.$$

3.2. If Y is *complete* then by [GA2], (3.10) holds. We briefly indicate the argument under the assumption that there exists $y \in Y$, such that ρ_y, the distance function from y is smooth; in the general case one uses regularization to obtain a smooth approximation to ρ_y. Let ϕ_n be as in Example 3.1, and set $f_n = \phi_n \circ \rho_y$. Then one checks that if $\alpha \in \text{dom } \bar{d}_j$, then $(f_n\alpha) \to \alpha$, $d_{i,0}(f_n\alpha) \to d_i\alpha$, which implies $\bar{d}_{i,0} = \bar{d}_i$.

In certain incomplete cases of interest below, one can show $\bar{d}_i = \bar{\delta}^*_{i+1}$ without the availability of a cutoff function. However, in the complete case using f_n, one can prove the strong additional property that $h \in L^2$, $\Delta h = 0$ implies $h \in \mathcal{H}^i$; (i.e. $dh = \delta h = 0$); see [DR], [AV]. In the incomplete case, this property is quite delicate. It is not an invariant of the quasi isometry class of the metric and can fail to hold even if $\bar{d} = \bar{\delta}^*$; e.g. it fails for the double cover of the punctured plane with the pulled back (flat) metric. This phenomenon is responsible for the difficulties of the incomplete case which were described in the introduction.

In the complete case, if $h \in L^2$, $\Delta h = 0$, we have, by Stokes' Theorem,

$$<\delta dh, f_n^2 h> - <f_n dh, 2df_n \wedge h> = <f_n dh, f_n dh> \tag{3.11}$$

$$<d\delta h, f_n^2 h> \pm <f_n \delta h, 2*(df_n \wedge *h)> = <f_n \delta h, f_n \delta h> \tag{3.12}$$

Since $|<a,b>| \leq \frac{1}{2}<a,a> + \frac{1}{2}<b,b>$

$$\frac{1}{2}<f_n dh, f_n dh> + 2<df_n \wedge h, df_n \wedge h> \geq |<f_n dh, 2df_n \wedge h>| \tag{3.13}$$

$$\frac{1}{2}<f_n \delta h, f_n \delta h> + 2<df_n \wedge *h, df_n \wedge *h> \geq |<f_n \delta h, 2*(df_n \wedge *h)>| . \tag{3.14}$$

Adding (3.11), (3.12) and using (3.13), (3.14) and $\Delta h = 0$, we get

$$\begin{aligned} &\frac{1}{2}(<f_n dh, f_n dh> + <f_n \delta h, f_n \delta h>) \\ &\leq 2(<df_n \wedge h, df_n \wedge h> + <df_n \wedge *h, df_n \wedge *h>) . \end{aligned} \tag{3.15}$$

As $n \to \infty$, it is easy to see that the right-hand side of (3.15) $\to 0$. Since $f_n \to 1$, it follows that $dh = \delta h = 0$.

To account for the possibility that range $\bar{d}_{i-1}$ may not be closed, it is customary to define the reduced L^2-cohomology by setting

$$\bar{H}^i_{(2)}(Y) = \ker \bar{d}_i / \overline{\text{range } d_{i-1}} . \tag{3.16}$$

If $\bar{d} = \bar{\delta}^*$, then automatically, $\bar{H}^i_{(2)}(Y) \simeq \mathcal{H}^i$. Suppose that one now takes as his objective to find *some* topological interpretation of the space $\mathcal{H}^i$ and then to derive properties of the resulting object from general properties of harmonic forms; e.g. if Y is *any* complete Kähler manifold then $\mathcal{H}^*$, (which might possibly be infinite dimensional for some i), satisfies the Kahler package. Then $\bar{H}^i_{(2)}(Y)$ can be viewed as a "bridge"; i.e. to interpret $\mathcal{H}^i$ it suffices to calculate $\bar{H}^i_{(2)}(Y)$. If in fact $\bar{H}^i_{(2)}(Y) = H^i_{(2)}(Y)$, one can calculate on open subsets and apply the usual exact cohomology sequences; see [C3]. But since $\bar{H}^*_{(2)}(Y)$ is not the cohomology of a complex of cochains, these sequences may not hold if $\bar{H}^i_{(2)}(Y) \neq H^i_{(2)}(Y)$; compare [APS].

3.3. We now describe the L^2-cohomology for the simplest singularity in the compact case, that of a metric cone. Let N^m be a riemannian manifold with metric $\tilde{g}$. The *metric cone* $C^*(N^m)$ is by definition the completion of the smooth incomplete riemannian manifold $C(N) = R^+ \times N$, with

$$g = dr \otimes dr + r^2 \tilde{g} \ . \tag{3.17}$$

We denote by $C_{r_0,r}(N^m)$ the subset $(r_0, r) \times N \subset C(N^m)$: Suppose that X^{m+1} is a compact metric space such that for some finite set of points $\{p_j\}$, $X - \bigcup_{j=1}^{N} p_j$ is a smooth riemannian manifold. We say that X^{m+1} has *isolated metrically conical singularities* if there exist smooth compact riemannian manifold N^m_j, and neighborhoods U_j of p_j, such that $U_j - p_j$ is isometric to $C_{0,r_j}(N^m)$. We say that X has *isolated conical singularities* if the metric on $X - Up_j$ is quasi isometric to a metric of the above type. We define $H^*_{(2)}(X)$ by

$$H^i_{(2)}(X) = H^i_{(2)}(X - \bigcup_{j=1}^{N} p_j) \ . \tag{3.18}$$

In [C3], it is shown that $H^i_{(2)}(X)$ does not change if further points are removed from $X - \bigcup_{j=1}^{N} p_j$. Thus $H^i_{(2)}(X)$ is well defined.

The Poincaré Lemma in this situation takes the following form.

$$H^i_{(2)}(C_{0,1}(N^m)) = \begin{cases} H^i(N^m) & i \leq m/2 \\ 0 & i > m/2 \ . \end{cases} \tag{3.19}$$

This corresponds to the calculation in §2.2, for IH_* of a truncated cone. When combined with the standard exact sequences, (3.19) yields the tabulation of $H^i_{(2)}(X^{m+1})$ given in §2.2 for $m+1$ even. In particular, $H^i_{(2)}(X^{m+1})$ is finite dimensional which implies that $\bar{d}_{i-1}$ has closed range.

If $m+1 = 2k$, or in case $m+1 = 2k+1$, if $H^k(N^{2k}, R) = 0$, then $\bar{d} = \bar{\delta}^*$. Thus in these cases, the Strong Hodge Theorem holds. If $m+1 = 2k+1$ and $\dim H^k(N^{2k}, R) > 0$, then $d_k \neq \delta^*_{k+1}$. Moreover a calculation based on (3.19) shows that Poincaré duality also fails in this case. These interesting phenomena demonstrate that the global topology of the link plays a significant role in the theory. However, they do *not* occur for algebraic varieties, and thus will not be discussed further here; see however [C1], [C2], [C3], [C5]. (The point here is that algebraic varieties admit a stratification by strata of even codimension.)

The calculation in (3.19) can be made intuitively plausible as follows. Let π_2 be the natural projection of $C_{0,1}(N^m)$ onto N^m; ($C_{0,1}(N^m)$ is topologically $(0,1) \times N^m$). If ϕ is an i-form on N^m, then the pointwise norm of $\pi_2^*(\phi)$ at a point (r, x) is r^{-i} times the norm of ϕ at x. Since the cross sectional area of $C_{0,1}(N^m)$ varies as r^m, the condition for $\pi_2^*(\phi)$ to define an element of $H^i_{(2)}(C_{0,1}(N^m))$; (i.e. for $\pi_2^*(\phi)$ to be in L^2), is just $m - 2i > -1$, or equivalently $i \leq \frac{m}{2}$.

The proof of (3.19) uses the following homotopy operators. Let $\theta(r, x) = \phi(r, x) + dr \wedge \omega(r, x)$ and let $a \in (0, 1)$. Set

$$\mathcal{K}\theta = \begin{cases} \displaystyle\int_a^r \omega & i \leq \frac{m}{2} + 1 \\ \displaystyle\int_0^r \omega & i > \frac{m}{2} + 1 . \end{cases} \tag{3.20}$$

Appropriate estimates show that the integrals converge and define bounded operators on L^2. A computation and some estimates show that

$$(\bar{d}\mathcal{K} + \mathcal{K}\bar{d})\,\theta = \begin{cases} \theta - \phi(a) & i < \frac{m+1}{2} \\ \theta & i \geq \frac{m}{2} + 1 . \end{cases} \tag{3.21}$$

The cases $i = \frac{m+1}{2}$, $i = \frac{m}{2} + 1$ are a little more delicate, but the end result is (3.19).

3.4. Let X^n be a pseudomanifold—i.e. a simplicial complex such that each point is contained in a closed n-simplex, each n–1 simplex is a face of exactly two n-simplices, and the n-simplices can be compatibly oriented. We assume X^n is embedded as a subcomplex of some piecewise linear triangulation of $\mathbf{R}^n$. Let Σ^i denote the closed i-skeleton of X^n. The induced metric on $X^n - \Sigma^{n-2}$, gives $X^n - \Sigma^{n-2}$ the structure of a *smooth* flat riemannian manifold; (a neighborhood of a point on an (n–1)-simplex is isometric to an open subset of $\mathbf{R}^n$). If X^n is now given the structure of a metric space in such a way that the induced metric on $X^n - \Sigma^{n-2}$ is quasi-isometric to the flat metric then this metric space will be called a *riemannian pseudo-manifold with conical singularities*; (e.g. the metric on X^n induced from the embedding in $\mathbf{R}^n$ obviously has this property).

Define $H^i_{(2)}(X) = H^i_{(2)}(X-\Sigma)$. The results on isolated conical singularities: $\overline{d} = \overline{\delta}^*$ and the strong Hodge theorem extend to this case. If X also has a stratification by even dimensional strata, then the axioms of §2.2 are satisfied by the complex of sheaves of locally L^2-differential forms on $X-\Sigma$ which are in dom(d). We conclude that $H^k_{(2)}(X) \cong IH_{n-k}(X) \cong \mathrm{Hom}(IH_k(X), \mathbf{C})$. This isomorphism is also constructed directly by integration in [C3]. Here it is most convenient to use a certain subcomplex ($\mathcal{S}^*$ of [C3], p. 136) of the L^2 forms, which has the same cohomology, but for which integration over chains which satisfy the "middle intersection group allowability condition" is always defined.

Finally we mention the following three general facts. The usual cohomology groups $H^*(X^n)$ can be represented by forms whose norms are uniformly bounded on $X^n - \Sigma^{n-2}$; e.g. forms on an open neighborhood of $X^n \subset \mathbf{R}^N$. From this it follows that $H^*_{(2)}(X^n)$ is a module over $H^*(X^n)$.

For each pseudomanifold X there is an associated space $\hat{X}$, the *normalization* of X [GM2]. Since X and $\hat{X}$ have the same nonsingular parts, $H^*_{(2)}(\hat{X}) \cong H^*_{(2)}(X)$. (In fact, $\hat{X}$ is topologically just the metric completion of the nonsingular part $X-\Sigma$.)

Consider a finite covering of $X-\Sigma^{n-2}$. If one pulls back and completes the Riemannian metric on $X-\Sigma$, one obtains a space X' and a map $\pi : X' \to X$ which is a (topological) *branched covering* (see [F]). There is a natural map $\pi^* : H^i_{(2)}(X) \to H^i_{(2)}(X')$. By Poincaré duality, π^* is an injection. (Note that analytic properties on X do not carry over automatically to properties on X'. For example, the torus T^2 is a branched cover of the sphere S^2, so $C(T^2)$ is a branched cover of $C(S^2) = \mathbf{R}^3$. But $\dim H^1(T^2, \mathbf{R}) \neq 0$ so $\bar{d}_1 \neq \delta_2^*$ on $C(T^2)$.)

3.5. Let X be a compact analytic variety which admits an embedding in some compact Kähler manifold (e.g. complex projective space). Any choice $\rho : X \to M$ of such an embedding determines a metric Ω_ρ on the nonsingular part $X-\Sigma$ of X by restriction of the Kähler metric on M.

PROPOSITION. *The quasi-isometry class of Ω_ρ is independent of the embedding ρ.*

As in §3.4 we define $H^i_{(2)}(X)$ to be $H^i_{(2)}(X-\Sigma)$ with the metric Ω_ρ. By the proposition, it is independent of ρ.

We note that we do not know the general fact about normalization of §3.4 in the analytic context because we do not know that the metric on the nonsingular part of $\hat{X}$ is quasi-isometric to that of X. A similar remark applies to branched coverings.

Let $V \subset \mathbf{C}^N$ be an analytic subvariety. We say that V is a cone at $p \in V$ if for some union W of affine complex lines through p and some neighborhood U of p, we have $V \cap U = W \cap U$. For example any subvariety of $\mathbf{C}^N$ defined by homogeneous equations is conical at 0.

DEFINITION. An analytic variety X is *locally analytically conical* if each point $p \in X$ has a neighborhood U and an analytic embedding $\rho : U \to \mathbf{C}$ such that $\rho(U)$ is a cone at $\rho(p)$.

EXAMPLES 1. Not all cones at 0 are locally analytically conical: for example $X^2Z = Y^3$ fails the test at $p = (0, 0.1)$.

2. Single condition Schubert varieties (see §5.2) are locally analytically conical.

PROPOSITION. *Locally analytically conical varieties are conical.*

In other words, the nonsingular part of a locally analytically conical variety is quasi-isometric to an open dense subset of a polyhedron. The converse is false: all algebraic curves are conical in the quasi-isometry sense.

COROLLARY. *For a locally analytically conical variety* X, $H^i_{(2)}(X) \simeq \mathrm{Hom}(IH_i(X), \mathbb{C})$.

As remarked in §1.7, this does not in itself imply the theorems of the "Kähler package" for locally analytically conical varieties. But it is known for other reasons: see the section on algebraically conical singularities of §6.2.

3.6. We close this section by noting the following construction of complete Kähler metrics. Let X be a projective algebraic variety. Then as is well known, there exists an algebraic map $\pi : X \to \mathbb{C}P^N$ which is a branched covering. Let $Z \subset \mathbb{C}P^N$ denote the branch locus, and let $\mathfrak{L}$ be the line bundle over $\mathbb{C}P^N$ corresponding to Z, equipped with a Hermition metric. Let Ω denote the Kähler form of the usual metric on $\mathbb{C}P^N$. Then, if σ is a holomorphic section of $\mathfrak{L}$ which vanishes on Z and ϕ is a smooth function such that $\phi|Z \equiv 1$, $\phi \equiv 0$ off and ϵ-tubular neighborhood of Z, then as in [CG], for small ϵ',

$$\epsilon' \sqrt{-1}\,(dd^c \log \log^2 \phi \|\sigma\|^2) + \Omega \tag{3.22}$$

defines the Kähler form of a complete Kähler metric on $\mathbb{C}P^N - Z$. We do not know if the L^2-cohomology of this metric is isomorphic to the usual cohomology of $\mathbb{C}P^N$. If so, it makes sense to ask if the L^2-cohomology of the pulled back metric on $X-\pi^{-1}(Z)$ is isomorphic to $IH^*(X)$. This would imply that the "Kähler package" holds for X.

§4. *Conjectures*

In this section X will denote a complex n-dimensional projective variety. Conjecture A states that the intersection homology groups $IH_*(X)$ satisfy the conditions of the "Kähler package." Conjectures B and C state that the De Rham and Hodge theorems hold for the L^2 differential forms on the nonsingular part of X.

Let Σ denote the singular set of X. We shall denote the L^2 cohomology of $X-\Sigma$ (with the metric induced from the embedding in projective space) by $H^*_{(2)}(X)$.

Conjecture A : The Kähler package

A.1. (*Pure Hodge* (p, q) *decomposition*). There is a natural direct sum decomposition

$$IH_k(X) \cong \bigoplus_{p+q=k} IH_{(p,q)}(X)$$

such that

$$IH_{(p,q)}(X) \cong \overline{IH_{(q,p)}(X)} .$$

This decomposition is compatible with f^* and f_* when f is a branched covering or is normally nonsingular. For example, if $f: Y \to X$ is normally nonsingular with relative dimension m then

$$f_* : IH_{(p,q)}(Y) \to IH_{(p,q)}(X)$$

and

$$f^* : IH_{(p,q)}(X) \to IH_{(p-m,q-m)}(Y) .$$

The map from cohomology $H^i(X) \to IH_{2n-i}(X)$ is a morphism of Hodge structures.

A.2. (*Hard Lefschetz*). Let H be a hyperplane in the ambient projective space, which is transverse to a Whitney stratification of X. Let $\Omega \in H^2(X)$ denote the cohomology class represented by $H \cap X$ (as in §2.1) and let $L : IH_i(X) \to IH_{i-2}(X)$ denote multiplication by this class. Then the map

$$L^k : IH_{n+k}(X) \to IH_{n-k}(X)$$

is an isomorphism for each k.

If we define the primitive intersection homology $P_{n+k}(X) = \ker(L^{k+1})$ then we have the Lefschetz decomposition $IH_m(X) = \bigoplus_k L^k(P_{m+2k}(X))$. This decomposition is compatible with the Hodge decomposition.

A.3. *Poincaré duality*. The intersection pairing

$$IH_i(X) \times IH_{2n-i}(X) \to \mathbf{C}$$

is nonsingular in complementary dimensions.

A.4. *Lefschetz Hyperplane Theorem*: If H is a hyperplane which is transverse to each stratum of X, then the homomorphism induced by inclusion

$$IH_k(X \cap H: \mathbf{Z}) \to IH_k(X; \mathbf{Z})$$

is an isomorphism for $k < n-1$ and a surjection for $k = n-1$.

A.5. *Hodge Signature Theorem*. For $\xi \in P_{(p,q)}$ a primitive intersection homology class of dimension $k = p+q$, we have

$$(\sqrt{-1})^{p-q}(-1)^{(n-k)(n-k-1)/2} L^k(\xi \cap \bar{\xi}) > 0 .$$

If $\sigma(X)$ denotes the signature of the intersection pairing (A.3) on $IH_n(X)$ then

$$\sigma(X) = \sum_{p+q \equiv 0 (\mathrm{mod}\ 2)} (-1)^p \dim IH_{(p,q)}(X) .$$

Conjecture A would follow from the stronger Conjectures B and C below. (In fact, for this implication, it would suffice to substitute $\bar{H}^*_{(2)}$ for $H^*_{(2)}$ in what follows.)

Conjecture B. The L^2 cohomology group $H^k_{(2)}(X)$ is finite dimensional and is isomorphic to the subspace $\mathcal{H}^k$ of $\Lambda^k \cap L^2$ which consists of the square summable differential k-forms which are closed and co-closed. Furthermore, the operator J preserves this subspace $\mathcal{H}^k$.

Conjecture C. For almost any chain $\xi \in C_k(X)$ and almost any differential form $\theta \in \Lambda^k \cap L^2$, the integral $\int_\xi \theta$ is finite, and $\int_{\partial\eta} \theta = \int_\eta d\theta$ whenever both sides are defined. The induced homomorphism

$$H^k_{(2)}(X) \xrightarrow{\iota} \mathrm{Hom}(IH_k(X), \mathbb{C})$$

is an isomorphism.

§5. EXAMPLES

In this section we consider two classes of examples: varieties with isolated conical singularities and single condition Schubert varieties. These examples will be used as illustrations throughout the rest of the paper.

In each case we will give a stratification and a resolution. We also calculate the cohomology and the intersection homology of these examples and verify that the intersection homology has a Hodge (p, q) decomposition.

5.1. *Isolated algebraically conical singularities*

We will treat the case where X has a unique algebraically conical singular point x. The case of several isolated algebraically conical singular points is entirely similar.

DEFINITION. The isolated singularity $x \in X$ is said to be algebraically conical if the Hopf blowup $\pi : \tilde{X} \to X$ of X at x is nonsingular and the exceptional division $D = \pi^{-1}(x)$ is nonsingular.

EXAMPLE. A locally analytically conical varieties (§3.5) with one singular point is algebraically conical.

Let $L \to D$ be the normal complex line bundle of D and let $S \to D$ be its circle bundle, which is the boundary of a tubular neighborhood of D in $\tilde{X}$. Consider the Gysin sequence

$$H_i(D) \xrightarrow{\cap C_1 L} H_{i-2}(D) \xrightarrow{\beta} H_{i-1}(S) \xrightarrow{\alpha} H_{i-1}(D) \xrightarrow{\cap C_1 L} H_{i-3}(D)$$

where $C_1 L$ denotes the first chern class of L. Since $D = \pi^{-1}(x)$ is the exceptional divisor it must satisfy the following

Blowing down condition: $C_1 L$ can be represented by a Kähler form on D, so $\cap C_1 L$ satisfies hard Lefschetz.

Since X is a space with isolated singularities we have:

$$IH_i(X) = \begin{cases} H_i(X) & \text{for } i > n \\ H_i(X-x) & \text{for } i < n \\ \operatorname{Im}(H_i(X-x) \to H_i(X)) & \text{for } i = n\ . \end{cases}$$

The column and rows in the following diagrams are exact:

I

$$\begin{array}{ccccccccc} & & & & & H_{i-1}(S) & & & \\ & & & & \nearrow & & \searrow \alpha & & \\ H_i(D) & \longrightarrow & H_i(\tilde{X}) & \longrightarrow & H_i(X) & \longrightarrow & H_{i-1}(D) & \longrightarrow & H_{i-1}(\tilde{X}) \end{array}$$

II

$$\begin{array}{ccccccccc} & & & & & H_{i-1}(S) & & & \\ & & & & \beta \nearrow & & \searrow & & \\ H_i(\tilde{X}-D) & \longrightarrow & H_i(\tilde{X}) & \longrightarrow & H_{i-2}(D) & \longrightarrow & H_{i-1}(\tilde{X}-D) & \longrightarrow & H_{i-1}(\tilde{X}) \\ H^{2n-i}(X) & & & & & & & & \end{array}$$

III

$$\begin{array}{ccccc} & & H_i(D) & & \\ & & \downarrow & \searrow \cap C_1 L & \\ H^{2n-i}(X) = H_i(\tilde{X}-D) & \longrightarrow & H_i(\tilde{X}) & \longrightarrow & H_i(\tilde{X}, \tilde{X}-D) = H_{i-2}(D) \\ & \gamma \searrow & & & \\ & & H_i(\tilde{X}, D) = H_i(X)\ . & & \end{array}$$

If $i > n$ it follows from the blowing down condition and the Gysin sequence that $\alpha = 0$. Therefore

$$IH_i(X) = \text{coker}\,(H_i(D) \to H_i(\tilde{X}))$$

which has a Hodge (p,q) decomposition induced from that on $\tilde{X}$.

Similarly if $i < n$ then $\beta = 0$ so

$$IH_i(X) = \ker\,(H_i(\tilde{X}) \to H_{i-2}(D))$$

which is also pure.

Finally, if $i = n$, the map $\cap C_1 L$ of diagram III is an isomorphism, so $IH_n(X) = \ker(H_n(\tilde{X}) \to H_{n-1}(D))$ which is pure.

5.2. *Single condition Schubert varieties*

Fix integers i, j, k, ℓ such that $j+k \le \ell$, and

$$\begin{matrix} & & j & & \\ & \le & & \le & \\ i & & & & \ell\,. \\ & \le & & \le & \\ & & k & & \end{matrix}$$

Let $F^j \subset \mathbb{C}^\ell$ denote a fixed j-dimensional subspace and let $G_k(\mathbb{C}^\ell)$ denote the Grassmann variety of k planes in $\mathbb{C}^\ell$. Define

$$\mathcal{S} = \{V^k \in G_k(\mathbb{C}^\ell) \,|\, \dim(V^k \cap F^j) \ge i\}\,.$$

Such an $\mathcal{S}$ is called a single condition Schubert variety. It has complex dimension $i(j-i)+(k-i)(\ell-k)$. Define

$$\tilde{\mathcal{S}} = \left\{ \text{partial flags } (W^i \subset V^k \subset \mathbb{C}^\ell) \,|\, \begin{matrix} & & F^j & & \\ & \subset & & \subset & \\ W^i & & & & \mathbb{C}^\ell \\ & \subset & & \subset & \\ & & V^k & & \end{matrix} \right\}.$$

The map $\pi : \tilde{\mathcal{S}} \to \mathcal{S}$ (given by $\pi(W^i \subset V^k) = V^k$) is a resolution of singularities. $\mathcal{S}$ is stratified by the single-condition Schubert subvarieties

$$\mathcal{S}_p = \{V^k \in G_k(\mathbb{C}^\ell) \,|\, \dim(V^k \cap F^j) \ge p\}$$

for $i \le p \le \min(j,k)$. The codimension of $\mathcal{S}_p$ in $\mathcal{S}$ is $C=(p-i)(p+i+\ell-j-k)$. If $x \in \mathcal{S}_p - \mathcal{S}_{p+1}$, then $\dim(\pi^{-1}(x)) = i(p-i)$. Since $i(p-i) \le \frac{1}{2}C$, $\tilde{\mathcal{S}}$ is a small resolution of $\mathcal{S}$ and consequently $IH_*(\mathcal{S}) \cong H_*(\tilde{\mathcal{S}})$ (see [GM3]). It follows that $IH_*(\mathcal{S})$ inherits a Hodge (p,q) decomposition from that of $\tilde{\mathcal{S}}$. (It is known that $H_{p,q}(\tilde{\mathcal{S}}) = 0$ unless $p=q$ so the same is true of $IH_{p,q}(\mathcal{S})$.)

We now give the Poincaré polynomials for these spaces. Define $P_n(t) = 1(1+t^2)(1+t^2+t^4)\cdots(1+t^2+\cdots+t^{2n-2})$. The Poincaré polynomial $Q_\ell^k(t)$ for $G_\ell(\mathbb{C}^k)$ is

$$Q_\ell^k(t) = \frac{P_\ell(t)}{P_k(t)\,P_{\ell-k}(t)}\,.$$

The Poincaré polynomial for $IH_*(\mathcal{S})$ is $Q_i^k(t)\,Q_{j-i}^{\ell-i}(t)$. The Poincaré polynomial for $H^*(\mathcal{S})$ is

$$\sum_{i \le p \le \min(j,k)} Q_p^k(t)\,Q_{j-p}^{\ell-k}(t)\,t^{2j(j-p)}\,.$$

(Each term in this sum is a contribution arising from a stratum.)

The map $H^*(\mathcal{S}) \to H^*(\tilde{\mathcal{S}}) \cong IH_*(\mathcal{S})$ is an injection. One may verify from the Poincaré polynomials that it is in general far from being a surjection.

§6. *Status of the conjectures*

In this section, we present the currently available evidence for the conjectures of Chapter 4.

In §6.1 we discuss the parts of the Kähler package that have been proved for all complex projective varieties. The methods of proof are topological or algebraic: they do not use L^2 analysis. The main gap at present is that there is no general proof of the existence of a pure Hodge (p, q) decomposition.

Even in the nonsingular case, establishing a conjugation invariant (p, q) decomposition requires analysis. Two possible approaches to the singular case are to reduce it to a related nonsingular space—e.g. a resolution of singularities, or to develop L^2 analysis directly on the singular variety. As for the first possibility there is a conjecture that the intersection homology of a variety X is always a direct summand of the homology of any resolution $\tilde{X}$ and that the inclusion of $IH_*(X)$ into $H_*(\tilde{X})$ respects the Hodge decomposition of $H(\tilde{X})$ (see [GM3]). This conjecture of course would imply the existence of a pure Hodge decomposition on X, but the conjecture itself appears to be very difficult in general. In §6.2 we give a number of special cases in which this conjecture can be established.

As for analysis on X itself, there are again two approaches as indicated in §1.8 and §3. One is to use the incomplete metric on $X-\Sigma$ induced by the inclusion $X \subset \mathbb{C}P^N$, and the other is to fabricate a complete metric. In both cases the analysis is extremely delicate and it depends on as yet unexplored aspects of the metric structure of X near a singularity. To carry this out for general X may well be even more difficult than the procedure using $\tilde{X}$, but the resulting understanding of the differential geometry of the singularities of X would be extremely interesting in itself. The progress to date on this is sketched in §6.4.

6.1. *Results for general varieties*

Poincare Duality: As mentioned in §2 the Poincaré duality theorem ($IH_i(X) \times IH_{2n-i}(X) \to \mathbb{C}$ is nonsingular) is true for all projective varieties X.

Weak Lefschetz: The weak Lefschetz theorem (§4.A4) has been proven for all complex projective varieties. There is a sheaf-theoretic proof [GM3] following ideas of Artin [A]. There is also a proof using the Morse-theoretic techniques of [GM4].

Hard Lefschetz: We have been informed by O. Gabber that he has found a proof of the hard Lefschetz theorem (4.A.2) for the ℓ-adic analogue of $IH_*(X)$ when X is a variety defined over a field of characteristic p. This implies the same result in characteristic 0.

Purity in characteristic p: The proof of Gabber also shows that the ℓ-adic analogues of $IH_*(X)$ are pure in the sense of Deligne (all eigenvalues of the Frobenius action have the same absolute value). According to the heuristic dictionary of Deligne [D1] §3, this is the characteristic p analogue of conjecture 4.A.1. But it is moral evidence only. It does not imply the existence of a (p,q) decomposition.

Mixed Hodge structures: Verdier has informed us of the existence of sheaf-theoretic techniques which may be used to put a mixed Hodge structure on $IH_*(X)$. It would remain to show that this mixed Hodge structure is pure.

6.2. *Special classes of varieties*

Curves and surfaces: The Hodge (p,q) decomposition conjecture (4.A.1) is true for varieties X with complex dimension 1 or 2. For curves this is true because $IH_*(X) = IH_*(\hat{X})$ where $\hat{X}$ is the normalization of X, which is always nonsingular. For surfaces the (p,q) decomposition of $IH_*(X)$ may be deduced from that of $H_*(\tilde{X})$ where $\tilde{X}$ is the minimal resolution of X. The proof is similar to that in §5.1 but uses Grauert's blowing down condition in place of the ampleness condition on the normal bundle of the exceptional divisor.

Schubert varieties: For the single condition Schubert varieties of §5.2, $IH_*(X)$ is isomorphic to the cohomology of the small resolution $\tilde{X}$ and therefore has a Hodge (p,q) decomposition.

Algebraically conical singularities:

DEFINITION. A monoidal transformation $\pi : \tilde{Y} \to Y$ with center $Z \subset Y$ is called a *clean blowup* if

1. Z is nonsingular.
2. $D \to Z$ is a topological fibration, where $D = \pi^{-1}(Z)$ is the exceptional divisor.
3. The inclusion $D \subset \tilde{Y}$ is normally nonsingular (§2).

A variety X is said to have *algebraically conical singularities* if it can be desingularized by a sequence of clean blowups. (The sequence of centers Z may be chosen so as to have increasing dimension.) For example, a locally analytically conical variety (§3.5) has algebraically conical singularities, but not vice-versa.

For varieties with algebraically conical singularities, the (p, q) decomposition of $IH_*(X)$ is induced from that of $IH_*(\tilde{X})$ where $\tilde{X}$ is the resolution produced by the sequence of clean blowups. This may be proven in a manner analogous to the case of isolated conical singularities (§5.1) using as ingredients the hard Lefschetz theorem of Gabber, and Deligne's criterion for the degeneration of a spectral sequence [D6] related to the fibrations $D \to Z$.

Rational homology manifolds: An example of a rational homology manifold is the quotient of any smooth compact manifold by the smooth action of any finite group.

Suppose X is an algebraic variety which is a rational homology manifold. Since (for any rational homology manifold) the cohomology coincides with the intersection homology [GM2], it suffices to find a Hodge (p, q) decomposition of the cohomology. However Deligne shows this exists by observing that Poincaré duality on X implies that the cohomology of X injects into the cohomology of any resolution of X, and therefore inherits a Hodge structure from the resolution.

6.3. *Results in special dimensions*

The first Betti number: For all projective varieties X, $\dim(IH_1(X))$ is even (as would be predicted by the existence of a conjugation invariant (p, q) decomposition). This follows from inductive application of the weak Lefschetz theorem and a direct verification for surfaces (see §6.2). It is always true that $IH_1(X) \cong H_1(\hat{X})$ where $\hat{X}$ is the normalization of X [GM2]. Therefore we obtain the following corollary, which was pointed out to us by Horrocks:

COROLLARY. *For any normal projective variety* X, *the first Betti number of* X *(for ordinary homology) is even.*

Varieties with small singular sets: Suppose the singular set of X has dimension $\leq p$. Then if we iterate the process of taking a generic hyperplane section $p+1$ times, we obtain a nonsingular variety. By repeated application of the Lefschetz hyperplane theorem, we have

THEOREM. $IH_k(X)$ *has a pure Hodge* (p,q) *decomposition for all* $k < n-p-1$ *and all* $k > n+p+1$.

REMARKS 1. If we also use the fact that $IH_*(X)$ has an appropriately natural mixed Hodge structure (see §6.1), this theorem can be extended to the cases $k = n-p-1$ and $k = n+p+1$.

2. The same idea shows that the hard Lefschetz map $L^k : IH_{n+k}(X) \to IH_{n-k}(X)$ is an isomorphism for all $k > p+1$.

6.4. L^2-*cohomology*

As explained in §3, for compact spaces with conical singularities $H^*_{(2)}(X) \simeq IH^*(X)$ and the Strong Hodge Theorem holds. However, if in addition the metric on the nonsingular part of X is Kähler this is still not enough to imply the "Kähler package" because the almost complex structure J may not preserve the space $\mathcal{H}^i$; (we still conjecture that J does preserve $\mathcal{H}^i$ if the singularities are conical in a suitable complex analytic sense, e.g. if X is an algebraic variety with metric induced from its embedding in $\mathbf{C}P^N$; see §4). At present, there are two cases when J can be shown to preserve $\mathcal{H}^i$, see [C4] for details.

Isolated metrically conical singularities

Let $C(N^m)$ be a metric cone, where $m = 2k-1$ is odd. Then it can be shown that $h \in L^2$, $\Delta h = 0$, implies $h \in \mathcal{H}^i$, with the possible exception of the cases $i = \frac{m-1}{2}, \frac{m+1}{2}, \frac{m+3}{2}$. Thus if the metric on $C(N^m)$ is Kähler, $J(\mathcal{H}^i) = \mathcal{H}^i$ except possibly in these dimensions. Now assume further that the complex structure is invariant under the 1-parameter group of dilations of $C(N^m)$; e.g. suppose $C(N^m)$ is a complex affine cone. Then it can be checked directly that J preserves the space of forms θ such that $\theta, d\theta, \delta\theta, d\delta\theta, \delta d\theta \in L^2$. This suffices to show that for compact Kähler manifolds with isolated metrically conical singularities, such that J commutes with dilations, $J(\mathcal{H}^i) = \mathcal{H}^i$. More generally, the same follows if the metric and complex structure satisfy these conditions to sufficiently high order at the singular point.

Piecewise flat spaces

The arguments in the example above can be generalized to certain piecewise flat spaces by induction, and "the method of descent"; (compare [CT], example 4.5). Rather than giving a general definition of this class of spaces we will indicate how to construct some examples. Let Y be a compact Kähler manifold such that the metric g is flat and let Z be an arbitrary union of compact totally geodesic complex hypersurfaces. Let $\pi : X \to Y$ be a finite branched covering of Y, branched along Z. Then the completion of the metric $\pi^*(g)$ on $X - \pi^{-1}(Z)$ gives X the structure of a piecewise flat space with metrically conical singularities, and $J(\mathcal{H}^i) = \mathcal{H}^i$ on X. More generally, Y and Z might be quotients of piecewise flat spaces in this construction. For example, let Y be the space obtained by dividing $\overbrace{\mathbf{C} \times \cdots \times \mathbf{C}}^{n}$ by the group generated by the standard lattice, together with multiplication by -1 in each factor and permutations of the factors. Then Y is homeomorphic to $\mathbf{C}P^n$.

Note that in both of the above cases, it is only the Kähler property that is relevant. Thus X need not be an algebraic variety.

Complete metrics (see [M], [ZU1], [ZU2])

In [ZU2],* Zucker considers $H^*_{(2)}(\Gamma\ X)$, the L^2 cohomology of of quotients of symmetric spaces by arithmetic groups, for which the natural metric is complete and has finite volume. In the Hermitian cases, the metric is Kähler. He shows that in certain cases $H^*_{(2)}(\Gamma\backslash X)$ is naturally isomorphic to $IH^*(\Gamma\backslash X^*)$, where $\Gamma\backslash X^*$ is the Baily Borel compactification of $\Gamma\backslash X$.

---

*Other strong evidence is provided by [ZU1].

§7. *Relations with mixed Hodge theory*

In [D1], [D2], [D3] Deligne defines a mixed Hodge structure on the cohomology of any algebraic variety X. This gives a filtration

$$w_0 \subset w_1 \subset \cdots \subset w_{2i} = H^i(X)$$

such that w_j/w_{j-1} has a Hodge (p,q) decomposition with $p+q = j$ ("w_j/w_{j-1} has pure weight j"). He shows:

(7.1) $$X \text{ compact} \Longrightarrow w_i = w_{i+1} = \cdots = w_{2i}$$

(7.2) $$X \text{ smooth} \Longrightarrow w_0 = w_1 = \cdots = w_{i-1} = 0\,.$$

In §7.1 we give a (conjectural) relation between the mixed Hodge structure on the cohomology of X and the (conjectured) pure Hodge structure on $IH_*(X)$. In §7.2 we deduce both structures from the pure Hodge structure of a resolution of X, when X has isolated singularities. We find that an additional criterion is needed for the procedure to work with intersection homology. This criterion is sharpened in §7.3 and gives rise to new (conjectural) necessary conditions for blowing down.

7.1. *Conjecture.* $w_{i-1}(H^i(X)) = \ker(H^i(X) \to IH_{2n-i}(X))$ for compact X. Notice that the kernel always contains w_{i-1} if the Hodge (p,q) decomposition conjecture (4.A.1) is true (because the map is strictly compatible with the filtration [D2] 2.3.5). A consequence of this conjecture is that the (conjectural) pure Hodge structure on $IH_{2n-i}(X)$ determines the one from mixed Hodge theory on w_i/w_{i-1}. The reverse is not true. For single condition Schubert varieties (§5.2) the map $w_i/w_{i-1} \to IH_{2n-i}(X)$ is not surjective.

Conjecture 7.1 is true for the examples of §5 by direct calculation.

Deligne has suggested [D5] the existence of a technique whereby the Hodge structures on the other w_j/w_{j-1} could be similarly determined using the pure Hodge structures on other intersection homology groups.

This technique would apply to the hypercohomology of complexes of algebraic sheaves (as well as to the ordinary cohomology) thereby extending mixed Hodge theory to such hypercohomology groups.

7.2. In this section we describe Deligne's construction of the weight filtration on the cohomology of a space with an isolated singularity. This induces a mixed Hodge structure on intersection homology.

Let D be any compact subvariety of a nonsingular n-dimensional compact complex variety $\tilde{X}$. We first construct a mixed Hodge structure on the cohomology of $\tilde{X}/D$ (the space obtained by collapsing D to a point). In the case that $\tilde{X}/D$ admits the structure of an algebraic variety X (compatibly with the projection $\tilde{X} \to X$), this calculation gives the mixed Hodge structure on X. Consider the exact sequence of the pair (diagram II of 5.1):

$$\to H^{i-1}(X) \to H^{i-1}(\tilde{X}) \overset{\theta}{\to} H^{i-1}(D) \to H^i(X) \to H^i(\tilde{X}) \overset{\theta}{\to} H^i(D) \to .$$

Here, $w_i = H^i(X)$

$w_{i-1} = \operatorname{coker}(H^{i-1}(X) \to H^{i-1}(D))$

$w_j = w_j(H^{i-1}(D))$ for $j < i-1$.

One can see directly from the exact sequence (and the fact that each homomorphism is strictly compatible with the filtration) that w_j/w_{j-1} has a pure Hodge (p,q) decomposition of weight j.

Since $H^i(X) \cong IH_{2n-i}(X)$ for $i > n$ and $\operatorname{Hom}(H^i(X), \mathbb{C}) \cong IH_i(X)$ for $i < n$, we obtain mixed Hodge structures on $IH_j(X)$ for all $j \neq n$. (However (7.1) is not satisfied even though X is compact.) It is easy to see from diagram III of §5.1 that $IH_n(X)$ has a pure Hodge structure of weight n. Note: this mixed Hodge structure on $IH_*(X)$ depends only on the algebraic structure of $\tilde{X} - D$, the nonsingular part of X.

This gives the following result:

PROPOSITION. *A necessary and sufficient condition that* $IH_*(X)$ *have a pure Hodge structure, is that the map*

$$\theta : H^i(\tilde{X}) \to H^i(D)$$

be a surjection for all $i \geq n$, *or equivalently*

$$H_j(D) \to H_j(X) \quad \textit{is an injection for all} \quad j \geq n\ .$$

Observe that in the example of §5.1, this condition is guaranteed by the blowing down condition. However even in this example, the cohomology has only a *mixed* Hodge structure. Thus, to prove that the intersection homology of a variety with an isolated singularity has a pure Hodge structure, one must verify the above condition on any resolution. We do not know how to do this in general, although the preceding construction of the *mixed* Hodge structure (on H^* and IH_*) requires no further condition. Thus the existence of a pure Hodge structure on $IH_*(X)$ appears to involve more subtle structure of the variety than does the existence of a mixed Hodge structure on cohomology.

7.3. We now turn the question around and ask what blowing down conditions are implied by these ideas.

Let D be an arbitrary (compact) subvariety of a nonsingular compact n-dimensional variety $\tilde{X}$ and suppose $X = \tilde{X}/D$ is algebraic.

Conjecture. $H_j(D) \to H_j(\tilde{X})$ is an injection for all $j \geq n$ and this holds for local reasons near D, i.e. if T is a tubular neighborhood of D in $\tilde{X}$, with boundary S, then $H_j(T) \to H_j(T, S)$ is an injection for all $j \geq n$. (This conjecture is a consequence of the "direct sum conjecture" in [GM3].)

REMARKS. The local condition is stronger than the global condition because of the factorization

$$\begin{array}{ccccc} H_j(D) \cong H_j(T) & \longrightarrow & H_j(\tilde{X}) & \longrightarrow & H_j(\tilde{X}, \tilde{X}-D) \\ & \searrow & & & \wr\!\!\| \\ & & & & H_j(T, S) \end{array}$$

This conjecture has two interesting consequences:

Consequence 1. For all $j \geq n$ the mixed Hodge structure on

$$H_j(D) = \ker(H_j(\tilde{X}) \to IH_j(X)) \text{ is actually pure.}$$

Consequence 2. The map given by pushing into $\tilde{X}$ and then restricting to D,

$$H_i(D) \cong H_i(T) \longrightarrow H_i(T,S) \xrightarrow{\cong} H^{2n-i}(T) \cong H^{2n-i}(D)$$

is an injection.

Consequence 2 is part of the blowing down condition from the example in §5.1 since the map $H_i(D) \to H^{2n-i}(D) \to H_{i-2}(D)$ coincides with $\cap C_1 L$.

If $\tilde{X}$ is a surface and D is a divisor with normal crossings, Grauert's necessary and sufficient blowing down criterion is that the intersection form be negative definite. Our necessary condition (2) is that the intersection form be nonsingular.

BIBLIOGRAPHY

[A] M. Artin, Théorème de finitude pour un morphism propre: dimension cohomologique des schemas algébriques affines, EGA4 expose XIV, Springer Lecture Notes in Math No. 305, Springer Verlag, New York (1973).

[APS] M. Atiyah, V. Patodi, and I. Singer, Spectral asymmetry and Riemannian geometry I, II, and III, Math. Proc. Cam. Phil. Soc. *77* (1975), 43-69, *78*(1976), 405-432, *79*(1976), 315-330.

[AV] A. Andreotti and E. Vesentini, Carleman estimates for the Laplace-Beltrami equation on complex manifolds. Publ. Math. I.H.E.S. *25* (1965), 81-130.

[C1] J. Cheeger, On the Spectral Geometry of Spaces with Cone-like Singularities, Proc. Nat. Acad. Sci., *76*(1979), pp. 2103-2106.

[C2] ———, Spectral Geometry of Spaces with Cone-like Singularities, preprint 1978.

[C3] ———, On the Hodge Theory of Riemannian Pseudomanifolds. Proc. of Symp. in Pure Math. *36*(1980), 91-146.

[C4] ———, Hodge Theory of Complex Cones, Proc. Luminy Conf., "Analyse et Topologie sur les Espaces singuliers," (to appear).

[C5] ———, Spectral Geometry of Singular Riemannian Spaces (to appear).

[CT] J. Cheeger and M. Taylor, On the Diffraction of Waves by Conical Singularities, C.P.A.M. (to appear).

[CG] M. Cornalba and P. Griffiths, Analytic cycles and vector bundles on noncompact algebraic varieties, Inv. Math. *28*(1975), 1-106.

[D1] P. Deligne, Théorie de Hodge I, Actes du Congrès international des mathematicians, Nice 1970.

[D2] ———, Théorie de Hodge II, Publ. Math. I.H.E.S. *40*(1971), 5-58.

[D3] ———, Théorie de Hodge III, Publ. Math. I.H.E.S. *44*(1974), 5-78.

[D4] ———, La conjecture de Weil I, Publ. Math. I.H.E.S. *43*(1974), 273-307.

[D5] ———, letter to D. Kazhdan and G. Lusztig (spring 1979).

[D6] ———, Théorème de Lefschetz et critères de dégénérescence de suites spectrales, Publ. Math. I.H.E.S. *35*(1968), 107-126.

[DR] G. De Rham, Variétés Differentiables, 3rd ed. Hermann, Paris, 1973.

[F] R. Fox, Covering Spaces with Singularities, in *Algebraic Geometry and Topology* (Symposium in honor of S. Lefschetz) Princeton Univ. Press 1957.

[FM] W. Fulton and R. MacPherson, Bivariant theories (preprint, Brown University 1980).

[GA1] M. Gaffney, The harmonic operator for differential forms, Proc. Nat. Acad. Sci. *37*(1951), 48-50.

[GA2] ———, A special Stokes theorem for Riemannian manifolds, Ann. of Math. *60*(1954), 140-145.

[GO] M. Goresky, Whitney stratified chains and cochains to appear in Trans. Amer. Math. Soc.

[GM1] M. Goresky and R. MacPherson, La dualité de Poincaré pour les espaces singuliers, C.R. Acad. Sci. *284* Série A(1977), 1549-1551.

[GM2] ———, Intersection Homology Theory, Topology *19*(1980), 135-162.

[GM3] ———, Intersection Homology Theory II, to appear.

[GM4] ———, Stratified Morse Theory (preprint, Univ. of B.C. 1980).

[H] R. Hardt, Topological properties of subanalytic sets, Trans. Amer. Math. Soc. *211*(1975), 57-70.

[L] S. Lefschetz, *Topology*. Amer. Math. Soc. Colloq. Publ. XII New York, 1930.

[MA] J. Mather, Stratifications and mappings, in *Dynamical Systems* (M.M. Peixoto, ed.) Academic Press, New York (1973).

[MC] C. McCrory, Poincaré Duality in spaces with singularities. Thesis, Brandeis Univ. (1972).

[M] W. Müller, Manifolds with cusps, and a related trace formula (preprint, 1980).

[O] A. Ogus, Local cohomological dimension of algebraic varieties. Annals of Math. *98*(1973), 327-365.

[T] R. Thom, Ensembles et stratifies, Bull. Amer. Math. Soc. *75*(1969), 240-284.

[ZE] C. Zeeman, Dihomology I and II, Proc. London Math. Soc. (3) *12* (1962), 609-638, 639-689.

[ZU1] S. Zucker, Hodge theory with degenerating coefficients: L^2 cohomology in the Poincaré metric. Ann. of Math. *109*(1979), 415-476.

[ZU2] ———, L^2 Cohomology, Warped Products, and Arithmetic groups (preprint, 1980).

FUNCTION THEORY OF NONCOMPACT KÄHLER MANIFOLDS OF NONPOSITIVE CURVATURE

R. E. Greene*

The principle which underlies the facts discussed in this article is that the function theory of complete noncompact Kähler manifolds is controlled by the curvature properties of their Kähler metrics. For the sake of clarity, it is appropriate initially to separate the purely function theoretic problems from the topological ones by studying the function theoretic properties of complete Kähler manifolds which are diffeomorphic as real manifolds to Euclidean spaces $\mathbb{R}^{2n}$. The case $n = 1$ is made comparatively simple by the uniformization theorem. But, for $n \geq 2$, there are a vast variety of biholomorphically distinct complex structures on $\mathbb{R}^{2n}$, of course. The most familiar are the $\mathbb{C}^n$ structure and the bounded domains in $\mathbb{C}^n$ which are real diffeomorphic to $\mathbb{R}^{2n}$. These latter form already an infinite-dimensional family of biholomorphically distinct structures ([5], [9]). Unbounded domains in $\mathbb{C}^n$ can have unexpected function theoretic properties if one forms expectations by analogy with the unit disc-complex plane dichotomy ([7]). And if the Kähler condition were not imposed, complex structures on $\mathbb{R}^{2n}$ with such pathological properties as that all holomorphic functions are constant would have to be considered

*Research supported by an Alfred P. Sloan Foundation Fellowship, the National Science Foundation (U.S.A.), the Institute for Advanced Study, Princeton, and the University of Bonn, Sonderforschungsbereich "Theoretische Mathematik."

Seminar on Differential Geometry
0-691-08268-5/82/000341-17$00.85/0 (cloth)
0-691-08296-0/82/000341-17$00.85/0 (paperback)

([6]: The examples there have no Kähler metrics. But whether every complete Kähler structure on R^{2n} has nonconstant holomorphic functions is unknown). In view of this being unknown and of the general multiplicity of possibilities, it is reasonable to restrict the considerations further by imposing not only the condition of the existence of a complete Kähler metric but also curvature hypotheses of considerable strength.

A reasonable general restriction is to complete simply connected Kähler manifolds of nonpositive sectional curvature. (It is a special case of the classical Cartan-Hadamard theorem that such a Kähler manifold is diffeomorphic as a real manifold to R^{2n}; thus explicit restriction to manifolds real diffeomorphic to R^{2n} can be omitted.) There are many biholomorphically distinct complete Kähler manifolds of this type, in addition to the obvious two constant holomorphic sectional curvature (zero and negative) examples (which are biholomorphically C^n or the ball by [25] and [28]): Every C^∞-boundary domain in C^n that is sufficiently C^∞ close to the ball has on it a complete Kähler metric of everywhere negative sectional curvature; such a metric can be constructed directly ([11]). A more delicate analysis ([11] and [12]) shows that the Bergman metric for such domains is itself a complete Kähler metric of negative sectional curvature and in fact its curvature is globally close to that of the ball. Any C^∞ neighborhood of the ball contains infinite-dimensional families of biholomorphically distinct domains ([5]; see also the discussion in [11] and [12]); thus there are infinite-dimensional families of complete simply connected Kähler manifolds of negative curvature. (Another example of such a Kähler manifold is constructed in [36]; it cannot be realized as a domain in C^n with smooth boundary, according to [38].) Various general results on the function theory of complete simply connected Kähler manifolds of nonpositive curvature are discussed in §1; and in §2 some proof techniques related to these results are discussed.
A second general curvature condition to consider is positivity of sectional curvature. Every complete Riemannian manifold of positive sectional curvature that is noncompact is diffeomorphic to a Euclidean space ([23];

see also [18] for a proof related to the Kähler manifold considerations). Although a constant holomorphic sectional curvature example cannot exist in this case, it is possible to exhibit explicitly a complete positive curvature Kähler metric on $\mathbb{C}^n$ ([31]). Any complete noncompact Kähler manifold of positive curvature is a Stein manifold ([13]; [15]; [18]). Also, such manifolds have no bounded nonconstant holomorphic functions ([46]) and no L^2 holomorphic $(n,0)$ forms, $n = \dim_{\mathbb{C}} M$, except the 0-form ([13]; see also [46]). In fact, such a manifold has no L^2 holomorphic $(p,0)$ forms, $p > 0$, except the 0-form (same references); but, for $p < n$, this fact is in part a metric and not a purely function theoretic conclusion since only in case $p = n$ is the L^2 inner product independent of metric choice. On the basis of all this evidence, it has been conjectured that every complete noncompact Kähler manifold of positive curvature is biholomorphic to $\mathbb{C}^n$ ([16]). Another conjecture involving positivity of some curvature is that every complete noncompact Kähler manifold of positive holomorphic bisectional curvature should be a Stein manifold ([40]; [45]). On any such manifold, there are strictly plurisubharmonic functions ([17]; see also [44]); furthermore, global holomorphic functions separate points and give local coordinates (unpublished observation of Y. T. Siu and S. T. Yau; see [40]; proof is sketched in the last section of the present paper). The positive curvature results and conjectures will be discussed further only incidentally in this paper.

§1. *The role of curvature at infinity*

If M and N are Stein manifolds of complex dimension at least two and if there are compact sets K_1 in M and K_2 in N such that $M - K_1$ is biholomorphic to $N - K_2$, then M is biholomorphic to N (this follows from the results in [39] by a brief argument). From this generalized Hartogs' phenomenon, an expectation arises that a complete simply connected Kähler manifold of nonpositive curvature, which is necessarily a Stein manifold ([43]), has its function theory controlled not only by its curvature behavior everywhere but even by just the behavior of its curvature at large distances from a fixed point ("at infinity"). Results regarding this type of determination of function theory by curvature fall in general into three classes: (i) conditions for biholomorphism to the ball; (ii) conditions for function theoretic similarity to (or biholomorphism to) general bounded domains in $\mathbb{C}^n$; and (iii) conditions for biholomorphism to $\mathbb{C}^n$. The corresponding results in Riemann surface theory (e.g. [20], [34], [1], [27], [4]) do not of course distinguish between (i) and (ii), but the curvature conditions arising in the Riemann surface case are very suggestive of the appropriate conditions for the (ii)-(iii) dichotomy in higher dimensions. In particular, it is suggested by the Riemann surface case that curvature's going rapidly to zero with increasing distance from a fixed point is connected with function theoretic resemblance to $\mathbb{C}^n$, while curvature's going slowly to (or not at all) to zero is connected with resemblance to bounded domains (cf. [16]).

(i) *Biholomorphism to the ball*

If a complete simply-connected Kähler manifold of everywhere nonpositive sectional curvature has constant negative holomorphic sectional curvature outside some compact set, then it is biholomorphic to the ball ([10]). To prove this, let p be a point of the manifold M and be a number so large that on $M - \overline{B}(p:r)$ the holomorphic sectional curvature is constant negative. (Here $\overline{B}(p;r)$ = the closed metric ball of radius r around p.) Then since $M - \overline{B}(p;r)$ is simply connected, a local holomorphic isometry to the ball (with a suitable multiple of the Bergman metric) around a point q in $M - \overline{B}(p;r)$ can be continued to a holomorphic locally isometric map of $M - \overline{B}(p;r)$ onto a subregion of the ball. By the Hartogs' phenomenon mentioned this map extends to be a holomorphic map of M into the ball. It can then be seen that this map is biholomorphic (see [10] for details).

This type of result does not genuinely require actual constancy of holomorphic sectional curvature (outside the compact set). For example if the metric structure of a complete Kähler Stein manifold real diffeomorphic to $\mathbb{C}^n$ converges C^k (suitable finite k) to that of the ball sufficiently rapidly with increasing distance from a fixed point, then the manifold is still necessarily biholomorphic to the ball, in dimension ≥ 3 ([10]). In this case, one finds an expanding sequence of regions analogous to spherical shells with increasing outer and inner radii which are C^k nearly isometric to and so have complex structure C^{k-1} close to that of geometric spherical shells in $\mathbb{C}^n$. By Hamilton's theorem ([24]; see also [12] for a treatment related to the present situation), each is then biholomorphic to a domain in $\mathbb{C}^n$ obtained by perturbation of the corresponding spherical shell. The Hartogs' phenomenon again applies to show that the (filled-in) interior of the shell-like region is biholomorphic to a successively smaller perturbation of the ball in $\mathbb{C}^n$. Using a normal families argument yields a limit map of the manifold to the ball. Finally, application of the generalized Schwarz lemma of [47] shows that the maps may be chosen so that the limit map is nondegenerate and hence biholomorphic. This argument, which used primarily very general principles, naturally applies to a much wider variety of cases than just the ball; in effect, it shows that in a suitable sense Stein manifold complex structures are all metrically isolated (in dimension ≥ 3). Further details and the generalizations are given in [10] (see also the remarks in [12] on the related question of abstract isolation of complex structures, in terms of their structure tensors).

That some form of quite rapid convergence to constant negativity of holomorphic sectional curvature is genuinely necessary, not just technically needed, in the previous discussion is shown by the curvature behavior of the Bergman metric of domains C^∞ near the ball ([12]) mentioned earlier.

(ii) *Function-theoretic similarity to bounded domains*

It is natural to ask whether there is a condition on curvature that would imply that a manifold was biholomorphic to a domain in $\mathbb{C}^n$ without being so restrictive as to apply only in case the manifold was biholomorphic to the ball. A number of conjectures in this direction have been made along with some related conjectures on existence of bounded nonconstant holomorphic functions (e.g. [16], [20]). One natural condition (on a complete simply-connected Kähler manifold of nonpositive curvature) to consider in this connection is that the curvature be bounded above and below by negative constants; the bound above rules out the $\mathbb{C}^n$ structure while the bound below is technically indispensable in many of the known methods of investigation in this subject. But it remains unknown whether the existence of such curvature bounds implies the presence of bounded, nonconstant holomorphic functions. A number of suggestive results, of interest in themselves, have been obtained with this and other related curvature conditions as hypotheses. First, there is the well-known result, that follows from Ahlfors' Schwarz lemma ([2]), that a Hermitian manifold with holomorphic sectional curvature bounded above by a negative constant is hyperbolic (see [33] for proof and further references) and various refinements of this ([20]). Second, there are quite general conditions on a complete simply-connected Kähler manifold of nonpositive curvature under which the Bergman metric constructed from L^2 holomorphic $(n,0)$ forms exists and is positive definite, and even complete ([20]): if, outside some compact set, sectional curvature $\leq -A/r^2 (\log r)^{1-\varepsilon}$ (A, ε positive constants, r = distance from a fixed point), then the Bergman kernel is nonvanishing on the diagonal and the Bergman metric is positive definite; and if (in addition) $-B \leq$ sectional curvature $\leq -C$, (B, C positive constants) or $-B/r^2 \leq$ sectional curvature $\leq -C/r^2$ outside some compact set (B, C again positive constants) then the Bergman metric is complete. (The original definition of the Bergman metric in the manifold, $(n,0)$ form context is in [32] and [42]: see also the discussion in [20] preliminary to the specific results just quoted.) By consideration of the Riemann surface case, the results on the nonvanishing of the kernel and positive definiteness of the metric are the best possible results of this general type. The proof technique of [20] for these results is discussed briefly in the next section.

(iii) *Biholomorphism to* $\mathbb{C}^n$

Since $\mathbb{C}^n$ has no obvious family of variations analogous to the C^∞ variations of the ball, it might be hoped that the conditions on curvature at infinity required to characterize the biholomorphism type of $\mathbb{C}^n$ would not have to be as stringent as those required in the case of the ball. This hope would be justified: biholomorphic equivalence to $\mathbb{C}^n$ is implied by just sufficiently rapid going to zero of curvature. For instance, it is proved in [20] that if M is a complete simple-connected Kähler manifold of nonpositive sectional curvature and if, for some point $p \in M$, the (non-negative) function $k, k : [0, +\infty) \to \mathbf{R}$, defined by $k(s) = -$ infimum of the sectional curvatures of M at all points of distance exactly s form p, has the properties (1) $\int sk(s) < +\infty$ and (2) $s^2 k(s) \to 0$ as $s \to +\infty$, then M is biholomorphic to $\mathbb{C}^n$ ([the formally stated result, Theorem J in [20], is slightly different but the result just stated is also proved there). That M is biholomorphic to $\mathbb{C}^n$ if $k(s) > -A/(1+s^2)^{1+\varepsilon}$ is proved in [4] using the L^2 $\bar\partial$-theory; the proof of the just stated, more general result (in [20]) uses many of the same techniques.

Recent work of N. Mok, Y. T. Siu and S. T. Yau [The Poincaré-Lelong equation on complete Kähler manifolds, to appear, Andreotti Memorial Volume] has put the just stated results on biholomorphism to $\mathbb{C}^n$ in a new perspective: First, they have shown that if M is a complete Kähler manifold and if there is a point $p \in M$ such that $\exp_p : TM_p \to M$ is a diffeomorphism and such that there are suitable positive constants C, ε with, for all $q \in M$, each sectional curvature at q between $-C/(1+r^{2+\varepsilon})$ and $C/(1+r^{2+\varepsilon})$, $r =$ distance from p to q, then M is biholomorphic to $\mathbb{C}^n$. Second, they have shown that if such a Kähler manifold has the further properties that its complex dimension is at least two and that its sectional curvature is either everywhere nonpositive, or everywhere nonnegative, then it is (biholomorphically) isometric to $\mathbb{C}^n$. (Lead by this second result to consider the Riemannian case, the author and H. Wu have obtained similar results for Riemannian manifolds of dimension greater than two. [On a new gap phenomenon in Riemannian geometry, to appear, Proc. Nat.

Acad. Sciences, U.S.A.].) Thus the hypothesis of nonpositive curvature in the nonpositive-curvature-of-faster-than-quadratic-decay theorem is not only unnecessary but in fact is illustrated only by certain metrics on $\mathbb{C}$ and by $\mathbb{C}^n$ with the standard metric, $n \geq 2$. On the other hand, examples abound of complete Kähler metrics on $\mathbb{C}^n$ which are not isometric to the standard metric but do have curvature of faster than quadratic decay and have the exponential map a diffeomorphism at some point (as noted, both positive and negative curvatures must occur in this case). For instance, they can be obtained as the Levi form of a small perturbation of the potential of the Euclidean metric.

§2. *Some basic proof techniques*

Some basic techniques used in the proofs of the theorems stated in the previous section on the Bergman metric (and in the original proofs of the biholomorphic equivalence to $\mathbb{C}^n$ theorems) will be discussed in this section. For detailed references, full generality of statement, and complete proofs, see [20], on which the present discussion is based.

(i) *Plurisubharmonicity and convexity*

A C^2 function on a Kähler manifold which is convex (i.e., has nonnegative second derivatives along geodesics) is plurisubharmonic (i.e., has nonnegative definite Levi form); and a C^2 strictly convex function (positive second derivatives along geodesics) is strictly plurisubharmonic (positive definite Levi form); these statements follow from direct calculations (see e.g., [14]). The corresponding statements also hold for convex functions, and in a suitable sense strictly convex functions, which are not necessarily C^2 ([14], [18], and [19]); this is shown by construction of suitable smooth local approximations. Nonpositivity of curvature is associated to convexity: on a complete simply-connected Riemannian manifold of nonpositive sectional curvature the function $p \to \mathrm{dis}^2(p,q)$, q fixed, dis = Riemannian distance is a C^∞ strictly convex function. Since completeness implies also that it is an exhaustion function, a complete simply-connected Kähler manifold of nonpositive curvature is a Stein manifold ([43]), the function $p \to \mathrm{dis}^2(p,q)$ then being a C^∞ strictly plurisubharmonic exhaustion function. In the case of noncompact complete manifold of positive sectional curvature, the dis^2 function need not be convex, or in the Kähler case plurisubharmonic. But on any complete noncompact Riemannian manifold of positive sectional curvature, there is a C^∞ strictly convex exhaustion function ([18]); hence a complete noncompact Kähler manifold of positive curvature is a Stein manifold ([13], [15], [18]).

For more detailed analysis of the nonpositive curvature case, estimates are needed on the convexity of functions of distance. The general principle here is that more negativity of curvature yields more convexity (i.e., larger second derivatives on geodesics) of increasing functions of distance. The specific result needed for the biholomorphism to $\mathbf{C}^n$ theorem is that if $f:[0,+\infty) \to \{-\infty\} \cup \mathbf{R}$ is a nondecreasing function, finite-valued and C^∞ on $(0,+\infty)$ and such that $f((\Sigma|z_i|^2)^{1/2})$ is plurisubharmonic on $\mathbf{C}^n$ then on a complete simply-connected Kähler manifold of nonpositive curvature the function $p \to f(\mathrm{dis}(p,q))$ q fixed, is plurisubharmonic. For instance, the function $\log r$, $r =$ distance from a fixed point, is plurisubharmonic on such a manifold.

(ii) *The* $L^2 - \bar\partial$ *method*

The fundamental method of constructing holomorphic functions and forms on a complex manifold is the solution of suitable $\bar\partial$ problems. In the present context, the appropriate specific result to be used is the following version of the L^2 method with weight factors developed in [3] and [26] (see [21] for a convenient form of the relevant complex Laplacian calculation, originally carried out by Kodaira).

(*) If M is a Stein manifold with a complete Kähler metric g, if λ_2 is a plurisubharmonic function on M, and λ_1 is a C^∞ function on M, then

(a) if the Levi form L_{λ_1} is $\geq cg$ for some positive continuous function c on M and if f is a $C^\infty(n,1)$ form on M with $\bar\partial f = 0$, then there exists a $C^\infty(n,0)$ form u on M such that $\bar\partial u = f$ and

$$\int_M |u|^2 e^{-\lambda_1-\lambda_2} \leq \int_M c^{-1}|f|^2 e^{-\lambda_1-\lambda_2} \ (\leq +\infty) .$$

(b) if $L_{\lambda_1} + \mathrm{Ric} \geq cg$ for some positive continuous function c, where Ric is the Ricci form of g, and if f is a C^∞ (0,1) form on M with $\bar\partial f = 0$ then there exists a C^∞ function u on M such that $\bar\partial u = f$ and

$$\int_M |u|^2 e^{-\lambda_1-\lambda_2} \leq \int_M c^{-1}|f|^2 e^{-\lambda_1-\lambda_2} .$$

A typical type of application of (*) is to the following situation: If $F: U \to C$ is a holomorphic function on an open subset of a manifold M (satisfying the hypotheses of (*)), if $p \in U$, and if $b: U \to \mathbf{R}$ is a non-negative function which is identically 1 near p but has compact support in U, then for any choice of λ_1 and any λ_2 that is continuous, finite-valued except perhaps at p, the integral $\int_M c^{-1}|\bar\partial(bF)|^2 e^{-\lambda_1-\lambda_2}$ is finite. Thus there is a solution u of $\bar\partial u = \bar\partial(bF)$ with $\int_M |u|^2 e^{-\lambda_1-\lambda_2}$ finite. If λ_2 and hence $\lambda_1+\lambda_2$ are sufficiently singular, with value $-\infty$, at p, then necessarily u will vanish to a high order at p and $bF - u$ will equal to high order at p the function F. A similar procedure applies to holomorphic $(n,0)$ forms. Also, even if F is not holomorphic but is only known to have $\bar\partial F$ vanishing to a certain order at p, then a solution u of $\bar\partial u = \bar\partial(bF)$ can still be found with u having forced vanishing at p. Thus global holomorphic objects (of the form $bF - u$) can be found with specified behavior at p.

In statement (*), the role of M being a Stein manifold is two-fold: First, M being a Stein manifold guarantees a solution of $\bar\partial u = f$ when $\bar\partial f = 0$, independently of f satisfying any estimate. Second, it makes certainly possible the approximation of λ_2 by a monotone nonincreasing sequence of C^∞ plurisubharmonic functions (see [37] and [19] for a general discussion of the C^∞ approximation question). The first point is just for convenience in the statement and only the second is essential. If the integral $\int c^{-1}|f|^2 e^{-\lambda_1-\lambda_2}$ is assumed finite and if suitable C^∞ approximations of λ_2 are known to be possible, then the conclusions of (*) hold, even if M is not a Stein manifold. In case λ_2 is C^∞, finite-valued except at a finite number of points, then suitable approximations always exist if there is a C^∞ strictly plurisubharmonic function on M; this is obvious near the singular points of λ_2 by local coordinate smoothing and can be seen globally by a patching construction ([37] and [19]).

The observations of the previous paragraph can be used to demonstrate the existence of enough global holomorphic functions to separate points and give local coordinates on any complete noncompact Kähler manifold of positive holomorphic bisectional curvature ([40], on unpublished work of Y.-T. Siu and S.-T. Yau), or in fact on a complete noncompact Kähler manifold with holomorphic bisectional curvature nonnegative and positive outside some compact set. On such a manifold, there always exists a C^∞ strictly plurisubharmonic function ([17] and, for the outside a compact set case [44]; cf., also [8]), $\phi : M \to \mathbf{R}$. Also, the Ricci tensor of M is nonnegative so that $L_\phi + \mathrm{Ric} > cg$ for some positive continuous function c. If $(z_1, \cdots, z_n)$ is a local coordinate system around a point $p \in M$ with p corresponding to $(0, \cdots, 0)$, then, for any positive number c_1, $c_1 \log(\Sigma|z_i|^2)$ is plurisubharmonic in a neighborhood of p. Multiplying this locally defined function by a nonnegative C^∞ function that is identically 1 near p and 0 away from p (a "bump" function) and adding a suitable (large positive) constant multiple of ϕ yields a global plurisubharmonic function with the same singularity at p as $c_1 \log(\Sigma|z_i|^2)$ and no other singularities (it is C^∞ on $M - \{p\}$). Setting λ_2 = a finite sum of such plurisubharmonic functions, $\lambda_1 = \phi$, and $f = \bar{\partial}$ applied to a sum of products of locally defined holomorphic functions around the finite number of singular points with bump functions around the points yields by the same reasoning as before a global holomorphic function with a specified finite-order holomorphic jet at each of the finite number of singular points of λ_2; thus there is a global holomorphic function with specified finite-order holomorphic jets at an arbitrary finite set of points. This procedure yields the same conclusion, in fact, on any complete Kähler manifold which has nonnegative Ricci curvature and on which there is a strictly plurisubharmonic function.

(iii) *Sub-mean-value theorems and uniform estimates*

The fact that if f is holomorphic on a region in $\mathbb{C}^n$ then $|f|^2$ is subharmonic on the region yields the estimate that if f is holomorphic on a ball B in $\mathbb{C}^n$ with center p then $|f(p)|^2 \le [\text{volume }(B)]^{-1} \int_B |f|^2$. It is important to have corresponding estimates on manifolds of the sort under consideration, but of course the fact that the Kähler metric need not be of zero curvature must be taken into account. On any Kähler manifold, $|f|^2$ is subharmonic if f is holomorphic, no matter what the curvature of the Kähler metric is: this can be checked by direct calculation. If p is a point of a complete simply-connected Riemannian manifold of nonpositive sectional curvature and if F is a nonnegative subharmonic function on the ball B of radius r around p, then $F(p) \le [V(r)]^{-1} \int_B F$, where $V(r)$ = the volume of the ball of radius r in Euclidean space of (real) dimension = dimension of the manifold ([13]). (In this statement, the assumptions on the manifold can also be localized to B, but this generality is not needed in the present context.) Combined with the subharmonicity of $|f|^2$, this statement implies that if f is holomorphic on the ball B of radius r around the point p in a complete simply-connected Kähler manifold of nonpositive sectional curvature, then $|f(p)|^2 \le [V(r)]^{-1} \int_B |f|^2$.

It is useful to obtain similar estimates for local holomorphic sections of Hermitian vector bundles. The norm^2 of a local holomorphic section will necessarily be subharmonic if the curvature of the bundle is nonpositive in the sense of [22]. In this case, the required estimate follows as before. In practice, it is of course necessary on occasion to modify the metric of the vector bundle so as to obtain this curvature condition. This can always be done locally, which suffices to yield an estimate of the type indicated; and the modification will introduce a controllable constant factor at most.

(iv) *Geometric construction of almost holomorphic objects*

If $z_1 = x_1 + iy_1, \cdots, z_n = x_n + iy_n$ are unitary complex linear coordinates on the tangent space of a Kähler manifold M at p (i.e. $J\left(\frac{\partial}{\partial x_j}\right) = \frac{\partial}{\partial y_j}$, $j = 1, \cdots, n$), then the associated functions $z_j \circ \exp_p^{-1}$ defined in a neighborhood of p in M and also denoted by z_j are not in general holomorphic since $\exp_p$ is not in general a holomorphic map. But $\bar{\partial} z_j$ does vanish to second order at p for each $j = 1, \cdots, n$. If M is a complete simply-connected Kähler manifold of nonpositive curvature, then the functions z_j are globally defined, and $\bar{\partial} z_j$ can be estimated in terms of the curvature of M. For fixed positive r and $p \in M$, there is a constant C_r such that $|(\bar{\partial} z_j)(q)| \leq C_{r,p} \operatorname{dis}^2(p,q)$ if $\operatorname{dis}(p,q) < r$. If there is a global (constant) lower bound on the sectional curvature of M, then $C_{r,p}$ can be taken independent of p for fixed r and can also be taken to depend on r in a predictable way. Similarly, forms with controllable deviation from holomorphicity can be constructed from the exponential map.

As an example of the application of these techniques the proof of the Bergman metric results will be briefly described. The nonvanishing of the Bergman kernel on the diagonal and the positive definiteness of the Bergman metric are of course connected to the existence of L^2 holomorphic $(n,0)$ forms; for nonvanishing at a point $p \in M$, there should be a L^2 holomorphic $(n,0)$ form ω with $\omega(p) \neq 0$ and for positive definiteness there should be an L^2 holomorphic $(n, 0)$ form with specified 1-jet at p. To find such forms, the L^2-$\bar{\partial}$ technique is used. Careful application of the comparison principle described under (i), the comparison being in this case with certain metrics on the unit disc and unit ball, yields a C^∞ function $\phi : M \to \mathbf{R}$ such that $\phi(M) \subset [0, 1)$, ϕ is of the order of $\operatorname{dis}^2(\ , p)$ at p, $\log \phi$ is plurisubharmonic and ϕ strictly plurisubharmonic, $\phi^{-1}((-\infty, a])$ is compact for all $a \in [0, 1)$ and $\phi^{-1}(\{0\}) = p$. The function ϕ is analogous to $\Sigma|z_i|^2$ on the unit ball. Taking, in statement (*), $\lambda_1 = \phi$, λ_2 = a suitable positive multiple of $\log \phi$ and $f = \bar{\partial}(b\omega_1)$, ω_1 a local holomorphic $(n, 0)$ form around p and b a bump function at p, yields a global holomorphic $(n, 0)$ form $b\omega_1 - u$ with the jet of a specific order of this form equal to that of ω_1, i.e., arbitrary. This form $b\omega_1 - u$ is in fact L^2 without weight factors: since $e^{-\lambda_1 - \lambda_2}$ is bounded below, u is L^2, and $b\omega_1$ has compact support. This construction establishes the nonvanishing of the Bergman kernel on the diagonal and the positive definiteness of the Bergman metric. The lower bound on the Bergman metric needed to prove the completeness results is obtained by carrying out this type of procedure not using local coordinates to generate ω_1 (which would not yield controllable estimates as p varied) but rather using an almost holomorphic form in place of ω_1, this form being constructed via the exponential map as discussed in (iv). The lower bounds on curvature are used in the applications of (iii) as well as in the construction indicated using (iv).

The basis idea of the proof of the biholomorphism to $\mathbf{C}^n$ theorems is to identify in intrinsic terms the functions on the manifold that correspond to linear functions on $\mathbf{C}^n$ by finding the holomorphic functions of slowest possible growth among the nonconstant holomorphic functions. These functions are most easily obtained by first finding the (one-dimensional) space of holomorphic $(n, 0)$ forms of slowest possible growth, corresponding to $\{c\, dz_1 \wedge \cdots \wedge dz_n | c \in \mathbf{C}\}$ on $\mathbf{C}^n$; then finding the next fastest growing family, corresponding to $\{L\, dz_1 \wedge \cdots \wedge dz_n | L = \Sigma a_j z_j, a_j \in \mathbf{C}\}$ on $\mathbf{C}^n$;

of course, the analogue of $dz_1 \wedge \cdots \wedge dz_n$ must be shown to be nowhere zero (see [29] for a refined result on this point). The finding of these families of holomorphic $(n, 0)$ forms is again by the techniques (i) - (iv). Naturally, many technical difficulties must be disposed of to complete this program in detail, but the basic idea ([41]) as noted has a pleasing directness, and the techniques have wide applicability.

DEPARTMENT OF MATHEMATICS
UNIVERSITY OF CALIFORNIA, LOS ANGELES
LOS ANGELES, CALIF. 90024

REFERENCES

[1] Ahlfors, L. V., Sur le type d'une surface de Riemann, C. R. Acad. Sci. Paris 201 (1935), 30-32.

[2] ________, An extension of Schwarz's lemma, Trans. Amer. Math. Soc. 43 (1938), 359-364.

[3] Andreotti, A. and Vesentini, E., Carleman estimates for the Laplace-Beltrami equation on complex manifolds, Instit. Hautes Etudes Sci., Pub. Math. 25 (1965), 81-138.

[4] Blanc, C. and Fiala, F., Le type d'une surface et sa courbure totale, Comment. Math. Helv. 14 (1941-42), 230-233.

[5] Burns, D., Shnider, S; and Wells, R. O., On deformations of strictly pseudoconvex domains, Inv. Math. 46 (1978), 237-253.

[6] Calabi, E. and Eckmann, B., A class of compact complex manifolds which are not algebraic. Ann. Math. 58 (1953), 494-500.

[7] Diederich, K. and Sibony, N., Strange complex structures on Euclidean space, J. reine angew. Math. 311/312 (1979), 397-407.

[8] Elencwajg, G., Pseudoconvexité locale dans les variétés Kählériennes. Ann. l'Ins. Fourier 25 (1975), 295-314.

[9] Fefferman, C., The Bergman kernel and biholomorphic mappings of pseudoconvex domains, Inv. Math. 26 (1974), 1-65.

[10] Greene, R. E., Metric determination of complex structures, to appear.

[11] Greene, R. E. and Krantz, S., Stability properties of the Bergman kernel and curvature properties of bounded domains, to appear in Proc. Conference in Sev. Complex Var. 1979, Princeton.

[12] ________, Deformations of complex structures, estimates for the $\bar{\partial}$ equation, and stability of the Bergman kernel, to appear.

[13] Greene, R. E., and Wu, H., Curvature and complex analysis, I, II and III. Bull. Amer. Math. Soc. 77 (1971), 1045-1049; *ibid.* 78 (1972), 866-870; *ibid.* 79 (1973), 606-608.

[14] ————, On the subharmonicity and plurisubharmonicity of geodesically convex functions, Indiana Univ. Math. J. 22 (1973), 641-653.

[15] ————, A theorem in complex geometric function theory, Value Distribution Theory, Part A, Dekker, New York, 1974, 145-167.

[16] ————, Some function theoretic properties of noncompact Kähler manifolds, Proc. Sym. Pure Math. Vol. 27, Part II, Amer. Math. Soc., Providence, R. I. 1975, 33-41.

[17] ————, On Kähler manifolds of positive bisectional curvature and a theorem of Hartogs, Abh. Math. Sem. Univ. Hamburg, Kähler Jubilee Volume, Vol. 47, April, 1978.

[18] ————, C^∞ convex functions and manifolds of positive curvature, Acta Math. 137 (1976), 209-245.

[19] ————, C^∞ approximations of convex, subharmonic and plurisubharmonic functions, Ann. scient. Ec, Norm. Sup., 4e serie, t. 12 (1979), 47-84.

[20] ————, *Function Theory on Manifolds Which Possess a Pole*, Lecture Notes in Math. No. 699, Springer-Verlag, Berlin-Heidelberg-New York, 1979.

[21] ————, Harmonic forms on noncompact Riemannian and Kähler manifolds, to appear, Michigan Math. J.

[22] Griffiths, P. A., Hermitian differential geometry, Chern classes and positive vector bundles, Global Analysis, Univ. of Tokyo Press, Tokyo, 1969, 183-251.

[23] Gromoll, D. and Meyer, W., On complete open manifolds of positive curvature. Ann. Math. 90 (1969), 75-90.

[24] Hamilton, R., Deformation of complex structures on manifolds with boundary. I: the stable case, J. Diff. Geom. 12 (1977), No. 1, 1-46.

[25] Hawley, N. S., Constant holomorphic curvature, Can. J. Math. 5 (1953), 53-56.

[26] Hörmander, L., *An Introduction to Complex Analysis in Several Variables*, Second Edition, North-Holland, Amsterdam-London, 1973.

[27] Huber, A., On subharmonic functions and differential geometry in the large, Comm. Math. Helv. 32 (1957), 13-72.

[28] Igusa, J., On the structure of a certain class of Kähler manifolds, Amer. J. Math. 76 (1954), 669-678.

[29] Kasue, A. and Ochiai, T., On holomorphic sections with slow growth of Hermitian line bundles on certain Kähler manifolds with a pole, to appear.

[30] Klembeck, P., Kähler metrics of negative curvature, the Bergman metric near the boundary and the Kobayashi metric on smooth bounded strictly pseudoconvex domains, Indiana Univ. Math. J. 27 (1978), No. 2, 275-282.

[31] ________, A complete Kähler metric of positive curvature on $\mathbb{C}^n$, to appear.

[32] Kobayashi, S., Geometry of bounded domains, Trans. Amer. Math. Soc. 92 (1959), 267-290.

[33] ________, *Hyperbolic Manifolds and Holomorphic Mappings*, Dekker, New York, 1970.

[34] Milnor, J., On deciding whether a surface is parabolic or hyperbolic, Amer. Math. Monthly, 84 (1977), 43-46.

[35] Moser, J., On Harnack's theorem for elliptic differential equations, Comm. Pure Apply. Math. 14 (1961), 571-591.

[36] Mostow, G. and Siu, Y.-T., A compact Kähler surface of negative curvature not covered by the ball, to appear.

[37] Richberg, R., Stetige streng pseudoconvexe Funktionen, Math. Ann. 175 (1968), 257-286.

[38] Rosay, J., to appear.

[39] Shiffman, B., Extension of holomorphic maps into Hermitian manifolds, Math. Ann. 194 (1971), 249-258.

[40] Siu, Y. T., Pseudoconvexity and the problem of Levi, Bull. Amer. Math. Soc. 84 (1978), no. 4, 481-512.

[41] Siu, Y. T. and Yau, S. T., Complete Kähler manifolds with nonpositive curvature of faster than quadratic decay, Ann. Math. 105 (1977), 225-264. (Errata, 109 (1979), 621-623.)

[42] Weil, A., *Introduction à l'Etude des Varietés Kähleriennes*, Hermann, Paris, 1958.

[43] Wu, H., Negatively curved Kählerian manifolds, Notices Amer. Math. Soc. 14 (1967), Abstract Nr. 675-327, 515.

[44] ________, An elementary method in the study of nonnegative curvature, Acta. Math. 142 (1978), 57-78.

[45] ________, Open Problems in geometric function theory, Proc. Sym. in Math. of the Taniguchi Foundation 1978.

[46] Yau, S. T., Some function theoretic properties of Riemannian manifolds and their applications to geometry. Indiana Univ. Math. J. 25 (1976), No. 71, 659-670.

[47] ________, A general Schwarz lemma for Kähler manifolds, Amer. J. Math. 100 (1978), No. 1, 197-204.

KÄHLER MANIFOLDS WITH VANISHING FIRST CHERN CLASS

M. L. Michelsohn

The purpose of this note is to demonstrate the following theorem.

THEOREM. *Let* X *be a compact, simply-connected Kähler manifold with* $c_1(X) = 0$. *Then* $\mathrm{Todd}(X) = 0$ *or* 2^k *for some* k. *Furthermore,* X *is a product of simply-connected Kähler manifolds with vanishing first Chern classes*:

$$X = X_1 \times \cdots \times X_k$$

such that

$$\mathrm{Todd}(X_i) = \begin{cases} 0 & \text{if } \dim(X_i) \text{ is odd.} \\ 2 & \text{if } \dim(X_i) \text{ is even.} \end{cases}$$

It should be noted that this theorem is part of a larger work [5] in which Clifford algebras are used to develop a cohomology for Kähler manifolds, as well as a cohomology for spinor bundles and twisted spinor bundles over Kähler manifolds. Several Weitzenbock formulas are developed from among which the Lichnerowicz Theorem [4] is retrieved.

In particular, we show the following. Let $S = \mathcal{S} \otimes E$ where $\mathcal{S}$ is the spinor bundle and E is any holomorphic hermitian bundle with the canonical connection. Then there is a decomposition $S = \bigoplus_{r \geqq 0} S^r$, and there are operators

Seminar on Differential Geometry
0-691-08268-5/82/000359-3 $00.50/0 (cloth)
0-691-08296-0/82/000359-3 $00.50/0 (paperback)

$$\xrightarrow{\mathcal{D}} \Gamma(S^{r-1}) \xrightarrow{\mathcal{D}} \Gamma(S^r) \xrightarrow{\mathcal{D}}$$

with $\mathcal{D}^2 = 0$, which form an elliptic complex of index $\{\mathrm{chE}\,\hat{A}(X)\}([X])$. The resulting cohomology groups are denoted $H^r(X;S)$ for $r \geqq 0$. We also establish formulas of the type:

$$4(\mathcal{D}\bar{\mathcal{D}} + \bar{\mathcal{D}}\mathcal{D}) = \nabla^*\nabla + \mathcal{R}^S \tag{1}$$

where $\bar{\mathcal{D}}$ denotes the formal adjoint of $\mathcal{D}$, where ∇^* denotes the formal adjoint of the connection $\nabla : \Gamma(S) \to \Gamma(T^*X \otimes S)$, and where $\mathcal{R}^S$ is a zero-order operator which is explicitly described in terms of the curvatures of X and E and by using Clifford multiplication. When E is a fractional power of the canonical bundle, the operator $\mathcal{R}^S$ depends only on the Ricci transformation of X.

We suppose now that $c_1(X) = 0$. By Yau [6], [7] we know that we can endow X with a Ricci-flat Kähler metric. Then, after letting E be trivial, the Weitzenböck formula (1) becomes simply

$$4(\mathcal{D}\bar{\mathcal{D}} + \bar{\mathcal{D}}\mathcal{D}) = \nabla^*\nabla ,$$

which implies that every harmonic section is parallel. Furthermore, in this case, there is a connection-preserving bundle isomorphism $\mathcal{S}^r \cong \Lambda^{0,r}$, inducing the isomorphism $H^r(X;\mathcal{S}) \cong H^r(X;\mathcal{O})$. (Here $\Lambda^{0,r}$ denotes the bundle of differential forms of bidegree $(0, r)$.)

Consequently, any harmonic $(0, r)$-form is parallel. Of course $\Lambda^{0,0}$ and $\Lambda^{0,n}$ are already known to be flat. Therefore, if $\hat{A}(X) = \mathrm{Todd}(X) = \Sigma(-1)^r \dim H^r(X;\mathcal{O}) \neq 1 + (-1)^n$, then there exist parallel $(0, r)$ forms with $r \neq 0, n$. However, this would imply that the holonomy group G of X is properly contained in SU_n. If G is not a product of two non-trivial groups, then G belongs to the list of Berger [1], and so $G = Sp_{n/2}$, for n even and > 2. This case has been recently ruled out by Bogomolov [2]. We conclude that G is a product of two non-trivial groups. This implies that X is a non-trivial Riemannian product of Ricci flat Kähler manifolds with $c_1 = 0$. Iterating this argument gives the theorem.

We note that simple-connectedness is a reasonable assumption in light of the work of Cheeger and Gromoll [3] which shows that the universal covering of a compact Ricci-flat Kähler manifold splits as $\mathbb{C}^k \times X_0$ where X_0 is a compact, simply-connected Ricci-flat manifold and where $\mathbb{C}^k$ is flat.

REFERENCES

[1] M. Berger, *Sur les groupes d'holonomie homogène des variétiés à connexion affine et des variétiés Riemanniennes*, Bull. Soc. Math., France, 83(1955), 279-330.

[2] F. A. Bogomolov, *Hamiltonian Kähler manifolds*, Sov. Math. Dokl., *19* (1978), 1462-1465.

[3] J. Cheeger and D. Gromoll, *The splitting theorem for manifolds of non-negative Ricci curvature*, J. Diff. Geom., 6(1971), 119-128.

[4] A. Lichnerowicz, *Spineurs harmoniques*, C. R. Acad. Sci., Paris, Ser. A-B *257*(1963), 7-9.

[5] M. L. Michelsohn, *Clifford and spinor coholomogy for Kähler manifolds*, Amer. J. Math. *102*(1980), 1083-1146.

[6] S. T. Yau, *Calabi's conjecture and some new results in algebraic geometry*, Proc. Nat. Acad. Sci. U.S.A. *74*(1977), 1798-1799.

[7] ———, *On the Ricci curvature of a compact Kähler manifold and the complex Monge Ampere equation I*, Comm. Pure and Appl. Math., *31*(1978), 339-441.

COMPACTIFICATION OF NEGATIVELY CURVED COMPLETE KÄHLER MANIFOLDS OF FINITE VOLUME

Yum-Tong Siu and Shing-Tung Yau*

The compactification of quotients of bounded symmetric domains with finite volume was obtained by Satake [12], Baily-Borel [3], and Andreotti-Grauert [1]. In this paper we investigate the problem of compactification of negatively curved complete Kähler manifolds of finite volume. It can be regarded as the generalization of the compactification result of quotients of bounded symmetric domains with finite volume in the case of rank 1.

MAIN THEOREM. *Let* M *be a complete Kähler manifold whose sectional curvature is bounded between two negative numbers. If the volume of* M *is finite, then* M *is biholomorphic to an open subset* M′ *of a projective algebraic subvariety* X *such that* X − M′ *is an exceptional set of* X *which can be blown down to a finite number of points.*

The method of the proof is as follows. We first show that for a ray γ in M the Busemann function $B_{\tilde{\gamma}}$ on the universal covering $\tilde{M}$ of M associated to the lifting $\tilde{\gamma}$ of γ is plurisubharmonic. Then we show that for c sufficiently large, when restricted to $B_{\tilde{\gamma}} < -c$, the function $B_{\tilde{\gamma}}$ descends to M. Moreover, the minimum of a finite number of such

*Research partially supported by NSF grants.

Seminar on Differential Geometry
0-691-08268-5/82/000363-18$00.90/0 (cloth)
0-691-08296-0/82/000363-18$00.90/0 (paperback)

descended functions forms an exhaustion function of M. By using the canonical line bundle of M we embed M into a projective algebraic subvariety as an open subset. Finally we use the Schwarz-Pick lemma to show that the complement of the image of M is an exceptional set in the projective algebraic subvariety which can be blown down to a finite number of points.

We should mention that P. Eberlein has independently obtained the results of §2. They appear in Annals of Mathematics 111 (1980), 435-476.

Table of Contents

§1. Plurisuperharmonicity of Busemann function
§2. Negatively curved Riemannian manifolds of finite volume
§3. Projective embedding by canonical line bundle
§4. Use of the Schwarz-Pick lemma
Appendix

§1. *Plurisubharmonicity of Busemann function*

Let M be a complete Kähler manifold and $\gamma:[0,\infty)\to M$ be a *ray* in M (parametrized by arc-length), i.e. $d(\gamma(t_1),\gamma(t_2)) = |t_1-t_2|$ for $t_1, t_2 \in [0,\infty)$, where $d(\cdot,\cdot)$ is the distance function of M.

For $p \in M$, define the Busemann function associated to γ by

$$B_\gamma(p) = \lim_{t\to\infty} (d(p,\gamma(t))-t) .$$

PROPOSITION 1. *If* M *is simply-connected and if the sectional curvature of* M *is negative, then* B_γ *is strictly plurisubharmonic.*

Proof. Fix $p_0 \in M$. For $0 \le t < \infty$ denote by $r_t(p)$ the distance between p and $\gamma(t)$. Let X be a tangent vector at p_0. We want to calculate

$$H(r_t)(X,X) + H(r_t)(JX,JX) ,$$

where $H(r_t)$ is the Hessian of the function r_t. Let $\alpha_t:[0,\ell_t]\to M$ be the shortest geodesic parametrized by arc-length joining $\gamma(t)$ to p_0 with

$\alpha_t(0) = \gamma(t)$ and $\alpha_t(\ell_t) = p_0$. Decompose X into the tangential and normal components to α_t

$$X = e\frac{\partial}{\partial r_t} + X'.$$

By parallel translating X, X' along α_t, we obtain vector fields $X_t(s)$, $X'_t(s)$, $0 \leq s \leq \ell_t$, along α_t. Let $Y_t(s)$ be the Jacobi field along α_t such that $Y_t(\ell_t) = X'$ and $Y_t(0) = 0$. By considering the exponential map at p_0, we can extend $Y_t(s)$ to a vector field $\tilde{Y}$ in a neighborhood of α_t such that $\left[\tilde{Y}, \frac{\partial}{\partial r_t}\right] = 0$.

$$\begin{aligned}
H(r_t)(X,X) &= H(r_t)(X',X') \\
&= \tilde{Y}\tilde{Y}\, r_t - (\nabla_{\tilde{Y}}\tilde{Y})\, r_t \\
&= \tilde{Y}\left\langle \tilde{Y}, \frac{\partial}{\partial r_t}\right\rangle - \left\langle \nabla_{\tilde{Y}}\tilde{Y}, \frac{\partial}{\partial r_t}\right\rangle \\
&= \left\langle \tilde{Y}, \nabla_{\tilde{Y}}\frac{\partial}{\partial r_t}\right\rangle \\
&= \left\langle \tilde{Y}, \nabla_{\frac{\partial}{\partial r_t}}\tilde{Y}\right\rangle \\
&= \int_0^{\ell_t} \frac{d}{dr_t}\left\langle \tilde{Y}, \nabla_{\frac{\partial}{\partial r_t}}\tilde{Y}\right\rangle dr_t \\
&= \int_0^{\ell_t}\left(\left|\nabla_{\frac{\partial}{\partial r_t}}\tilde{Y}\right|^2 + \left\langle \tilde{Y}, \nabla_{\frac{\partial}{\partial r_t}}\nabla_{\frac{\partial}{\partial r_t}}\tilde{Y}\right\rangle\right) dr_t \\
&= \int_0^{\ell_t}\left(\left|\nabla_{\frac{\partial}{\partial r_t}}\tilde{Y}\right|^2 - \left\langle R\left(\tilde{Y}, \frac{\partial}{\partial r_t}\right)\frac{\partial}{\partial r_t}, \tilde{Y}\right\rangle\right) dr_t .
\end{aligned}$$

Let a be a positive number. Then as the sectional curvature of M is negative, there exists a positive number c such that

$$\left\langle R\left(\tilde{Y}(s), \frac{\partial}{\partial r_t}\right)\frac{\partial}{\partial r_t}, \tilde{Y}(s)\right\rangle$$
$$\leq -c\|\tilde{Y}(s)\|^2$$

when the distance of $a_t(s)$ from p_0 is $\leq a$.

By the Rauch comparison theorem [12],

$$\|Y(t)\| \geq \frac{t}{\ell_t} \|X'\| .$$

Hence

$$H(r_t)(X, X) \geq \frac{ca(\ell_t - a)^2}{\ell_t^2} \|X'\|^2 .$$

Similarly

$$H(\gamma_t)(JX, JX) \geq ca \frac{(\ell_t - a)^2}{\ell_t^2} \|e\|^2 .$$

Adding these two inequalities together, we have

$$H(\gamma_t)(X, X) + H(\gamma_t)(JX, JX) \geq ca \frac{(\ell_t - a)^2}{\ell_t^2} \|X\|^2 .$$

Letting $\ell_t \to \infty$, it follows that, in the sense of distributions,

$$H(B_\gamma)(X, X) + H(B_\gamma)(JX, JX) \geq ca\|X\|^2 .$$

§2. *Negatively curved Riemannian manifolds of finite volume*

In this section we discuss the structure of Riemannian manifolds with finite volume with sectional curvatures bounded between two negative numbers. We are going to prove that such a manifold has only a finite number of "infinite ends." Through each point in an end there is a unique geodesic ray in that end. The lifting of such a geodesic ray in the universal covering of the manifold defines a Busemann function on the universal covering. This Busemann function descends down to the infinite end to define an exhaustion function on that end. When the manifold is Kähler, this exhaustion function is strictly plurisuperharmonic by §1. The above results will be proved by using the lemmas of Margulis and Gromov given below.

LEMMA 1 (Margulis [11]). *Let* M *be an* n*-dimensional simply-connected complete Riemannian manifold whose sectional curvature* K *satisfies* $-1 \leq K \leq 0$. *Let* Γ *be a discrete group of isometries of* M. *Then there exists a positive number* ε_n *depending only on* n *such that for every* $x \in M$ *and* $0 < \varepsilon \leq \varepsilon_n$ *the group* $\Gamma_\varepsilon(x)$ *generated by* $\{\gamma \in \Gamma | d(x, \gamma x) < \varepsilon\}$ *is almost nilpotent in the sense that it contains a nilpotent subgroup of finite index.*

A complete proof of the Margulis lemma is given by Gromov [7, 8] where instead of the condition $-1 \leq K \leq 0$ Gromov used the following weaker condition $-1 \leq K \leq 1$ and M has no closed geodesic of length < 1 (see [7, p. 225]).

Before we state Gromov's lemma we introduce some definitions.

DEFINITION. Let M be a simply-connected complete Riemannian manifold of nonpositive sectional curvature. Two geodesic rays γ_1, γ_2 parametrized by arc-lengths are called *equivalent* if $d(\gamma_1(t), \gamma_2(t))$ is bounded for $t > 0$. The set of all equivalence classes of all geodesic rays parametrized by arc-length is denoted by $M(\infty)$. $M \cup M(\infty)$ with the cone topology [6] (i.e. a subbase for the topology is the set of all open cones of geodesic rays) is a compact topological space homeomorphic to a cell. Every isometry of M has a natural homeomorphic extension to $M \cup M(\infty)$. An isometry of M is called *elliptic* if it has a fixed point in M. An isometry of M is called *parabolic* if it has no fixed point in M and exactly one fixed point in $M(\infty)$. An isometry of M is called *hyperbolic* if it has no fixed point in M and exactly two fixed points in $M(\infty)$. It is proved in [6, Th. 6.5] that if the sectional curvature of M is bounded from above by a negative number, then the group of isometries of M are divided into the three mutually disjoint classes of elliptic, parabolic, and hyperbolic elements.

LEMMA 2 (Gromov [7, p. 226, Corollary 3.5]). *Let* M *be a simply-connected complete Riemannian manifold whose sectional curvature* K *satisfies* $-1 \leq K < 0$. *Let* Γ *be a fixed-point-free discrete subgroup of isometries of* M *such that the volume of* M/Γ *is finite. Then there exists a positive number* ε' *depending on* M *and* Γ *such that if* $\gamma \in \Gamma$ *and* $d(x, \gamma x) < \varepsilon'$ *for some* $x \in M$, *then* γ *is not hyperbolic.*

For the rest of this §2 we assume that M is an n-dimensional simply-connected complete Riemannian manifold whose sectional curvature is bounded between two negative numbers and Γ is a fixed-point-free discrete subgroup of isometries of M such that the volume of M/Γ is finite.

For $x \in M(\infty)$ let Γ_x be the set of all parabolic elements of Γ having x as the unique fixed point. Since a parabolic element of Γ and a hyperbolic element of Γ cannot have any common fixed point [6, p. 75, Prop. 6.8], it follows that Γ_x is either empty or equal to the set of all elements of Γ having x as a fixed point.

The set of all parabolic elements of Γ can be written as a disjoint union of subsets Γ_{x_i}, where $x_i \in M(\infty)$. Let ε be the minimum of the two constants ε_n and ε' from Lemmas 1 and 2. Let

$$A_i = \{x \in M \mid \min_{\gamma \in \Gamma_{x_i}} d(x, \gamma x) < \varepsilon\}$$

and

$$D = \{x \in M \mid \min_{\gamma \in \Gamma} d(x, \gamma x) \geq \varepsilon\} .$$

By Lemma 2, $M = D \cup (\bigcup_i A_i)$.

Let $\pi : M \to M/\Gamma$ be the natural projection. The set $\pi(D)$ is compact. Otherwise there is a sequence of points $\{y_\nu\} \subset M/\Gamma$ so that the geodesic balls with centers y_ν and radius $\min(\varepsilon_n, \varepsilon')$ are disjoint and with volume bounded from below, contradicting the finiteness of the volume of M/Γ.

LEMMA 3. *Let* $x \in A_i$ *and* $\gamma \in \Gamma$. *If* $\gamma x \in A_j$, *then* $\gamma x_i = x_j$.

Proof. Since $x \in A_i$, $d(x, \gamma' x) < \varepsilon$ for some $\gamma' \in \Gamma_{x_i}$. Since $\gamma x \in A_j$, there exists $\gamma'' \in \Gamma_{x_j}$ such that $d(\gamma x, \gamma'' \gamma x) < \varepsilon$, that is, $d(x, \gamma^{-1}\gamma''\gamma x) < \varepsilon$. It follows that both γ' and $\gamma^{-1}\gamma''\gamma$ belong to $\Gamma_\varepsilon(x)$. By Lemma 1, $\Gamma_\varepsilon(x)$ contains a nilpotent subgroup N of finite index. Hence the center of N contains an element $\sigma \neq 1$. Since Γ is fixed point free, it contains no elliptic element. By Lemma 2, σ is parabolic and has some unique fixed point $y \in M(\infty)$. For some positive integer k, $(\gamma')^k \in N$. Since no parabolic element is of finite order, $(\gamma')^k \in \Gamma_{x_i} - \{1\}$. It follows from

$$\sigma x_i = \sigma(\gamma')^k x_i = (\gamma')^k \sigma x_i$$

that $\sigma x_i = x_i$ and $y = x_i$. For some positive integer ℓ, $(\gamma^{-1}\gamma''\gamma)^\ell \in N$. It follows from

$$\sigma(\gamma^{-1}\gamma''\gamma)^\ell x_i = (\gamma^{-1}\gamma''\gamma)^\ell \sigma x_i = (\gamma^{-1}\gamma''\gamma)^\ell x_i$$

that $(\gamma^{-1}\gamma''\gamma)^\ell x_i = x_i$. Since $\gamma^{-1}\gamma''\gamma$ is parabolic by Lemma 2, we have $\gamma^{-1}\gamma''\gamma x_i = x_i$ and $\gamma''\gamma x_i = \gamma x_i$. Hence $\gamma x_i = x_j$. Q.E.D.

COROLLARY. *If* $\pi(A_i) \cap \pi(A_j) \neq \emptyset$, *then* $\pi(A_i) = \pi(A_j)$.

Proof. For some $\gamma \in \Gamma$ and $x \in A_i$ we have $\gamma x \in A_j$. By Lemma 3, $\gamma x_i = x_j$. It follows that $\gamma \Gamma_{x_i} \gamma^{-1} = \Gamma_{x_j}$ and $\gamma A_i = A_j$. Q.E.D.

LEMMA 4. $M/\Gamma - \pi(D)$ *is covered by a finite number of* $\pi(A_i)$.

Proof. Let η be a positive number which is smaller than the injectivity radius of the points x in $\pi(D)$. Then $d(x, \gamma x) \geq \eta$ for $\gamma \in \Gamma \setminus \{1\}$ and

$x \in D$. We claim that for each i there exists $y_i \in A_i$ such that $d(y_i, \gamma y_i) \geq \frac{1}{2} \min(\varepsilon, \eta)$ for all $\gamma \in \Gamma_{x_i}$. For we can take $y'_i \in A_i$ and assume that

$$\min_{\gamma \in \Gamma_{x_i}} d(y'_i, \gamma y'_i) < \frac{1}{2} \min(\varepsilon, \eta) .$$

Then on any path joining x to y'_i there exists y_i such that

$$\min_{\gamma \in \Gamma_{x_i}} d(y_i, \gamma y_i) = \frac{1}{2} \min(\varepsilon, \eta) .$$

By the definition of A_i, $y_i \in A_i$. Hence the projection π restricted to the ball with center y_i and radius $\frac{1}{2} \min(\varepsilon, \eta)$ is injective. Thus the volume of each $\pi(A_i)$ is bounded from below independent of i. The lemma now follows from the finiteness of the volume of M/Γ and the corollary to Lemma 3. Q.E.D.

We can select a finite number of pairwise disjoint $\pi(A_i)$, say $\pi(A_1)$, say $\pi(A_1), \cdots, \pi(A_m)$ so that they cover $M/\Gamma - \pi(D)$. Each point x of A_i can be joined to x_i by a geodesic ray $\sigma(t)$, $0 \leq t < \infty$. We claim that the geodesic ray σ lies in A_i. Since the Busemann function associated to σ is invariant under the action of Γ_{x_i} [6, p. 83, Prop. 7.8], by the lemma of the Appendix $d(\sigma(t), \gamma\sigma(t))$ is a decreasing function of t for $\gamma \in \Gamma_{x_i}$ (see also [10, Th. 2.4]). It follows that σ lies in A_i. Suppose Y is a compact subset of M and for any $y \in Y$, $\sigma_y(t)$ is the geodesic ray parametrized by its arc-length which joins y to x_i. Then by the lemma of the Appendix for $\gamma \in \Gamma_{x_i}$, $d(\sigma_y(t), \gamma\sigma_y(t))$ decreases to zero as $t \to \infty$ uniformly in $y \in Y$. Hence for some t_0, $\sigma_y(t_0) \in A$ for all $y \in Y$. By applying this to the case where Y is a curve in M joining two points of A, we conclude that A is connected. Therefore $\pi(A_1), \cdots, \pi(A_m)$ are the components of $M/\Gamma - \pi(D)$.

LEMMA 5. *For each point* $x \in A_i$ *there is exactly one geodesic ray* $\sigma(t)$, $t \geq 0$, *in* A_i *issued from* x *and it is the ray joining* x *to* x_i. *Moreover, the geodesic line* $\sigma(t)$, $-\infty < t < \infty$, *must intersect* D.

Proof. Suppose σ_1, σ_2 are two distinct geodesic rays in A_i issued from x such that σ_1 is the geodesic joining x and x_i. Let the point represented by σ_2 in $M(\infty)$ is x'_i. Then x_i and x'_i can be joined by a unique geodesic τ (see [6]). Let $y \in M$ be a point on τ.

Let t_j be an increasing sequence of real numbers going to ∞. There exists $\gamma_j \in \Gamma_{x_i}$ such that $d(\sigma_2(t_j), \gamma_j \sigma_2(t_j)) < \varepsilon$. The number of distinct γ_j is infinite, otherwise some γ_j has x'_i as fixed point. By passing to a subsequence, we can assume without loss of generality that $\gamma_j(x)$ converges to a point x''_i in $M(\infty)$. Since for any discrete subgroup of parabolic isometries of M its limit point set in $M \cup M(\infty)$ is a singleton which equals its common fixed point [6, p. 89, Prop. 8.9 P], it follows that $x''_i = x_i$.

Since the geodesic segment $\gamma_j \sigma_2(t), 0 \leq t \leq t_j$, approaches τ, there exists $0 \leq t'_j \leq t_j$ such that $\gamma_j \sigma_2(t'_j)$ approaches y. Clearly $t'_j \to \infty$. From

$$\begin{aligned} d(\sigma_2(t_j), x) &\leq d(\sigma_2(t_j), \gamma_j\sigma_2(t_j)) + d(\gamma_j \sigma_2(t_j), y) + d(y, x) \\ &< \varepsilon + d(\gamma_j \sigma_2(t_j), y) + d(y, x) \end{aligned}$$

and

$$\begin{aligned} d(\sigma_2(t_j), x) &= d(\sigma_2(t_j), \sigma_2(t'_j)) + t'_j \\ &= d(\gamma_j \sigma_2(t_j), \gamma_j \sigma_2(t'_j)) + t'_j , \end{aligned}$$

it follows that

$$\begin{aligned} t'_j &\leq \varepsilon + d(\gamma_j \sigma_2(t_j), y) - d(\gamma_j \sigma_2(t_j), \gamma_j \sigma_2(t'_j)) + d(y, x) \\ &\leq \varepsilon + d(\gamma_j \sigma_2(t'_j), y) + d(y, x) \end{aligned}$$

which yields a contradiction when $j \to \infty$. Hence there cannot be two distinct geodesic rays in A_i issued from x.

Suppose the geodesic line $\sigma(t)$, $-\infty < t < \infty$, does not intersect D. Then the connected curve $\pi\sigma(t)$, $-\infty < t < \infty$, must lie completely in $\pi(A_i)$. It follows that $\sigma(t)$, $-\infty < t < \infty$, must lie completely in A_i, which is not possible, because then there are two distinct geodesic rays $\sigma(t)$, $t \geq 0$, and $\sigma(t)$, $t \leq 0$ in A_i issued from x. G.E.D.

Clearly A_i is invariant under Γ_{x_i}. On the other hand, if $\gamma \in \Gamma$ leaves A_i invariant, then $\gamma \in \Gamma_{x_i}$. For, if γ maps x_i to another point x_i' of $M(\infty)$ and σ is a geodesic joining a point x of A_i to x_i, then $\gamma\sigma$ is a geodesic ray in A_i issued from γx which is different from the geodesic ray in A_i joining γx to x_i.

Take a geodesic ray σ_i in A_i and let B_i be the Busemann function on M associated to σ_i. By [5, p. 83, Prop. 7.8], B_i is invariant under Γ_{x_i}. There exists a function ϕ_i on $\pi(A_i)$ such that $B_i = \phi_i \circ \pi$ on A_i. Take $\eta > 0$ and let

$$D_\eta = \{x \in M \mid d(x, D) < \eta\}.$$

We claim that for all $c \in \mathbf{R}$ the set

$$L_c(\phi_i) = \{y \in \pi(A_i - D_\eta) \mid \phi_i(y) \geq -c\}$$

is compact.

Since $\pi(A_1), \cdots, \pi(A_m)$ are the components of $M/\Gamma - \pi(D)$ and $\pi(D_\eta)$ is a relatively compact open subset of M/Γ, it follows that the boundary E of $\pi(A_i - D_\eta)$ in M/Γ is a compact subset of $\pi(A_i)$. Let μ be the supremum of ϕ_i on E. Take $y \in L_c(\phi_i)$. Let $x \in A_i - D_\eta$ such that $\pi(x) = y$. Let $\sigma(t)$, $t \geq 0$, be the geodesic ray joining x to x_i. By Lemma 5, the geodesic line $\sigma(t)$, $-\infty < t < \infty$, intersects the closure $\overline{D_\eta}$ of D_η. Let t_0 be the largest value of t such that $\sigma(t_0) \in \overline{D_\eta}$. Then $\pi\sigma(t_0) \in E$. Since $\phi_i(y) = \phi_i(\pi\sigma(t_0)) + t_0$, it follows that $t_0 \geq c - \mu$. Clearly t_0 is a negative number. Hence $d(y, E) \leq |c-\mu|$ and $L_c(\phi_i)$ is compact.

Similar but different results on complete Riemannian manifolds with negative curvature and finite volume were proved by Heintze [9]. Though the results in this section are rather direct easy consequences of the lemmas of Margulis and Gromov, we present them there with complete proofs, because such complete proofs for these results in the form needed cannot be found yet in the literature and they are not completely trivial. Eberlein announced some related results in [5].

§3. *Projective embedding by canonical line bundle*

Let M be a complete Kähler manifold of complex dimension ≥ 2 whose sectional curvature is bounded between two negative numbers. Assume that the volume of M is finite. By applying the results of §1 and §2, we obtain a real-valued function ϕ on M which approaches $-\infty$ at the boundary of M and which is plurisubharmonic outside some compact subset of M. Let L be the canonical line bundle of M.

LEMMA 6. $\bigcup_{\nu=0}^{\infty} \Gamma(M, L^{\nu})$ *separates points and gives local coordinates of* M.

Proof. For $x \in M$, let r_x be the distance function on M measured from x. Take two distinct points x_1, x_2 of M. There exists a neighborhood U_i of x_i in M such that $\log r_{x_i}$ is plurisubharmonic on U_i and smooth on $U_i - x_i$. Let ρ_i be a nonnegative smooth function on U_i with compact support which is identically $2n+2$ on some open neighborhood of x_i. Let Ω be the curvature form of L and $\psi = \sum_{i=1}^{2} \rho_i \log r_{x_i}$. There exists a positive integer ν_0 such that for $\nu \geq \nu_0$, $\sqrt{-1}\,\partial\bar{\partial}\psi + \nu\Omega$ is strictly positive on M and therefore dominates a positive-valued function c times the Kähler form. Since M is complete, from the L^2 estimates of $\bar{\partial}$ it follows that for $\nu > \nu_0$ and for any C^{∞} L^{ν}-valued $\bar{\partial}$-closed $(0,1)$-form g on M with

$$\int_M |g|^2 c^{-1} e^{-\psi} < \infty$$

there exists a C^∞ cross section u of L^ν over M such that $\bar\partial u = g$ and

$$\int_M |u|^2 e^{-\psi} \leq \int_M |g|^2 c^{-1} e^{-\psi} .$$

Take a nowhere zero holomorphic cross section s of L^ν over some open neighborhood W of x_1 for some $\nu > \nu_0$. Let σ be a nonnegative smooth function on W with smooth support which is identically 1 on some open neighborhood of x_1. Apply the above result of L^2 estimates of $\bar\partial$ to $g = (\bar\partial\sigma)s$. Then the holomorphic cross section $\sigma s - u$ of L^ν over M is nonzero at x_1 and zero at x_2. Hence $\bigcup_{\nu=0}^{\infty} \Gamma(M, L^\nu)$ separate x_1 and x_2.

Now we want to show that $\bigcup_{\nu=0}^{\infty} \Gamma(M, L^\nu)$ gives local coordinates of M at x_1. By shrinking W, we can assume that there is a holomorphic coordinate system $z_1, \cdots, z_n$ on W with $z_\mu(x) = 0$, $1 \leq \mu \leq n$. Fix $1 \leq \mu \leq n$ and apply the above result of L^2 estimates of $\bar\partial$ to $g = (\bar\partial\sigma)z_\mu s$ and we obtain u. Let $s_\mu = \sigma z_\mu s - u$. Then $s_1, \cdots, s_n$ are holomorphic cross sections of L^ν over M and $\frac{s_1}{s}, \cdots, \frac{s_n}{s}$ form a local coordinate system of M at x_1 with x_1 as the origin. Q.E.D.

It follows from the existence of ϕ, Lemma 6, and a theorem of Andreotti-Tomassini [2, p. 97, Th. 2] that M is biholomorphic to an open subset of a projective algebraic manifold X.

§4. *Use of the Schwarz-Pick Lemma*

Let M be as in §3. By §2 for some compact subset E of M, $M-E$ has a finite number of components $A'_1, \cdots, A'_m$, each of which is an end of M. Moreover, the function ϕ which approaches $-\infty$ at the boundary of M is strictly plurisubharmonic on $M-E$. By §3 we can assume that M is an open subset of a projective algebraic manifold X. Because of ϕ, some open neighborhood of $X-M$ is holomorphically convex. By blowing down the exceptional set of this holomorphically convex neighborhood, we can assume that M is an open subset of a complex space X', possibly with isolated singularities, such that $X'-M$ admits a Stein neighborhood W in $X'-E$. We can extend ϕ to be a plurisubharmonic function on $X'-E$ by defining it to be $-\infty$ on $X'-M$. Since $X'-M$ is the $-\infty$ set of a plurisubharmonic function on X', it follows that for any connected open subset G of X', $G \cap M$ is still connected.

LEMMA 7. *Let* $x_1 \in X'-M$ *and* U *be a relatively compact Stein open neighborhood of* x_1 *in* W *such that* $U \cap M$ *is contained in some* A'_i. *Then* $U-M = \{x_1\}$.

Proof. Suppose besides x_1 there is another point $x_2 \in U-M$. We can find a holomorphic map $f: W \to \mathbb{C}^N$ embedding W as a closed complex subvariety of $\mathbb{C}^N$. Let B be a bounded open ball in $\mathbb{C}^N$ which contains $f(U)$ and $f(X'-M)$. Let $\delta(\cdot,\cdot)$ be the distance function on $f^{-1}(B)$ defined by the metric induced by an invariant metric of B. Let a be a positive number such that $a < \delta(x_1, x_2)$ and $a < \delta(x_j, f^{-1}(B)-U)$, $j = 1, 2$. Take a point x'_j in $U \cap M$ such that $\delta(x_j, x'_j) < \frac{a}{4}$, $j = 1, 2$. Let y_j be a point on $U-M$ which is closest to x'_j. Then $y_1 \neq y_2$. For $j = 1, 2$ there exists a curve $\sigma_j: [0,1] \to U$ into U such that $\sigma_j(t) \in U \cap M$ for $0 \leq t < 1$ and $\sigma_j(1) = y_j$. (We can take σ_j to be the geodesic joining x'_j and y_j.)

Let β be a positive number such that $\beta < \inf_E \phi$ and $\beta < \phi(\sigma_j(0))$, $j = 1, 2$. Let s_ν be a decreasing sequence in $(-\infty, \beta]$ such that $s_\nu \to -\infty$ as $\nu \to \infty$. Since $\phi(\sigma_j(t)) \to -\infty$ as $t \to 1$ and $\beta < \phi(\sigma_j(0))$, by the continuity of ϕ there exists $0 \leq t_{j,\nu} < 1$ such that $\phi(\sigma_j(t_{j,\nu})) = s_\nu$.

Let $F_s = A_i \cap \{\phi = s\}$. Since $\beta < \inf_E \phi$, each F_s is compact for $s \leq \beta$. By considering the universal covering $\pi: \tilde{M} \to M$ of M and applying

Lemma 5, we conclude that, if x, x' are two distinct points of F_s for some $s \leq \beta$, then there are two geodesic rays in A_i issued from x, x' which can be extended to geodesics intersecting F_β and which can be lifted up to geodesic rays in $\tilde{M}$ having the same limit point in $M(\infty)$. Since the sectional curvature of M is bounded from above by a negative number, it follows that the diameter of F_s (calculated with respect to the Kähler metric of M) approaches zero as $s \to -\infty$ (see the lemma in the Appendix).

Consider the holomorphic map $f^{-1}(B) \cap M \to B$ defined by f. Since $M - f^{-1}(B)$ is a compact subset of the complete Kähler manifold M whose sectional curvature is bounded from below and since B with the invariant metric is a complete Kähler manifold whose sectional curvature is bounded from above by a negative number, it follows from the Schwarz-Pick Lemma [13] that there exists a positive constant α such that $\alpha\delta(x,y) \leq d(x,y)$ for $x, y \in f^{-1}(B) \cap M$, where $d(\cdot,\cdot)$ is the distance function on M calculated with respect to the Kähler metric of M. (More precisely, the Schwarz-Pick Lemma is applied to the holomorphic map $\tilde{M} \to B$ which is the extension, by Hartogs' theorem of the holomorphic map $f \cdot \pi$.) Let γ be the distance of $f(y_1)$ and $f(y_2)$ with respect to the invariant metric of B. Then

$$\begin{aligned}&\liminf_{\nu\to\infty} d(\sigma_1(t_{1,\nu}), \sigma_2(t_{2,\nu}))\\ &\geq \alpha \liminf_{\nu\to\infty} \delta(\sigma_1(t_{1,\nu}), \sigma_2(t_{2,\nu}))\\ &\geq \alpha\gamma > 0\,.\end{aligned}$$

Since $\sigma_j(t_{j,\nu}) \in F_{s_\nu}$ for $j = 1, 2$, this contradicts the fact that the diameter of F_{s_ν} approaches 0 as $\nu \to \infty$. Q.E.D.

Clearly $X' - M$ can be covered by a finite number of relatively compact connected Stein open subsets $U_1, \cdots, U_\ell$ of W. We can assume that $U_j \cap (X' - M) \neq \emptyset$ for $1 \leq j \leq \ell$. Since $U_j \cap M$ is a connected subset of $M-E$, $U_j \cap M$ is contained in some A'_i. Let G_i be the union of all U_j

such that $U_j \cap M \subset A'_i$. By Lemma 7, $G_i - M$ consists of a single point x_i. This concludes the proof of the Main Theorem. Moreover, we have shown that this compactification X' of M is obtained by adding one point x_i to each end A'_i of M.

Appendix

Distance between points on geodesic rays with the same infinity point

In this Appendix we present the proof of a lemma used in the paper. After receiving a preprint of an earlier version of this paper, Eberlein informed us that this lemma is a consequence of [10, Th. 2.4]. The statement in [10, Th. 2.4] is stronger, but the proof given here is more straightforward.

Let M be a complete Riemannian manifold of sectional curvature $\leq -\kappa^2 < 0$. Take a 2-dimensional linear subspace H in the tangent space of M at some point Q of M. Let $P(r,\theta)$ denote the point in H with polar coordinates (r,θ). Let f be the exponential map at Q. Take $0 < \tau < t_0$ and $\theta_0 > 0$. Let

$$\begin{aligned} A &= f(P(t_0, 0)) \\ B &= f(P(t_0, \theta_0)) \\ C &= f(P(t_0 - \tau, 0)) \\ D &= f(P(t_0 - \tau, \theta_0)) . \end{aligned}$$

Let $\ell(A,B)$ (respectively $\ell(C,D)$) be the length of the image under f of the circular arc $P(t_0,\theta)$, $0 \leq \theta \leq \theta_0$ (respectively the circular arc $P(t_0-\tau,\theta)$, $0 \leq \theta \leq \theta_0$). We now fix A, B, and C (so that τ is fixed) and let $t_0 \to \infty$ (so that the point Q recedes to the boundary of M). Since A, B, and C are fixed, $\theta_0 = \theta_0(t_0)$ is a function of t_0. Let

$$\ell_0(A,B) = \lim_{t_0 \to \infty} \ell(A,B)$$

$$\ell_0(C,D) = \lim_{t_0 \to \infty} \ell(C,D) .$$

LEMMA 8. $\dfrac{\ell_0(A, B)}{\ell_0(C, D)} \geq e^{\kappa\tau}$.

Proof. Let $V(r, \theta)$ be the image of $\frac{\partial}{\partial\theta}$ under df. Then for fixed θ along the geodesic ray $f(P(r, \theta))$, $0 \leq r < \infty$, $V(r, \theta)$ is a Jacobi field which is 0 at Q and is perpendicular to the geodesic ray at every point. Since

$$\ell(A, B) = \int_{\theta=0}^{\theta_0} \|V(t_0, \theta)\| d\theta$$

$$\ell(C, D) = \int_{\theta=0}^{\theta_0} \|V(t_0 - \tau, \theta)\| d\theta ,$$

it follows that

$$\frac{\ell(A, B)}{\ell(C, D)} = \frac{1}{\ell(C, D)} \int_{\theta=0}^{\theta_0} \frac{\|V(t_0, \theta)\|}{\|V(t_0 - \tau, \theta)\|} \|V(t_0 - \tau, \theta)\| d\theta$$

$$\geq \inf_{0 \leq \theta \leq \theta_0} \frac{\|V(t_0, \theta)\|}{\|V(t_0 - \tau, \theta)\|}$$

$$= \inf_{0 \leq \theta \leq \theta_0} \exp \int_{r=t_0-\tau}^{t_0} \frac{\partial}{\partial r} \log \|V(r, \theta)\| dr .$$

To finish the proof, it suffices to show that

$$\frac{\partial}{\partial r} \log \|V(r, \theta)\| \geq \kappa .$$

To prove this, we let $V_r = \frac{\partial}{\partial r} V$ and $V_{rr} = \frac{\partial^2}{\partial r^2} V$ and suppress the variable θ after we fix it.

$$\frac{\partial}{\partial r} \log \|V(r)\| = \frac{\langle V(r), V_r(r)\rangle}{\langle V(r), V(r)\rangle}$$

$$= \frac{\langle V(r), V_r(r)\rangle - \langle V(0), V_r(0)\rangle}{\langle V(r), V(r)\rangle - \langle V(0), V(0)\rangle}$$

$$= \frac{\langle V_r(r'), V_r(r')\rangle + \langle V(r'), V_{rr}(r')\rangle}{2\langle V(r'), V_r(r')\rangle}$$

(by the mean value theorem, where $0 < r' < r$)

$$= \frac{\langle V_r(r'), V_r(r')\rangle - \langle R(V(r'), T(r'))T(r'), V(r')\rangle}{2\langle V(r'), V_r(r')\rangle}$$

(by the Jacobi field equation, where $T(r')$ is the unit vector of the geodesic ray $f(P(r, \theta))$, $0 \le r < \infty$, at the point $r = r'$)

$$\geq \frac{\langle V_r(r'), V_r(r')\rangle + \kappa^2\langle V(r'), V(r')\rangle}{2\langle V(r'), V_r(r')\rangle}$$

which is $\geq \kappa$ by the inequality

$$2\kappa ab \leq a^2 + \kappa^2 b^2 .$$

Q.E.D.

YUM-TONG SIU
DEPARTMENT OF MATHEMATICS,
STANFORD UNIVERSITY
STANFORD, CALIFORNIA 94305

SHING-TUNG YAU
SCHOOL OF MATHEMATICS,
INSTITUTE FOR ADVANCED STUDY
PRINCETON, NEW JERSEY 08540

REFERENCES

[1] A. Andreotti and H. Grauert, Algebraische Körper von automorphen Funktionen, Nachr. Akad. Wiss. Göttingen, math.-phys. Klasse, 1961, 39-48.

[2] A. Andreotti and G. Tomassini, Some remarks on pseudoconcave manifolds, *Essays on Topology and Related Topics*, dedicated to G. de Rham, ed. by A. Haefliger and R. Narasimhan, 85-104, Springer-Verlag 1970.

[3] W. L. Baily, Jr. and A. Borel, Compactification of arithmetic quotients of bounded symmetric domains, Ann. Math. *84*(1966), 442-528.

[4] J. Cheeger and Ebin, D., Comparison theorems in Riemannian geometry, North-Holland Mathematical Library, 9(1975).

[5] P. Eberlein, Lattices in spaces of nonpositive curvature.

[6] P. Eberlein and B. O'Neill, Visibility manifolds, Pacific J. of Math. *46*(1973), 45-109.

[7] M. Gromov, Manifolds of negative curvature, J. Diff. Geom. *13*(1978), 223-230.

[8] ________, Almost flat manifolds, J. Diff. Geom. *13*(1978), 231-241.

[9] E. Heintze, Mannifaltigkeiten negativer Krümmung.

[10] E. Heintze and H. C. Im Hof, Geometry of horospheres, J. Diff. Geom. *12*(1977), 481-491.

[11] R. A. Margulis, On connections between metric and topological properties of manifolds of nonpositive curvature, Proc. of the VI Topological Conf. 1972, p. 83, Tbilisi, USSR (Russian).

[12] I. Satake, On compactifications of the quotient spaces for arithmetically defined discontinuous groups. Ann. Math. *72*(1960), 555-580.

[13] S. T. Yau, A general Schwarz lemma for Kähler manifolds, Amer. J. Math. *100*(1978), 197-203.

LOCAL ISOMETRIC EMBEDDINGS

H. Jacobowitz

In §1 a proof of the Cartan-Janet theorem is outlined and a new result is presented on the local isometric deformation of analytic submanifolds. In §2 the isometric embedding problem for non-analytic metrics is related to hyperbolic differential equations. This is contrasted with the observation that elliptic techniques probably cannot be used in this problem.

§1. Let U be an open neighborhood of the origin in R^n and let $g = g_{ij}(x)$ be a Riemannian metric on U. A map $f : U \to E^N$ of U into some Euclidean space is an isometric immersion if it satisfies the equations

$$(1) \qquad \sum_{\alpha=1}^{N} \frac{\partial f^\alpha}{\partial x_i} \frac{\partial f^\alpha}{\partial x_j} = g_{ij}(x), \qquad 1 \le i, j \le n .$$

Any solution to (1) is necessarily an immersion and thus an embedding upon being restricted to a smaller neighborhood of the origin. If $N < (n+1)/2$, then (1) is an over-determined system and for almost all choices of g (even if g is real analytic) this system has no solution. For $N > n(n+1)/2$ the system is underdetermined. One expects solutions to always exist. It is a consequence of the work of Nash that for $N \ge n(n+3)/2$ solutions do exist even if g is only finitely differentiable. See Nash [13], Greene [5], and Gromov and Rokhlin [6].

Seminar on Differential Geometry
0-691-08268-5/82/000381-13 $00.65/0 (cloth)
0-691-08296-0/82/000381-13 $00.65/0 (paperback)

In this paper we are interested in the determined case and we shall always take $N = n(n+1)/2$. The fundamental result for this dimension is for real analytic metrics and is due to Janet [11] and Cartan [2].

CARTAN-JANET THEOREM. *Let* g_{ij} *be real analytic in some neighborhood of the origin. Then in a possibly smaller neighborhood of the origin the system (1) has a real analytic solution.*

Thus a real analytic Riemannian manifold of dimension n always has a local isometric embedding into E^N, $N = n(n+1)/2$. This number is called the Cartan-Janet dimension.

We wish to sketch a simple version of Janet's proof. Details may be found in [9]. The idea is to apply the Cauchy-Kowalewski theorem. But since (1) cannot be put into a form suitable for the Cauchy-Kowalewski theorem, we must consider a related second order system.

We start by considering an extension problem. Let M^n be a Riemannian manifold and H^{n-1} a co-dimension one submanifold. Assume that $h: H \to E^N$ is an isometric embedding. We seek an isometric embedding $f: M \to E^N$ which extends h. That is, we seek to solve (1) with the initial condition $f(x) = h(x)$ for all $x \in H$. That such an extension need not always exist is clear. For consider a non-geodesic curve $H \subset M^2$ and an isometric embedding $h: H \to \Gamma$ where Γ is a straight line in E^3. Assume that an extension $f: U \to E^3$ exists where U is some open set in M^2 which includes in its interior some arc of H. The minimal geodesic γ between two sufficiently close points p and q of this arc lies entirely in U. So $f(\gamma)$ is a curve in E^3 between the points $f(p)$ and $f(q)$. Since f is isometric, γ and $f(\gamma)$ have the same length. Also the distance d between p and q measured along H must be the same as the distance between $f(p)$ and $f(q)$ measured along Γ. So $L(\gamma) = L(f(\gamma))$, $L(\gamma) < d$, $L(f(\gamma)) > d$. This is impossible.

It is not hard to see that a necessary condition for extending a map $h: H^1 \to E^3$ is that the geodesic curvature of H at each point is no greater than the curvature of the space curve h(H) at the corresponding point. The strict inequality is a sufficient condition for extending (Darboux [3, p. 274]).

A similar situation is true in higher dimensions. Let L_M be the second fundamental form of H in M and L_E be the second fundamental form of H in E. Here we have identified H with its image h(H). Note that L_M is real valued while L_E takes values in the normal bundle of H in E. If an extension f exists, let ν be a unit vector field tangent to f(M) and orthogonal to h(M). Then ν satisfies

1- $\langle L_E(X,Y), \nu\rangle = L_M(X,Y)$ all $X, Y \in T(H)$

2- $\nu \perp h(H)$

3- $\|\nu\| = 1$.

Thus a necessary condition for there to exist an extension of $h: H \to E$ is that there exist a unit vector field along H satisfying 1, 2, and 3. Generically, this condition is also sufficient as we now show.

DEFINITION. A map $h: H^{n-1} \to E^N$ is non-degenerate at p if the osculating space of h(H) at p has dimension N-1.

Recall that the osculating space of h(H) is the space spanned by the first and second derivatives of h. Introduce coordinates $(x_1, \cdots, x_{n-1}, y)$ on M so that $H = \{(x,y) | y = 0\}$. Then h is nondegenerate if $\{h_{x_i}, h_{x_j x_k} | 1 \le i, j, k \le n-1\}$ is a linearly independent set of vectors in E^N. In the generic case ν given above does not lie in the osculating space so that its projection into this osculating space satisfies 1 and 2 but has norm strictly less than one.

THEOREM 1 [9]. *Let* M^n *be a real analytic Riemannian manifold,* H^{n-1} *a real analytic submanifold,* p *a point of* H *and* $h: H \to E^N$, $N = n(n+1)/2$, *a real analytic isometric embedding. If* h *is nondegenerate at* p *and if there exists a vector* $\nu \in T_{h(p)}E$ *which satisfies*

1 - $$<L_E(X,Y),\nu> = L_M(X,Y) \quad \textit{all} \ X, Y \in T_pH$$

2 - $$\nu \perp h(H) \quad \textit{at} \ p$$

3′- $$\|\nu\| < 1$$

then there exists a neighborhood U *of* p *in* M *and a real analytic isometric embedding* $f: U \to E$ *with* $f = h$ *on* $U \cap H$.

Proof. Again we take $H = \{(x,y) | y = 0\}$ where y is identified with x_n. Without loss of generality we may assume that the coordinates were chosen so that $g_{nn} = 1$ and $g_{in} = 0$, $1 \le i \le n-1$ in a neighborhood in M of p. We will use the notation f_i to denote $\frac{\partial f}{\partial x_i}$ where $1 \le i \le n-1$, etc. In the next lemma h is isometric, so $h_i h_j = g_{ij}(x,0)$. Of course g_{ij} does not denote differentiation so we use $g_{ij,y}$ for differentiation.

LEMMA. *Any solution to the initial value problem*

$$(2) \qquad \begin{aligned} f_i \cdot f_{yy} &= 0 \\ f_y \cdot f_{yy} &= 0 \\ f_{ij} \cdot f_{yy} &= -\tfrac{1}{2}(g_{ij})_{yy} + f_{yi} \cdot f_{yj} \\ f(x,0) &= h(x) \\ f_y(x,0) &= \mu(x) \end{aligned}$$

also solves the initial value problem

$$(3) \qquad \begin{aligned} f_i \cdot f_j &= g_{ij} \\ f_i \cdot f_y &= 0 \\ f_y \cdot f_y &= 1 \\ f(x,0) &= h(x) \end{aligned}$$

provided $\mu(x)$ *satisfies*

$$\begin{aligned} <L_E(X,Y),\mu> &= L_M(X,Y) \\ \mu &\perp h(H) \\ \|\mu\| &= 1 . \end{aligned}$$

Proof. Since $\frac{\partial}{\partial y}(f_y \cdot f_y) = 2f_y \cdot f_{yy} = 0$ while $f_y \cdot f_y = \|\mu\|^2$ at $y = 0$, the uniqueness theorem for ordinary differential equations gives $f_y \cdot f_y \equiv 1$. Also, $\frac{\partial}{\partial y}(f_i \cdot f_y) = f_i \cdot f_{yy} + f_{iy} \cdot f_y = \frac{1}{2}\frac{\partial}{\partial x_i}\|f_y\|^2 = 0$ while $f_i \cdot f_y = 0$ at $y = 0$. So $f_i \cdot f_y \equiv 0$. From these two results we derive that $f_{ij} \cdot f_y = -f_i f_{jy}$ and that $f_{ijy} \cdot f_y = -f_{jy} \cdot f_{iy}$. Now let $s(x,y) = f_i \cdot f_j$. Then $s_y = -2f_{ij} \cdot f_{yy}$ and $s_{yy} = 2f_{ij} \cdot f_{jy} - 2f_{ij} \cdot f_{yy} = 2f_{iy} \cdot f_{jy} - 2\left(-\frac{1}{2}(g_{ij})_{yy} + f_{iy} \cdot f_{jy}\right) = (g_{ij})_{yy}$. Now $s(x,0) = h_i \cdot h_j = g_{ij}(x,0)$ and $s_y(x,0) = -2h_{ij} \cdot \mu$. But $\langle L_E\left(\frac{\partial}{\partial x_i}, \frac{\partial}{\partial x_j}\right), \mu\rangle = (h_{ij})^{\perp} \cdot \mu = h_{ij} \cdot \mu$, where $h_{ij}^{\perp}$ is the part of $\frac{\partial^2 h}{\partial x_i \partial x_j}$ which is perpendicular to $h(H)$. And $L_M\left(\frac{\partial}{\partial x_i}, \frac{\partial}{\partial x_j}\right) = \Gamma_{ij}^n = -\frac{1}{2} g_{ij,y}$. Thus from $\langle L_E, \mu\rangle = L_M$ we see that $s_y(x,0) = g_{ij,y}$. It then follows that $f_i \cdot f_j \equiv g_{ij}$.

Now to prove the theorem. Let ν satisfy 1, 2, and 3′ at p. At p, h is nondegenerate and so h is nondegenerate in a neighborhood of p. This implies that for each q close to p there is a unique vector $\nu_1(q)$ in the osculating space of $h(H)$ at q satisfying 1 and 2. Further $\nu_1(q)$ is a real analytic function of q. Because $\nu_1(p)$ is the projection of ν into the osculating space we have that $\|\nu_1\| < 1$ at p and hence in a neighborhood of p. Let $\nu_2(q)$ be orthogonal to the osculating space at q and chosen in a real analytic manner so that $\|\nu_1 + \nu_2\| = 1$. Finally let $\mu = \nu_1 + \nu_2$. By the lemma, it is enough to show that the initial value problem (2) has a solution. We have arranged that μ has a non-zero component in the direction orthogonal to the osculating space of $h(H)$. Thus on H the N vectors f_i, f_y, f_{ij} are linearly independent. Thus near this initial data the coefficient in (2) of f_{yy} is invertible. Hence the Cauchy-Kowalewski theorem may be applied to show that (2) has a real analytic solution. This establishes our theorem. This theorem can be taken as the basis for an inductive proof of the Cartan-Janet Theorem. See [9] or [16, pp. 216-225].

This theorem can also be used to obtain information about isometric deformations. Let $f: M^n \to E^N$ be a given real analytic and isometric embedding.

DEFINITION. f has a non-trivial isometric deformation if there exists a family f_t of real analytic maps of $M \to E$, depending smoothly upon t, with

1) $f_0 = f$
2) each f_t isometric
3) there does not exist a family $g_t : E \to E$ of rigid motions with $f_t = g_t \circ f$.

We say a submanifold $M \subset E$ has a non-trivial isometric deformation if the associated $f : M \to E$ does.

Let $H = \{(x, y) | y = 0\}$ be a hypersurface of M, p a point of H, and $f : M^n \to E^N$ an isometric embedding.

DEFINITION. The hypersurface H is asymptotic at p if at p the set of vectors $\{f_y, f_i, f_{jk} | 1 \le i, j, k \le n-1\}$ is linearly dependent.

Thus H is a non-asymptotic hypersurface at p if the linear span of $\{f_y, f_i, f_{jk} | 1 \le i, j, k \le n-1\}$ is of dimension N. This is a special case of a definition in [8]. We note that for $n = 2$ our definition agrees with the classical definition of an asymptotic curve. Whether H^{n-1} is asymptotic at p or not depends only on the tangent space to H at p —the curvature of H is irrelevant. Finally, if H is nonasymptotic at p then there exists a vector ν as in the hypothesis of Theorem 1.

THEOREM 2 [10]. *Let* M^n *be a real analytic submanifold of* E^N, $N = \frac{1}{2} n(n+1)$, *and* p *a point of* M. *If* M *has a non-asymptotic hypersurface at* p *then some neighborhood of* p *in* M *has a non-trivial isometric deformation.*

Let H^{n-1} be a real analytic hypersurface which is non-asymptotic at p. Pick a sequence $H^1 \subset H^2 \cdots \subset H^{n-1}$ of real analytic submanifolds such that H^i is a non-asymptotic hypersurface in H^{i+1}. H^1 is a curve and so has a non-trivial isometric deformation. The proof of Theorem relies on repeated use of a strengthened form of Theorem 1 applied to the manifolds

$H^2, \cdots, H^{n-1}, M$. Theorem 2 improves the result of Tenenblat [19] where it was shown that such an M always has a non-trivial infinitesimal isometric deformation.

It might be thought that Theorem 2 is true even without the hypothesis that there is a non-asymptotic hypersurface through p. However, at least for the case $n=2$, this hypothesis cannot be omitted. Efimov has shown that there exist real analytic surfaces in E^3 which are rigid in any neighborhood of a distinquished point. See [4] and the references cited there. In particular the surface $(x, y, x^9 + \lambda x^7 y^2 + y^9)$, where λ is any transcendental number, has no real analytic isometric deformations in any neighborhood of the origin except for those deformations induced by rigid motions in E^3.

It should be emphasized that the results of this section are for the Cartan-Janet dimension and have been proved only in the real analytic category. If the ambient dimension is taken to be somewhat larger than $n(n+1)/2$ then all these results hold even for C^k mappings and metrics. The analogue of the Cartan-Janet theorem is the Nash embedding theorem [13] which asserts the existence of a global isometric embedding for any C^3 metric. The analogue of the extension result (Theorem 1) may be found in Theorems 5.1 and 5.2 of [9]. The analogue of the deformation result occurs on page 306 of [9] where it is shown that the deformation can even be chosen so as to leave the given hypersurface point-wise fixed.

In the next section we consider results for non-analytic metrics in the Cartan-Janet dimension. As we shall see, very little is known except for $n=2$.

§2. We now consider non-analytic metrics. Known results for $n=2$ will lead us to a conjecture for $n > 2$. We first replace (1) by its linearization. So let $(x_1, x_2, z(x_1, x_2))$ be a surface in E^3 with induced metric $\overline{g}_{ij}(x)$. Given some nearby metric $g = \overline{g} + \varepsilon h$ we look for a solution to (1) of the form $f = (x_1, x_2, z(x)) + \varepsilon(u(x), v(x), w(x))$. Disregarding terms quadratic in ε, we are left with

$$
(4) \qquad \begin{aligned}
2(u_x + z_x w_x) &= h_{11} \\
u_y + v_x + z_y w_x + z_x w_y &= h_{12} \\
2(v_y + z_y w_y) &= h_{22} .
\end{aligned}
$$

Here it is convenient to use (x,y) in place of (x_1, x_2). Let L be the operator $Lw = z_{yy} w_{xx} - 2z_{xy} w_{xy} + z_{xx} w_{yy}$.

LEMMA [7]. *System (4) has a smooth solution if and only if the equation* $Lw = -\frac{1}{2}(h_{11,yy} + h_{12,xx}) + h_{12,xy}$ *has a smooth solution.*

The operator L is of interest because its characteristics $\phi(x,y) =$ constant are given by the solutions to $z_{yy}\phi_x^2 - 2z_{xy}\phi_x\phi_y + z_{xx}\phi_y^2 = 0$ and so L has real characteristics precisely when $z_{xx}z_{yy} - z_{xy}^2 \leq 0$. Recall that the curvature of the original surface is given by $K = (z_{xx}z_{yy} - z_{xy}^2) \cdot (1 + z_x^2 + z_y^2)^{-2}$. Thus L is elliptic wherever $K > 0$ and L is hyperbolic wherever $K < 0$. This means that the linearized system (4) can be solved in the neighborhood of any point where the curvature is not zero.

THEOREM. *Let* M *be a smooth Riemannian manifold of dimension* 2. M *has a smooth local isometric embedding into* E^3 *in the neighborhood of any point at which the Gaussian curvature is nonzero.*

For points of positive curvature the existence of this local isometric embedding may be proved using the linearized system (4), see [7]. For negative curvature it is best to work with asymptotic coordinates and quasi-linear hyperbolic equations, Poznyack [15]. In the theorem smooth can mean sufficiently differentiable or C^∞. The existence of isometric embeddings near a point at which K is zero is still unproved in general. There are counter-examples with low differentiability. That is, there are metrics of class $C^{2,1}$ with no C^2 isometric embeddings into E^3 (Pogorelov [14]). This can be modified to yield a $C^{3+\alpha}$ metric with no C^3 isometric embedding into E^3. (But every $C^{3+\alpha}$ metric has a $C^{3+\alpha}$ isometric embedding into E^4 [7].)

Although there are no results for $n>2$, we do have the following conjecture which seems to be widely believed.

CONJECTURE. There exists an open set of C^∞ metrics defined on some small open set in R^n such that each metric can be isometrically embedded in E^N, $N=\frac{1}{2}n(n+1)$.

Further we may expect this set of metrics to have an intrinsic characterization involving curvature inequalities. The idea behind this conjecture is that it should be possible in some cases to reduce (1) or its linearization to some standard form, e.g. an elliptic or hyperbolic system.

We have several strong clues as to how to proceed. We have seen that when $n=2$ the system (1) may be studied using the operator L. The characteristics of L can easily be seen to be the asymptotic curves on the surface $(x,y,z(x,y))$. In §1 we saw that certain results for $n=2$ and non-asymptotic curves carry over to $n>2$ and non-asymptotic hypersurfaces. Thus we may hope that the asymptotic hypersurfaces are the characteristic hypersurfaces of some simpler system equivalent to (1). (Note that every hypersurface is characteristic for (1). This is why we could not apply the Cauchy-Kowalewski Theorem directly to (1).) Further evidence of this is found in the work of Tenenblat [18] where it is shown that the characteristics in the sense of Cartan for the differential system corresponding to (1) for an embedded manifold are precisely the asymptotic hypersurfaces.

A word of caution here. It is natural to now call a submanifold $U^n \subset E^N$ "elliptic at a point p" if there are no hypersurfaces in U^n which are asymptotic at the point p and to call U "elliptic" if it is elliptic at each of its points. For any such elliptic submanifold the system (1) or its linearization should be equivalent to a system which is elliptic in the usual sense and the conjecture should then hold in a neighborhood of the metric induced on U. But we have the following result which is essentially due to Tanaka [17, page 146].

LEMMA. *If* $U^n \subset E^N$, $N=\frac{1}{2}n(n+1)$, *is elliptic at some point* p *then* $n=2$ *and* $N=3$.

Proof. Choose bases for the tangent and normal spaces to U at p. Then the second fundamental form is given by $N-n$ symmetric $n\times n$ matrices $A_1, \cdots, A_r$, $r = N-n$. It is not hard to show that if U is elliptic at p then every nonzero matrix in the linear span of $\{A_1, \cdots, A_r\}$ has at least two eigenvalues of the same sign. But according to Kaneda and Tanaka [12, page 11] it then follows that either $r < n(n-1)/2$ or $r=1$ and $n=2$. In our case $r = n(n-1)/2$ so $n=2$ and $N=3$. (The result of Kaneda and Tanaka is based on the work of Adams, Lax, and Phillips [1].)

So it seems that elliptic techniques will not be useful in this problem. However, it is possible to find submanifolds such that the linearization of (1) is equivalent to a hyperbolic system. Unfortunately the system is only weakly hyperbolic. Since the theory of such systems is not well developed, there remains much work to do on the conjecture. We describe the results for $n=3$, $N=6$.

Let $u: M^3 \to E^6$ be given. The Cauchy problem for the linearized system is

$$\sum_{\alpha=1}^{6} \left(\frac{\partial u^\alpha}{\partial x_i} \frac{\partial v^\alpha}{\partial x_j} + \frac{\partial u^\alpha}{\partial x_j} \frac{\partial v^\alpha}{\partial x_i} \right) = g_{ij} \qquad 1 \leq i, j \leq 3 \tag{5}$$

$$v(x_1, x_2, 0) = h(x_1, x_2)$$

where h and g are assumed to satisfy the compatibility conditions $u_i h_j + u_j h_i = g_{ij}$, $1 \leq i, j \leq 2$.

As before we take $y \equiv x_3$ and $H = \{(x,y) | y = 0\}$ with H nonasymptotic at the origin. Now we define $\nu(x)$ along H to be the vector field satisfying

$$u_y \cdot \nu = \frac{1}{2} g_{33}$$

$$u_i \cdot \nu = g_{i3} - u_y \cdot h_i$$

$$u_{ij} \cdot \nu = \frac{1}{2} (g_{j3,i} + g_{i3,j} - g_{ij,y}) - u_y \cdot h_{ij} .$$

Such a vector field exists and is unique since $\{u_y, u_i, u_{ij} | 1 \le i, j \le 2\}$ is linearly independent. We assume that $u(x_1, x_2, y) = (u^1(x, y), \cdots, u^6(x, y))$ with $u^i = x_i$, $u^3 = y$. This means that we are working with a special non-asymptotic hypersurface rather than a general one. Set $\tilde{h} = (h^4, h^5, h^6)$ and $\tilde{\nu} = (\nu^4, \nu^5, \nu^6)$.

LEMMA. *The Cauchy problem (5) may be solved if the following Cauchy problem has a solution*

$$(6) \qquad \begin{aligned} &\sum_{\alpha=4}^{6} (u^\alpha_{ij} v^\alpha_{yy} + u^\alpha_{yy} v^\alpha_{ij} - u^\alpha_{iy} v^\alpha_{jy} - u^\alpha_{jy} v^\alpha_{iy}) = \gamma_{ij} \qquad (1 \le i, j \le 2) \\ &v(x, 0) = \tilde{h}(x) \\ &v_y(x, 0) = \tilde{\nu}(x) . \end{aligned}$$

Here $\gamma_{ij} = \frac{1}{2}(g_{j3,iy} + g_{i3,jy} - g_{ij,yy} - g_{33,ij})$.

The point of the lemma is that system (6) is not as degenerate as (5). Indeed we will find conditions which guarantee that (6) is of hyperbolic type.

Let $\sigma(\xi_1, \xi_2, \tau)$ be the symbol of the system in (6). It is easily seen that $\sigma = \tau^3 q(\xi_1, \xi_2, \tau)$ where q is a cubic polynomial in τ.

DEFINITION. $u : U^3 \to E^6$ is hyperbolic with respect to the non-asymptotic hypersurface H if for each $(\xi_1, \xi_2) \ne (0, 0)$ the equation $q(\xi_1, \xi_2, \tau) = 0$ has three real roots in τ.

Hyperbolic submanifolds do exist.

EXAMPLE. The submanifold

$$u(x_1, x_2, y) = \left(x_1, x_2, y, \frac{1}{2} x_1^2 + \alpha x_1 y + \frac{1}{2} y^2, x_1 x_2, \frac{1}{2} x_2^2\right)$$

is hyperbolic at all its points if and only if $\alpha \ge 1$.

When $\alpha = 1$ the cubic $q(\xi_1, \xi_2, \tau)$ has repeated roots in τ for each $(\xi_1, \xi_2) \neq (0,0)$. When $\alpha > 1$ the roots are distinct for all $(\xi_1, \xi_2) \neq (0,0)$ except for $(0, \xi_2)$, $\xi_2 \neq 0$. Unfortunately there do not exist submanifolds with distinct roots for all $(\xi_1, \xi_2) \neq (0,0)$.

We relate hyperbolicity and asymptotic hypersurfaces. Let $u(x_1, x_2, y) = (x_1, x_2, y, u(x,y), v(x,y), w(x,y))$ be a submanifold passing through the origin with u, v, w all of second order at the origin. Fix some $(\xi_1, \xi_2) \neq (0,0)$ and identify this pair with the differential form $\xi_1 dx_1 + \xi_2 dx_2$. We seek to determine when the surface in U annihilated by $\xi_1 dx_1 + \xi_2 dx_2 + \tau dy$ is asymptotic. Again let u stand for the 3-vector (u, v, w). The annihilator of $\xi_1 dx_1 + \xi_2 dx_2 + \tau dy$ is spanned by $A = \xi_2 \frac{\partial}{\partial x_1} - \xi_1 \frac{\partial}{\partial x_2}$ and $B = \tau\xi_1 \frac{\partial}{\partial x_1} + \tau\xi_2 \frac{\partial}{\partial x_2} - (\xi_1^2 + \xi_2^2) \frac{\partial}{\partial y}$. We identify the first vector with $\xi_2 u_1 - \xi_1 u_2$ and the second with $\tau\xi_1 u_1 + \tau\xi_2 u_2 - (\xi_1^2 + \xi_2^2) u_y$. The space $\{A, B\}$ is asymptotic when the vectors $A \vdash A$, $A \vdash B$, $B \vdash B$ are linearly dependent, that is, when their determinant vanishes.

LEMMA. $$\det \begin{pmatrix} A \vdash A \\ A \vdash B \\ B \vdash B \end{pmatrix} = (\xi_1^2 + \xi_2^2)^3 q(\xi_1, \xi_2, \tau) .$$

The lemma may be reworded as follows.

LEMMA. U *is hyperbolic at the point* p *with respect to the nonasymptotic hypersurface* H *if and only if through each* $v \in T_pH$ *there are three asymptotic* 2-*planes (counted with multiplicity).*

RUTGERS UNIVERSITY
CAMDEN, NEW JERSEY

REFERENCES

[1] J.F. Adams, P.D. Lax, R.S. Phillips, On matrices whose real linear combinations are non-singular, Proc. Amer. Math. Soc. 16(1965), 318-322.

[2] E. Cartan, Sur la possibilité de plonger un espace riemannian donné dans un espace euclidien, Ann. Soc. Pol. Math. 6(1927), 1-7.

[3] G. Darboux, *Lecons sur la théorie générale des surfaces, Vol. 3*, Gauthier-Villars, Paris, 1894.

[4] N. V. Efimov, Qualitative problems of the theory of deformation of surfaces, in *Differential Geometry and Calculus of Variations*, Translations Series 1, Vol 6, Amer. Math. Soc., Providence, Rhode Island, 1962 (Translated from Uspeki Mat. 3, 2(24)(1948), 47-168).

[5] R. Greene, *Isometric embedding of Riemannian and pseudo-Riemannian manifolds*, Amer. Math. Soc. Memoir 97, Providence, 1970.

[6] M. L. Gromov and V. A. Rokhlin, Embeddings and immersions in Riemannian Geometry, Russian Math. Surveys 25(1970), 1-57.

[7] H. Jacobowitz, Local isometric embeddings of surfaces into Euclidean four space, Indiana University Math. J. 21(1971), 249-254.

[8] ———, Deformations leaving a hypersurface fixed, in *Proc. Symposia Pure Mathematics, Vol XXIII*, Amer. Math. Soc., Providence, 1973.

[9] ———, Extending isometric embeddings, J. Differential Geometry 9(1974), 291-307.

[10] ———, Local analytic isometric deformations, to appear in Indiana University Math. J.

[11] M. Janet, Sur la possibilité de plonger un espace riemannian donné dans un espace euclidien, Ann. Soc. Polon. Math. 5(1926), 38-43.

[12] E. Kaneda and N. Tanaka, Rigidity for isometric embeddings, J. Mathematics of Kyoto University 18(1978), 1-70.

[13] J. Nash, The embedding problem for Riemannian manifolds, Ann. of Math. 63(1956), 20-64.

[14] A. V. Pogorelov, An example of a two-dimensional Riemannian metric not admitting a local realization in E_3, Doklady Akad. Nauk USSR 198(1971), 42-43.

[15] E. G. Poznyack, Regular realization in the large of two-dimensional metrics of negative curvature, Soviet Math. Dokl. 7(1966), 1288-1291.

[16] M. Spivak, *Differential Geometry, Vol 5*, Publish or Perish, Boston, 1975.

[17] N. Tanaka, Rigidity for elliptic isometric embeddings, Nagoya Math. J. 51(1973), 137-160.

[18] K. Tenenblat, On characteristic hypersurfaces of submanifolds in Euclidean space, Pacific J. Math. 74(1978), 507-517.

[19] ———, On infinitesimal isometric deformations, Proc. Amer. Math. Soc. 75(1979), 269-275.

YANG-MILLS THEORY: ITS PHYSICAL ORIGINS AND DIFFERENTIAL GEOMETRIC ASPECTS

Jean Pierre Bourguignon* and H. Blaine Lawson, Jr.**

One of the most intriguing developments in mathematical physics over the past quarter century, at least for many of us who work in differential geometry, has been the discovery of the fundamental role played by bundles, connections and curvature in expressing the basic laws of nature.

Certain aspects of this theory, called *Yang-Mills theory*, have recently attracted widespread attention from mathematicians. This has been due largely to a series of articles and lectures by M. F. Atiyah, I. M. Singer and others. (See [3], [4], [5] and [12].) In these articles certain difficult mathematical problems in the theory are solved by beginning with fundamental ideas of R. Penrose [22] and ultimately by using quite recent results in algebraic geometry. Our emphasis here will be of a more differential geometric nature.

We have several rather distinct objectives in writing this paper. The first is to present a brief, general introduction to this physical theory and its underlying mathematical models. This is done in Part One. The second is to discuss some specific contributions that we have made to the mathe-

*Supported in part by NSF Grant MCS 77-18723(02).
**Supported in part by NSF Grant MCS 77-23579.

Seminar on Differential Geometry
0-691-08268-5/82/000395-27 $01.35/0 (cloth)
0-691-08296-0/82/000395-27 $01.35/0 (paperback)

matical side of the theory. These contributions are summarized in Part One and their proofs are outlined in Part Two. (Full details appear in [10].) Finally, in Part Three we prove some new results. In particular, we present a detailed analysis of weakly stable SU_2-, U_2-, SU_3- and SO_4-Yang-Mills fields over compact homogeneous 4-manifolds.

This work extended over almost a two-year period and benefited from conversations with many mathematicians and physicists. We take this opportunity to thank them all. Special thanks are due to D. Friedmann for useful comments on Part One.

Table of Contents

Part One: Geometry and physics of Yang-Mills fields
- A) *The development of gauge physics*
- B) *A mathematical description of gauge theories*
- C) *Special aspects of dimension four*
- D) *Some mathematical problems and results*
- E) *Some comments on the tangent bundle*

Part Two: An overview of the stability theorems
- A) *Picking appropriate variations of the connection*
- B) *Using an averaging procedure*
- C) *Introducing some elementary algebraic lemmas*
- D) *The basic stability theorem*
- E) *The* SU_2*-stability theorem over* S^4

Part Three: Stability results for homogeneous 4-manifolds
- A) *Yang-Mills fields of least action*
- B) *The* U_N *set-up*
- C) *The* SO_4 *set-up*
- D) *The specialized stability theorems*

Part One: Geometry and physics of Yang-Mills fields

A) *The development of gauge physics*

In the middle of last century, when Maxwell wrote down the laws of electromagnetism, the electric and magnetic fields were thought of as distinct entities. Then Lorentz, at the turn of the century, pointed out that if space-time $R^3 \times R$ is equipped with a pseudo-riemannian metric (which we then call $R^{3,1}$), these two fields can be naturally combined as components of an exterior 2-form ω on $\mathbf{R}^{3,1}$, called the *electromagnetic field,* and then Maxwell's (vacuum) equations take the beautiful and condensed form

$$d\omega = 0 , \quad \delta\omega = 0 .$$

These equations are invariant under the *Lorentz group* of linear isometries of $\mathbf{R}^{3,1}$. The Lorentz argument was an early instance where an appropriate invariance relation under a group simplified and unified a theory.

Also at the turn of the century Einstein pointed out that it was necessary to associate particles (or *photons*) to electromagnetic interactions in order to explain phenomena observed in the photoelectric effect. This duality between waves and particles led to the development of quantum mechanics and eventually to the tremendously successful theory of *quantum electrodynamics.*

During the same period of time, two new types of interactions were discovered: the *strong* nuclear *interactions* which explain the cohesion of the nucleus, and *weak interactions,* introduced by Fermi to explain β-radioactivity. Physicists now had to face four fundamental interactions (strong, weak, electromagnetic, and gravitational) with different strengths, ranges, and groups of symmetries. To make matters worse, not all particles are subject to all types of interactions. (For example, electrons, muons and neutrinos have no strong interactions.)

Because of their range and nature, the phenomena of the weak and strong interactions can only be modeled quantum mechanically. However, a theory for these interactions which is analogous to quantum electrodynamics, requires a model at the "classical mechanical level." Such a theory was proposed in 1954 by C. N. Yang and R. Mills [26], and over the past 25 years their theory has attracted much attention and has eventually proved to be physically relevant.

The key feature of the theory of Yang and Mills is again the invariance of the physics under a group, but in this case an infinite-dimensional group. To understand this we return to the classical electromagnetic field ω mentioned above. Notice that since $d\omega = 0$ we may express ω as

$$\omega = d\alpha$$

where α is a 1-form on $\mathbf{R}^4$ called the *electromagnetic potential.* The form α is defined only up to an exact form, i.e., we may replace α by $\alpha + df$ where f is any smooth function on $\mathbf{R}^4$. Such a replacement is called a *change of gauge* or a *gauge transformation.* The insensitivity of the Physics to the group of gauge transformations lies at the heart of matters. It is called the "*principle of local invariance.*"

In an attempt to describe strong interactions at the classical level, C. N. Yang and R. Mills proposed in [26] that the Lagrangian of the interaction should involve a potential with values in the Lie algebra of the non-abelian group SU_2 (describing the degrees of freedom of isotopic spin, the first quantum number to be understood in relation with strong interactions). Moreover, this Lagrangian should be invariant under the group of local internal symmetries, again called gauge transformations. Such theories have become known as *gauge theories.* (See [25].) They have spread to other areas of physics like solid state physics, statistical mechanics, etc.

One of the striking features of the Yang-Mills proposal was that the potential α (an $\mathfrak{su}_2$-valued 1-form) was required to transform *like a connection.* Specifically, a gauge transformation is here defined as a map $g : \mathbf{R}^{3,1} \to SU_2$, and the transformed potential is given by

$$\alpha^g = g\alpha g^{-1} + g dg^{-1} .$$

Furthermore, in this terminology, the proposed density was just $\|\Omega\|^2$ where $\Omega = d\alpha + \frac{1}{2}[\alpha, \alpha]$ was the curvature of the connection α. This connection, of course, lives on the trivialized principal SU_2-bundle over $\mathbf{R}^{3,1}$. The gauge transformation g simply amounts to a change of the trivialization (or, if you prefer, a principal bundle automorphism).

We note that the invariance under gauge transformations together with certain homogeneity requirements actually force the Lagrangian density to be $\|\Omega\|^2$.

The Yang-Mills formulation above can be considered a strict analogue of electromagnetic theory as follows. For electromagnetic theory, the relevant Lie group is U_1. The potential $i\alpha$ (with values in $i\mathbb{R} = \mathfrak{u}_1$) can be considered as a connection on a trivialized principal U_1-bundle over $\mathbb{R}^{3,1}$. A gauge transformation is then a smooth map $g : \mathbb{R}^4 \to U_1$ which can be written in the form $g(x) = \exp(-if(x))$. The transformed connection is $\alpha^g = \alpha + df$ (as above), and the associated field $\omega = d\alpha$ is just the curvature of the connection. By d'Alembert's Principle, the field equations are obtained by setting $\delta L = 0$ where $L \equiv \frac{1}{2} \int \|\omega\|^2$ is the total energy. Let $\alpha_t = \alpha + t\beta$ be a family of potentials, and note that

$$\frac{d}{dt} L_t \Big|_{t=0} = \int < d\alpha, d\beta >$$

$$= \int < \omega, d\beta >$$

$$= \int < \delta\omega, \beta > = 0 .$$

Thus, the condition $\delta L_\alpha = 0$ implies Maxwell's equations for the field ω.

The principal bundle formalism of Yang and Mills is certainly very suggestive to a geometer. It is natural to ask, for example, whether there is a physical interpretation of the connection, i.e., of the gauge potential. For many years it was thought that the electromagnetic potential was merely a mathematical artifice, convenient but physically meaningless. Then in 1959 an experiment suggested by Y. Aharonov and D. Bohm (cf. [1]), and performed for the first time by Chambers, revealed that in the absence of the field, the electromagnetic potential does play a role. In this experiment, one reflects a coherent beam of electrons in a closed path encircling a solenoid. This solenoid is considered as an infinite, perfectly insulated tube. Although the field outside the tube is zero, the phase shift caused by the self-interaction of the beam is observed to vary with the intensity of the current in the tube.

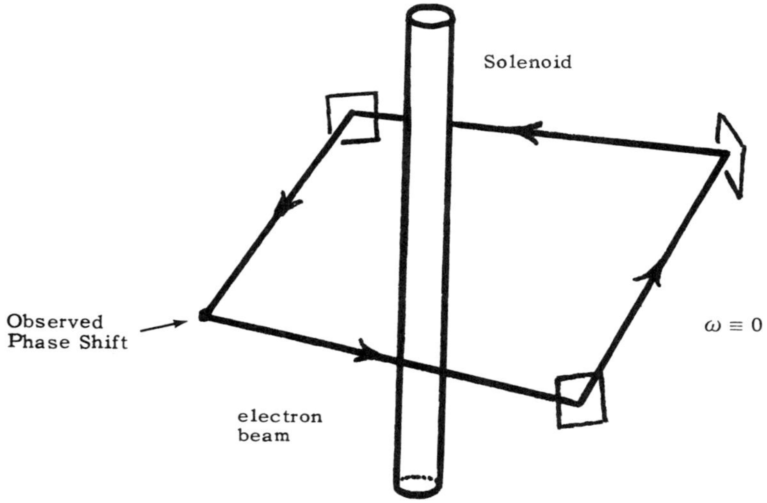

(This phase shift is simply interpreted as the holonomy transformation of the flat bundle, generated by parallel translation around the closed path.)

One might also wonder whether, over topologically non-trivial space-times, the principal bundle models of field theory should be topologically trivial. In 1930, Dirac introduced the notion of a *magnetic monopole*, an electromagnetic field with an isolated singularity in space. He observed that the integral of the field over a 2-sphere surrounding the singularity (properly normalized) could take on non-zero integer values. These integers, of course, come from the first Chern class of the underlying U_1-bundle, and their non-vanishing proves the non-triviality of this bundle. The existence of magnetic monopoles still remains conjectural. Nevertheless, the value of the principal bundle formalism in electromagnetic theory is evident.

Recently gauge theories gained considerable interest by providing a *renormalizable theory* for the coupled weak and electromagnetic interactions.

In this theory, due to S. Weinberg and A. Salam, the potential to consider takes its values in the nonabelian group U_2. (The center or "U_1-factor" is associated with the electromagnetic interaction and the "SU_2-factor" with the three directions of the so-called *intermediate bosons* which should be the exchange particles of the weak interactions.) Another application of an SU_3-gauge theory, called *quantumchromodynamics*, to strong interactions is expected to explain why the quarks, suspected to be more fundamental constituents of elementary particles, remain confined and as a result have not yet been isolated. (For a nice discussion of these ideas by a physicist, see [18].)

B) *A mathematical description of gauge theories*

Taking now a more axiomatic approach, we describe the characteristic features of gauge theories in differential geometric terms. Over space-time we consider a *vector bundle* E whose fibre variables are *acted upon by the group* G *of internal symmetries* of the interaction under study. (For example, in electromagnetism we have $G = U_1$ rotating the polarization of photons. For strong interactions we can have $G = SU_2$, SU_3, or SU_4 acting on the degrees of freedom of isotopic spin together with strangeness, or strangeness and charm, respectively.)

The shift from the principal bundle point of view developed in section A) to the vector bundle point of view that we are presenting is mathematically merely a matter of taste, but corresponds physically to having focused our attention on a specific particle (or collection of particles). Recall that *an elementary particle* subject to a certain interaction is associated with (and sometimes identified with) an irreducible representation of the internal group of symmetries of the interaction in question. The corresponding wave functions are then just sections of the bundle associated with this representation.

From a purely mathematical point of view, we may consider any vector bundle E with structure group G (a compact Lie group) over a general manifold M. A *potential* is then a G-connection on the bundle. This can be defined as a linear differential operator ∇ from sections of E to differential 1-forms with values in E with the property that, for any function f on M and any section s of E,

$$\nabla(fs) = f\nabla s + df\otimes s .$$

This operator (naturally extended) must also annihilate the tensor fields defining the G-structure.

The *field* associated with the potential is just the curvature R^∇ of the connection ∇. Many equivalent definitions of the curvature are available. For example, let d^∇ denote exterior differentiation on the space $\Omega^k(M,E)$ of E-valued exterior differential k-forms given on α in $\Omega^k(M,E)$ by the formula

$$(d^\nabla \alpha)_{x_0,\dots,x_k} = \sum_{i=0}^{k} (-1)^i \nabla_{x_i}(\alpha_{x_0,\dots,\hat{x}_i,\dots,x_k})$$

$$+ \sum_{i<j} (-1)^{i+j} \alpha_{[x_i,x_j],x_0,\dots,\hat{x}_i,\dots,\hat{x}_j,\dots,x_k} .$$

Then we have

$$R^\nabla = d^\nabla \circ d^\nabla$$

acting on 0-forms, i.e., on sections of E. From this identity it follows easily that R^∇ satisfies the *second Bianchi identity*

$$d^\nabla R^\nabla = 0 .$$

The *gauge group* $\mathcal{G}_E$ of E is just the group of automorphisms of E, that is, an element of $\mathcal{G}_E$ is a G-bundle isomorphism $g: E \to E$. Note that $\mathcal{G}_E$ is just the group of cross-sections of the bundle $G_E \subset \mathrm{Hom}(E,E)$, whose fibre at a point m is the group of G-isomorphisms of the fibre of E at m. Associated to G_E is the bundle of Lie algebras $\mathfrak{g}_E \subset \mathrm{Hom}(E,E)$ whose cross-sections will be called the *gauge algebra* of E.

Suppose that P is the principal G-bundle underlying E. Then the bundles above are isomorphic to the following associated bundles

$$G_E \cong P \times_{Ad} G ; \quad \mathfrak{g}_E \cong P \times_{ad} \mathfrak{g} .$$

Notice that the curvature R^∇ is a $\mathfrak{g}_E$-valued differential 2-form on M.

For classical physics it would be reasonable to consider our base space M to be a 4-dimensional Lorentzian manifold. However, for technical reasons, internal to current quantization methods, it is important to study fields over 4-dimensional *riemannian* manifolds. (For example, one must replace Minkowski space $\mathbf{R}^{3,1}$ by flat euclidean space $\mathbf{R}^4$.)

We suppose now that M is a riemannian manifold equipped with a G-bundle E, and we consider the space $\mathcal{C}_E$ of G-connections on E. On this space we have defined the *Yang-Mills functional* or *action*:

$$\mathcal{YM}(\nabla) \equiv \frac{1}{2} \int_M \|R^\nabla\|^2$$

where the norm $\|\cdot\|$ is defined naturally in terms of the riemannian metric on M and a G-invariant norm in $\mathfrak{g}_E$. The difference of two G-connections on E is easily seen to be a 1-form with values in $\mathfrak{g}_E$. Consequently, the space $\mathcal{C}_E$ *is simply an affine space with* $\Omega^1(M, \mathfrak{g}_E)$ *as the vector group of translations*. Furthermore, the gauge group $\mathcal{G}_E$ acts naturally on $\mathcal{C}_E$ as follows. Given g in $\mathcal{G}_E$ and ∇ in $\mathcal{C}_E$, we define the transformed connection to be

$$\nabla^g \equiv g \circ \nabla \circ g^{-1} .$$

The associated curvature has the form $R^{\nabla^g} = g \circ R^\nabla \circ g^{-1}$, from which we see that $\mathcal{YM}(\nabla) = \mathcal{YM}(\nabla^g)$, i.e., *the Yang-Mills functional is invariant under the gauge group.*

A connection ∇ in $\mathcal{C}_E$ with finite action is said to be a *Yang-Mills potential,* and R^∇ is called a *Yang-Mills field,* if ∇ represents a critical point in $\mathcal{C}_E$ among (potentials with finite action). Given a family of connections $\nabla^t = \nabla + A^t$, where A^t lies in $\Omega^1(M, \mathfrak{g}_E)$ and $A^0 = 0$ the corresponding curvature can be written as $R^{\nabla^t} = R^\nabla + d^\nabla A^t + \frac{1}{2}[A^t \wedge A^t]$. From here one sees that ∇ is a Yang-Mills potential if and only if

$$\delta^\nabla R^\nabla = 0$$

where $\delta^\nabla : \Omega^2(M, \mathfrak{g}_E) \to \Omega^1(M, \mathfrak{g}_E)$ is the adjoint of d^∇. In terms of a local orthonormal frame field $(e_1, \cdots, e_n)$ on M, these equations can be written as

$$(\delta^\nabla R^\nabla)_Y = -\sum_{j=1}^{n} (\nabla_{e_j} R^\nabla)_{e_j, Y} .$$

Combining this with the Bianchi Identity above, we see that the Yang-Mills potentials are exactly the connections with *harmonic curvature*. In fact, if the group G is U_1, or more generally, any compact abelian group, then we can appeal to the standard Hodge theory for 2-forms. In this sense, the general Yang-Mills theory can be considered a "*non-abelian Hodge Theory.*"

C) *Special aspects of dimension four*

For physical reasons the Yang-Mills functional is of particular interest over manifolds of dimension four, and the theory has certain features which are special to this dimension. One of the most basic is that in dimension four, the Yang-Mills functional is *conformally invariant*, i.e., it only depends on the conformal class of the metric on M. To see this, we recall that the norm of a 2-form ω can be given by $|\omega|^2 * 1 = \omega \wedge *\omega$ where $*$ is the Hodge star-operator. If $(e_1, \cdots, e_4)$ is an oriented orthonormal basis of T^*M, then $*1 = e_1 \wedge \cdots \wedge e_4$ and $*(e_i \wedge e_j) = e_k \wedge e_\ell$, where (i, j, k, ℓ) is an even permutation of $(1, 2, 3, 4)$. On 2-forms, this operator is clearly conformally invariant.

This invariance has the following consequence. Recall that flat euclidean space $\mathbf{R}^4$ is conformally equivalent to the standard 4-sphere

S^4 punctured at one point. Consequently, bundles with connection over $\mathbf{R}^4$ which are appropriately asymptotically trivial, can be extended to S^4. Moreover, it has been proved by Uhlenbeck [24] that any bundle with a Yang-Mills connection of finite action over $\mathbf{R}^4$ automatically extends to S^4. Hence, there is special interest in bundles over the standard sphere. Notice that the topological class of the bundle obtained over S^4 depends on the conditions of asymptotic triviality chosen for the connection on $\mathbf{R}^4$.

Over any compact oriented 4-manifold M every G-bundle E has characteristic numbers. If G is simple, there is only one, the first Pontrjagin number $p_1(E)$ (sometimes referred to by physicists as the "*topological quantum number*" or "*instanton number*"). We *denote* its absolute value by k. Given a connection ∇ on E, this number can be expressed, via Chern-Weil Theory, in terms of the curvature by the formula

$$4\pi^2 p_1(E) = \int_M \langle R^\nabla \wedge R^\nabla \rangle$$

where $\langle \cdot, \cdot \rangle$ stands for the inner product in $\mathfrak{g}_E$.

On an oriented 4-manifold the 2-forms decompose as $\Lambda^2 = \Lambda^+ \oplus \Lambda^-$ where $\Lambda^\pm$ are the ± 1 eigenspaces under the involution $*$. This gives a decomposition $R^\nabla = R_+^\nabla + R_-^\nabla$ where $*R_\pm^\nabla = \pm R_\pm^\nabla$, and one can easily see that:

$$4\pi^2 p_1(E) = \int_M \{\|R_+^\nabla\|^2 - \|R_-^\nabla\|^2\}$$

and

$$\mathcal{YM}(\nabla) = \int_M \{\|R_+^\nabla\|^2 + \|R_-^\nabla\|^2\}.$$

It follows immediately that

$$4\pi^2 k \leqq \mathcal{YM}(\nabla)$$

for any connection ∇ on E. Furthermore, this lower bound for $\mathcal{YM}(\nabla)$ is achieved if and only if $R_\varepsilon^\nabla \equiv 0$ where $\varepsilon k = -p_1(E)$. Fields for which R_-^∇ (resp. R_+^∇) vanishes are called *self-dual* (resp. *anti-self-dual*).

When $G = SU_2$, the first explicit examples of these special fields over S^4 (called *instantons*) were given in [8]. Later G. 't Hooft gave in [17] solutions depending on $5k$ parameters, which can be thought of as the positions and sizes of k *pseudoparticles*. M. Atiyah, N. Hitchin, and I. M. Singer proved (cf. [4]) that in fact the space of instanton connections is $(8k-3)$-dimensional. An almost complete description of these fields is now available (cf. [13]) and relies on algebro-geometric techniques. The basic idea, due to R. S. Ward (cf. [7]) is to use Penrose's approach (cf. [22]) to convert the field equations into complex analytic geometry on complex projective 3-space $\mathbb{C}P^3$, which is viewed as the total space of a bundle over S^4 with fibres $\mathbb{C}P^1$. By pulling back to $\mathbb{C}P^3$ both the bundle and the self-dual connection, one automatically gets an analytic (hence by J. P. Serre's theorem algebraic) bundle with connection over $\mathbb{C}P^3$. It is a piece of good fortune that the bundles arising in this way turn out to be precisely of the restricted type algebraic geometers have recently been able to classify, namely they are algebraically trivial along a 4-parameter family of lines in $\mathbb{C}P^3$ and have a symplectic structure. Hence they belong to the family of *stable* bundles (cf. [15]). The final construction may be described by using complexes of special analytic sheaves on $\mathbb{C}P^3$, the so-called *monads*. (See [13]; people interested in more explicit formulas should consult [11].) The construction carries over for any simple group G.

If the local nature of the instanton space is now clear, its global structure is still largely unknown (compare [6]): for example, it is only for $k = 1, 2$ that it is known to be connected.

Since the space of G-connections is an affine space (and hence, contractible), and since the Yang-Mills functional is invariant under the gauge group $\mathcal{G}$, this functional gives rise to a function on the classifying space

$B\mathcal{G}$ for $\mathcal{G}$. It has been shown in [6] when $M = S^4$ and $G = SU_2$, the space of (equivalence classes of) self-dual connections carries a large "initial part" of the topology of $B\mathcal{G}$.

D) *Some mathematical problems and results*

One of the outstanding mathematical problems in the theory at the moment is the following:

PROBLEM I. *For* SU_2-, SU_3- *and* U_2-*bundles over* S^4, *determine whether critical points of* $\mathcal{YM}$ *other than absolute minima, exist. In particular, determine whether there exist Yang-Mills fields which are not self-dual or anti-self-dual.*

The tangent bundle of S^4 with its canonical riemannian connection is a Yang-Mills field with group SO_4. It is not self-dual, however it does minimize the functional. This can be understood as follows. The group SO_4 is not simple and the Euler number of the bundle E appears as a new constraint on the curvature via the Chern-Gauss-Bonnet formula. (See Part Three of this paper.) In particular for the tangent bundle of S^4, this new constraint supercedes the Pontryagin constraint to give a topological lower bound on $\mathcal{YM}$. Of course the Euler class is an unstable invariant. Thus, suppose one enlarges the group of the tangent frame bundle P of S^4 by setting $\tilde{P} = P \times_\rho G$ for some non-trivial representation $\rho : SO_4 \to G$ where G is large (of rank $\geqq 3$, say). Then the topological lower bound to $\mathcal{YM}$ given by the Euler class disappears. Furthermore, the canonical connection on P has a natural extension to a Yang-Mills connection on $\tilde{P}$. It is a very nice observation of M. Itoh [19], that (when $\operatorname{rank}(G) \geqq 3$) *this connection on* $\tilde{P}$ *is an unstable critical point*, i.e., there is a smooth variation of this connection which decreases the functional.

Progress on Problem I has recently been made by C. Taubes (cf. [23]) who proved self-duality for Yang-Mills fields with an axial-symmetry condition.

The authors have obtained the desired conclusions under a stability assumption. Namely, we say that a Yang-Mills connection ∇ is *weakly stable* if the second variation of $\mathcal{YM}$ at ∇ is non-negative. (A local minimal is always weakly stable.)

STABILITY THEOREM I (cf. [10]). *Any weakly stable Yang-Mills field over* S^4 *with group* SU_2, SU_3 *or* U_2 *is self-dual or anti-self-dual.*

An analogous result holds for bundles with group SO_4. Here the conclusion is of *two-fold self-duality.* (See Part Three.)

On the sphere S^n, $n \geqq 5$, there are no weakly stable Yang-Mills fields. This result was proved by J. Simons and formed the starting point of our work on the theorem above.

The proof of the Stability Theorem I will be outlined in Part Two.

Another important general problem is the following:

PROBLEM II. *Determine the structure of Yang-Mills fields on general compact riemannian 4-manifolds, and in particular, on homogeneous ones.*

In Part Three we give a detailed discussion of minimizing fields on homogeneous spaces and prove the following result.

STABILITY THEOREM II. *Let* X *be a compact oriented homogeneous riemannian 4-manifold. Then any weakly stable Yang-Mills field on* X *with group* SU_2, SU_3, U_2 *or* SO_4 *(or abelian) is absolutely minimizing. (See the statement in Part Three for more details.)*

It is a nice observation due to C. H. Gu that non-stable fields exist on the homogeneous space $S^1 \times S^3$ (the tangent bundle is trivial but the standard Levi-Civita product connection which is riemannian symmetric and hence Yang-Mills, is not flat).

Also in connection with problem II, the authors have proved that over sufficiently positively curved riemannian manifolds, the absolute minima of $\mathcal{YM}$ are isolated from other critical points. As a particular case we have

ISOLATION THEOREM ($n \geqq 5$). *Any Yang-Mills connection* ∇ *over the standard* n*-sphere* S^n, $n \geqq 5$, *such that* $\|R^\nabla\|^2 < \frac{1}{2}\binom{n}{2}$ *pointwise, is trivial (i.e.,* $R^\nabla \equiv 0$ *).*

Note: For this explicit estimate we take the norm on $\mathfrak{g}_E$ to be $\|A\|^2 = \frac{1}{2}\,\mathrm{trace}(A^t \circ A)$.

ISOLATION THEOREM ($n = 4$). *Let* R^∇ *be a Yang-Mills connection on a bundle* E *over* S^4 *which satisfies the pointwise condition* $\|R^\nabla\|^2 \leqq 3$. *Then either* E *is flat or* $E = E_0 \oplus S$ *where* E_0 *is flat and* S *is one of the (two) 4-dimensional bundles of tangent spinors with the canonical riemannian connection.*

Furthermore, if R^∇ *satisfies the pointwise condition* $\|R_+^\nabla\|^2 < 3$ *(or* $\|R_-^\nabla\|^2 < 3$ *), then* $R_+^\nabla \equiv 0$ *(resp.* $R_-^\nabla \equiv 0$ *).*

ISOLATION THEOREM ($n = 3$). *Let* R^∇ *be a Yang-Mills field on a bundle* E *over* S^3 *which satisfies the pointwise condition* $\|R^\nabla\|^2 \leqq \frac{3}{2}$. *Then either* E *is flat or* $E = E_0 \oplus S$ *where* E_0 *is flat and* S *is the 4-dimensional tangent spin bundle with the canonical riemannian connection.*

Note that neither the topological type nor the structure group of the bundle enters the statements of the isolation theorems. For proofs of these results and some refinements of them, the reader is referred to [10].

E) *Some comments on the tangent bundle*

Among bundles over a 4-dimensional manifold M, the tangent bundle plays a special role. It is the setting for gravitational theory. It has the mathematical property that the diffeomorphism group of M acts naturally on it, and it is this group which plays the role in gravitational theory that the gauge group plays in ordinary gauge theory.

Of course one can do standard gauge theory on the tangent (SO_4-) bundle. In this case, its special features make the theory interesting. For example, on TM one can consider the special class of *torsion-free* or *symmetric* connections, whose interplay with the Yang-Mills equations is still largely unknown. (See [9], however, for a result in this direction.)

An interesting fact is that the Levi-Civita connection of an Einstein metric on M induces a self-dual connection on the bundle Λ^+M of $*$-invariant 2-forms, and an anti-self-dual connection on the $*$-anti-invariant 2-forms Λ^-M (cf. [5]), whence the name *gravitational instanton* sometimes given to such structures. The total space of the unit sphere bundle in Λ^-M (or Λ^+M) carries a natural complex structure if it is *half-conformally flat*. (See [5] again.) (This is an important step in Penrose's program.) A large class of half-conformally flat Einstein manifolds is provided by S. T. Yau's solution of Calabi's conjecture [27], which guarantees the existence of Ricci-flat Kähler metrics on K3 surfaces. An explicit construction of such metrics is still unknown. Quite recently some physicists have been interested in such manifolds since they provide renormalizable supersymmetric models (cf. [14]). These manifolds are on the other hand the only non-locally symmetric examples of half conformally flat non-conformally flat spaces. Notice that in the case of positive scalar curvature N. Hitchin has recently proved in [16] that an Einstein half conformally flat manifold is indeed S^4 or $\mathbb{C}P^2$ with its standard metric.

Part Two: An overview of the stability theorems

In this part we explain the structure of the proof of the 4-dimensional stability theorems. We shall emphasize the ideas involved rather than the computations. (The only technical details presented in this part are the ones we need in Part Three.)

The proof decomposes naturally into two stages. In the first stage we work with an arbitrary compact Lie group G and an arbitrary compact homogeneous riemannian orientable 4-manifold (later called a CHROM). In the second stage we shall restrict the size of the group and the nature of the base.

A) *Picking appropriate variations of the connection*

We shall use the stability assumption via the *second variation formula*: if A is an infinitesimal variation of a Yang-Mills connection ∇ on the

G-bundle $\pi : E \to M$, then

$$\text{(2.1)} \quad \frac{d^2}{dt^2} \mathcal{YM}(\nabla + tA)\Big|_{t=0} = (\mathcal{S}^\nabla A, A) = \int_M \Big(\delta^\nabla d^\nabla A + 2\sum_{i=1}^n [R_{e_i,\cdot}, A_{e_i}], A\Big) .$$

The first step in the proof is then to construct A adapted to the geometry of the situation. Such variations will be suggested to us by the *enlarged gauge group* $\tilde{\mathcal{G}}$, i.e., the group of diffeomorphisms of E which preserve the G-structure and which cover an isometry of M in general dimensions or a conformal transformation if M is 4-dimensional. These diffeomorphisms preserve the Yang-Mills functional as one easily sees.

Elements of the Lie algebra $\tilde{\mathcal{G}}$ can be described as follows. Let $\tilde{Y}$ be the Δ-horizontal lift in E of a Killing field Y on M (or a conformal vector field Y if M is 4-dimensional), then $\tilde{Y}$ belongs to $\tilde{\mathcal{G}}$. Moreover the infinitesimal variation of ∇ arising from $\tilde{Y}$ for the action of $\tilde{\mathcal{G}}$ on $\mathcal{C}_E$ is nothing but $i_Y R^\nabla$, the contraction of the curvature with the vector field Y. It then follows directly from the invariance of $\mathcal{YM}$ that $\mathcal{S}^\nabla(i_Y R^\nabla) = 0$. It should be no surprise that when one writes explicitly what such equations say about ∇, one gets a pure tautology.

However if, guided by the preceding calculations, one takes a conformal vector field Y on S^n and considers variations $i_Y R^\nabla$, then one has the identity

$$\text{(2.2)} \qquad \mathcal{S}^\nabla(i_Y R^\nabla) = (4-n)\, i_Y R^\nabla .$$

This leads to J. Simons' theorem (cf. [10]) about the non-existence of stable Yang-Mills fields on S^n for $n \geq 5$.

If Y is a Killing field on a CHROM X, one can consider the variations $i_Y R_+^\nabla$ for which one can prove the identity

$$\text{(2.3)} \qquad \mathcal{S}^\nabla(i_Y R_+^\nabla). = \sum_{i=1}^4 \{[R^\nabla_{+e_i,Y}, R^\nabla_{-e_i,\cdot}] + [R^\nabla_{+e_i,\cdot}, R^\nabla_{-e_i,Y}]\} .$$

On the right-hand side notice the symmetry between Y and "·".

To establish formulas (2.2) and (2.3) it is convenient to use Bochner-Weitzenböck formulas expressing the Hodge-de Rham Laplacian on 1- and 2-forms in terms of the operator $\nabla^*\nabla$. This operator has in particular the advantage of being defined on general tensor fields and not merely on exterior differential forms.

B) *Using an averaging procedure*

One important consequence of formulas (2.2) and (2.3) is that the second variation of $\mathcal{YM}$ is an algebraic expression in the curvature for our special variations. Moreover, because of the symmetry pointed out in (2.3), the average of these second variations over (the sphere in) the Lie algebra $\mathfrak{I}$ of Killing fields of a CHROM X vanishes. To see this, one evaluates it as a trace using an appropriately chosen basis of $\mathfrak{I}$ for each point of X.

If the connection ∇ is weakly stable, this forces the second variation for each $i_Y R_+^\nabla$ (Y in $\mathfrak{I}$) to vanish. It follows then that the $i_Y R_+^\nabla$'s lie in the kernel of $\mathcal{S}^\nabla$.

C) *Introducing some elementary algebraic lemmas*

The above discussion establishes the following identity which holds for all tangent vectors V and W

$$\sum_{i=1}^{4} [R^\nabla_{+\,e_i,V}, R^\nabla_{-\,e_i,W}] = 0 . \tag{2.4}$$

We point out that the term above is automatically symmetric in V and W (this is a consequence of the following purely algebraic fact: the tensor product of the SO_4-modules $\Lambda^+\mathbf{R}^4$ and $\Lambda^-\mathbf{R}^4$ is isomorphic to the SO_4-module $S_0^2\mathbf{R}^4$ of traceless symmetric 2-tensors).

It is another elementary algebraic fact that the identity (2.4) is equivalent to the following one: for all tangent vectors V, W, Y and Z

$$[R^\nabla_{+\,V,W}, R^\nabla_{-\,Y,Z}] = 0 . \tag{2.5}$$

It will be better to free our discussion from the vectors V, W, Y and Z. For that purpose we introduce the algebras $\mathfrak{a}_{+m}$ (resp. $\mathfrak{a}_{-m}$) generated in $\mathfrak{g}_{E,m}$ by the transformations $R^{\nabla}_{+\,V,W}$ (resp. $R^{\nabla}_{-\,V,W}$) for all tangent vectors V, W at m. We set $\mathfrak{a}_{\pm} = \bigcup_{m \in M} \mathfrak{a}_{\pm m}$.

D) *The basic stability theorem*

We can restate (2.5) as our

BASIC STABILITY THEOREM. *Any weakly stable Yang-Mills field over a CHROM* X *has the property*

$$[\mathfrak{a}_+, \mathfrak{a}_-] \equiv 0$$

on X.

E) *The* SU_2*-stability theorem over* S^4

We come now to the second stage of the proof of the theorem which requires special assumptions on the group or on the base manifold.

We consider here the case $G = SU_2$ over the 4-sphere.

If $\mathfrak{g} = SU_2$, the centralizer of every non-trivial element is reduced to the line generated by this element. Then at each point either $\mathfrak{a}^+$ or $\mathfrak{a}^-$ is reduced to 0 or they are equal and 1-dimensional. The last case can be ruled out by coming back to the Bochner-Weitzenböck formula of R^{∇}. Consequently one of the two subalgebras is reduced to 0 on an open set and hence on all of X since the harmonic field $R^{\nabla}_{\pm}$ behaves like an analytic field (by the Aronszajn theorem, cf. [2]).

In Part Three we shall deal with the new complications arising from the larger groups U_2, SU_3 and SO_4 and from the more complicated topology of general CHROM's.

Part Three: Stability results for homogeneous 4-manifolds

In this part we shall conduct a detailed analysis of weakly stable fields over certain 4-dimensional manifolds. This will lead to a Specialized Stability Theorem over any 4-dimensional compact homogeneous riemannian manifold (or CHROM) X. This class of manifolds includes S^4, CP^2, $S^2 \times S^2$, $S^1 \times (S^3/\Gamma)$ (with any left-invariant metric on S^3), $S^2 \times T^2$ and T^4 (where T^2 and T^4 can be arbitrary flat tori). When we want to insist that a particular statement is true for general 4-manifolds, we shall denote the manifold by M instead of X.

A) *Yang-Mills fields of least action*

In this section we discuss the geometry associated with certain special groups G. To begin we recall what happens when G is abelian. In this case R is an ordinary closed 2-form. Any other connection $\nabla' = \nabla + A$ on the same bundle, has curvature $R' = R + dA$. Conversely, any 2-form which differs from R by an exact form, is the curvature 2-form of a connection on this bundle. The given connection is Yang-Mills if and only if R is harmonic in the sense of ordinary Hodge theory. We see that $\mathcal{YM}(\nabla') = \int_M \|R'\|^2 = \int_M (\|R\|^2 + \|dA\|^2) \geqq \mathcal{YM}(\nabla)$, since $\int_M \langle R, dA\rangle = \int_M \langle \delta R, A\rangle = 0$. Hence, *in the abelian case, a Yang-Mills field represents the unique minimum of the functional and every bundle carries such a field.*

It is interesting to note that a compact homogeneous 4-manifold carrying non-trivial harmonic 2-forms turns out to be symmetric, hence on it every harmonic 2-form, i.e., every abelian Yang-Mills field, is parallel. (To see this, check case-by-case.)

It is useful to understand when a given field reduces to an abelian one. A valuable criterion is provided by the following.

PROPOSITION 3.1. (See [10, 3.15].) *Suppose ϕ in $\Omega^2(M, \mathfrak{g}_E)$ is harmonic and takes its values in a 1-dimensional sub-bundle of $\mathfrak{g}_E$. If moreover $[R, {}^\nabla\phi] = 0$, then $\phi = \phi_0 \otimes e$ where ϕ_0 is a harmonic scalar-valued 2-form and where e is a parallel section of $\mathfrak{g}_E$.*

COROLLARY 3.2. *Let R be a Yang-Mills field with group G such that at each point m of M the dimension of the space $\{R_{V,W} : V, W \in T_m M\}$ is $\leqq 1$. Then R reduces to an abelian field, i.e., there exists a principal U_1-bundle P_{U_1} with connection and a homomorphism $\rho : U_1 \hookrightarrow G$ so* *that the naturally constructed principal* G*-bundle with connection,* $P_G = P_{U_1} \times_\rho G$, *is equivalent to the given one.*

For G abelian, the minimizing Yang-Mills field is unique as we saw. It is self-dual or anti-self-dual if and only if the cohomology class it represents is self-dual or anti-self-dual. (It is so on $\mathbb{C}P^2$ and for the diagonal or antidiagonal classes on $S^2 \times S^2$, but not so for the other classes on $S^2 \times S^2$, for example.)

When G is simple, the Pontryagin constraint $4\pi^2 k = 4\pi^2 |p_1(E)|$ gives a topological lower bound for $\mathcal{YM}$ and only the absolute minima are self-dual or anti-self-dual as we saw in Part One.

B) *The* U_N *set-up*

The unitary group U_N represents an interesting mixture of these two cases.

Let E be a complex hermitian N-plane bundle over M with a unitary connection, and let $\mathfrak{u}_E$ be the associated bundle of skew hermitian endomorphisms of E. Then there is a natural splitting

$$\mathfrak{u}_E = \mathfrak{c}_E \oplus \mathfrak{su}_E \tag{3.3}$$

where $\mathfrak{c}_E$ denotes the center of $\mathfrak{u}_E$ at each point. The curvature transformation of $\mathfrak{u}_E$ (with its induced connection) is given by $R^{\mathfrak{u}_E}(\sigma) = [R^E, \sigma]$ for σ in $\mathfrak{u}_E$. It follows immediately that *the bundle* $\mathfrak{c}_E$ *is flat.*

With respect to the decomposition (3.3) we can write the curvature of E as

$$R^E = R^0 + R^1 , \tag{3.4}$$

for

$$R^0 = c \otimes \tau \tag{3.5}$$

where c is a real-valued 2-form and where $\tau : E \to E$ denotes scalar multiplication by $\sqrt{-1}$. The form c is closed, and the de Rham cohomology class of $(N/2\pi)c$ in $H^2(M, \mathbb{R})$ is the (real) *first Chern class* $c_1(E)$ *of* E.

A straightforward calculation now shows that R^E is harmonic (i.e., Yang-Mills) if and only if both R^0 and R^1 are harmonic. Of course, R^0 is harmonic if and only if c is harmonic. Since the splitting (3.3) is orthogonal, the Yang-Mills density can be written as

$$\begin{aligned}\|R^E\|^2 &= \|R^0\|^2 + \|R^1\|^2 \\ &= N\|c\|^2 + \|R^1\|^2 .\end{aligned}$$

As we showed in the abelian case, the integral of $\|R^0\|^2$ is minimized when R^0 is harmonic. Furthermore, from Part One we see that

$$\begin{aligned}\int_M \langle R^1 \wedge R^1 \rangle &= \int_M (\|R^1_+\|^2 - \|R^1_-\|^2) \\ &= 4\pi^2 p_1(E) - \int_X R^0 \wedge R^0 \\ &= 4\pi^2 \left[p_1(E) - \frac{1}{N}\, c_1^2(E)\right] .\end{aligned}$$

The last term is independent of the connection. From this and our observation above we conclude the following.

PROPOSITION 3.6. *Any unitary Yang-Mills field* R *on* M *whose* SU_N*-piece* R^1 *is self-dual or anti-self-dual, gives an absolute minimum of the Yang-Mills functional.*

It is interesting to note that the components of R in the decomposition (3.4) can be changed independently while preserving harmonicity. To see this we recall that given connections ∇ and ∇' on vector bundles E and E′ over M, there are canonically induced connections $\nabla \oplus \nabla'$ on $E \oplus E'$ and $\nabla \otimes \nabla' \equiv \nabla \otimes 1 + 1 \otimes \nabla'$ on $E \otimes E'$. (See [20] or [21].) If E and E′ are hermitian and if ∇ and ∇' are unitary, the tensor product can be taken over the complex numbers. A straightforward calculation proves the following useful and interesting fact.

PROPOSITION 3.7. *The direct sum and the (real or complex) tensor product of Yang-Mills connections are again Yang-Mills connections.*

Suppose now that E is a complex N-plane bundle with unitary Yang-Mills connection, and let Λ be any complex line bundle. We can equip Λ with a Yang-Mills connection and thereby construct a Yang-Mills connection on $E \otimes_{\mathbb{C}} \Lambda$. Note that $c_1(E \otimes_{\mathbb{C}} \Lambda) = c_1(E) + Nc_1(\Lambda)$. Thus, if $c_1(E)$ is N times an integral class, we can choose Λ so that $c_1(E \otimes \Lambda) = 0$. The corresponding Yang-Mills field will have $R^0 = 0$. If M is simply connected, this means that the connection has been reduced to an SU_N-connection.

C) *The* SO_4 *set-up*

We now confine our attention to the case where E is a real 4-dimensional bundle and $G = SO_4$. The curvature of an SO_4-connection on E can be viewed as a bundle map $R : \Lambda^2 TM \to \Lambda^2 E$. There are decompositions: $\Lambda^2 TM = \Lambda^+ TM \oplus \Lambda^- TM$ and $\Lambda^2 E = \Lambda^+ E \oplus \Lambda^- E$ into $+1$ and -1 eigenbundles under the respective Hodge operators. This leads to a decomposition

$$R^\nabla = R_+^+ + R_+^- + R_+^- + R_-^- \tag{3.8}$$

where $R_\pm^\nabla = R_\pm^+ + R_\pm^-$ are the components appearing in Part One. The bundle E has characteristic invariants, the Pontrjagin number $p_1(E)$ discussed above and the Euler number $\chi(E)$. These can be expressed in terms of the curvature R^∇ of any SO_4-connection Δ as follows

$$\tag{3.9} \begin{cases} 4\pi^2 p_1(E) = \displaystyle\int_M (\|R_+^+\|^2 + \|R_+^-\|^2 - \|R_-^+\|^2 - \|R_-^-\|^2) \\ \\ 8\pi^2 \chi(E) = \displaystyle\int_M (\|R_+^+\|^2 - \|R_+^-\|^2 - \|R_-^+\|^2 + \|R_-^-\|^2) . \end{cases}$$

Comparison with the Yang-Mills functional proves the following.

PROPOSITION 3.10. *Let* E *be a 4-dimensional oriented vector bundle with an* SO_4*-connection* ∇ *over* M *and set* $p = 4\pi^2 p_1(E)$ *and* $\chi = 8\pi^2\chi(E)$. *Then* ∇ *minimizes the Yang-Mills functional on* E *if the following condition holds*:

$$(3.11)\qquad \begin{cases} R_-^+ = R_-^- = 0 & \text{if} \quad p \geqq |\chi| \\ R_+^+ = R_+^- = 0 & \text{if} \quad -p \geqq |\chi| \\ R_-^+ = R_+^- = 0 & \text{if} \quad \chi \geqq |p| \\ R_+^+ = R_-^- = 0 & \text{if} \quad -\chi \geqq |p| \; . \end{cases}$$

An SO_4-connection whose curvature satisfies (3.11) will be called *two-fold self-dual.* Such connections exist in each equivalence class of SO_4-bundles over S^4 (cf. [10]). Furthermore, the riemannian connection on TM for any 4-dimensional Einstein manifold M is two-fold self-dual, since the Einstein condition is equivalent to $R_-^+ = R_+^- = 0$. (For further discussion of this case, see [9].)

D) *The specialized stability theorems*

We are now in a position to state and prove our second main result.

SPECIALIZED STABILITY THEOREM. *Let* R *be a weakly stable Yang-Mills field with group* G *over a CHROM* X *of dimension* 4.

(i) *If* $G = SU_2$ *or* SO_3, *then either* R *is self-dual or anti-self-dual or it reduces to an abelian field.*

(ii) *If* $G = U_2$, *then the* R^1*-component of* R *in the decomposition (3.4) is either self-dual or anti-self-dual, or* R *reduces to an abelian field.*

(iii) *If* $G = SO_4$, *then either* R *is two-fold self-dual or it reduces to a* U_2*-field.*

(iv) *If* $G = SU_3$, *then either* R *is self-dual or anti-self-dual or it reduces to a* U_2*-field.*

In particular, in each case above the field minimizes the Yang-Mills functional.

Note 1. We emphasize that when we say a field "reduces," we mean that the given connection reduces, i.e., the given connection is canonically induced from a connection on some topological reduction of the principal bundle. This reduced connection is, of course, itself a weakly stable Yang-Mills connection.

Note 2. The restriction of G to certain low dimensional Lie groups is essential for the conclusions of the Specialized Stability Theorem to hold. This follows from M. Itoh's observation [19] discussed in Part One, D).

Proof. From the Aronszajn Theorem [2] on unique continuation of solutions to elliptic systems, it suffices to establish our conclusions on a (non-empty) open subset of X. Many of the following statements are initially valid only on such a subset, but for simplicity we shall not make constant reference to this fact.

The important fact here is the Basic Stability Theorem which guarantees that $[\mathfrak{a}_+, \mathfrak{a}_-] \equiv 0$ on X.

i) Let $\mathfrak{g}$ be the Lie algebra of G and suppose $\mathfrak{g} \cong \mathfrak{su}_2$. It is an elementary fact that for any non-zero V in $\mathfrak{su}_2$, the only elements which commute with V are multiples of V. It therefore follows from Part One that either $\mathfrak{a}_+ \equiv 0$ or $\mathfrak{a}_- \equiv 0$ or $\mathfrak{a} = \mathfrak{a}_+ + \mathfrak{a}_-$ is abelian on X. In the third case, the proof is completed by applying Proposition 3.1.

ii) Suppose now that $\mathfrak{g} \cong \mathfrak{u}_2$. From the decompositions (3.3) and (3.4) we obtain decompositions $\mathfrak{a}_\pm = \mathfrak{a}_\pm^0 \oplus \mathfrak{a}_\pm^1$ (where $\mathfrak{a}_\pm^1$ are the subalgebras of $\mathfrak{su}_E$ generated by the transformations $(R_\pm^1)_{V,W}$ at each point). Arguments of the paragraph above now show that either $\mathfrak{a}_+^1 \equiv 0$ or $\mathfrak{a}_-^1 \equiv 0$ or $\dim(\mathfrak{a}_+^1 + \mathfrak{a}_-^1) \leqq 1$. In the third case the field reduces to an abelian one.

iii) Suppose that $\mathfrak{g} \cong \mathfrak{so}_4$ and let $\mathfrak{g}_E = \mathfrak{g}_E^+ \oplus \mathfrak{g}_E^-$ be the decomposition corresponding to the splitting $\mathfrak{so}_4 \equiv \mathfrak{su}_2 \oplus \mathfrak{su}_2$. This gives a corresponding splitting $\mathfrak{a}_\pm = \mathfrak{a}_\pm^+ \oplus \mathfrak{a}_\pm^-$, where $\mathfrak{a}_\pm^\pm$ are generated by the transformations $R^{\pm}_{\pm}$ from the curvature decomposition (3.8). The condition $[\mathfrak{a}_+, \mathfrak{a}_-] = 0$ implies that $[\mathfrak{a}^+_+, \mathfrak{a}^+_-] = [\mathfrak{a}^-_+, \mathfrak{a}^-_-] = 0$. Since the bundle $\mathfrak{g}^+_E$ has fibre $\mathfrak{su}_2$, the above arguments show that either $\mathfrak{a}^+_+ \equiv 0$ or $\mathfrak{a}^-_- \equiv 0$ or, by Proposition 3.1, there is a non-zero parallel cross-section τ of $\mathfrak{g}^+_E$. The analogous statement holds for $\mathfrak{g}^-_E$. We conclude that the field is two-fold self-dual or that there exists a parallel complex structure τ on E. (This structure may or may not be compatible with the given orientation on E.) In the latter case we have a reduction to a U_2-field.

iv) Suppose finally that $\mathfrak{g} \cong \mathfrak{su}_3$, and suppose that $\mathfrak{a}^+ \neq 0$ and $\mathfrak{a}^- \neq 0$. In this case, either $\mathfrak{a}^+$ or $\mathfrak{a}^-$ must be abelian everywhere on X. It then follows from Proposition 3.1 that $\mathfrak{g}_E$ admits a non-zero parallel cross-section, i.e., a parallel, skew hermitian bundle map $L : E \to E$ with $\operatorname{trace}(L) \equiv 0$. Then E decomposes into at least 2 distinct eigenbundles for L, and since L is parallel, this splitting is preserved by the connection. This gives the desired reduction to a U_2-field. This completes the proof. ■

REFERENCES

[1] Y. Aharonov, D. Bohm, Phys, Rev. 123(1961), 1511.

[2] N. Aronszajn, A unique continuation theorem for solutions of elliptic partial differential equations or inequalities of second order, J. Math. Pure et Appl. 35(1957), 235-249.

[3] M. F. Atiyah, Lecture given at the Int. Congress of Mathematicians, Helsinki (1978).

[4] M. F. Atiyah, N. Hitchin, I. M. Singer, Deformations of instantons, Proc. Nat. Acad. Sci. USA 74(1977), 2662-2663.

[5] ———, Self-duality in four-dimensional riemannian geometry, Proc. R. Soc. London, A 362(1978), 425-461.

[6] M. F. Atiyah, J. D. Jones, Topological aspects of Yang-Mills theory, Comm. Math. Phys. 61(1978), 97-118.

[7] M. F. Atiyah, R. S. Ward, Instantons and algebraic geometry, Comm. Math. Phys. 55(1977), 97-118.

[8] A. Belavin, A. Polyakov, A. Schwarz, Y. Tyupkin, Pseudoparticle solutions of the Yang-Mills equations, Phys. Lett. 59B(1975), 85-87.

[9] J. P. Bourguignon, Les variétés de dimension 4 à signature non nulle dont la courbure est harmonique sont d'Einstein, Preprint IAS, Princeton.

[10] J. P. Bourguignon, H. B. Lawson, Jr., Stability and isolation phenomena for Yang-Mills fields, to appear in Comm. Math. Phys.

[11] E. F. Corrigan, D. B. Fairlie, R. G. Yates, P. Goddard, The construction of self-dual solutions to SU_2-gauge theory, Comm. Math. Phys. 58(1978), 223-240.

[12] A. Douady, J. L. Verdier, Les équations de Yang-Mills, Séminaire E.N.S. 1977-1978, Astérisque No. 71-72, (1980).

[13] V. G. Drinfeld, Y. I. Manin, A description of instantons, Comm. Math. Phys. 63(1978), 177-192.

[14] D. Z. Friedmann, L. A. Gaume, Kähler geometry and the renormalization of supersymmetric σ-models, Preprint, ITP SUNY at Stony Brook.

[15] R. Hartshorne, Stable vector bundles and instantons, Comm. Math. Phys. 59(1978), 1-15.

[16] N. Hitchin, Kählerian twistor spaces, Preprint, Oxford.

[17] G. 'T Hooft, Phys, Rev. Letters 37(1977), 8-11.

[18] J. Iliopoulos, Unified theories of elementary particle interactions, Contemp. Phys. 21 (1980), 159-183.

[19] M. Itoh, Invariant connections and self-duality condition for Yang-Mills solutions, Preprint, Tsukuba University.

[20] H. B. Lawson, Jr., M. L. Michelson, Clifford bundles, immersions of manifolds and the vector field problem, to appear in J. of Differential Geometry.

[21] R. S. Palais, Seminar on the Atiyah-Singer index theorem, Annals of Math. Studies No. 57(1965), Princeton University Press.

[22] R. Penrose, The twistor programme, Rep. on Math. Phys. 12(1977), 65-76.

[23] C. H. Taubes, On the equivalence of the first and second order equations for gauge theories, Preprint, Lyman Lab. of Phys., Harvard.

[24] K. Uhlenbeck, Removable singularities in Yang-Mills fields, Preprint.

[25] T. T. Wu, C. N. Yang, Concept of nonintegrable phase factors and global formulation of gauge fields, Phys. Rev. D12(1975), 3845-3857.

[26] C. N. Yang, R. Mills, Phys. Rev. 96(1954), 191.

[27] S. T. Yau, On the Ricci curvature of a compact Kähler manifold and the complex Monge-Ampère equation I, Comm. Pure and Appl. Math. XXXI (1978), 339-411.

SYMMETRY AND ISOLATED SINGULARITIES OF CONFORMALLY FLAT METRICS AND OF SOLUTIONS OF THE YANG-MILLS EQUATIONS

Basilis Gidas

I. *Introduction and results*

The problems we study here have been motivated from a study of the particular equation

$$\Delta u + u^{\frac{n+2}{n-2}} = 0 \quad \text{in} \quad \mathbf{R}^n, \ n > 2 . \tag{1.1}$$

Here $u(x)$ is a non-negative function. As is well known, equation (1.1) has a Differential Geometric origin: Let g_{ij} be a conformally flat metric on $\mathbf{R}^n$, i.e., for some positive function $u(x)$,

$$g_{ij} = u(x)^{\frac{4}{n-2}} \delta_{ij} . \tag{1.2}$$

Then g_{ij} has scalar curvature $K(x)$ iff

$$\Delta u + \frac{n-2}{4(n-1)} K(x) u^{\frac{n+2}{n-2}} = 0 . \tag{1.3}$$

If K is constant and positive, then (1.3) reduces to (1.1) by an appropriate stretching of the coordinates. For $n = 4$, solutions of (1.1) give rise to solutions of the Euclidean Yang-Mills equations via 't Hooft's ansatz [1, 2].

Seminar on Differential Geometry
0-691-08268-5/82/000423-19 $00.95/0 (cloth)
0-691-08296-0/82/000423-19 $00.95/0 (paperback)

Another particular equation to which our results apply is

$$\Delta u + u^{\alpha} = 0 \quad \text{in} \quad \mathbf{R}^n, \ 1 < \alpha < \frac{n+2}{n-2}, \ n > 2 . \tag{1.4}$$

For $n = 3$ and $\alpha < 5$, this equation arises in Astrophysics. Our consideration of this equation has been motivated by a study of existence theorems for (1.1). The non-linearity in (1.1) is critical for the Sobolev inequality, and no Palais-Smale condition exists. Thus one would consider studying the problem with $\alpha < \frac{n+2}{n-2}$, and then letting $\alpha \to \frac{n+2}{n-2}$.

In order to motivate our general theorems below, we analyze the spherically symmetric solutions of (1.1) and (1.4) [3, 4]. This is conveniently achieved by setting

$$t = -\log r \tag{1.5a}$$

$$\psi(t) = r^{\frac{2}{\alpha-1}} u(r) . \tag{1.5b}$$

Then

$$\frac{d^2\psi}{dt^2} + \frac{n-2}{\alpha-1}\left(\frac{n+2}{n-2} - \alpha\right)\frac{d\psi}{dt} - \frac{2(n-2)}{(\alpha-1)^2}\left(\alpha - \frac{n}{n-2}\right)\psi + \psi^{\alpha} = 0 . \tag{1.6}$$

The solutions of (1.6) may be analyzed via phase-space analysis.

Case $\alpha = \frac{n+2}{n-2}$: In this case equation (1.6) reads

$$\psi'' - \left(\frac{n-2}{2}\right)^2 \psi + \psi^{\frac{n+2}{n-2}} = 0 . \tag{1.7}$$

Equation (1.7) is explicitly soluble. The positive spherically symmetric solutions (1.1) fall into two groups

a) regular solutions

$$u(r) = \left(\frac{\lambda\sqrt{n(n-2)}}{\lambda^2 + r^2}\right)^{\frac{n-2}{2}}, \quad \lambda > 0 , \tag{1.8}$$

b) singular solutions

$$u(r) = \frac{\psi(\log r)}{r^{\frac{n-2}{2}}} . \tag{1.9}$$

The simplest singular solution is

$$u(r) = \frac{\left(\frac{n-2}{2}\right)^{\frac{n-2}{2}}}{r^{\frac{n-2}{2}}} . \tag{1.10}$$

The function $\psi(\log r)$ in (1.9) is strictly positive with bounded oscillations about $\left(\frac{n-2}{2}\right)^{\frac{n-2}{2}}$. The functions (1.9) constitute a two-parameter family of solutions [3] which interpolate [4] between (1.10) and (1.8). Solution (1.8) is regular everywhere (including infinity), while (1.9) has two isolated singularities—one at the origin and one at infinity. Since equation (1.1) is conformally invariant, if $x \mapsto y$ is a conformal transformation with Jacobian $J(x)$, then the function

$$u(y) = (J(x))^{-\frac{n-2}{2n}} u(x) \tag{1.11}$$

is also a solution of (1.1). We use this fact to display the center of (1.8), and move around the singular points of (1.9). By a translation $x \mapsto x-a$, (1.8) becomes

$$u(x) = \left(\frac{\lambda\sqrt{n(n-2)}}{\lambda^2 + |x-a|^2}\right)^{\frac{n-2}{2}} , \quad \lambda > 0 . \tag{1.8$'$}$$

We employ the conformal transformation

$$x \mapsto |a-b|^2 \frac{x+a-b}{|x+a-b|^2} + b \tag{1.12}$$

which maps $x = 0$ into $x = a$, and $x = \infty$ into $x = b$. Its Jacobian is

$$J(x) = |a-b|^{2n}\,|x+a-b|^{-2n}\,.$$

Solution (1.9) transforms, under (1.12), into

$$u(x) = \frac{|a-b|^{\frac{n-2}{2}}\,\psi\!\left(\log\left[|a-b|\,\left|\frac{x-a}{x-b}\right|\right]\right)}{|x-a|^{\frac{n-2}{2}}\,|x-b|^{\frac{n-2}{2}}} \tag{1.9$'$}$$

Case $\alpha < \frac{n+2}{n-2}$: It is known [4] that equation (1.4) has no positive solutions (with or without singularities) for $1 < \alpha \le \frac{n}{n-2}$. There exists an explicit solution of (1.4), namely

$$u(r) = \frac{C_0}{r^{\frac{2}{\alpha-1}}} \tag{1.13a}$$

where

$$C_0 = \left\{\frac{2(n-2)}{(\alpha-1)^2}\left(\alpha - \frac{n}{n-2}\right)\right\}^{\frac{1}{\alpha-1}}\,. \tag{1.13b}$$

This is singular at the origin as well as at infinity. A phase-space analysis [3] of (1.6) yields the existence of another positive spherically symmetric solution $u(r)$ which is singular at the origin but regular at infinity. More precisely, for $\frac{n}{n-2} < \alpha < \frac{n+2}{n-2}$,

$$u(r) \sim \begin{cases} \dfrac{C_0}{r^{\frac{2}{\alpha-1}}} & \text{near } r = 0 \\[2ex] \dfrac{C}{r^{n-2}} & \text{near } r = +\infty \end{cases}\,. \tag{1.14}$$

Here C_0 is given by (1.13b), and C is an arbitrary parameter.

REMARK 1.1. It is worth noting that equation (1.1) has no spherically symmetric positive solutions with only one singular point (compare with

(1.14)), while equation (1.4) has no spherically symmetric regular solutions. The solutions (1.9), except the particular one (1.10), have no analogues in equation (1.4).

The questions that motivated the general theorems below, and other current work, are:

i) Are there regular positive (non-spherically symmetric) solutions of (1.1) other than (1.8) (or (1.8′))?

ii) Are there positive solutions of (1.1) with only one isolated singularity? Are there positive (non-spherically symmetric) solutions, with only one singularity, of (1.4), other than (1.14)?

iii) Besides (1.9) (or (1.9′)) for $\alpha = \frac{n+2}{n-2}$, and (1.13) for $\alpha < \frac{n+2}{n-2}$, are there other positive solutions of (1.1) or (1.4), with exactly two isolated singularities?

iv) Let $u(x)$ be a positive solution of (1.1) or (1.4) in $0 < |x| < R$. How does $u(x)$ behave near the singular point $x = 0$?

THEOREM 1. *Let* $u(x)$ *be a positive* C^2 *solution of*

$$\Delta u + f(u) = 0 \quad \text{in} \quad R^n, \ n > 2 . \tag{1.15}$$

Suppose that

$$u(x) = O\left(\frac{1}{|x|^m}\right) \text{ at infinity} \tag{1.16}$$

where $m > 0$ *and* $2(n-3)m > (n-4)(n-2)$. *Assume that* f *satisfies*

$$\frac{d}{dt}\left(t^{-\frac{n+2}{n-2}} f(t)\right) \le 0 \tag{1.17}$$

for all $t \ge 0$. *Then*

a) *If* $f(t)$ *is not identically equal to* const. $\cdot\, t^{\frac{n+2}{n-2}}$, *no solution exists.*

b) *If* $f(t) = t^{\frac{n+2}{n-2}}$, *then (1.8′) is the only positive solution.*

REMARK 1.2. The proof of Theorem 1, under a condition much weaker than (1.16), is given in [5]. The following theorem proven in [6], covers a much larger class of non-linearities than Theorem 1, but it assumes stronger decay at infinity.

THEOREM 2. *Let* $u(x)$ *be a positive* C^2 *solution of*

$$\Delta u + f(u) = 0 \quad \text{in} \quad \mathbf{R}^n, \quad n > 2 \tag{1.18}$$

with

$$u(x) = O\left(\frac{1}{|x|^{n-2}}\right) \text{ at infinity .} \tag{1.19}$$

Assume

i) $f(u) \geq 0$ *on* $0 \leq u \leq u_0 = \max u$, $f = f_1 + f_2$, *with* $f_1 \in C^1$, *and* f_2 *continuous and non-decreasing.*

ii) *For some* $\alpha > \frac{n+1}{n-2}$, $f(u) = O(u^\alpha)$ *near* $u = 0$. *Then* $u(x)$ *is spherically symmetric about some point in* $\mathbf{R}^n$, *and* $u_r < 0$ *for* $r > 0$, *where* r *is the radial coordinate about that point.*

Our next theorem (proven in [5]) is a strong version of Theorem 1 without any condition at infinity.

THEOREM 3. *Let* $u(x)$ *be a positive* C^2 *solution of*

$$\Delta u + u^\alpha = 0 \quad \text{in} \quad \mathbf{R}^n, \quad n \geq 3 \tag{1.20}$$

with $1 < \alpha \leq \frac{n+2}{n-2}$. *Then*

a) *If* $\alpha < \frac{n+2}{n-2}$, *then no solution exists.*

b) *If* $\alpha = \frac{n+2}{n-2}$, *then (1.8′) is the only positive solution.*

REMARK 1.3. This theorem holds [5] for a larger class of non-linearities. The interesting thing about this theorem is that there is no assumption about the behavior of $u(x)$ at infinity which could be a singular point. As a result we obtain

COROLLARY 1.

a) *Equation (1.1) has no positive solution with only one isolated singularity.*

b) *Equation (1.4) has no positive solution with only one isolated singularity located at infinity.*

REMARK 1.4. Because of the conformal invariance of (1.1), the isolated singularity in part a) of Corollary 1 could be located anywhere in R^n.

The following theorem [6] covers the uniqueness of (1.14). Its proof is similar to the proof of Theorem 5 below, which we give in Section II.

THEOREM 4. *Let* $u(x)$ *be a positive* C^2 *solution of*

$$\Delta u + f(u) = 0 \quad in \quad R^n - \{0\} \tag{1.21}$$

with

$$u(x) = O\left(\frac{1}{|x|^{n-2}}\right) \quad at \quad \infty \tag{1.22}$$

$$u(x) \to +\infty \quad as \quad x \to 0 . \tag{1.23}$$

Assume:

i) $f(u)$ *is continuous, non-decreasing in* u *for* $u \geq 0$, *and for some* $\alpha > \frac{n+1}{n-2}$, $f(u) = O(u^\alpha)$ *near* $u = 0$.

ii)
$$\varliminf_{u \to +\infty} \frac{f(u)}{u^p} \geq c > 0 \quad for\ some \quad p \geq \frac{n}{n-2} . \tag{1.24}$$

Then $u(x)$ *is spherically symmetric about the origin, and* $u_r < 0$ *for* $r > 0$.

REMARK 1.5.

i) We expect the result to remain true if $f(u) = O(u^\alpha)$ holds for $\alpha > \frac{n}{n-2}$.

ii) Condition (1.23) has recently been established in [5] for a class of non-linearities including $f(u) = u^\alpha$ (see Theorem 6 below).

The following theorem establishes the uniqueness of (1.9) and (1.13), for positive solutions of (1.1) and (1.4) respectively, with two isolated singularities.

THEOREM 5. *Let* $u(x)$ *be a positive* C^2 *solution of*

$$\Delta u + u^{\alpha} = 0 \quad in \quad R^n - \{0, \infty\}, \quad \frac{n+1}{n-2} < \alpha \le \frac{n+2}{n-2}, \quad n > 2 \tag{1.25}$$

with two isolated singularities at the origin and at infinity. More precisely assume

$$u(x) \to +\infty \quad as \quad x \to 0 \tag{1.26}$$

$$|x|^{n-2}\, u(x) \to +\infty \quad as \quad x \to \infty\,. \tag{1.27}$$

Then $u(x)$ *is spherically symmetric about the origin, and therefore it is equal to (1.9) or (1.13).*

REMARK 1.6. Theorem 5 is a slight generalization of Theorem 4 of [6] which treats only the conformal invariant equation $\alpha = \frac{n+2}{n-2}$. Its proof is given in Section II. For $\alpha < \frac{n+2}{n-2}$, conditions (1.26) and (1.27) have been proven in [5] from the positivity of $u(x)$ and the fact that $u(x)$ is regular outside 0 and ∞. Current work of Spruck and myself indicates that these conditions are also true for $\alpha = \frac{n+2}{n-2}$.

Concerning the classification of isolated singularities, Spruck and I have proven [5]:

THEOREM 6. *Let* $u(x)$ *be a positive* C^2 *solution of*

$$\Delta u + u^{\alpha} = 0 \quad in \quad 0 < |x| \le R \tag{1.28}$$

for $\frac{n}{n-2} < \alpha < \frac{n+2}{n-2}$. *Then the singularity at the origin is either removable, or there exist constants* C_1 *and* C_2 *such that*

$$\frac{C_1}{|x|^{\frac{2}{\alpha-1}}} \le u(x) \le \frac{C_2}{|x|^{\frac{2}{\alpha-1}}} \quad near \quad x = 0\,. \tag{1.29}$$

Furthermore

$$|x|^{\frac{2}{\alpha-1}} u(x) \xrightarrow[|x|\to 0]{} C_0 \tag{1.30}$$

where C_0 *is given by (1.13b).*

REMARK 1.7. Current work of Spruck and myself indicates that (1.29) holds also for $\alpha = \frac{n+2}{n-2}$.

We close this section by describing the implications of the above results for the Yang-Mills equations. Let $u(x)$ be a solution of

$$\Delta u + u^3 = 0 \quad on \quad R^4 \tag{1.31}$$

then a solution of the Euclidean Yang-Mills equations is obtained via 't Hooft's ansatz

$$A_\mu = i\sigma_{\mu\nu}\partial_\nu \log u(x) \tag{1.32}$$

where $\sigma_{\mu\nu}$ is defined in terms of Pauli matrices [1,2]. The results of Theorem 3 together with well-known results for $\Delta u - u^3 = 0$ on R^4, yield [2]:

THEOREM 7. *Let* $A_\mu(x)$ *be a solution of the Yang-Mills equations given by 't Hooft's ansatz (1.32). Suppose that* $A_\mu(x)$ *is regular everywhere except at one point (which could be infinity). Then the singularity at that point is removable, and* $A_\mu(x)$ *is self-dual (or antiself-dual). In particular any finite action solution of the Yang-Mills equations which is given by 't Hooft's ansatz is either self-dual or antiself-dual.*

REMARK 1.8. The number of Yang-Mills connections with Pontryagin number k one obtains via 't Hooft's ansatz is $5|k| + a(k)$, where $a(k) = 0$ for $|k| = 1$, $a(k) = 3$ for $|k| = 2$, and $a(k) = 4$ for $|k| \geq 3$.

Part a) of Corollary 1 motivates the following conjecture, currently under investigation:

CONJECTURE. There is no solution of the Yang-Mills equations which has, in some gauge, only one (non-gauge removable) isolated singularity.

II. *Proof of Theorem 5*

Let $e_n = (0, 0, \cdots, 1)$ and

$$y = \frac{x-e_n}{|x-e_n|^2} + e_n \tag{2.1}$$

$$v(y) = |x-e_n|^{n-2} u(x) . \tag{2.2}$$

Then $v(y)$ satisfies

$$\Delta v + |y-e_n|^{-(n-2)\left(\frac{n+2}{n-2} - \alpha\right)} v^{\alpha}(y) = 0 . \tag{2.3}$$

Thus $v(y)$ satisfies (2.3) in $\mathbf{R}^n - \{0, e_n\}$ and has two isolated singularities located at $y = 0$, and $y = e_n$. We shall prove that $v(y)$ is rotationally symmetric about the y_n-axis, i.e.,

$$\left(y_i \frac{\partial}{\partial y_j} - y_j \frac{\partial}{\partial y_i}\right) v(y) = 0 \quad \text{for} \quad i, j \neq n . \tag{2.4}$$

Equation (2.4) implies (via (2.1) and (2.2))

$$\left(x_i \frac{\partial}{\partial x_j} - x_j \frac{\partial}{\partial x_i}\right) u(x) = 0 \quad \text{for} \quad i, j \neq n . \tag{2.5}$$

Thus $u(x)$ is invariant under the x_n-axis. Since the choice of the axis was arbitrary, we conclude that $u(x)$ is spherically symmetric about the origin.

The following proposition asserts that the singular solution $v(y)$ is a distribution solution at the singular points $y = 0$, and $y = e_n$.

PROPOSITION 2.1. *Let*

$$D = \{y \in \mathbf{R}^n : |y| \leq R\}$$

$$D_{e_n} = \{y \in \mathbf{R}^n : |y - e_n| \leq R\}$$

and

$$f(y) \equiv |y - e_n|^{-(n-2)\left(\frac{n+2}{n-2} - \alpha\right)} v^{\alpha}(y) . \tag{2.6}$$

Let $v(y)$ *be a* C^2 *positive solution of (2.3) in* $\Omega_0 = D_0 - \{0\}$ *(or* $\Omega_{e_n} = D_{e_n} - \{e_n\}$*). Then* $f(y) \in L_1^{loc}(D_0)$ $(f(y) \in L_1^{loc}(D_{e_n}))$, *and* $v(y)$ *is a distribution solution of (2.3) at the origin (at the point* e_n*), and the integral equation*

$$v(y) = c \int d^n z \, \frac{f(z)}{|y-z|^{n-2}} \tag{2.7}$$

holds in the sense of distributions. Here $c = \dfrac{\Gamma(n/2)}{2(n-2)\,\pi^{n/2}}$.

Proof. Let D denote either D_0 or D_{e_n}, and D^{ε} denote either $D_0 - \{y \in \mathbf{R}^n : |y| \leq \varepsilon\}$, or $D_{e_n} - \{y \in \mathbf{R}^n : |y - e_n| \leq \varepsilon\}$. Let

$$\begin{aligned} \Delta\psi_{\varepsilon} &= -\lambda_{\varepsilon}\psi_{\varepsilon} && \text{in } D^{\varepsilon} \\ \psi_{\varepsilon} &= 0 && \text{on } \partial D^{\varepsilon} \\ \psi_{\varepsilon} &> 0 && \text{in } D^{\varepsilon} \\ \|\psi_{\varepsilon}\|_{L_2(D^{\varepsilon})} &= 1 \end{aligned}$$

i.e., λ_{ε} is the first eigenvalue of $-\Delta$ with zero Dirichlet data on ∂D^{ε}, and ψ_{ε} the corresponding eigenfunction. Let λ and ψ the corresponding objects of $-\Delta$ in D. Then [7]

$$\begin{aligned} \lambda_{\varepsilon} &\to \lambda \quad \text{as} \quad \varepsilon \downarrow 0 \\ \psi_{\varepsilon} &\to \psi \quad \text{in} \quad H_0^1(D) \end{aligned} \tag{2.8}$$

and $\psi_{\varepsilon} \to \psi$ in C^2 in a neighborhood of ∂D. From (2.3) we obtain by Green's theorem

$$\int_{\Omega_\varepsilon} f(y)\psi_\varepsilon(y) = -\int_{\Omega_\varepsilon} v(y)\Delta\psi_\varepsilon(y) - \int_{\substack{|y|=\varepsilon \\ \text{or } |y-e_n|=\varepsilon}} ds\, v\, \frac{\partial\psi_\varepsilon}{\partial r} + C\,.$$

By Hopf's boundary lemma $\frac{\partial\psi_\varepsilon}{\partial r} > 0$ on $|y| = \varepsilon$ or $|y-e_n| = \varepsilon$. Thus

$$\int_{\Omega_\varepsilon} f(y)\psi_\varepsilon(y) \le \lambda_\varepsilon \int_{\Omega_\varepsilon} v(y)\psi_\varepsilon(y) + C\,.$$

By (2.8), we quickly deduce that $f(y) \in L_1^{loc}(D)$. To prove that $v(y)$ is a distribution solution, we prove that for $\zeta \in C_0^\infty(D)$,

$$\int_D d^n y\,\{v\Delta\zeta + \zeta f(y)\} = 0\,. \tag{2.9}$$

Let $\tau_\varepsilon(r) = \tau\left(\frac{r}{\varepsilon}\right)$, where $\tau(r)$ is C^∞, zero in a neighborhood of the origin, and 1 for $r \ge R$ (here r is the radius from 0 or e_n). Integration by parts, and the equation (2.3) imply

$$\int_D \tau_\varepsilon(r)\{v(y)\Delta\zeta + \zeta f(y)\} = -\int_{\substack{|y|<\varepsilon \\ \text{or } |y-e_n|<\varepsilon}} \left\{ v\zeta\Delta\tau_\varepsilon + 2v\,\frac{\partial\zeta}{\partial y_i}\,\frac{\partial\tau_\varepsilon}{\partial y_i}\right\}. \tag{2.10}$$

Note that $|\Delta\tau_\varepsilon| \le \frac{C}{\varepsilon^2}$ and $\left|\frac{\partial\tau_\varepsilon}{\partial y_i}\right| \le \frac{C}{\varepsilon}$. Then

$$\int_{|y-e_n|\varepsilon} |v\zeta\Delta\tau_\varepsilon| \le C\left(\int_{|y-e_n|<\varepsilon} f(y)\right)^{\frac{1}{\alpha}} \left(\int_{|y-e_n|<\varepsilon} |y-e_n|^{\frac{n-2}{\alpha-1}\left(\frac{n+2}{n-2}-\alpha\right)}\right)^{\frac{\alpha-1}{\alpha}} \varepsilon^{-2}$$

$$\le C\left(\int_{|y-e_n|<\varepsilon} f(y)\right)^{\frac{1}{\alpha}} \varepsilon^{n\frac{\alpha-1}{\alpha}\,\frac{n-2}{\alpha}\left(\frac{n+2}{n-2}-\alpha\right)} \varepsilon^{-2} = o\left(\varepsilon^{\frac{2}{\alpha}}\right) \to 0 \text{ as } \varepsilon \downarrow 0$$

similarly for the other term in (2.9), and for the case $|y| < \varepsilon$. This establishes (2.9). The validity of (2.7) in the distributional sense is an easy consequence.

REMARK. The use of the eigenfunctions ψ_ε and ψ has been motivated from [7].

We now begin the proof of (2.4). We pick a direction, say the y_1-axis, perpendicular to the y_n-axis. For each $\lambda \geq 0$ and $y \in \mathbf{R}^n$, let $y^\lambda = (2\lambda - y_1, y_2, \cdots, y_n)$. The proof of (2.4) will be a consequence of the following three lemmas (stated for the y_1-direction, but they hold for any direction perpendicular to the y_n-axis).

LEMMA 2.1. *Suppose that* $u(x)$ *satisfies the assumptions of Theorem 5, and* $v(y)$ *is defined by (2.2). Then there exists a* $\lambda_0 > 0$ *such that for all* $\lambda \geq \lambda_0$

$$v(y) > v(y^\lambda), \quad \textit{for} \quad y_1 < \lambda . \tag{2.11}$$

LEMMA 2.2. *Under the same assumptions as in Lemma 2.1, the set of* $\lambda > 0$ *for which (2.11) holds is open.*

LEMMA 2.3. *Assume that for some* $\lambda > 0$

$$v(y) \geq v(y^\lambda) \qquad y_1 < \lambda \tag{2.12a}$$

$$v(y) \not\equiv v(y^\lambda) . \tag{2.12b}$$

Then

$$v(y) > v(y^\lambda) \quad \textit{for} \quad y_1 < \lambda \tag{2.13}$$

and

$$v_{y_1}(y) < 0 \quad \textit{on} \quad y_1 = \lambda . \tag{2.14}$$

Proof of Lemma 2.3. Let $w(y) = v(y^\lambda)$, $y_1 \leq \lambda$. Then

$$\Delta w + |y^\lambda - e_n|^{-(n-2)\left(\frac{n+2}{n-2} - \alpha\right)} w^\alpha(y) = 0 \tag{2.15}$$

subtracting (2.3) from (2.15) we have

$$\Delta(w-v) + C(y)(w-v) =$$

$$\left\{ |y-e_n|^{-(n-2)\left(\frac{n+2}{n-1}-\alpha\right)} - |y^\lambda - e_n|^{-(n-2)\left(\frac{n+2}{n-2}-\alpha\right)} \right\} v^\alpha(y) \geq 0 \ . \tag{2.16}$$

This together with $w-v \leq 0$, the maximum principle, and Hopf's boundary lemma, imply

$$w-v < 0 \quad \text{for} \quad y_1 < \lambda$$

and

$$0 < (w-v)_{y_1} = -2v_{y_1} \quad \text{on} \quad y_1 = \lambda \ .$$

These yield (2.13) and (2.14).

The proof of Lemmas 2.1 and 2.2 employs the integral equation (2.7). For simplicity we set $c = 1$.

Proof of Lemma 2.1. From (2.7)

$$v(y) - v(y^\lambda) = I_1 - I_2 \tag{2.17}$$

where

$$I_1 = \int_{z_1<\lambda} f(z) \left\{ \frac{1}{|y-z|^{n-2}} - \frac{1}{|y-z^\lambda|^{n-2}} \right\} \tag{2.18a}$$

$$I_2 = \int_{z_1>\lambda} f(z) \left\{ \frac{1}{|y-z^\lambda|^{n-2}} - \frac{1}{|y-z|^{n-2}} \right\} . \tag{2.18b}$$

For $y_1 < \lambda$, and λ sufficiently large we shall prove

$$I_1 \geq C_1 \frac{1}{|y^\lambda|^{n-1}} \frac{(\lambda-y_1)(\lambda-R)}{|y^\lambda|} \tag{2.19a}$$

$$I_2 \leq C_2 \frac{1}{|y^\lambda|^{n-1}} \frac{\lambda-y_1}{1} \frac{1}{\lambda^{\alpha(n-2)-(n+1)}} \tag{2.19b}$$

where C_1, C_2 and R are constants independent of y and λ. Estimates (2.19) quickly imply (2.11).

For $r \le s$ we use

$$(n-2)\frac{1}{s^{n-1}}(s-r) \le \frac{1}{r^{n-2}} - \frac{1}{s^{n-2}} \le (n-2)\frac{1}{r^{n-1}}(s-r) . \tag{2.20}$$

We also use

$$|y^\lambda - z| - |y-z| = \frac{4(\lambda-y_1)(\lambda-z_1)}{|y^\lambda - z + y - z|} \le 2(\lambda - y_1) . \tag{2.21}$$

Let R be a fixed number, using $|y^\lambda - z| \le |y^\lambda| + |z|$, and $|y^\lambda - z| \ge |y-z|$ valid for $z_1 < \lambda$, we obtain for large λ

$$I_1 \ge C(\lambda-y_1) \int\limits_{z_1<\lambda} f(z) \frac{1}{|y^\lambda - z|^{n-1}} \frac{\lambda - z_1}{|y^\lambda - z| + |y-z|}$$

$$\ge C(\lambda - y_1) \frac{1}{|y^\lambda|^{n-1}} \frac{1}{|y^\lambda|} \int\limits_{\substack{z_1<\lambda \\ |z|<R}} f(z)(\lambda - z_1)$$

$$\ge C \frac{1}{|y^\lambda|^{n-1}} \frac{(\lambda-y_1)(\lambda-R)}{|y^\lambda|} .$$

This establishes (2.19a). Next we establish (2.19b). Using (2.20) and (2.21) we obtain

$$I_2 \le I_2' + I_2'' + I_2''' \tag{2.22}$$

where

$$I_2' = C(\lambda - y_1) \int\limits_{\substack{z_1>\lambda \\ |z|>2|y^\lambda|}} f(z) \frac{1}{|y^\lambda - z|^{n-1}}$$

$$I_2'' = C(\lambda - y_1) \int_{\substack{z_1 > \lambda \\ |z| < 2|y^\lambda| \\ |z-y^\lambda| < \frac{|y^\lambda|}{2}}} f(z) \frac{1}{|y^\lambda - z|^{n-1}}$$

$$I_2''' = C(\lambda - y_1) \int_{\substack{z_1 > \lambda \\ |z| < 2|y^\lambda| \\ |y^\lambda - z| > \frac{|y^\lambda|}{2}}} f(z) \frac{z_1 - \lambda}{|y^\lambda - z|^n} .$$

To bound I_2', we use i) $|y^\lambda - z| > |y^\lambda|$ implied by $|z| > 2|y^\lambda|$, and ii) $f(z) = O(|z|^{-\alpha(n-2)})$ for large $|z|$. Then for large λ

$$\begin{aligned} I_2' &\le C \frac{1}{|y^\lambda|^{n-1}} (\lambda - y_1) \int_{\substack{z > \lambda \\ |z| > 2|y^\lambda|}} f(z) \\ &\le C \frac{1}{|y^\lambda|^{n-1}} \frac{\lambda - y_1}{} \frac{1}{|y^\lambda|^{\alpha(n-2)-(n-1)}} . \end{aligned} \tag{2.23}$$

To bound I_2'', we use $|z| > \frac{1}{2}|y^\lambda|$, and $f(z) = O(|z|^{-\alpha(n-2)})$. Then for large λ

$$I_2'' \le C \frac{1}{|y^\lambda|^{n-1}} (\lambda - y_1) \frac{1}{|y^\lambda|^{\alpha(n-2)-n}} . \tag{2.24}$$

For I_2''' we have $\frac{1}{2}|y^\lambda| < |y^\lambda - z| < 3|y^\lambda|$. Thus

$$\begin{aligned} I_2''' &\le C \frac{1}{|y^\lambda|^n} (\lambda - y_1) \int_{z_1 > \lambda} f(z)|z| \\ &\le C \frac{1}{|y^\lambda|^n} (\lambda - y_1) \frac{1}{\lambda^{\alpha(n-2)-(n+1)}} . \end{aligned} \tag{2.25}$$

Estimate (2.29b) is obtained by combining (2.22)-(2.25). This completes the proof of Lemma 2.1.

Proof of Lemma 2.2. Suppose that (2.11) holds for some $\lambda > 0$ but not for all λ in some neighborhood of λ. Then there exists a sequence of $\{\lambda^j\}$, $\lambda^j \to \lambda$, and a sequence $\{y^j\}$, with $y_1^j < \lambda^j$ such that

$$v(y^j) \le v(y^{j\lambda^j}) . \tag{2.26}$$

Then there is a subsequence (denoted again by y^j) such that either

$$y^j \to y \quad \text{with} \quad y_1 < \lambda$$

or

$$y^j \to \infty .$$

In the former case

$$v(y) \le v(y^\lambda) \tag{2.27}$$

and in view of (2.11) we must have $y_1 = \lambda$. But this is also impossible, because in this case (2.27) implies $v_{y_1}(y) \ge 0$ (on $y_1 = \lambda$) which contradicts (2.14). Next we show that $y^j \to \infty$ is also impossible. A long and tedious computation [6] yields

$$0 \ge \frac{|y^j|^n}{\lambda^j - y_1^j}\left\{v(y^j) - v(y^{j\lambda^j})\right\} \xrightarrow[|y^i| \to +\infty]{} 2(n-2)\int dz f(z)(\lambda - z_1)$$

$$= -2(n-2)\int dz f(z^\lambda)(\lambda - z_1) .$$

Thus

$$0 \ge \int dz (f(z) - f(z^\lambda))(\lambda - z_1) . \tag{2.28}$$

Since $v(y) > v(y^\lambda)$ and $|y^\lambda - e_n| > |y - e_n|$, for $y_1 < \lambda$, it follows that $f(y) > f(y^\lambda)$ for $y_1 < \lambda$. Thus (2.28) is impossible—a contradiction. This proves Lemma 2.2.

Completion of the proof of (2.4). By Lemmas 2.1 and 2.2, there exists a maximal open interval $0 \leq \mu < \lambda < +\infty$ such that (2.11) holds. We shall prove that $\mu = 0$. By Lemma 2.3

$$v_{y_1} < 0 \quad \text{for} \quad y_1 > \mu \tag{2.29}$$

and by continuity

$$v(y) \geq v(y^\mu) \quad \text{for} \quad y_1 < \mu . \tag{2.30}$$

If $\mu > 0$, then since $v(y) \to +\infty$ as $y \to 0$ or $y \to e_n$, we cannot have $v(y) \equiv v(y^\mu)$. Thus by Lemma 2.3 we must have

$$v(y) > v(y^\mu) \quad \text{for} \quad y_1 < \mu$$

$$v_{y_1} < 0 \quad \text{for} \quad y_1 = \mu .$$

If $y \neq \infty$, then this is impossible by Lemma 2.2. For a sequence $\{y^j\}$, $|y^j| \to +\infty$, we again reach a contradiction as in (2.28). This establishes (2.4), and the proof of Theorem 5 is completed.

SCHOOL OF MATHEMATICS
INSTITUTE FOR ADVANCED STUDY
PRINCETON, NEW JERSEY 08540

REFERENCES

[1] Jackiew, R., Nohl, C., and Rebbi, C.: Phys. Rev. D15 (1977), 1642.

[2] Gidas, B.: "Euclidean Yang-Mills and Related Equations," in *Bifurcation Phenomena in Mathematical Physics and Related Topics*, pp. 243-267, D. Reidel Publishing Co. (eds. C. Bardos and D. Bessis).

[3] Fowler, R.H.: Further Studies on Emden's and Similar Differential Equations, Quart. J. Math. 2 (1931), 259-288.

[4] Gidas, B.: "Symmetry Properties and Isolated Singularities of Positive Solutions of Nonlinear Elliptic Equations," in *Nonlinear Differential Equations in Engineering and Applied Sciences*, ed. R. L. Sternberg, Marcel Dekker, Inc. (1980).

[5] Gidas, B. and Spruck, J.: Global and Local Behavior of Positive Solutions of Nonlinear Elliptic Equations, to appear in Commun. Pure and Applied Math (1981).

[6] Gidas, B., Ni, W. M., and Nirenberg, L.: Symmetry of Positive Solutions of Nonlinear Equations in $\mathbf{R}^n$, to appear.

[7] Lions, P. L.: "Isolated Singularities in Semilinear Problems," preprint.

ON PARALLEL YANG-MILLS FIELDS

Gu Chaohao

As a class of special solutions of the Yang-Mills equations the parallel Yang-Mills fields on four-dimensional Riemannian manifolds are constructed on the basis of the tangential bundles and the Levi-Civita connections of some special Riemannian manifolds.

I. *Introduction*

The theory of Yang-Mills fields now is not only extremely important in physics, but also very attractive for mathematicians. One of the central questions is to obtain solutions of the Yang-Mills equations on a Riemannian manifold M_n (or on a Lorentz Manifold), i.e., to find principal bundles $P(M_n, G)$ (or vector bundles over M_n) and connections on the bundles such that the Yang-Mills equations be satisfied. Here G is a Lie group, being compact usually. A Yang-Mills field is called parallel [1] if the gauge derivatives of the field strength, i.e., the covariant derivatives of the curvature, are all zero. Obviously, such fields are solutions of the Yang-Mills equations. Especially, every solution of the Yang-Mills equations on a two-dimensional manifold M_2 is parallel. However, in the case of higher dimension only a few Riemannian manifolds admit non-trivial parallel Yang-Mills fields.

In this report we shall give a description of the parallel Yang-Mills fields, mainly, in the four-dimensional case. We suppose that the manifolds

Seminar on Differential Geometry
0-691-08268-5/82/000443-11 $00.55/0 (cloth)
0-691-08296-0/82/000443-11 $00.55/0 (paperback)

are oriented and connected. For completeness we begin with the two-dimensional case and construct the solutions explicitly. When the dimension of M_n is four we mainly consider the non-Abelian case, since the Abelian case is easy to deal with. In the non-Abelian case, the Riemannian manifold should be symmetric spaces or conformally half-flat Einstein spaces [1]. It is proved that all parallel Yang-Mills fields over these manifolds can be constructed on the basis of the tangential bundles and the Levi-Civita connections. It is also seen that the local analytic conformally half-flat Einstein metrics can be constructed by solving a system of partial differential equations and the solutions essentially depend on two arbitrary functions in three variables.

It should be noted that the parallel Yang-Mills fields partially coincide with the point-wisely spherically symmetric gauge fields studied by C.N. Yang [2], [3].

We use the following notations:

Let $\{U_\alpha\}$ be a covering of M_n and the restricted bundle over each U_α be trivial. Choosing a local section on each U_α the gauge potential is represented by g-valued 1-forms

$$\underset{\alpha}{b} = \underset{\alpha}{b}\lambda\, dx^\lambda \qquad (\lambda = 1, 2, \cdots, n) , \tag{1}$$

the field strength is

$$\underset{\alpha}{f} = \frac{1}{2}\underset{\alpha}{f}_{\lambda\mu} dx^\lambda \wedge dx^\mu = -d\underset{\alpha}{b} - \frac{1}{2}[\underset{\alpha}{b}, \underset{\alpha}{b}] \tag{2}$$

and the Yang-Mills equations are

$$g^{\mu\nu}\underset{\alpha}{f}_{\lambda\mu|\nu} = g^{\mu\nu}(\underset{\alpha}{f}_{\lambda\mu;\nu} + [\underset{\alpha}{b}_\nu, \underset{\alpha}{f}_{\mu\nu}]) = 0 , \tag{3}$$

where "|" and ";" are symbols for gauge derivatives and covariant derivatives with respect to the Levi-Civita connection, respectively, and $g^{\lambda\mu}$ is the metric tensor.

The transformation

$$\underset{\alpha}{b} \to \underset{\alpha}{b}' = (\mathrm{ad}\,\underset{\alpha}{\zeta})\underset{\alpha}{b} - (d\,\underset{\alpha}{\zeta})\underset{\alpha}{\zeta}^{-1} \tag{4}$$

is called a gauge transformation via the G-value function $\underset{\alpha}{\zeta}$, defined on $\underset{\alpha}{U}$. Under the gauge transformation we have

$$\underset{\alpha}{f} \to \underset{\alpha}{f}' = (\mathrm{ad}\,\underset{\alpha}{\zeta})\underset{\alpha}{f}\,, \qquad \zeta_{\beta\alpha} \to \zeta'_{\beta\alpha} = \zeta_\beta \zeta_{\beta\alpha} \zeta_\alpha^{-1}\,, \tag{5}$$

where $\zeta_{\beta\alpha}$ are transition functions, defined on $U_\alpha \cap U_\beta$.

The gauge potential and the transition function should satisfy the following consistency conditions

$$\underset{\beta}{b} = (\mathrm{ad}\,\zeta_{\beta\alpha})\underset{\alpha}{b} - (d\,\zeta_{\beta\alpha})\zeta_{\alpha\beta}\,. \tag{6}$$

2. *The case of* M_2

The two-dimensional case was treated by using Morse theory provided M_2 is compact [4]. However, the Yang-Mills equations can be solved directly and explicitly for each M_2.

Near an arbitrary point there exist a local coordinate system and a local gauge such that the metric of M_2 is

$$ds^2 = du_1^2 + a(u_1, u_2)\,du_2^2 \tag{7}$$

and the gauge potential satisfies

$$b_1(u_1, u_2) = 0\,, \qquad b_2(0, u_2) = 0\,. \tag{8}$$

Writing down the simplified Yang-Mills equations and integrating them, we obtain

$$b_1 = 0\,, \quad b_2 = c\sigma(u_1, u_2)\,, \quad f_{12} = c\sqrt{a}\,, \tag{9}$$

where c is an element of g and

$$\sigma = -\int_0^{u_1} \sqrt{a(\tau, u_2)}\, d\tau \, .$$

Thus, the field strength is proportional to the area form of M_2. Moreover, a global solution can be represented as follows: M_2 is covered by $\{U_\alpha\}$ in each U_α

$$b_\alpha = c_\alpha h_\alpha \, , \qquad f_\alpha = c_\alpha \cdot \text{Area form} \, ,$$

where $c_\alpha \in g$ and h_α are 1-forms, satisfying $-dh_\alpha = \text{Area form}$.

It is easily seen that via a gauge transformation we have

$$c_1 = c_2 = \cdots = c \, . \tag{10}$$

If M_2 is compact, c must satisfy the following quantization conditions:

(1) c generates a compact subgroup $G_1 \subset G$,

(2) The first Chern number is an integer, i.e.,

$$c = kc_1 \, , \qquad \frac{kA}{2\pi} = \text{integer} \, . \tag{11}$$

Here A is the area of M_2 and c_1 is a normalization of c such that $\exp(tc_1) \to e^{it}$ is an isomorphism.

If c is given, solutions of the Yang-Mills equation on M_2 can be determined by a homomorphism from the fundamental group $\pi_1(M_2)$ to the group $H = \{a \in G \mid (ad\, a)c = c\}$. This can be seen from the argument in [5], [6].

3. *The case of* M_4

If the gauge group G is Abelian, the underlying manifold M_4 admits a parallel 2-form and hence is Kählerian or decomposible to $M_2 \times M_2'$ at least locally. The parallel Yang-Mills fields can be obtained easily.

For the essentially non-Abelian case we have proved in [1] the following theorem:

THEOREM 1. *The Riemannian manifolds* M_4, *admitting essentially non-Abelian parallel Yang-Mills fields are:*

(I) *The spaces of nonvanishing constant curvature.*

(II) *The Kählerian manifolds of nonvanishing constant holomorphic sectional curvature.*

(III) *Other conformally half-flat Einstein spaces with* $R \neq 0$.

(IV) *The spaces* $M_1 \times M_3$, *where* M_3 *are spaces of nonvanishing constant curvature.*

Now we are going to construct the fields. Let η_{aij} and $\overline{\eta}_{aij}$ be the 't Hooft symbols defined by:

(i) $\eta_{abc} = \overline{\eta}_{abc} = \varepsilon_{abc}$ $(a, b = 1, 2, 3)$

(ii) η_{aij} and $\overline{\eta}_{aij}$ are anti-symmetric with respect to $ij\,(i, j = 1,2,3,4)$.

(iii) η_{aij} and $\overline{\eta}_{aij}$ are self-dual and anti-self-dual respectively, i.e.

$$\eta_{aij} = \frac{1}{2}\varepsilon_{ijk\ell}\eta_{ak\ell}, \qquad \overline{\eta}_{aij} = -\frac{1}{2}\varepsilon_{ijk\ell}\overline{\eta}_{ak\ell}. \tag{12}$$

Then the 4×4 matrices with i, j as row and column indices

$$Y_a = -\frac{1}{2}(\eta_{aij}), \quad Y_{a'} = -\frac{1}{2}(\overline{\eta}_{aij}) \quad (a' = a+3) \tag{13}$$

are a set of standard bases of the Lie algebra SO_4. Moreover, $\{Y_a\}$ and $\{Y_a'\}$ generate the normal subgroups SU_2^+ and SU_2^- of SO_4 respectively. We have

$$SO_4/SU_2^+ \simeq SO_3, \qquad SO_4/SU_2^- \simeq SO_3. \tag{14}$$

The tangential bundle of a Riemannian manifold is an SO_4 bundle with the Levi-Civita connection. From the natural homomorphism $SO_4 \to SO_4/SU_2^-$ we obtain an SO_3 bundle E^+. The Levi-Civita connection ω_{ij} under orthonormal frames is decomposed as the sum of self-dual part ω_{ij}^+ and anti-self-dual part ω_{ij}^-. We have an SO_3 connection ω_{ij}^+ on the bundle E^+. Similarly we have an SO_3 connection ω_{ij}^- on E^-.

Evidently, the Levi-Civita connection on the tangential bundle of a space of constant curvature is a parallel Yang-Mills field. Moreover, the connection ω^+ on E^+, ω^- on E^- and their sum are also parallel. It is also seen in [1] that the connection ω^+ on E^+ is parallel if M_4 is conformally half-flat and Einstein with $W^+ = 0$, where W^+ is the self-dual part of the Weyl tensor. If $W^- = 0$, the connection ω^- on E^- is parallel. For the manifolds of type IV, obviously, there are natural SO_3 parallel fields. For the manifold of type II there exist parallel $U_1 \times SO_3$ fields, the SO_3 part of which is ω^+ on E^+. Some of these connections can be lifted as $SU_2 \times SU_2$, SU_2 or $U_1 \times SU_2$ parallel fields. All these fields stated above are defined as basic non-Abelian Yang-Mills parallel fields.

THEOREM 2. *Each irreducible* SO_3 *or* SU_2 *parallel Yang-Mills field is equivalent to a basic non-Abelian parallel Yang-Mills field.*

Proof. We sketch the proof for the conformally half-flat Einstein spaces M_4 which cover the spaces of type (I), (II) and (III). The proof for the spaces of type (IV) is similar and easier.

Suppose F be a parallel Yang-Mills field with gauge group SO_3 on M_4 and $W^+ = 0$. Consider R_{ijkl} as a symmetric linear operator on the space of skew symmetric 2-tensors, then η_{akl} are eigen elements and $\eta_{ak\ell}$ span an invariant subspace. Referring to certain orthonormal frames the curvature tensor can be expressed as

$$R_{ijkl} = \sum_a h\, \eta_{aij}\, \eta_{akl} + \sum_a h_{a'} \zeta_{aij}\, \zeta_{akl} \quad (a' = a+3,\ a = 1,2,3)\,. \tag{15}$$

Here $\zeta_a = (\zeta_{aij})$ are suitable bases of su_2^- and h is a constant $(h \neq 0)$. The field strength of F can be expressed as

$$f_{ij} = \sum_a L_a\, \eta_{aij} + \sum_a L_{a'} \zeta_{aij}\,. \tag{16}$$

Here L_a and L_a' are valued in SO_3. Substituting (15) and (16) in the generalized Ricci identity

$$f_{ih}R_{hjkl} + f_{hj}R_{hikl} + [f_{kl}, f_{ij}] = f_{ij|kl} - f_{ij|lk}, \tag{17}$$

we find that

$$[L_a, L_b] = -h \sum_a \varepsilon_{abc} L_c, \qquad L_{a'} = 0, \tag{18}$$

if $W^- \neq 0$. For the space of constant curvature we may have (18) or

$$L_a = 0, \qquad [L_{a'}, L_{b'}] = -h \sum_a \varepsilon_{abc} L_{c'} \tag{19}$$

which can be treated in the same way.

Let $M = \cup U_\alpha$ and on each U_α the tangential frame bundle and the bundle of F are trivial. In each U_α because of (18) one can choose orthonormal frames and a certain gauge such that the field strength of F takes the canonical form

$$\underset{\alpha}{f}_{ij} = h \sum_a \eta_{aij} X_a \tag{20}$$

where X_a are standard bases of SO_3. Let $\xi_{\alpha\beta} = (\ell^i_j)$ be the transition functions of the orthonormal frames and $\zeta_{\alpha\beta}$ be the transition functions of F. From (20) it follows that

$$(\mathrm{ad}\, \zeta_{\alpha\beta}) \Big(\sum_a \eta_{aij} X_a \ell^i_h \ell^j_l \Big) = \sum_a \eta_{akl} X_a .$$

From the properties of the 't Hooft symbol it is seen that $\zeta_{\alpha\beta}$ are the images of $\xi_{\alpha\beta}$ in the natural homomorphism $SO_4 \to SO_4/SU_2^-$. Hence $\zeta_{\alpha\beta}$ are the transition functions of E^+. Moreover, the curvature of the connection ω^+ on E^+ is

$$\overset{+}{R}_{klij} = h \sum_a \eta_{akl} \eta_{aij} . \tag{21}$$

Comparing (21) with (20) and noting that for these parallel fields the potentials are determined by the curvature, we conclude that the field F is equivalent to ω^+ on E^+.

Because each SU_2 Yang-Mills field is a lift of a SO_3 field, the above result holds true for SU_2 parallel Yang-Mills fields.

REMARK 1. Obviously the above theorem holds true for $SO_3 \times SO_3$, SO_4, $SU_2 \times SU_2$ and $U_1 \times SO_3$ parallel Yang-Mills fields.

REMARK 2. (20) is the canonical form of the field strength of a pointwisely spherically symmetric SU_2 gauge field considered by C. N. Yang [2], [3]. He proved that the base manifolds of such fields should be conformal to conformally half-flat Einstein manifolds. The corresponding SU_2 fields were also found. In fact after a conformal deformation of metric of the base manifolds such fields also become parallel and belong to the type (I), (II) and (III).

Now let G be a non-Abelian compact Lie group and F be a parallel Yang-Mills field with gauge group G. From the proof of Theorem 2 it is seen that there is a basic parallel Yang-Mills field F_1 with gauge group G, such that under a suitable gauge the gauge potentials and field strengths of F and F_1 are equal, if G_1 is considered as some subgroup of G. Let

$$H = \{\alpha \in G \mid (\mathrm{ad}\,\alpha)\beta = \beta, \ \forall \beta \in G_1\}$$

and $\sigma : \ell \to \psi_\ell \in H$ be a homomorphism of $\pi_1(M_4)$ to H. Here ℓ are loops having a fixed point 0 as the starting and ending points. Suppose $\ell \to \underset{1}{\Phi}_\ell \in G$ be the loop phase factors (holonomy) of the field F_1 [7]. We have

THEOREM 3. *Each parallel Yang-Mills field* F *with gauge group* G *is the product of a basic parallel field* F_1 *and a homomorphism* $\pi_1(M_4) \to H$ *in the sense that* $\Phi_\ell = \underset{1}{\Phi}_\ell \psi_\ell$, *where* Φ_ℓ *are loop phase factors of* F.

Proof. Let $\zeta_{\beta\alpha}$ and $\xi_{\beta\alpha}$ be the transition functions for the fields F and F_1 respectively. From the definition of H and consistency conditions (6) for F and F_1 it is seen that if $\eta_{\beta\alpha} = \zeta_{\beta\alpha}\xi_{\alpha\beta}$, then

$$d\eta_{\beta\alpha} = 0, \qquad \eta_{\gamma\beta}\eta_{\beta\alpha} = \eta_{\gamma\alpha} . \tag{22}$$

Hence $\eta_{\beta\alpha} = \text{const.}$ in each connected part of $U_\alpha \cap U_\beta$. Moreover, the transition functions $\{\eta_{\beta\alpha}\}$ and the vanishing connection generate a homomorphism of $\pi_1(M_4)$ to H. Theorem 3 follows from the method of construction of loop phase factors [7], [8].

Thus we have constructed all parallel Yang-Mills fields if the base manifolds have been constructed.

4. *Conformally half-flat Einstein metric*

The only unknown metrics in Theorem 1 are those of type (III), though the global classification of complete hyperbolic manifolds is to be completed.

We have seen that such a metric admits an SO_3 parallel Yang-Mills field F and

$$f^a_{\lambda\mu} = h\eta_{aij}\ell^i_\lambda \ell^j_\mu \tag{23}$$

locally. Here $f^a_{\lambda\mu}X_a$ is the field strength and $\ell^i_\lambda dx^\lambda = \omega^i$ are bases of orthonormal coframes. In fact (23) is the coordinate form of (20). The metric is $ds^2 = \sum_i (\ell^i_\lambda dx^\lambda)^2$.

Choose a coordinate system such that $x^4 = \text{const.}$ be parallel surfaces and x^4 be the length of normal curves. There exists a gauge such that $b_4 = 0$. Change the orthonormal frame via group SU_2^- such that the fourth base vector is normal to $x^4 = \text{const.}$ This is possible since SU_2^- is transitive on S^3. Then we have

$$\ell^4_4 = 1, \quad \ell^4_a = \ell^a_4 = 0 \quad (a = 1,2,3) . \tag{24}$$

Taking $\lambda = b$, $\mu = 4$, from (23) we have

$$b^a_{b,4} = h\ell^a_b , \tag{25}$$

where "," is the symbol for partial derivatives. Let

$$\vec{b}_a = (b_a^1, b_a^2, b_a^3), \qquad \vec{f}_{bc} = (f_{bc}^1, f_{bc}^2, f_{bc}^3). \tag{26}$$

The other equations of (23) are

$$\vec{b}_{a,4} \times \vec{b}_{b,4} = h\vec{f}_{ab}, \tag{27}$$

or

$$\vec{b}_{a,4} = \pm\frac{1}{2D}\,\varepsilon_{abc}\,\vec{\alpha}_b \times \vec{\alpha}_c \tag{28}$$

where

$$\vec{\alpha}_a = \frac{1}{2}\,\varepsilon_{abc}\,\vec{f}_{bc}$$

$$D = \{h(\vec{\alpha}_1 \times \vec{\alpha}_2)\cdot\vec{\alpha}_3\}^{1/2}. \tag{29}$$

(28) is a system of non-linear partial differential equations with $\vec{b}_a$ as unknowns. All analytic local solutions of (28) can be obtained by using Kovalevski theorem. The solutions depend on nine analytic functions in three variables as the initial conditions subjected to $D \neq 0$. However, the choice of $x^4 = 0$ depends on one arbitrary function. The gauge on $x^4 = 0$ which remains arbitrary depends on three arbitrary functions. The coordinate system on $x^4 = 0$ depends on three arbitrary functions too. Hence the solutions of our question depend essentially on two arbitrary analytic functions in three variables.

THEOREM 4. *The local analytic metrics which are conformally half-flat and Einstein with non-vanishing scalar curvature exist and the degree of freedom is two arbitrary analytic functions in three variables.*

REMARK. If the scalar curvature is zero the metrics should be Kählerian and the global existence theorem is a consequence of Yau's theorem [9]. If the scalar curvature is not zero the metrics other than those of type (II) should not be Kählerian. Recently, N. Hitchin [10] proved that if the scalar curvature is positive and M_4 is compact then M_4 is CP^2 or S^4.

Acknowledgment

It is a great pleasure to express my thanks to Prof. C. N. Yang for the initiative in studying the problem and for valuable discussions. I am indebted to the State University of New York at Stony Brook for its hospitality.

INSTITUTE FOR THEORETICAL PHYSICS
STATE UNIVERSITY OF NEW YORK AT STONY BROOK
NEW YORK, NEW YORK 11794

and

DEPARTMENT OF MATHEMATICS
FUDAN UNIVERSITY
SHANGHAI, CHINA

REFERENCES

[1] C. H. Gu, On the Riemannian spaces admitting parallel Yang-Mills fields (in English), Chinese Annals of Math. 1 (1980), No. 2.

[2] C. N. Yang, Pointwise SO_4 Symmetry of the BPST pseudo-particle solution. F. Block Festschrift (to be published).

[3] The letter from C. N. Yang to I. M. Singer dated May 25, 1977.

[4] M. F. Atiyah and J. D. S. Jones, Topological Aspects of Yang-Mills Theory, Commun. Math. Phys. 61 (1978), 97-118.

[5] B. Kostant, Lecture Notes in Math. 170 (1970), Springer Verlag.

[6] C. H. Gu, Electromagnetic fields and global U_1 gauge fields. Scientia sinica (Chinese Edition) 1976, 320-328.

[7] ———, The loop phase-factor approach to gauge theories, Fudan Jour. 1976 No. 2 51-60; Phys. Energ. Fort et Phys. Nucl. 2 (1978), 98-108.

[8] T. T. Wu and C. N. Yang, Concept of nonintegrable phase factors and global formulation of gauge fields. Phys. Rev. D*12* (1975), 3845-3857.

[9] S.-T. Yau, On the Ricci curvature of a compact Kähler manifold and the complex Monge-Ampére equation. Commun. Pure Appl. Math. 31 (1978), 339-411.

[10] N. J. Hitchin, Kählerian twistor spaces, preprint 1980.

VARIATIONAL PROBLEMS FOR GAUGE FIELDS

Karen K. Uhlenbeck

This talk describes some of the similarities and differences between the application of direct methods to variational problems for gauge fields and results for other geometric variational problems. The integrals we are interested in contain as a term in the action $\int_M |F|^2 *1$. Here F is the curvature (field) of a connection in a bundle over a Riemannian manifold M. The geometric problems we shall discuss for comparison are the Yamabe problem and the harmonic map problem. The physics literature on gauge fields is not discussed here.

§1. *The modern calculus of variations*

The classical calculus of variations is almost entirely the study of single integral problems. The action (called J for the rest of this paper) has the form

$$J(s) = \int_a^b L(s(t), s'(t), t)\,dt .$$

Here $s : [a,b] \to R^\ell$ and the Euler-Lagrange equations form a second order system of ordinary differential equations. The accepted approach to the global study of the structure of the critical (stationary) curves of this

Seminar on Differential Geometry
0-691-08268-5/82/000455-10$00.50/0 (cloth)
0-691-08296-0/82/000455-10$00.50/0 (paperback)

integral is to push it down onto a finite dimensional calculus problem [10]. Given s, find a broken solution $\tilde{s}$ by solving the end point problem of the Euler-Lagrange equations for $\tilde{s}(t_j) = s(t_j)$, $t_i \leq t \leq t_{i+1}$, on small intervals $a = t_0 < t_1 <, \cdots, < t_{\ell+1} = b$. Then $f(s(t_0), s(t_1), \cdots, s(t_{\ell+1})) = J(\tilde{s})$ is a function of $y_i = s(t_i)$. Now study f as a function on $\{y_1, \cdots, y_\ell\} \in R^\ell$.

Multiple integral problems do not lend themselves to this approach. Generally one is forced to work directly in a function space. However, this talk is intended to convince you that the difficulties arise more from the variety of phenomena which can occur rather from lack of method. [See [1] page 2.]

There are at least three levels of success possible in finding critical points.

(A) *Existence of absolute minima for* J:

One can nearly always find these. Hilbert used the idea to solve the linear Dirichlet problem and Tonelli developed it further. Find a minimizing sequence and take a weak limit in the appropriate function space. Sometimes it is difficult to show the minimum is classical (smooth). This minimum is also often trivial in geometric problems [11].

(B) *Existence of minima subject to a topological or functional constraint*:

Here the problem is to show the preservation of the constraint under weak limits. The usual function spaces are Sobolev spaces

$$L_k^p(M) = \{\text{functions on } M \text{ with derivatives through order } k \text{ in } L^p\}.$$

The important theorem is the Sobolev embedding theorem: $L_k^p(M) \subset C^0(M)$ is a compact inclusion if M is compact and $pk > \dim M$. Usually one needs to be in the C^0 framework to preserve all topological constraints. This is perhaps the best explanation for success in the classical $\dim M = 1$ situation. The natural function space for a variational problem is often $L_1^2(M)$ which lies in $C^0(M)$ only when $\dim M = 1$.

(C) *Morse theory or Ljusternik-Schnirelmann theory*:

By this we mean a count of critical points or the topology of critical sets which agrees with the topology of the function space. These theorems are conceptually easiest to prove by verifying the Palais-Smale condition for J. This condition replaces the local compactness used in studying $f: R^\ell \to R$. Here instead of R^ℓ we have an infinite dimensional complete Banach manifold X^∞ of functions and $J: X^\infty \to R$ differentiable. Then J is said to satisfy the *Palais-Smale condition* if given a sequence $s_\alpha \in X^\infty$, $|J(s_\alpha)| \leq b$ and $\lim_{\alpha\to\infty} |dJ(s_\alpha)| \to 0$, then there exists a subsequence $s_{\alpha'} \to s$ in X^∞ [16].

The verification generally proceeds in a standard fashion. One uses the condition $|J(s_\alpha)| \leq b$ to get a weakly convergent subsequence $s_{\alpha'} \rightharpoonup s$ as in the usual existence A or B. The ellipticity of the Euler-Lagrange equations for J and $|dJ(s_\alpha)| \to 0$ will then imply strong convergence $s_\alpha \to s$. The difficulties arise again more because X^∞ is often prescribed by the nature of the problem to be $L_1^2(M)$; here J is not differentiable, the object $L_1^2(M)$ is not a Hilbert manifold, and since $L_1^2(M) \not\subset C^0(M)$, topological constraints are lost in taking weak limits [14, Chapter 19].

§2. *Geometric variational problems*

When can a Riemannian metric on M be changed by a conformal (length) factor to have constant scalar curvature? This question is known as the Yamabe problem [24]. The non-linear partial differential equation which arises is

$$\Delta\phi - K\phi + \lambda\phi^{(n+2)/(n-2)} = 0 .$$

Here $K \in C^\infty(M)$ is the known scalar curvature of the metric on M and $\phi \in C^\infty(M)$, $\lambda \in R$ are unknowns. The special case when $M = R^4$

$$\Delta\phi + \lambda\phi^3 = 0$$

is a well-known special reduced form of the Yang-Mills equations.

There are several variational formulations possible. Let

$$J(\phi) = \int_M (|d\phi|^2 + K\phi^2) * 1$$

and minimize subject to the non-linear constraint $\int_M \phi^q * 1 = 1$. The Euler-Lagrange equations are then

$$\Delta\phi - K\phi + \lambda\phi^{q-1} = 0 .$$

The range $q < 2n/(n-2)$ is the range in which §1.C, the Palais-Smale condition, holds. This is a typical non-linear eigenvalue problem [15] with a large number of solutions. At the critical exponent $q = 2n/(n-2)$ or $q-1 = (n+2)/(n-2)$ of the Yamabe problem, the Sobolev embedding $L_1^2 M \subset L^q(M)$ fails to be compact. The Palais-Smale condition fails, but one hopes §1.B is true. The difficulty lies in preserving the constraint $\int_M \phi^q * 1$ under weak limits in $L_1^2(M)$. The support of a minimizing sequence could concentrate over isolated points; in that case the weak limit is zero and the functional constraint $\int_M \phi^q * 1 = 1$ is lost. This problem has been solved in many cases [2].

The harmonic map problem has also been extensively studied. This problem concerns critical maps of the energy integral defined on maps $s : M \to N$ between two Riemannian manifolds.

$$J(s) = E(s) = \int_M |ds|^2 * 1 .$$

Unfortunately the critical dimension for this problem is $n = 2$ and the energy integral is in the Palais-Smale range only for $\dim M = n = 1$ and borderline for $\dim M = n = 2$. In both cases, this integral is very important in geometry. For $\dim M = 1$, the critical points are geodesics and for $\dim M = 2$, the harmonic maps are closely connected with minimal surfaces. If we were allowed to change the problem, the integral $J(s) = \int_M |ds|^p *1$ is Palais-Smale for $p > \dim M$ [14].

When $\dim M = 2$, the harmonic map problem seems to be typical of variational problems depending on the conformal structure of M. The absolute minimum is taken on by trivial maps to a point. In minimizing energy in a homotopy class $\Lambda \in [M, N]$, the topological type of the weak limit of a minimizing sequence in $L_1^2(M, N)$ may change. However, the change occurs because small neighborhoods of isolated points in M may dilate to cover spheres in N. Therefore, existence theory is limited [17]. For example, if we minimize E in the space of maps of degree k from S^2 to S^2, we know from complex analysis that the meromorphic functions of degree k take on the minimum. To date there is no direct variational method which can produce this minimizing set unless $k = 1$. This is also the situation for the known Yang-Mills instanton fields over S^4.

Partial results show that one minimal solution $s : S^2 \to N$ exists if there is any topological reason for it [17]. In addition, Lemaire, Schoen and Yau [9], [18] show that the topological behavior of maps on π_1 is conserved under weak limits in $L_1^2(M, N)$. Minimizing harmonic maps exist under the topological constraint of behavior on π_1, although a minimum may not be taken on in every homotopy class in $[M, N]$.

There is an important, famous result on the existence of harmonic maps for M of arbitrary dimension. Eells, Hartman and Sampson [4], [7] show that the topological implications of Palais-Smale theory hold for energy whenever N has non-positive sectional curvature. (The Palais-Smale condition is not true unless $\dim M = 1$.) Whenever a geometric condition on a variational problem corresponding to the non-positive sectional curvature of N for harmonic maps can be identified, we expect good results. The term in the Yang-Mills equations corresponding to sectional curvature can be identified, but its sign cannot be prescribed by geometric conditions (it is the cubic term $(F \cdot [F, F])$ of the Weizenboch formula).

The relevance of the harmonic map problem for gauge theory occurs specifically in the question of the global existence of Lorentz (Hodge) gauges for connections. Given a trivial bundle over M and a connection $d+\tilde{A}$, when is $d+\tilde{A}$ gauge equivalent to a connection $d+A$ with $d^*A=0$? A gauge change is a map $s: M \to G$ where G is the Lie group. The connection form $\tilde{A}$ changes by $A = s^{-1}ds + s^{-1}\tilde{A}s$. The integral for this variational problem is

$$J(s) = \int_M |A|^2 *1 = \int_M |s^{-1}ds + s^{-1}\tilde{A}s|^2 *1$$

over $s: M \to G$. The Euler-Lagrange equations $d^*A = 0$ agree with the equation for harmonic maps $s: M \to G$ in the highest term. The local existence theory is now well understood [22], [23]. Because the group G will tend to have positive curvature, one does not expect global results as good as the Eells-Sampson existence theory for harmonic maps. However, the absolute minimum, under no constraints whatsoever, does exist in a weak sense and it is quite possible it is actually smooth (classical). In the harmonic map case, the corresponding minimum is trivial and therefore smooth.

§3. *Field theory*

Assume η is a vector bundle over a Riemannian manifold M with structure group G. Denote by $\mathrm{Ad}\,\eta$ and $\mathrm{Aut}\,\eta$ the associated adjoint and automorphism bundles. The space of connections in η is an affine space $\mathfrak{A}$. Given a base connection D_0, this space may be described as

$$\mathfrak{A} = \{D = D_0 + A : A \in C^\infty(M, T^*M \otimes \mathrm{Ad}\,\eta)\} .$$

For analytical purposes, one deals with spaces of Sobolev connections

$$\mathfrak{A}_k^p = \{D = D_0 + A : A \in L_k^p(M, T^*M \otimes \mathrm{Ad}\,\eta)\} .$$

A field is the curvature of a connection and is given by

$$F(D) = D^2 \in C^\infty(M, T^*M \wedge T^*M \otimes \mathrm{Ad}\,\eta)\,.$$

The curvature of a Sobolev connection behaves as expected when $p(k+1) \geq \dim M$. If $D \in \mathfrak{A}_k^p$,

$$F(D) = D^2 \in L_{k-1}^p(M, T^*M \wedge T^*M \otimes \mathrm{Ad}\,\eta)\,.$$

A number of important equations in physics arise as the Euler-Lagrange equations for integrals containing the term $\int_M |F|^2 *1$. If this term is the entire action, the Euler-Lagrange equations are the pure Yang-Mills equations. If the action has the form $\int_M |F|^2 *1 + \int_M |D\phi|^2 *1$ the Euler-Lagrange equations form a pair of coupled systems. One equation is a bundle Laplacian $\Delta\phi = 0$, i.e. linear. The second is a Yang-Mills equation coupled with a non-linear term in ϕ [8]. There are a number of disconcerting ways in which these integrals differ from the preceding geometric problems.

The first observation is the unusual existence of a very large (infinite dimensional) group of symmetries, the gauge group $\mathcal{G} = C^\infty(M, \mathrm{Aut}\,\eta)$. This has to be divided out in the variational approach. The gauge group for Sobolev connections $\mathfrak{A}_k^p$ has one more derivative: $\mathcal{G}_{k+1}^p = L_{k+1}^p(M, \mathrm{Aut}\,\eta)$. If we wish to preserve the entire topological structure of the bundle η under natural gauge changes, it is necessary to work in the Sobolev range $\mathcal{G}_{k+1}^p = L_{k+1}^p(M, \mathrm{Aut}\,\eta) \subset C^0(M, \mathrm{Aut}\,\eta)$ or $p(k+1) > \dim M$. Since curvature is essentially the exterior derivative of a connection, an integral containing the term $\int_M |F|^2 *1$ forces $X^\infty = \mathfrak{A}_1^2$, the space of L_1^2 connections. The natural gauge group is then $\mathcal{G}_2^2 = L_2^2(M, \mathrm{Aut}\,\eta)$, and the natural dimension constraint $4 > \dim M$. Of course, if one were willing to include a term $\int_M |F|^p *1$, then the dimension constraint is $2p > \dim M$. However, this topological constraint range suggests that the Palais-Smale range for Yang-Mills type equations over compact Riemannian manifolds is $\dim M < 4$ and $\dim M = 4$ is borderline. Recall from Section 1 the importance of weak convergence. The following theorem should be useful in verifying this conjecture, which is partially proved by Atiyah-Bott in $\dim M = 2$ [1], [22].

THEOREM. *Let* $2p > \dim M$ *and* $D(\alpha) \in \mathfrak{A}_1^p$ *be a sequence of connections in* η. *Assume also that* M *and* G *are compact, and* $\int_M |F(D)|^p * 1 \leq B$. *Then there exists a subsequence* $D(\alpha')$ *which is gauge equivalent by a sequence of elements in* $\mathcal{G}_2^p \subset C^0(M, \mathrm{Aut}\,\eta)$ *to a weakly convergent sequence in* $\mathfrak{A}_1^p$.

A not unrelated problem consists of the fact that the Yang-Mills type equations themselves are not elliptic. If the group G is abelian, the pure Yang-Mills equations are $d^*dA = 0$ for A a Lie algebra valued 1-form. In general, we will have a non-linear version of this equation to work with. However, the problem of choosing a good gauge *locally* in which $d^*A = 0$ is now understood. (This was described at the end of the previous section.) This added equation with the Euler-Lagrange equations form an elliptic system with very simple non-linearities.

Over compact manifolds of dimension less than 4, one expects good results for $\int_M |F|^2 * 1$ even with coupling. However, in dimension 4, we run across the typical phenomena associated with conformal variational problems described for the Yamabe and harmonic map problems. Typically, if a minimizing sequence of connections in $\mathfrak{A}_1^2$ is chosen, certain topological features of the bundle may change in the weak limit. Twists in the bundle can collect over isolated points and disappear in the limit; the second chern class of a bundle is clearly not conserved under all weak limits of connections in $\mathfrak{A}_1^2$. However, it is probably true that the first chern classes are conserved for weak limits in $\mathfrak{A}_1^2$ and a limited existence theory may be available here. Just as we do not really understand the harmonic map problem from S^2 to S^2 from a variational point of view, there is little to be said about the Yang-Mills fields over S^4 from this same point of view. (Note only the second chern class occurs.) The known solutions have not been found by variational methods.

The most serious gap in the variational theory occurs in the assumption that M is compact. The Yamabe problem and the harmonic map problem naturally occur for compact M; also the elliptic estimates used require compact M. However, gauge field theory problems require $M = R^2$, R^3 or R^4. For $M = R^4$, this is not terribly serious. The conformal invariance allows the problem over R^4 to be restated over $S^4 = R^4 \cup \{\infty\}$. The coupled problem is harder [12]. However, the integral $\int_{R^n} |F|^2 * 1$ for $n = 2$ or 3 behaves very differently than an integral over S^2 or S^3, especially when coupled with a potential term

$$J(D, \phi) = \int_{R^n} (|F|^2 + |D\phi|^2) * 1 .$$

In addition, unusual boundary conditions occur, such as the degree of $\phi/|\phi|$ on the sphere at ∞. Taubes [20], [21] has studied these coupled equations successfully, but the variational approach remains completely open.

REFERENCES

[1] M. Atiyah and R. Bott, On the Yang-Mills equations over Riemann surfaces (preprint).

[2] T. Aubin, Équations différentielles non linéaires et problème de Yamabe concernant la Courbure scalaire, J. Math. pures et appl. *55* (1976), 269-296.

[3] J. P. Bourguignon, H. B. Lawson, Jr., and J. Simons, Stability and gap phenomena for Yang-Mills fields, Proc. Nat. Acad. Sci. U.S.A. *76* (1979), 1550-1553.

[4] J. Eells and J. H. Sampson, Harmonic mappings of Riemannian manifolds, Amer. J. Math. *86*(1964), 109-160.

[5] J. Eells and L. Lemaire, A report on harmonic maps, Bull. London Math. Soc. 10(1978), 1-68.

[6] C. Gu, On the structure of classical Yang-Mills fields, Report on Mexico workshop (December 1979).

[7] P. Hartman, On homotopic harmonic maps, Can. J. Math. *19*(1967), 673-687.

[8] A. Jaffe, Introduction to gauge theories (International congress of mathematicians, Helsinki, (August 1978).

[9] L. Lemaire, Applications harmoniques de surfacés riemanniennes, J. Diff. Geo. *13*(1978), 51-78.

[10] J. Milnor, Morse theory, Ann. Math. Studies *51*(1970), Princeton Univ. Press.

[11] C. B. Morrey, Multiple integrals in the calculus of variations, Grundlehren *130*, Springer (1966).

[12] T. Parker, (Ph.D. thesis, Stanford).

[13] R.S. Palais, Ljusternik-Schnirelmann theory on Banach manifolds, Topology *5*(1966), 115-132.

[14] ———, Foundations of global non-linear analysis, Benjamen, New York (1968).

[15] ———, Critical point theory and the minimax principle, AMS Proc. of Sym. in Pure Math. XV (1970), 185-212.

[16] R.S. Palais and S. Smale, A generalized Morse theory, Bull. Amer. Math. Soc. 70(1964), 165-171.

[17] J. Sacks and K. Uhlenbeck, The existence of minimal 2-spheres, Annals of Math *113*(1981), 1-24.

[18] R. Schoen and S. T. Yau, Existence of incompressible minimal surfaces and the topology of 3-dimensional manifolds with non-negative scalar curvature, Ann. of Math. *110*(1979), 127-142.

[19] J. Schwartz, Generalizing the Ljusternik-Schnirelmann theory of critical points, Comm. Pure App. Math. *17*(1974), 307-315.

[20] C. Taubes, Arbitrary N-vortex solutions to the first order Landau-Ginzburg equations, Commun. Math. Phys. *72* #3 (1980), 277-292.

[21] A. Jaffe and C. Taubes, Vortices and Monopoles, Progress in Physics 2, Birkhauser, Boston (1980).

[22] K. Uhlenbeck, Removable singularities in Yang-Mills fields, to appear in Commun. Math. Phys.

[23] ———, Connections with L^p bounds on curvature, to appear in Commun. Math. Phys.

[24] H. Yamabe, On a deformation of Riemann structures on compact manifolds, Osaka Math. J. *12*(1960), 21-37.

Added in proof. Several relevant manuscripts have become available since this article was written.

R. Schoen and K. Uhlenbeck, Regularity of minimizing harmonic maps (U.C. Berkeley and U.I. Chicago Circle).

S. Sedlocek, A direct method for minimizing the Yang-Mills functional over 4-manifolds (Northwestern University).

C. Taubes, Self-dual Yang-Mills Connections on Non-self-dual 4-manifolds (Harvard University).

SOME GEOMETRICAL ASPECTS OF INTEGRABLE NONLINEAR EVOLUTION EQUATIONS

Hsing-Hen Chen and Yee-Chun Lee

Recently, the discovery of "soliton" solutions in many nonlinear evolution equations that has physical applications has aroused great interest and attention among physicists and mathematicians [1-5]. The word "soliton" was coined by Martin Kruskal [2] and Norman Zabusky to name the permanent nonlinear wave packet solution, indestructible even after collision among each other, that represents the nonlinear normal mode in nonlinear evolution systems. Kruskal and Zabusky first discovered the solitons through a numerical investigation of the Korteweg-deVries equation

$$q_t + 6qq_x + q_{xxx} = 0 , \tag{1}$$

an equation with many applications in various areas of physics. A typical one-soliton of this equation is given by

$$q = k^2 \operatorname{sech}^2 k(x - 4k^2 z - x_0) ,$$

where k and x_0 are arbitrary free parameters representing the strength (velocity) and the initial position of the soliton. Soon after the discovery of the solitons, Gardner, Greene, Kruskal and Miura [2] developed the inverse scattering method which gives the complete solution to the Cauchy problem of the Korteweg-deVries equation.

Seminar on Differential Geometry
0-691-08268-5/82/000465-18$00.90/0 (cloth)
0-691-08296-0/82/000465-18$00.90/0 (paperback)

The inverse scattering method involves a transformation of the nonlinear evolution equation into a pair of linear equations, also called the Lax equation [3],

$$L\psi = \lambda\psi \quad \text{and} \quad A\psi = \psi_t , \tag{2a}$$

where L and A are linear operators, depending on q, acting on the wavefunction ψ in a Hilbert space. The eigenvalue λ will be constant if the compatibility of the set in Equation (2a) is satisfied; that is,

$$L_t = [A, L] , \tag{2b}$$

This operator equation (2b), when written out explicitly, usually yields the nonlinear evolution equation under studying.

The presence of the eigenvalue λ in Equations (2) is crucial to the success of the inverse scattering method. The constancy of λ implies that the spectrum of the operator L does not change and are constants of motion [3]. The eigenvalues turn up in actual calculations as parameters that characterize the solitons. The constancy of these eigenvalues therefore guarantee the indestructibility of solitons.

The working of the inverse scattering problem is best illustrated in the following diagram

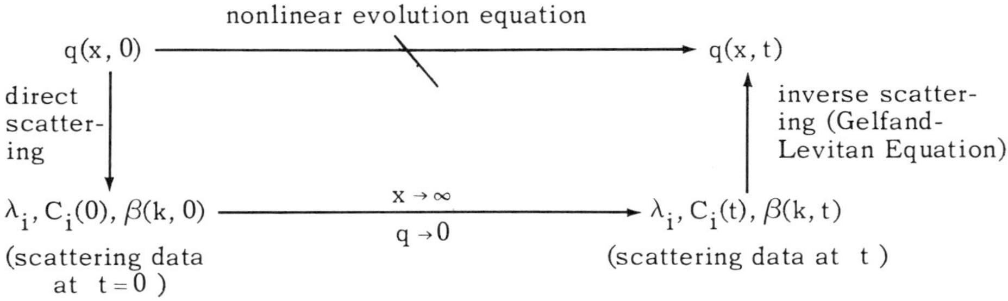

To go from q(x, 0) to q(x, t) through the solving of the nonlinear evolution equation directly is avoided. Instead, we take a detour by first solving the direct scattering problem $L\psi = \lambda\psi$ using q(x, 0) as the potential. The result is a set of initial scattering data, including the discrete eigenvalues

λ_i $(i=1,2,\cdots,\lambda)$, a set of normalization constants C_i associated with the bound states, and also $\beta(k, 0)$ the reflection coefficient associated with the continuum spectrum. The scattering data evolve in time very simply because the evolution can be carried out in the asymptotic region $(x \to \pm\infty)$ where $q(x,t) \to 0$ (boundary conditions of q) and is linear. We therefore obtain λ_i, $C_i(t)$ and $\beta(k,t)$. Then there is a way to reconstruct the potential function $q(x,t)$ from the scattering data at t. The so-called inverse scattering problem was solved a long time ago by Gelfand, Levitan [6], Marchenko [7], and others. It requires the solution of only a linear integral equation called the Gelfand-Levitan-Marchenko equation and the complete solution of the Cauchy problem is obtained by linear techniques only.

Through the above discussions, it is obvious that the presence of an eigenvalue λ in the Lax equation (2), is crucial to the success of the inverse scattering method. It was pointed out by Lund and Regge [8] that differential geometers [9, 10] had discovered the relation between nonlinear evolution equations (Gauss-Codazi equation) and a set of linear equation (Gauss-Weingarten equation) in the description of surfaces embedded in a three-dimensional Euclidean space long time ago. The important ingredients lacking in the Gauss-Weingarten equation to solve the nonlinear Gauss-Codazi equation is the eigenvalue λ. This may be one of the reasons why the mathematicians did not discover the inverse scattering problem at that time.

A classical example of the nonlinear evolution equation in the study of differential geometry is

$$\left(\frac{\partial^2}{\partial u^2} - \frac{\partial^2}{\partial v^2}\right) \omega = \sin\omega \cos\omega , \tag{3}$$

coined the sine-Gordon equation [11] by physicists. It first appeared in the literature in about 1875 [9]. Because of its many applications in physics, it received wide attentions recently [12]. Even though the complete solution of the Cauchy problem of this equation was not known then. The amazing fact is that mathematicians were already aware of a way to construct the so-called multi-soliton solutions through a method called the Bäcklund transformation [9, 13].

In the following, we will describe in detail the differential geometrical meaning of the nonlinear evolution equations as clear as possible.

When a two-dimensional surface is embedded in a three-dimensional Eucliding space [10], the construction of the surface can be described through the rotation of a trihedral $(\vec{x}_1, \vec{x}_2, \vec{x}_3)$ moving along two independent curves on the surface. The vectors $\vec{x}_1$ and $\vec{x}_2$ are usually taken as the tangent vectors along the two coordinate curves (u = const., v = const.) on the surface. And $\vec{x}_3$ will be the unit normal to the surface. We therefore have

$$\vec{x}_1 = \frac{\partial\vec{x}}{\partial u}, \quad \vec{x}_2 = \frac{\partial\vec{x}}{\partial u}, \quad \text{and} \quad \vec{x}_3 = \vec{x}_1 \times \vec{x}_2 / |\vec{x}_1 \times \vec{x}_2| \, , \tag{4}$$

where $\vec{x}(u, v)$ is the position vector of points on the surface.

Associated with the surface, there are two sets of invariants. The first set is associated with the first fundamental form or the metric of a curve on the surface.

$$ds^2 = E\, du^2 + 2F\, dudv + G\, dv^2 = \sum_{i,k=1}^{2} g_{ik} du^i du^k \, , \tag{5}$$

where $E = |\vec{x}_1|^2$, $F = \vec{x}_1 \cdot \vec{x}_2$, and $G = |\vec{x}_2|^2$ and $g_{ik} = \vec{x}_i \cdot \vec{x}_k$. Another set of invariants is associated with the so-called second fundamental form

$$Ldu^2 + 2M\, dudv + Ndv^2 = \sum_{i,k=1}^{2} L_{ik} du^i du^k \, . \tag{6}$$

describing the departure of a surface from its tangent plane. The Gauss-Weingarten equation then describes the rate of change of the trihedral as it moves along curves on the surface.

$$\begin{cases} \dfrac{\partial \vec{x}_i}{\partial u^k} = \Gamma^{\ell}_{ik}\vec{x}_{\ell} + L_{ik}\vec{x}_3 \quad i,k = 1,2 \\[2ex] \dfrac{\partial \vec{x}_3}{\partial u_j} = -L^k_i \vec{x}_k \ , \end{cases} \tag{7}$$

where Γ^{ℓ}_{ik} is the Christoffel symbol related to the metric tensor g_{ik} by

$$\begin{aligned} \Gamma^{\ell}_{ik} &= g^{\ell m}\Gamma_{ikm} \\ &= \frac{1}{2} g^{\ell m}\left(\frac{\partial g_{im}}{\partial u^k} + \frac{\partial g_{mk}}{\partial u^i} - \frac{\partial g_{ki}}{\partial u^m}\right) . \end{aligned} \tag{8}$$

The Gauss-Weingarten equation determines a surface once the two sets of invariants g_{ik} and L_{ik} are given. However, not arbitrary values of g_{ik} and L_{ik} can be chosen to yield a surface. The Gauss-Weingarten equation is an overdetermined set of linear equations for $\vec{x}_i$ and $\vec{x}_3$. We need a compatibility condition $(\partial^2 \vec{x}_i/\partial u^k \partial u^j = \partial^2 \vec{x}_i/\partial u^j \partial u^k)$ to guarantee solutions to this set of equations. The compatibility condition was given by Gauss and Codazi as

and
$$\begin{cases} R^n_{ijk} \equiv \dfrac{\partial \Gamma^n_{ik}}{\partial u^j} - \dfrac{\partial \Gamma^n_{ij}}{\partial u^k} + \Gamma^{\ell}_{ik}\Gamma^n_{ij} - \Gamma^{\ell}_{ij}\Gamma^n_{\ell k} = L_{ik}L^n_j - L_{ij} \\[2ex] \dfrac{\partial L_{ik}}{\partial u^j} - \dfrac{\partial L_{ij}}{\partial u^k} + \Gamma^{\ell}_{ik}L_{\ell j} - \Gamma^{\ell}_{ij}L_{\ell k} = 0 \ . \end{cases} \tag{9}$$

Equation (9) contains a lot of trivial and redundant equations. It can be reduced to essentially the following three equations.

$$\frac{\partial L_{12}}{\partial u^1} - \frac{\partial L_{11}}{\partial u^2} + \Gamma^{\ell}_{12}L_{\ell 1} - \Gamma^{\ell}_{11}L_{\ell 2} = 0$$

$$\frac{\partial L_{22}}{\partial u^1} - \frac{\partial L_{21}}{\partial u^2} + \Gamma^{\ell}_{22}L_{\ell 1} - \Gamma^{\ell}_{21}L_{\ell 2} = 0 \tag{10a}$$

known as the Codazi equation and

$$R_{2121} = L_{11} L_{22} - L_{12}^2 \tag{10b}$$

known as the Gauss equation.

The Gauss-Codazi equation is a set of complicated nonlinear partial differential equations. They are perhaps the most important equations occurred in the classical differential geometry. For example, the Gauss equation means that the Gaussian curvature of a surface is actually an intrinsic property of the surface independent of how the surface is embedded into the exterior space. The Gauss-Codazi equation is also important to us because it is the compatibility condition of a set of linear equations, the Gauss-Weingarten equation. It resembles very much the Lax equations in the inverse scattering method in solving nonlinear evolution equations. The difference is the absence of a free parameter (not present in the Gauss-Codazi equation) capable of being considered as an eigenvalue in the Gauss-Weingarten equation.

For example, the sine-Gordon equation describes surfaces with constant negative Gaussian curvature [9] (or pseudospheres). In this case, the curves $u = \text{const}$ and $v = \text{const}$ are chosen to be orthogonal. $F = g_{12} = \vec{x}_1 \cdot \vec{x}_2 = 0$ and the first fundamental form becomes

$$ds^2 = \cos^2\omega \, du^2 + \sin^2\omega \, dv^2 . \tag{11}$$

A variant form of the Gauss equation (10b) when $F = 0$ is given by [10]

$$K = -\frac{1}{2\sqrt{EG}} \left[\frac{\partial}{\partial v} \left(\frac{1}{\sqrt{EG}} \frac{\partial}{\partial v} E \right) + \frac{\partial}{\partial u} \left(\frac{1}{\sqrt{EG}} \frac{\partial}{\partial u} G \right) \right] ,$$

where K is the Gaussian curvature. If we put $E = \cos^2\omega$ and $G = \sin^2\omega$ into the above equation, we obtain

$$K = -\frac{1}{\sin\omega \cos\omega} \left(\frac{\partial^2}{\partial u^2} \omega - \frac{\partial^2}{\partial v^2} \omega \right) . \tag{12}$$

It is therefore clear that the solutions of the sine-Gordon equation (3) describe surfaces with Gaussian curvature $K = -1$. In fact, both the Gauss-Codazi and the Gauss-Weingarten equations for pseudospherical surfaces also depend on the invariants of the second fundamental form. They are given by

$$\begin{aligned} L_{11} &= -\sin\omega\cos\omega = -L_{22} \\ L_{12} &= 0 . \end{aligned} \tag{13}$$

After substituting both (11) and (13) into the Gauss-Codazi equation (10), we find (10a) trivially satisfied and (10b) reduced to (12). Therefore, the sine-Gordon equation is the Gauss-Codazi equation of pseudospherical surfaces. Furthermore, its associated linear Gauss-Weingarten equation is given by

$$\frac{\partial}{\partial u}\begin{pmatrix} \hat{e}_1 \\ \hat{e}_2 \\ \hat{e}_3 \end{pmatrix} = \begin{pmatrix} 0 & \frac{\partial\omega}{\partial v} & -\sin\omega \\ -\frac{\partial\omega}{\partial v} & 0 & 0 \\ \sin\omega & 0 & 0 \end{pmatrix} \begin{pmatrix} \hat{e}_1 \\ \hat{e}_2 \\ \hat{e}_3 \end{pmatrix} \tag{14a}$$

and

$$\frac{\partial}{\partial v}\begin{pmatrix} \hat{e}_1 \\ \hat{e}_2 \\ \hat{e}_3 \end{pmatrix} = \begin{pmatrix} 0 & \frac{\partial\omega}{\partial u} & 0 \\ -\frac{\partial\omega}{\partial u} & 0 & \cos\omega \\ 0 & \cos\omega & 0 \end{pmatrix} \begin{pmatrix} \hat{e}_1 \\ \hat{e}_2 \\ \hat{e}_3 \end{pmatrix} \tag{14b}$$

where $\hat{e}_1$, $\hat{e}_2$, and $\hat{e}_3$ are normalized vectors for $\vec{x}_1$, $\vec{x}_2$ and $\vec{x}_3$. They constitute an orthonormal set and Equation (14) tells us how the trihedral rotates as it moves in the u and the v directions.

Although very close in form to the linear Lax equations (2a) as the scattering equations for the sine-Gordon equation, the Gauss-Weingarten equation is handicapped by the absence of an eigenvalue. It is therefore unsuitable in the present form to be considered as the inverse scattering problem for the sine-Gordon equation (3). However, differential geometers a hundred years ago came up with an ingeneous method called the "Bäcklund transformation" [13], which enabled them to construct the hirachical multisoliton solutions of the sine-Gordon equation. The derivation of the Bäcklund transformation [14] can be proceeded as the following. Equation (14) is a set of equations for the nine components of three mutually orthogonal unit vectors. We can project these vectors to any arbitrary direction and obtain a set of equations for three real scalars satisfying $e_1^2 + e_2^2 + e_3^2 = 1$ (we still use the same notation e_i for their scalar components). We then introduce the transformation,

$$\phi = \frac{e_2 + ie_3}{1 - e_1}$$

and obtain from Equation (14) the following set of Ricatti equations.

$$\begin{cases} \dfrac{\partial\phi}{\partial u} = \dfrac{\partial\omega}{\partial v}(1+\phi^2) + i\sin\omega\,(1-\phi^2) \\[2ex] \dfrac{\partial\phi}{\partial v} = \dfrac{\partial\omega}{\partial u}(1+\phi^2) + 2\,i\cos\omega\,\phi\,. \end{cases}$$

Then, letting $\zeta = -2\,\mathrm{tg}^{-1}\phi$, we obtain the Bianchi transformation [9]

$$\begin{cases} \dfrac{\partial\zeta}{\partial u} + \dfrac{\partial\omega}{\partial v} = -i\sin\omega\cos\zeta \\[2ex] \dfrac{\partial\zeta}{\partial v} + \dfrac{\partial\omega}{\partial u} = -i\cos\omega\sin\zeta\,. \end{cases} \tag{15}$$

It is interesting to note that ζ in Equation (15) also satisfies the sine-Gordon equation

$$\frac{\partial^2\zeta}{\partial u^2} - \frac{\partial^2\zeta}{\partial v^2} = \sin\zeta\cos\zeta\,.$$

The usefulness of the Bianchi transformation is limited, however, because of the lack of a free parameter. A free parameter can be introduced into (15) by realizing that the sine-Gordon equation has a Lorentz symmetry.

It is invariant under the Lorentz transformation

$$\begin{aligned} u \to u' &= -i \cos\alpha\, u + i \csc\alpha\, v \\ v \to v' &= -i \csc\alpha\, u + i \cot\alpha\, v \,. \end{aligned} \tag{16}$$

A free parameter α is thus introduced into the Bianchi transformation (15). The resulting transformation with the parameter α was first discovered by A. V. Bäcklund [13] in 1875 and is now bearing his name—the Bäcklund transformation.

$$\begin{cases} \sin\alpha \left(\dfrac{\partial\zeta}{\partial u} + \dfrac{\partial\omega}{\partial v}\right) = \sin\zeta \cos\omega - \cos\alpha \cos\zeta \sin\omega \\ \sin\alpha \left(\dfrac{\partial\zeta}{\partial v} + \dfrac{\partial\omega}{\partial u}\right) = -\cos\zeta \sin\omega + \cos\alpha \sin\zeta \cos\omega \,. \end{cases} \tag{17}$$

The usefulness of the Bäcklund transformation (17) is its ability to generate multi-soliton solutions [15]. Starting with the trivial solution $\omega_0 = 0$ and an arbitrarily chosen value of the parameter $\alpha = \alpha_1$, Equation (17) yields a single-soliton solution.

$$\zeta(\alpha_1) = 2 \tan^{-1} e^{\csc\alpha_1 (u + \cos\alpha_1 v)} \,. \tag{18}$$

It represents a kink which assumes value 0 as $u \to -\infty$ and π as $u \to +\infty$ with an inflection point at $u = -\cos\alpha\, v$. Using this new solution in Equation (17) and putting in a new value of the parameter α_2, we then obtain from the Bäcklund transformation a second solution $\omega(\alpha_1, \alpha_2)$ depending on two parameters (also called a two-soliton solution). Obviously, when $\alpha_1 = \alpha_2$, the two-soliton solution reduces to the trivial solution $\omega(\alpha, \alpha) = 0$. This fact that no two solitons with identical parameters can exist at the same time reminds us of the analogous property of the Fermi particles in quantum statistical mechanics. In order to obtain explicit expressions of the two (or multi-) soliton solutions, it seems that we still have to solve a set of complicated differential equations. In reality, this difficulty is usually avoided with the aid of an identity obtained by

Bianchi [16]. He argued that the order of two successive Bäcklund transformations applied to a solution ω_0 (any solution of the nonlinear equation) could be irrelevant. In other words, there always exists a common solution $\omega(a_1, a_2) = \omega(a_2, a_1)$. The consequence of the permutability is a superposition formula enabling us to construct multi-solitons algebraically.

To derive this nonlinear superposition formula, we use only the first half of the Bäcklund transformation in Equation (17). Obviously, we have

$$\frac{\partial \zeta_1}{\partial u} + \frac{\partial \omega_0}{\partial v} = \frac{1}{\sin a_1} [\sin\zeta_1 \cos\omega_0 - \cos a_1 \cos\zeta_1 \sin\omega_0] \tag{19a}$$

$$\frac{\partial \zeta_2}{\partial u} + \frac{\partial \omega_0}{\partial v} = \frac{1}{\sin a_2} [\sin\zeta_2 \cos\omega_0 - \cos a_2 \cos\zeta_2 \sin\omega_0] \tag{19b}$$

$$\frac{\partial \zeta_1}{\partial u} + \frac{\partial \omega_{12}}{\partial v} = \frac{1}{\sin a_2} [\sin\zeta_1 \cos\omega_{12} - \cos a_2 \cos\zeta_1 \sin\omega_{12}] \tag{19c}$$

$$\frac{\partial \zeta_2}{\partial u} + \frac{\partial \omega_{21}}{\partial v} = \frac{1}{\sin a_1} [\sin\zeta_2 \cos\omega_{21} - \cos a_1 \cos\zeta_2 \sin\omega_{21}] . \tag{19d}$$

Now since $\omega_{12} = \omega_{21}$, we may eliminate all the derivatives on the left side by subtracting (a) and (d) from the sum of (b) and (c). After some simplifications, the result is given by

$$\tan\left(\frac{\omega_{12} - \omega_0}{2}\right) = \frac{\sin\left(\frac{a_1 + a_2}{2}\right)}{\sin\left(\frac{a_1 - a_2}{2}\right)} \tan\left(\frac{\zeta_1 - \zeta_2}{2}\right) . \tag{20}$$

Since Equation (20) does not contain derivatives of ω_{12}, the construction of ω_{12} involves only algebraic operations. Using this superposition formula, the hierarchy of multi-solitons can be constructed. The process of constructing more and more complicated solutions can be illustrated by the Lamb diagram [15]

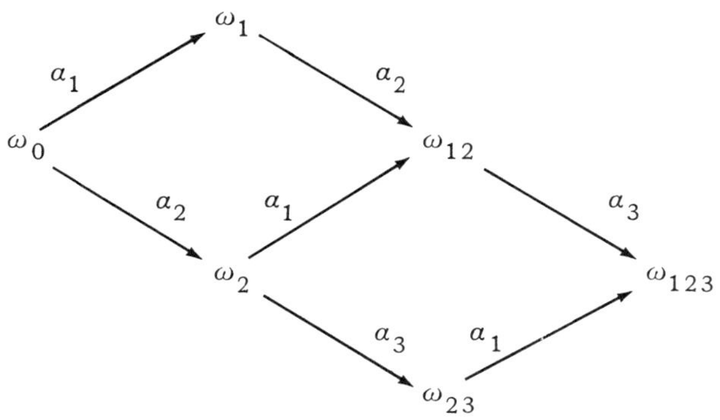

Each arrow indicates an application of the Bäcklund transformation with the specified parameters indicated. Each parallelogram represents an independent unit in which the Bianchi superposition formula (20) can be applied. Since the diagram can be extended to the right indefinitely, solutions with ever larger numbers of parameters (multi-soliton solutions) can be constructed with ease.

The Bäcklund transformation method illustrated above though ingenuous and useful in constructing multi-soliton solutions, nevertheless does not give complete solutions to the sine-Gordon equation. The only way to solve completely the Cauchy problem of the sine-Gordon equation is to apply the inverse scattering method. So far, we have associated with the nonlinear sine-Gordon equation a set of compatible linear equations, the Gauss-Weingarten equation, which is unsuitable for the inverse scattering method because of the absence of a free parameter. On the other hand, we derived from this set of equations a Bianchi transformation and subsequently introduced a free parameter to it through the Lorentz symmetry [17] the sine Gordon equation possessed. In fact, the Lorentz symmetry can be applied directly to the Gauss-Weingarten equation to generate an eigenvalue. However, another drawback of considering the Gauss-Weingarten equation as the scattering equation is the following: The vectors $\hat{e}_1$, $\hat{e}_2$ and $\hat{e}_3$ are mutually orthogonal and of unit length. Therefore, the projection of them along any fixed direction would satisfy the constraint

$$(\hat{n}\cdot\hat{e}_1)^2 + (\hat{n}\cdot\hat{e}_2)^2 + (\hat{n}\cdot\hat{e}_3)^2 = 1 \,. \tag{21}$$

This constraint makes the analysis of the scattering problem more difficult. However, we note that because of (21) the Gauss-Weingarten equation is actually a linear equation of not a three-component column but a real 3×3 matrices with its elements in each column satisfying (21). These matrices belong to the rotation group $O(3)$ and the matrix on the right side of Equation (14) belongs to the generators of the group which forms the $O(3)$ algebra. It is well known that an equivalent representation of the $O(3)$ algebra is the $SU(2)$ algebra. It is therefore possible to reduce the Gauss-Weingarten equation from three to two dimensions.

$$\begin{aligned} i\,\frac{\partial}{\partial u}\begin{pmatrix}\psi_1\\ \psi_2\end{pmatrix} &= \frac{1}{2}\left(\frac{\partial\omega}{\partial v}\,\sigma_1 - \sin\omega\;\sigma_2\right)\begin{pmatrix}\psi_1\\ \psi_2\end{pmatrix}\\ i\,\frac{\partial}{\partial v}\begin{pmatrix}\psi_1\\ \psi_2\end{pmatrix} &= \frac{1}{2}\left(\frac{\partial\omega}{\partial u}\,\sigma_1 - \cos\omega\;\sigma_3\right)\begin{pmatrix}\psi_1\\ \psi_2\end{pmatrix}\end{aligned} \tag{22}$$

where

$$\sigma_1 = \begin{pmatrix}0 & 1\\ 1 & 0\end{pmatrix}, \quad \sigma_2 = \begin{pmatrix}0 & -i\\ i & 0\end{pmatrix}, \quad \text{and} \quad \sigma_3 = \begin{pmatrix}1 & 0\\ 0 & -1\end{pmatrix}$$

are the Pauli spin matrices. The eigenvalue can then be introduced by applying the Lorentz transformation (16) to (22). We get finally

$$\left\{\begin{aligned} i\,\frac{\partial}{\partial u}\begin{pmatrix}\psi_1\\ \psi_2\end{pmatrix} &= \frac{1}{2}\left(\frac{\partial\omega}{\partial v}\,\sigma_1 + i\cot\alpha\,\sin\omega\;\sigma_2 - i\csc\alpha\,\cos\omega\;\sigma_3\right)\begin{pmatrix}\psi_1\\ \psi_2\end{pmatrix}\\ i\,\frac{\partial}{\partial v}\begin{pmatrix}\psi_1\\ \psi_2\end{pmatrix} &= \frac{1}{2}\left(\frac{\partial\omega}{\partial u}\,\sigma_1 - i\cot\alpha\,\cos\omega\;\sigma_3 + i\csc\alpha\,\sin\omega\;\sigma_2\right)\begin{pmatrix}\psi_1\\ \psi_2\end{pmatrix}\end{aligned}\right. \tag{23}$$

The solution of this scattering problem was given before in reference (18).

Recently, we have proposed a direct method [17] to test the integrability by inverse scattering method of a given nonlinear evolution equation. One of the key ingredients in our approach is to choose the linearized equation of the original nonlinear evolution equation as one of the two linear Lax equations needed to carry out the inverse scattering method. In case of the sine-Gordon equation (3), letting $\omega = \omega + \varepsilon\psi$, and collecting terms to first order in ε, we get

$$\frac{\partial^2\psi}{\partial u^2} - \frac{\partial^2\psi}{\partial v^2} = (\cos^2\omega - \sin^2\omega)\,\psi\ . \tag{24}$$

It is interesting to note that the Gauss-Weingarten equation, when $\hat{e}_1 = \hat{e}_2$ are eliminated, reduces to

$$\frac{\partial^2\hat{e}_3}{\partial u^2} - \frac{\partial^2\hat{e}_3}{\partial v^2} = (\cos^2\omega - \sin^2\omega)\hat{e}_3$$

exactly the same as the linearized equation (24).

This turns out to be not just a coincidence. There is a whole set of equations that can be solved through a scattering problem, the so-called Zakharov-Shabat [4] (or AKNS) [5] scattering problem.

$$\frac{\partial}{\partial u}\begin{pmatrix}\psi_1\\ \psi_2\end{pmatrix} = \begin{pmatrix}\lambda & q\\ r & -\lambda\end{pmatrix}\begin{pmatrix}\psi_1\\ \psi_2\end{pmatrix}. \tag{25}$$

Well-known examples are the KdV equation (1) when $r = 1$, the mKdV equation [18]

$$\frac{\partial q}{\partial v} \pm 6\, q^2\frac{\partial q}{\partial u} + \frac{\partial^3 q}{\partial u^3} \tag{26}$$

when $r = \pm q$, and the nonlinear Schrodinger equation [4]

$$i\,\frac{\partial q}{\partial v} + \frac{\partial^2 q}{\partial u^2} \pm 2\, qq^*q = 0 \tag{27}$$

when $r = \pm q^*$.

If we let

$$A = \psi_1^2 , \quad B = \psi_2^2 , \quad \text{and} \quad C = \psi_1 \psi_2 , \tag{28}$$

then Equation (25) becomes

$$\begin{aligned} A_u &= 2\lambda A + 2qC \\ B_u &= 2rC - 2\lambda B \\ C_u &= qB + rA . \end{aligned} \tag{29}$$

Obviously, from Equation (28) there is a constraint on A, B, and C; namely,

$$AB = C^2 . \tag{30}$$

This constraint is similar to the constraint (21) of the vectors in the Gauss-Weingarten equation. Actually, the constraint (30) is too strong. Equation (29) itself only requires $AB - C^2 = \text{constant}$. We may choose this constant to be unity and let

$$e_1 = (A + iB)/(1+i)\sqrt{2}$$

$$e_2 = (B + iA)/(1-i)\sqrt{2}$$

and

$$e_3 = iC . \tag{31}$$

Obviously, $e_1^2 + e_2^2 + e_3^2 = 1$ is satisfied if $C^2 - AB = 1$. Furthermore, we derive from Equations (29) and (31) the following equations

$$\frac{\partial}{\partial u}\begin{pmatrix} e_1 \\ e_2 \\ e_3 \end{pmatrix} = \begin{pmatrix} 0 & -2i\lambda & -(q+ir)\varepsilon \\ 2i\lambda & 0 & -(r+iq)\varepsilon \\ \varepsilon(q+ir), & \varepsilon(r+iq), & 0 \end{pmatrix} \begin{pmatrix} e_1 \\ e_2 \\ e_3 \end{pmatrix} \tag{32}$$

where $\varepsilon = \dfrac{1+i}{\sqrt{2}}$ is a square root of i.

Equation (32) looks like a member of the Gauss-Weingarten Equation (14). If it is so, then a second member is needed to complete the Gauss-

Weingarten equation for a surface associated with a given nonlinear evolution equation solvable by the Zakharov-Shabat scattering problem. Since the additional equation should characterize the particular equation to be solved, it is not surprising to find out that it is nothing but the linearized equation that we proposed earlier. To see this explicitly, let us consider the coupled nonlinear Schrodinger equation

$$\begin{cases} i\dfrac{\partial q}{\partial v} + \dfrac{\partial^2 q}{\partial u^2} - 2rq^2 = 0 \\[2ex] i\dfrac{\partial r}{\partial v} - \dfrac{\partial^2 r}{\partial u^2} - 2qr^2 = 0 \end{cases} \tag{33}$$

as an example. When $r = \pm q^*$, we recover Equation (27). The linearized equation of (33) is easily found to be

$$i\frac{\partial}{\partial v}\begin{pmatrix} A \\ \\ B \end{pmatrix} = \begin{pmatrix} \dfrac{\partial^2}{\partial u^2} + 4rq\,, & 2r^2 \\[2ex] 2q^2\,, & -\dfrac{\partial^2}{\partial u^2} + 4rq \end{pmatrix}\begin{pmatrix} A \\ \\ B \end{pmatrix}. \tag{34}$$

We have denoted the linearized quantities as A and B the same used in Equation (29) because direct checks on Equations (27) and (34) show that they are compatible if q and r satisfy the coupled nonlinear Schrodinger equation (33). Furthermore, the transformation (31) with the help of Equation (29) brings Equation (34) to a form of the second member of the Gauss-Weingarten equation (14b). The result is

$$\frac{\partial}{\partial v}\begin{pmatrix} e_1 \\ e_2 \\ e_3 \end{pmatrix} = \begin{pmatrix} 0 & h & -g\varepsilon \\ -h & 0 & -\ell\varepsilon \\ \varepsilon g & \varepsilon\ell & 0 \end{pmatrix}\begin{pmatrix} e_1 \\ e_2 \\ e_3 \end{pmatrix} \tag{35}$$

where

$$h = -4\lambda^2 - 2qr$$

$$g = 2(iq - r)\lambda + \frac{\partial}{\partial u}(iq + r)$$

$$\ell = 2(ir - q)\lambda - \frac{\partial}{\partial u}(q + ir)$$

and

$$\varepsilon = \frac{1+i}{\sqrt{2}} .$$

The recognition of Equations (32) and (35) as the Gauss-Weingarten equation of a certain surface is not completely right though. The two matrices in Equations (32) and (35) are in general complex while $\hat{e}_1$, $\hat{e}_2$ and e_3 are real vectors. The restriction $r = q^*$ and $\lambda = 0$ would make the matrices real, however. It implies that the geometrical space associated with the coupled nonlinear Schrodinger equation is, in general, complex if the restriction $r = q^*$ is not made. Since the generalization is straightforward, we would not pursue it further here.

As a final remark to the geometrical interpretation of nonlinear evolution equations, we would like to mention that nonlinear evolution equations do not always correspond to the compatibility equation (Gauss-Codazi equation). Sometimes, they would correspond to the Gauss-Weingarten equation directly. As an example, we would like to mention the one-dimensional spin-wave equation

$$\frac{\partial}{\partial t}\vec{S} = \vec{S} \times \frac{\partial^2}{\partial x^2}\vec{S} \tag{36}$$

describing the nearest neighbor interactions of spin vectors with unit length through a self-consistent magnetic field. It is easily shown that the unit vector $\hat{e}_3$ in the Gauss-Weingarten equations (32) and (35) of the nonlinear Schrodinger equation (27) with $r = -q^*$ satisfies Equation (36). That is,

$$\frac{\partial}{\partial v}\hat{e}_3 = \hat{e}_3 \times \frac{\partial^2}{\partial u^2}\hat{e}_3$$

is satisfied provided $\hat{e}_1$, $\hat{e}_2$ and $\hat{e}_3$ form a right-handed orthonormal system. Interestingly, the Gauss-Weingarten equation (14) associated with the Sine-Gordon equation (3) also yields a nonlinear vector evolution equation

$$\begin{cases} \dfrac{\partial \vec{x}_1}{\partial u} - \dfrac{\partial \vec{x}_2}{\partial v} = 0 \\[2ex] \dfrac{\partial \vec{x}_2}{\partial v} - \dfrac{\partial \vec{x}_1}{\partial u} = \vec{x}_1 \times \vec{x}_2 , \end{cases} \tag{37}$$

with $\vec{x}_1 = \hat{e}_1 \cos\omega$ and $\vec{x}_2 = \hat{e}_2 \sin\omega$. It was discovered by Lund and Regge [8] in connection with motions of a vortex line immersed in a fluid.

In this paper we have reviewed the relation between a nonlinear evolution equation and the embedding of a surface in higher spaces. The associated surfaces are shown to be obtained from linearizing the nonlinear evolution equations though caution has to be made for possible nonlinear evolution equations that should be associated with the Gauss-Weingarten equation itself instead of the compatibility conditions. The Bäcklund transformation, which is a way to construct the multi-soliton solutions, is also derived from the Gauss-Weingarten equation. In order to apply the inverse scattering method, an eigenvalue should be introduced into the Gauss-Weingarten equation which is usually possible by inspecting the continuous symmetry of the nonlinear evolution equation.

Although the above geometrical interpretation for nonlinear partial differential equations works very well. There are also nonlinear integro-differential equations that are integrable with soliton solutions. For example, the Benjamin-Ono equation [20, 21]

$$\frac{\partial q}{\partial t} + 2q\,\frac{\partial q}{\partial x} + \frac{\partial^2}{\partial x^2}\,\frac{p}{\pi} \int_{-\infty}^{\infty} \frac{q(y)}{y-x}\, dy = 0$$

is shown recently to have algebraic solitons. Their geometrical associations should be an interesting subject.

Acknowledgments. This work is supported by the Office of Naval Research (Contract N00014-79-C-0665) and the U. S. Department of Energy (Contract DE-AS05-77DP40032).

DEPARTMENT OF PHYSICS AND ASTRONOMY
UNIVERSITY OF MARYLAND
COLLEGE PARK, MARYLAND 20742

REFERENCES AND NOTES

[1] M. Kruskal and N. Zabusky, Phys. Rev. Lett. *15*, 240 (1965). See also A. Scott, F. Chu and D. McLauglin, Proc. IEEE *61*, 1443 (1973).

[2] C. S. Gardner, J. M. Greene, M. Kruskal and R. M. Miura, Phys. Rev. Lett. *19*, 1095 (1967); Comm. Pure Appl. Math. *27*, 97 (1974).

[3] P. Lax, Comm. Pure Appl. Math. *21*, 647 (1968).

[4] V. E. Zakharov and A. B. Shabat, Soviet Phys. J. E.T.P. *34*, 62 (1972).

[5] M. Ablowitz, D. J. Kaup, A. Newell and H. Segur, Phys. Rev. Lett. *31*, 125 (1973). See also Stud. Appl. Math. *53*, 249 (1974).

[6] I. M. Gelfand and B. M. Levitan, Am. Math. Soc. Trans. Ser. 2, *1*, 253 (1955).

[7] V. A. Marchenko, Dokl. Akad. Nauk S.S.S.R. *72*, 47 (1950).

[8] F. Lund and T. Regge, Phys. Rev. D. *14*, 1524 (1976).

[9] L. P. Eisenhart, "A Treatise on the Differential Geometry," Ginn and Company (1909).

[10] J. J. Stoker, Differential Geometry, Wiley-Interscience (1969).

[11] T. Skyrme, Proc. R. Soc. Lond. *A262*, 237 (1961).

[12] R. Rajaraman, Phys. Report *21C*, 227 (1975).

[13] A. V. Bäcklund, Lund Universitets Arsskrift *10* (1875).

[14] See also H. H. Chen, Phys. Rev. Lett. *33*, 925 (1974).

[15] G. E. Lamb, Jr., Rev. Mod. Phys. *43*, 99 (1971).

[16] L. Bianchi, Lezioni di Geometria Differenziale V.II, p. 418, Pisa (1902).

[17] S. Lie, Archiv for Mathematik og Naturvidenskab Vol. IV, p. 150 (1829). See also Ref. [8].

[18] F. Lund, Phys. Rev. Lett. *38*, 1175 (1977).

[19] H. H. Chen, C. S. Liu and Y. C. Lee, Physica Scripta *20*, 490 (1979).

[20] H. H. Chen, Y. C. Lee and N. Pereira, Phys. Fluds *22*, 187 (1979).

[21] H. H. Chen and Y. C. Lee, Phys. Rev. Lett. *43*, 264 (1979).

THE CONFORMALLY INVARIANT LAPLACIAN AND THE INSTANTON VANISHING THEOREM

R. O. Wells, Jr.*

§1. *Introduction*

In the recent solution of the self-dual Yang-Mills equations on S^4 due to Atiyah, Drinfeld, Hitchin and Manin ([2], see [1] for a full discussion of this theorem with proofs and references) a vanishing theorem for certain holomorphic vector bundles on $\mathbf{P}_3(\mathbf{C})$ plays a crucial role. Namely if $\pi : \mathbf{P}_3(\mathbf{C}) \to S^4$ is the natural mapping of projective 3-space to the 4-sphere obtained by putting a quaternionic structure on $\mathbf{P}_3$ (see §2 below), then a holomorphic vector bundle $E \to \mathbf{P}_3$ is an *instanton bundle* if $E = \pi^* F$, where F is a real-analytic G-bundle on S^4, for a compact semi-simple Lie group G, equipped with a connection ω which satisfies the self-dual Yang-Mills equations on S^4, and the holomorphic structure on E is induced in a natural manner from the differential-geometric data on F. If $H \to \mathbf{P}_3$ is the hyperplane section bundle, i.e., $c_1(H) = 1$, then the vanishing theorem of Atiyah-Drinfeld-Hitchin-Manin asserts that, if E is an instanton bundle, then

$$H^1(\mathbf{P}_3, \mathcal{O}_{\mathbf{P}_3}(E \otimes H^{-2})) = 0 .$$

*This research is supported by the National Science Foundation, the Vaughn Foundation, and the Institute for Advanced Study.

Seminar on Differential Geometry
0-691-08268-5/82/000483-16 $00.80/0 (cloth)
0-691-08296-0/82/000483-16 $00.80/0 (paperback)
For copying information, see copyright page.

The purpose of this paper is to give a short proof of this theorem, using results concerning the general Penrose transform developed in [5]. Namely, in [5], there is developed a general method for transforming cohomology groups on open subsets of $\mathbf{P}_3$ to solutions of differential equations on open subsets of a natural complexification $\mathbf{M}$ of compactified real Minkowski space. In §2 we show how S^4 is naturally embedded in $\mathbf{M}$. In §3 we develop the Penrose transform and show that it maps $H^1(\mathbf{P}_3, \mathcal{O}_{\mathbf{P}_3}(E \otimes H^{-2}))$ naturally and isomorphically onto solutions of a second order conformally invariant differential equation $D\phi = 0$ on $S^4 \subset \mathbf{M}$. Then we identify D with $\nabla^*\nabla + R/6$, by using a characterization of this conformally invariant operator due to Hitchin [7]. Here ∇ is covariant differentiation on F, and R is the scalar curvature of S^4. Then a Bochner-type vanishing argument (used in all of the proofs of this theorem) concludes the proof. In §4 we give a brief comparison of the notions of conformal weights as used in [5] and [7], and include a generalization of the well-known formula $K_{\mathbf{P}_n} \cong H^{-(n+1)}$ (for the canonical bundle of projective space) to a similar formula on Grassmannian manifolds.

§2. *Twistor geometry and the embedding of* S^4 *in complexified Minkowski space*

We will first recall the basic elements of twistor geometry (referring to [5] and [12] for further details and references). The space of twistors $\mathbf{T}$ is, by definition, a 4-dimensional complex vector space with an Hermitian form Φ of signature $++--$. We let

$$\mathbf{P} := \mathbf{F}_1, \quad \mathbf{M} := \mathbf{F}_2, \quad \text{and} \quad \mathbf{F} := \mathbf{F}_{12},$$

be the complex flag manifolds of 1-dimensional subspaces of $\mathbf{T}$, 2-dimensional subspaces of $\mathbf{T}$, and pairs of nested 1 and 2-dimensional subspaces of $\mathbf{T}$, respectively. Then there is a natural double fibration

(2.1)
$$\begin{array}{ccc} & \mathbf{F} & \\ {}^{\mu}\swarrow & & \searrow^{\nu} \\ \mathbf{P} & & \mathbf{M} \end{array}$$

where μ and ν are the natural maps $\mu(L_1, L_2) = L_1$, $\nu(L_1, L_2) = L_2$. Thus we have, for instance, that

$$\mathbf{M} = \{L \subset T : \dim L = 2\}$$

is a complex Grassmannian, while $\mathbf{P} \cong P_3(C)$ is a 3-dimensional complex projective space. Moreover, $\mathbf{F}$ is a 5-dimensional flag manifold fibred over both $\mathbf{P}$ and $\mathbf{M}$ with 2 and 1-dimensional projective spaces as fibres, respectively.

If L is a 2-dimensional subspace of T, then we say that $\Phi(L) = 0$ if and only if $\Phi(v) = 0$ for all $v \in L$. Thus we let

(2.2) $$M = \{L \in \mathbf{M} : \Phi(L) = 0\}$$

and one sees that the group $SU(2,2)$ acts transitively on M, since it preserves the form Φ, and one can verify that M is the natural (conformal) compactification of real flat Minkowski space (R^4 with a Minkowski metric of signature of type $+---$). In fact, $\mathbf{M}$ is a complexification of M, and is sometimes referred to as complexified, compactified Minkowski space. In [5] and [12] differential equations relating to mathematical physics on Minkowski space were studied in terms of holomorphic data on $\mathbf{P}$, using the double fibration as a means of defining the Penrose transform mapping holomorphic data on $\mathbf{P}$ to solutions of certain differential equations on open subsets of $\mathbf{M}$, which, in turn, can be restricted to M, giving real-analytic solutions of these equations on real Minkowski space (weak solutions are obtained as boundary values of appropriate holomorphic solutions, cf. [13] and [14]).

The manifold M is diffeomorphic to $S^1 \times S^3$, and the definition (2.2) realizes $S^1 \times S^3$ in a specific manner as a totally real submanifold of the 4-dimensional complex manifold (i.e., M is a real submanifold of $\mathbf{M}$ such that $T_x(M) \cap JT_x(M) = \{0\}$, for any $x \in M$, where J is the real-linear mapping of $T_x(\mathbf{M}) \to T_x(\mathbf{M})$ corresponding to multiplication by i, when $T_x(\mathbf{M})$ is considered as a complex vector space). In Euclidean quantum field theory one wants to consider certain differential equations on $\mathbf{R}^4$ with the Euclidean metric, or on S^4, its conformal compactification with the usual Riemannian metric. We want to relate S^4 to twistor geometry in a natural manner.

Choose coordinates (Z^0, Z^1, Z^2, Z^3) for $\mathbf{T}$, so that $\mathbf{T} \cong \mathbf{C}^4$ and consider a quaternionic structure on $\mathbf{T}$ given in the following manner (cf. Hartshorne [6]). Let $\mathbf{H}$ be the ring of quaternions, that is, a 4-dimensional real vector space with the standard basis $\{1, i, j, k\}$ where $i^2 = j^2 = k^2 = -1$, $ij = k$, etc. Define a 2-dimensional quaternionic structure $\mathbf{H}^2$ on the space $\mathbf{C}^4$ by letting

$$
(2.3) \qquad \begin{aligned} W^0 &= Z^0 + Z^1 j \\ W^1 &= Z^2 + Z^3 j . \end{aligned}
$$

Now let $\sigma : \mathbf{H}^2 \to \mathbf{H}^2$ be left multiplication by j, and we see that the induced mapping on $\mathbf{C}^4$ is given by

$$
\sigma(Z^0, Z^1, Z^2, Z^3) = (-\bar{Z}^1, \bar{Z}^0, -\bar{Z}^3, \bar{Z}^2)
$$

a conjugate-linear mapping of $\mathbf{C}^4$ to $\mathbf{C}^4$. This induces a mapping $\sigma : \mathbf{P}_3(\mathbf{C}) \to \mathbf{P}_3(\mathbf{C})$ by letting σ act on homogeneous coordinates of $\mathbf{P}_3(\mathbf{C})$. Then let $\mathbf{P}_1(\mathbf{H})$ be the projective line over $\mathbf{H}$, i.e., the equivalence classes of ordered pairs (W^0, W^1), $W^i \in \mathbf{H}$, under the equivalence relation

$$
(W^0, W^1) \sim (aW^0, aW^1), \quad 0 \neq a \in \mathbf{H} .
$$

Then $\mathbf{P}_1(\mathbf{H})$ is diffeomorphic to S^4, and the natural mapping

$$
\mathbf{C}^4 \to \mathbf{H}^2
$$

given by (2.3) induces

$$\text{(2.4)} \qquad \begin{array}{ccc} P(C^4) & \longrightarrow & P(H^2) \\ \cong & & \cong \\ \pi : P_3(C) & \longrightarrow & P_1(H) \cong S^4 \end{array}$$

and one can check that the fibres of π are 1-dimensional complex projective subspaces of $P_3(C)$. Now one of the elementary properties of the double fibration (2.1) is that if we let $\tau^{-1}(p) = \mu \circ \nu^{-1}(p)$, for $p \in \mathbf{M}$, (cf. [5], [12]), then $\tau^{-1}(p) \cong P_1(C)$ and moreover, $\mathbf{M}$ is the parameter space for *all* 1-dimensional projective subspaces of $P_3(C)$ (this is the *Klein correspondence*). Now from the fibration $\pi : \mathbf{P} \to S^4$ given by (2.4) we see that S^4 is the parameter space for a distinguished family of disjoint projective lines in $\mathbf{P}$. Thus we have an embedding of S^4 in $\mathbf{M}$, and a diagram of the following sort

$$\text{(2.5)} \qquad \begin{array}{ccccc} & \overset{\mu}{\swarrow} & F & \overset{\nu}{\searrow} & \\ \mathbf{P} & & & & \mathbf{M} \hookleftarrow M \\ & \underset{\pi}{\searrow} & & \nearrow & \\ & & S^4 \hookrightarrow & & \end{array}$$

where, as we will see below, S^4 is also embedded in $\mathbf{M}$ in a totally real manner.

Now, using the coordinates (Z^0, Z^1, Z^2, Z^3) as coordinates for $\mathbf{T}$ as above, we see that $SL(4, C)$ acts transitively on each of the three flag manifolds above. If we let

$$G_1 = \{T \in SL(4, C) : {}^t\bar{g}\,\Phi g = \Phi\}$$

then $G_1 \cong SU(2,2)$, and M can be identified with a closed orbit of G_1 acting on $\mathbf{M}$, and the action of G_1 on M induces the action of the (Lorentzian) conformal group acting on compactified Minkowski space (cf. [12]). In a similar manner, we note that the fibres of π are fixed under the action of σ, and if we let

$$G_2 = \{T \in SL(4, \mathbb{C}) : T\sigma = \sigma T\}$$

then $S^4 \hookrightarrow \mathbf{M}$ can be identified with a closed orbit of G_2, and $G_2 \cong$ Spin(5, 1), a 2-1 covering group of SO(4, 1), the (Euclidean) conformal group acting on the conformal compactification of Euclidean $\mathbf{R}^4$ (cf. [1]). Since both Spin(5, 1) and SU(2, 2) are real forms of $SL(4, \mathbb{C})$, it follows that M and S^4 are both totally real embeddings in (2.5).

§3. *The Penrose transform for instanton bundles*

In this section we will present the Penrose transform which maps cohomology on **P** to solutions of field equations on **M**. This will then be applied to the specific case of field equations of helicity zero. We will be following principally the development of the Penrose transform as given in [5].

Suppose that $U \subset \mathbf{M}$ is open, then let $U' = \nu^{-1}(U)$, and $U'' = \mu \circ \nu^{-1}(U)$ be derived open subsets of **F** and **P** respectively. Then we have the following diagram:

$$\begin{array}{ccccccc} & U' & & & & \mathbf{F} & \\ \mu\swarrow & & \searrow\nu & \underset{\text{open}}{\hookrightarrow} & \swarrow & & \searrow \\ U'' & & U & & \mathbf{P} & & \mathbf{M}. \end{array}$$

We now let

$$0 \longrightarrow \mu^{-1}\mathcal{O}_{\mathbf{P}} \longrightarrow \Omega^0_\mu \xrightarrow{d_\mu} \Omega^1_\mu \xrightarrow{d_\mu} \Omega^2_\mu \longrightarrow 0$$

be the exact sequence of holomorphic relative differential forms for the surjective holomorphic mapping μ. That is

$$\Omega^0_\mu = \Omega^0_{\mathbf{F}}$$

$$\Omega^1_\mu = \Omega^1_{\mathbf{F}}/\mu^*\Omega^1_{\mathbf{P}}$$

$$\Omega^2_\mu = \Omega^2_{\mathbf{F}}/\mu^*\Omega^1_{\mathbf{P}} \wedge \Omega^1_{\mathbf{F}}$$

and d_μ is the induced relative exterior differentiation operator. We can tensor this sequence with $\mu^{-1}\mathcal{O}_{\mathbf{P}}(V)$, where $V \to \mathbf{P}$ is a holomorphic vector bundle, which then yields the exact sequence

$$0 \longrightarrow \mu^{-1}\mathcal{O}_{\mathbf{P}}(V) \longrightarrow \Omega^0_\mu(V) \xrightarrow{d_\mu} \Omega^1_\mu(V) \xrightarrow{d_\mu} \Omega^2_\mu(V) \to 0 .$$

Here we have let

$$\Omega^p_\mu(V) = \Omega^p_\mu \otimes_{\mu^{-1}\mathcal{O}_{\mathbf{P}}} \mu^{-1}\mathcal{O}_{\mathbf{P}}(V) ,$$

and we note that d_μ is well defined on these bundle-valued forms since the transition functions of $\mu^* V$ are constant on the fibres of μ, and are thus annihilated by d_μ.

We now introduce spinor bundles on $\mathbf{M}$. Recalling that $\mathbf{M}$ is a Grassmannian manifold, let

$$S_- \to \mathbf{M}$$

be the universal bundle over $\mathbf{M}$, and consider the exact sequence

$$0 \to S_- \to \mathbf{M} \times \mathbf{T} \to S^+ \to 0 ,$$

which defines S^+ as the quotient bundle. Then we define

$$S^- = (S_-)^*$$
$$S_+ = (S^+)^*$$

as the $\mathbf{C}$-linear dual bundles. The tensor algebra of bundles generated by these bundles are the *spinor bundles* on $\mathbf{M}$ (cf., [5] where a different notation for the sheaves of sections of these bundles was employed). We have the following fundamental relation between these bundles and the tensor algebra generated by the tangent bundle:

(3.1)

a) $S^+ \otimes S^- \cong T(\mathbf{M})$

b) $(\Lambda^2 S_-)^4 \cong K_{\mathbf{M}} := \Lambda^4 T^*(\mathbf{M})$.

To see that (3.1a) holds we simply note that

$$S^+ \otimes S^- = S^+ \otimes (S_-)^* \cong \mathrm{Hom}(S_-, S^+) \cong T(\mathbf{M})$$

where the last isomorphism is the standard characterization of the tangent bundle of a Grassmannian manifold (cf. [8]). The second relation (3.1b) is discussed and a proof is given in §4.

It is easy to see that as holomorphic bundles on $\mathbf{M}$, $S_- \cong S_+$, and $S^+ \cong S^-$. Thus we find that

$$(\Lambda^2 S_+)^4 \cong K_{\mathbf{M}}.$$

It is sometimes important to distinguish between conformal weights defined in terms of S_- and S_+, which amounts to distinguishing between T and T^*, as was done in [5]. In this paper we will not concern ourselves with this distinction for the purpose of defining conformal weights. We will simply set, for any bundle F

$$F[p+q] := F \otimes (\Lambda^2 S_-)^p \otimes (\Lambda^2 S_+)^q \tag{3.3}$$

where $d = p+q$ is called the *conformal weight* attached to the bundle F. Thus conformal weights amount to positive powers of the 1/4-root of the canonical bundle (3.1b), which is the fundamental quantity in arguments involving integration. This concept is discussed further in §4.

Now we want to consider $V = E \otimes H^n$, where E is an instanton bundle. An *instanton bundle* E is a holomorphic vector bundle $E \to \mathbf{P}$ of the form $E = \pi^* F$, where F is a real-analytic G-bundle over S^4 of a specified rank, where G is a real semi-simple Lie group (e.g. $G = SU(2)$). Moreover, F is equipped with a connection ω satisfying the self-dual Yang-Mills equations (i.e. $\nabla\Omega = \nabla^*\Omega = 0$ and $*\Omega = \Omega$, where ∇ is the covariant differentiation associated to ω, Ω is the curvature, and $*$ is the Hodge $*$-operator associated with the spherical metric on S^4). The differential-geometric data on F induces the holomorphic structure on the bundle E (Ward [10], cf. [1], [9]). One can give intrisic characterizations of holomorphic vector bundles on $\mathbf{P}$ which are instanton bundles (cf. [3], [6]), but that won't concern us here.

Suppose that E is an instanton bundle on $\mathbf{P}$, then one can compute the following direct image sheaves near $S^4 \subset \mathbf{M}$ (noting that the bundle F corresponding to E has a unique holomorphic extension to a neighborhood of S^4):

$$\text{a)} \quad \nu_*^1 \Omega_\mu^0(E \otimes H^{-2}) \cong \mathcal{O}_{\mathbf{M}}(F)[1]$$

$$\text{(3.4)} \quad \text{b)} \quad \nu_*^0 \Omega_\mu^2(E \otimes H^{-2}) \cong \mathcal{O}_{\mathbf{M}}(F)[3]$$

c) all other direct images of $\Omega_\mu^p(E \otimes H^{-2})$ vanish.

This follows from standard cohomology calculations on the fibres of ν which are all isomorphic to $\mathbf{P}_1(\mathbf{C})$, and the fact that F is trivial on these fibres by construction (cf. [5]).

We now have two spectral sequences associated with the above geometric data. The first is the spectral sequence of the differential sheaf

$$\text{(3.5)} \quad 0 \longrightarrow \mu^{-1}\mathcal{O}_{\mathbf{P}}(E \otimes H^{-2}) \longrightarrow \Omega_\mu^0(E \otimes H^{-2}) \xrightarrow{d_\mu} \Omega_\mu^1(E \otimes H^{-2}) \xrightarrow{d_\mu} \Omega_\mu^2(E \otimes H^{-2}) \longrightarrow 0$$

on the open set U′:

$$\text{(3.6)} \qquad E_1^{pq} = H^p(U', \Omega_\mu^q(E \otimes H^{-2})) \Longrightarrow H^r(U', \mu^{-1}\mathcal{O}_{\mathbf{P}}(E \otimes H^{-2}))$$

where we note that (3.5) is not acylic on U′. Secondly, there is the Leray spectral sequence for the mapping ν, and for fixed j:

$$\text{(3.7)} \qquad H^p(U, \nu_*^q \Omega_\mu^j(E \otimes H^{-2})) \Longrightarrow H^r(U', \Omega_\mu^j(E \otimes H^{-2}))\ .$$

By (3.4c) we see that for each fixed j, all direct images sheaves but one vanish, thus insuring the complete degeneration of (3.7). Therefore we obtain the isomorphisms:

$$H^1(U', \Omega^0_\mu(E \otimes H^{-2})) \cong H^0(U, \mathcal{O}_{\mathbf{M}}(F[1]))$$

$$H^0(U', \Omega^2_\mu(E \otimes H^{-2})) \cong H^0(U, \mathcal{O}_{\mathbf{M}}(F[3]))$$

$$H^p(U', \Omega^q_\mu(E \otimes H^{-2})) = 0, \quad p \geq 0, \quad q \neq 0, 2 .$$

It then follows that the spectral sequence (3.6) degenerates to second order, and we obtain:

$$\begin{array}{ccc} H^1(U', \Omega^0_\mu(E \otimes H^{-2})) & \xrightarrow{d_2} & H^0(U', \Omega^2_\mu(E \otimes H^{-2})) \\ \wr\Vert & & \wr\Vert \\ H^0(U, \mathcal{O}_{\mathbf{M}}(F[1])) & \xrightarrow{D} & H^0(U, \mathcal{O}_{\mathbf{M}}(F[3])) \end{array} \tag{3.8}$$

where D is the operator on sections of F induced by the spectral sequence operator d_2. On the other hand, we see that

$$E^{10}_\infty \cong E^{10}_2 \cong \operatorname{Ker} d_2 \text{ in } (3.8)$$

while $E^{01}_\infty = E^{01}_2 = 0$. Thus we find that

$$H^1(U', \mu^{-1}\mathcal{O}_{\mathbf{P}}(E \otimes H^{-2})) \cong E^{10}_\infty \oplus E^{01}_\infty \cong E^{10}_2 \cong \operatorname{Ker} d_2 \text{ in } (3.8) .$$

We also have a mapping

$$H^1(U'', \mathcal{O}_{\mathbf{P}}(E \otimes H^{-2})) \to H^1(U', \mu^{-1}\mathcal{O}_{\mathbf{P}}(E \otimes H^{-2}))$$

given by the usual pullback, and as was shown in [5] by an elementary topological argument along the fibres of μ, this is an isomorphism if the fibres of $U' \xrightarrow{\mu} U''$ are 1-connected. However, if we consider a fundamental neighborhood system $\{U\}$ of $S^4 \subset \mathbf{M}$, then we see that $U'' = \mathbf{P}$ and the fibres of $U' \xrightarrow{\mu} \mathbf{P}$ are contractible and reduce to a single point in the limit. Putting all of this information together we obtain a sequence of isomorphisms:

$$H^1(\mathbf{P}, \mathcal{O}_{\mathbf{P}}(E \otimes H^{-2})) \xrightarrow{\cong} H^1(\nu^{-1}S^4, \mu^{-1}\mathcal{O}_{\mathbf{P}}(E \otimes H^{-2}))$$

$$\xrightarrow{\cong} \text{Ker } d_2 : H^1(\nu^{-1}(S^4), \Omega^0_\mu(E \otimes H^{-2})) \to H^0(\nu^{-1}(S^4), \Omega^2_\mu(E \otimes H^{-2}))$$

$$\xrightarrow{\cong} \text{Ker } D : H^0(S^4, \mathcal{O}_{\mathbf{M}}(F[1])) \to H^0(S^4, \mathcal{O}_{\mathbf{M}}(F[3])) .$$

Therefore we have constructed an isomorphism

$$H^1(\mathbf{P}, \mathcal{O}_{\mathbf{P}}(E \otimes H^{-2})) \underset{\cong}{\xrightarrow{\mathcal{P}}} \text{Ker } D \tag{3.9}$$

which is the Penrose transform in this geometric setting.

We note that D is invariant under the action of G_2, which acting on its orbit S^4, is locally the action of the Euclidean conformal group. It is known that there is essentially a unique second order conformally invariant differential operator of this type (cf. Hitchin [7]). We formulate this in our case as follows.

LEMMA 3.1. *The operator* D *in (3.8) can be identified with the conformally invariant operator*

$$D_0 = \nabla^*\nabla + R/6 : \Gamma(S^4, F[1]) \to \Gamma(S^4, F[3])$$

where ∇ *is the covariant derivative with respect to the connection of* F, ∇^* *is the adjoint with respect to the spherical metric on* S^4, *and* R *is the scalar curvature of* S^4.

Proof. We see that D and D_0 act on and map into the same conformally weighted spaces of sections of F (cf. Hitchin [7], for a good discussion of D_0 above). One can check readily that D and D_0 have the same symbol (cf. [5], §6). Knowing that the symbols agree, it then suffices, by the arguments in Hitchin [7] involving jet bundles and formal neighborhoods, to check that D annihilates the image of $H^1(U'', \mathcal{O}_{\mathbf{P}}(E \otimes H^{-2}))$ under the Penrose transform. This is however quite clear from our construction, as D is precisely the annihilator of the image of this cohomology group.

This yields immediately the following theorem, which was first announced in [2], with proofs in [4], [7] and [9].

THEOREM 3.2. *Let* E *be an instanton bundle on* $\mathbf{P}$, *then*

$$H^1(\mathbf{P}, \mathcal{O}(E \otimes H^{-2})) = 0 .$$

Proof. As in the papers cited above, one uses the isomorphism (3.4), Lemma 3.1 and the Bochner vanishing argument, noting that $R/6$ is positive implies that $\mathrm{Ker}(\nabla^* \nabla + R/6)$ acting on conformally weighted sections of F must be zero.

We also have the following generalization of (3.9).

THEOREM 3.3. *Let* U_0 *be an open set in* S^4, *then the Penrose transform*

$$\mathcal{P} : H^1(\pi^{-1}(U_0), \mathcal{O}(E \otimes H^{-2})) \longrightarrow \mathrm{Ker}\ D : \Gamma(U_0, \mathcal{O}_{\mathbf{M}}(F)[1]) \longrightarrow \Gamma(U_0, \mathcal{O}_{\mathbf{M}}(F)[3])$$

is an isomorphism.

This was first proved by Rawnsley [9], and follows readily from the arguments used above, which didn't depend on the tubular neighborhoods being neighborhoods of all of S^4. One can take simply neighborhoods of U_0 in $\mathbf{M}$. This isomorphism is, in fact an isomorphism of Frechet spaces, both spaces being naturally endowed with Frechet space topologies.

§4. *Conformal weights*

In this section we want to briefly compare the concepts of conformal weights as used in [5] and [7]. In [5] the concept of conformal weight used was the following definition. A holomorphic vector bundle E on $\mathbf{M}$ has conformal weight n means that E is of the form

$$E \otimes (\det U_2)^{-n}$$

where $U_2 \to G_{2,4}(\mathbf{C})\ (\cong \mathbf{M})$ is the universal or tautological bundle on $G_{2,4}(\mathbf{C})$, i.e., the fibre of U_2 at a point $p \in \mathbf{M}$ is the subspace of $\mathbf{C}^4$

defining the point $p \in \mathbf{M}$. The reason for assigning a negative weight to $\det U_2$ is that $\det U_2$ is a negative bundle in the sense of Chern classes and algebraic geometry, so that positive conformal weights correspond to powers of line bundles with positive first Chern class. In [5] we distinguished between $U_2 \to G_2(T)$ and $U_2 \to G_2(T^*)$ giving the notions of "primed and unprimed conformal weights." We can ignore that distinction here, and the sum of the "primed and unprimed conformal weights" in [5] will be simply the negative of the conformal weight introduced in §3. So, for instance, we have the operator D appearing in §3

$$D : \Gamma(S^4, F[1]) \to \Gamma(S^4, F[3]) .$$

On the other hand, Hitchin uses the more standard notion that E has conformal weight n on a k-dimensional real manifold X means that E is of the form $E \otimes (K^{1/k})^n$ where $K = \det T^*X$ is the canonical bundle of X. It is not true, in general, that $K^{1/k}$ is a well-defined line bundle, even though conformal weights are still well-defined concepts (see [7] for alternative definitions). On a complex manifold of complex dimension k, one would use powers of the k-th root of the holomorphic canonical bundle to define conformal weight from this point of view. The following theorem shows that for Grassmannian manifolds one can always take certain roots of the canonical bundle, yielding the fact that Hitchin's notion of conformal weight agrees with that used in §3 and differs from that used in [5] by a minus sign, as is desired. It is also of independent interest, being a generalization of well-known fact on projective space which doesn't seem to have been noted before.*

THEOREM 4.1. *Let* $U_m \to G_{m,n}$ *be the universal bundle over* $G_{m,n}$, *the Grassmannian manifold of* m*-planes in* n*-space (over any field), then*

$$\det(T^*G_{m,n}) \cong [\det U_m]^n .$$

*This theorem and its elementary proof developed out of conversations with A. Borel and J. Milnor, to whom I'd like to express my appreciation for their help on this point.

Proof. Let W be an m-dimensional vector space over a field k, and let, as usual, $W^* = \mathrm{Hom}(W, K)$, and $\det(W) = \Lambda^m W$. If

$$0 \to U \to V \to W \to 0$$

is a short exact sequence of vector spaces, then note that

$$\det(V) \cong \det(U) \otimes \det(W) .$$

Also, one can check easily that

$$\det(W^* \otimes W) \cong k$$

canonically. If $W \subset k^n$, then it follows from the exact sequence

$$0 \to W^* \otimes W \to W^* \otimes k^n \to W^* \otimes (k^n/W) \to 0$$

and the fact that $\det(W^* \otimes k^n) \cong [\det(W^*)]^n$, that we have

$$\det(W^* \otimes (k^n/W)) \cong \det(\mathrm{Hom}(W, k^n/W))$$

is canonically isomorphic to

$$[\det(W^*)]^n .$$

Here $[\ ]^n$ denotes the n-fold tensor product. Recalling that the tangent bundle of the Grassmannian is given by

$$\begin{aligned} TG_{m,n} &\cong \mathrm{Hom}(U_m, k^n/U_m) \\ &\cong U_m^* \otimes (k^n/U_m) \end{aligned}$$

(cf. [8]), we have that

$$\det(TG_{m,n}) \cong [\det U_m^*]^n$$

or

$$\det(T^*G_{m,n}) \cong [\det U_m]^n$$

as desired.

COROLLARY 4.2. $\det T^*G_{2,4}(\mathbf{C}) = (\det U_2)^4$.

This is what is needed for the comparison of conformal weights on $\mathbf{M}$.

COROLLARY 4.3. $K_{\mathbf{P}_n} = \Lambda^n T^*\mathbf{P}_n \cong (U_1)^{n+1} \cong (H)^{-n+1}$.

This last formula is well known in algebraic geometry (cf. e.g. [11], Example VI. 2.3).

RICE UNIVERSITY
and
THE INSTITUTE ON ADVANCED STUDY

REFERENCES

[1] Atiyah, M. F., *Geometry of Yang-Mills Fields*, Lezioni Fermione, Acad. Naz. dei Lincei: Scuola Normale Sup., Pisa, 1979.

[2] Atiyah, M. F., Hitchin, N. J., Drinfeld, V. G., Manin, Yu. I., "Construction of instantons," Physics Letters *65A* (1978), 185-187.

[3] Atiyah, M. and Ward, R., "Instantons and algebraic geometry," Comm. Math. Phys. *55*(1977), 111-124.

[4] Drinfeld, V. G. and Manin, Yu. I., "Instantons and sheaves on CP^3," Funct. Anal. and its Appl., *13*, No. 2 (1979), 59-74 (Eng. trans. 1979, pp. 124-134).

[5] Eastwood, M., Penrose, R., and Wells, R. O., Jr., "Cohomology and Massless fields," Comm. Math. Phys. *78*(1981), 305-351.

[6] Hartshorne, R., "Stable vector bundles and instantons," Comm. Math. Phys., *59*(1978), 1-15.

[7] Hitchin, N. T., "Linear field equations on self-dual spaces," Proc. Royal Soc. Lond. A. *370*(1980), 173-191.

[8] Milnor, J. and Stasheff, J., *Characteristic Classes*, Princeton Univ. Press, Princeton, N. J., 1974.

[9] Rawnsley, J. H., "On the Atiyah-Hitchin-Drinfeld-Manin vanishing theorem for cohomology groups of instanton bundles," Math. Ann. *241* (1979), 43-56.

[10] Ward, R., "On self-dual gauge fields," Physics Letters, Vol. 61A, (1977), 81-82.

[11] Wells, R. O., Jr., *Differential Analysis on Complex Manifolds*, Springer-Verlag, New York-Heidelberg-Berlin, 1980.

[12] Wells, R. O., Jr., "Complex manifolds and mathematical physics," Bull. Amer. Math. Soc. (New Series), 1 (1979), 296-336.

[13] ———, "Cohomology and the Penrose transform," in: *Complex Manifold Techniques in Theoretical Physics* (edited by D. E. Lerner and P. D. Sommers), Pitman, San Francisco, London, Melbourne, 1979; pp. 92-114.

[14] ———, "Hyperfunction solutions of the zero nest-mass field equations," Comm. Math. Phys. *78*(1981), 567-700.

CAUSALLY DISCONNECTING SETS, MAXIMAL GEODESICS AND GEODESIC INCOMPLETENESS FOR STRONGLY CAUSAL SPACE-TIMES

John K. Beem and Paul E. Ehrlich*

In [10, p. 538], Hawking and Penrose established the following theorem.

HAWKING-PENROSE THEOREM. *No space-time* (M, g) *of dimension* ≥ 3 *can satisfy all of the following three requirements together*:

(1) (M, g) *contains no closed timelike curves,*

(2) *every inextendible nonspacelike geodesic contains a pair of conjugate points,*

(3) (M, g) *contains a future or past trapped set* S.

Thus a chronological space-time of dimension ≥ 3 with everywhere nonnegative nonspacelike Ricci curvatures which satisfies the generic condition (cf. [9, p. 266]) and contains a future or past trapped set (cf. [9, p. 267]) is nonspacelike geodesically incomplete.

The purpose of this note is to explain how the Lorentzian distance function may be used to obtain a generalization of the Hawking-Penrose Theorem to the class of causally disconnected space-times. In section 1, we review the basic properties of space-times and the Lorentzian distance function needed for this purpose. In section 2, we introduce the concept

*Partially supported by NSF Grant MCS 77-18723(02).

Seminar on Differential Geometry
0-691-08268-5/82/000499-7 $00.50/0 (cloth)
0-691-08296-0/82/000499-7 $00.50/0 (paperback)

of causal disconnection. Finally in section 3, we discuss the geodesic completeness of causally disconnected space-times and indicate how our Theorem 3.1 implies the Hawking-Penrose Theorem.

1. *Preliminaries*

A Lorentzian manifold (M, g) is a smooth connected manifold with a countable basis together with a smooth Lorentzian metric g of signature $(-, +, \cdots, +)$. A space-time is a Lorentzian manifold which has been given a time orientation. With our signature convention, a nonzero tangent vector $v \in TM$ is said to be timelike (resp. nonspacelike, null, spacelike) according to whether $g(v, v) < 0$ (resp. ≤ 0, $= 0$, > 0). We will use the standard notations $p \ll q$ if there is a future directed timelike curve from p to q and $p \leq q$ if there is a future directed nonspacelike curve from p to q. The chronological future of p is defined by $I^+(p) = \{q \in M; p \ll q\}$ and the causal future of p is defined by $J^+(p) = \{q \in M; p \leq q\}$. The past sets $I^-(p)$ and $J^-(p)$ are defined dually.

The space-time (M, g) is said to be *chronological* if $p \notin I^+(p)$ for all p. It is *strongly causal* if each point has arbitrarily small neighborhoods such that no nonspacelike curve which leaves one of these neighborhoods ever returns. Finally, a strongly causal space-time is *globally hyperbolic* if $J^+(p) \cap J^-(q)$ is compact for all $p, q \in M$. If two Lorentzian metrics g and g' on M are globally conformal, then a curve γ is timelike (resp. null, spacelike) for (M, g) iff it is timelike (resp. null, spacelike) for (M, g'). Thus conformal metrics induce the same causal structure on M. We will denote by $C(M, g)$ the set of all Lorentzian metrics for M globally conformal to g.

The Lorentzian arc length functional is given by

$$L_g(\gamma) = L(\gamma) = \int_a^b \sqrt{|g(\gamma'(t), \gamma'(t))|}\, dt$$

and is upper semi-continuous for strongly causal space-times in the C^0-topology on curves (cf. [12, p. 49]).

The Lorentzian distance function $d = d(g) : M \times M \to \mathbf{R} \cup \{\infty\}$ may be defined as follows, cf. [9, p. 215]. Given $p \in M$, set $d(p,q) = 0$ if $q \notin J^+(p)$ and for $q \in J^+(p)$, let $d(p,q)$ be the supremum of lengths of all future directed nonspacelike curves from p to q. The distance function satisfies the reverse triangle inequality $d(p,q) \geq d(p,r) + d(r,q)$ whenever $p \leq r \leq q$. Also, the distance is lower semicontinuous where it is finite and if $p_n \to p$, $q_n \to q$ and $d(p,q) = \infty$, then $\lim d(p_n, q_n) = \infty$. However, the distance function is *not* upper semicontinuous for general space-times.

If $p \in I^+(p)$, then $d(p,p) = \infty$. Even strongly causal space-times such as the Reissner-Nordström space-times with $e^2 = m^2$ (cf. [9, p. 160]) may contain pairs of points with $d(p,q) = \infty$. But at least for strongly causal space-times, the distance function is finite and continuous on a neighborhood of the diagonal, cf. [3, pp. 368-9]. We say that (M, g) satisfies the *finite distance condition* if $d(g)(p,q) < \infty$ for all $p, q \in M$. If (M, g) is globally hyperbolic, then d(g) is continuous and satisfies the finite distance condition, cf. [9, p. 215]. In fact, globally hyperbolic space-times may be characterized among strongly causal space-times using this condition as follows, cf. [6, p. 95]: the strongly causal space-time (M, g) is globally hyperbolic iff (M, g') satisfies the finite distance condition for all $g' \in C(M, g)$.

We will define a future directed nonspacelike curve $\gamma : [a,b] \to (M,g)$ to be *maximal* if $L(\gamma) = d(\gamma(a), \gamma(b))$. If γ is maximal, then γ may be reparametrized to be a smooth (distance realizing) geodesic, cf. [4, p. 166]. Also if (M, g) is globally hyperbolic, then any pair of points with $p \leq q$ may be joined by at least one maximal geodesic, cf. [9, p. 213]. However, this property does *not* hold for arbitrary strong causal space-times, cf. [12, p. 7], Figure 7.

We caution that in many papers in General Relativity, "maximal" is used in a different sense. Namely, a timelike geodesic γ from p to q is sometimes said to be "maximal" if the index form is negative semi-definite; i.e., if γ is not "maximal," there is a small variation of γ which yields a longer curve from p to q, cf. [9, p. 110].

Finally, we recall that a geodesic γ is said to be *complete* if $\gamma(t)$ is defined for all positive and negative values of some (and hence any) affine parameter. For timelike and spacelike geodesics, this means that the geodesic has infinite length in both directions. A space-time is *timelike* (resp. *nonspacelike, null, spacelike*) *geodesically complete* if all inextendible timelike (resp. nonspacelike, null, spacelike) geodesics are complete. These forms of geodesic completeness are logically inequivalent, cf. [11], [8], [1]. In particular, there are globally hyperbolic space-times which are timelike and spacelike geodesically complete, but null geodesically incomplete. Geodesically incomplete space-times are often said to be *singular* in General Relativity.

2. *Causally disconnecting sets*

We now define causally disconnecting sets, cf. [4, p. 171], [2]. Using a construction different from the one given in [4], the condition "$d(p_n, q_n) < \infty$" may be omitted from Definition 6.1 of [4, p. 171].

DEFINITION 2.1. A space-time (M, g) is said to be *causally disconnected by a compact set* K if there exist sequences $\{p_n\}$, $\{q_n\}$ diverging to infinity such that for each n, $p_n \leq q_n$, $p_n \neq q_n$ and all future directed nonspacelike curves from p_n to q_n meet K.

Note first that if $k \neq n$, then p_k is not necessarily causally related to q_n or p_n. Also the compact set K may be quite different from a Cauchy surface and non-globally hyperbolic, strongly causal space-times may be causally disconnected even though they contain no Cauchy surfaces, e.g., a Reissner-Nordström space-time with $e^2 = m^2$. It is immediate that if (M, g) is causally disconnected, then so is (M, g') for all $g' \in C(M, g)$. Thus causal disconnection is a global conformal invariant.

THEOREM 2.2 ([2], [4, p. 173]). *Let* (M, g) *be a strongly causal space-time which is causally disconnected by a compact set* K. *Then* (M, g) *contains an inextendible nonspacelike geodesic which intersects* K *and is maximal between any pair of its points.*

It may also be shown that strongly causal space-times contain past directed and future directed nonspacelike geodesic rays issuing from every point, (cf. [2]). Also if (M, g) is strongly causal and $p \in M$ is *not* the origin of any future directed null geodesic ray, then (M, g) is causally disconnected by the future horismos $E^+(p) = J^+(p) \setminus I^+(p)$ of p. This last result and Theorem 2.2 imply that if (M, g) is a strongly causal space-time which contains no future directed null geodesic rays, then (M, g) contains a timelike geodesic line. This type of geodesic behavior occurs on the Einstein static universe $R \times S^1$, $g = -dt^2 + d\theta^2$.

3. *Geodesic incompleteness*

Using Theorem 2.2, we may prove the following generalization of the Hawking-Penrose Theorem (cf. [4, p. 172], [2]).

THEOREM 3.1. *No space-time of dimension* ≥ 3 *can satisfy all of the following three requirements together*:

(1) (M, g) *contains no closed timelike curves,*

(2) *every inextendible nonspacelike geodesic contains a pair of conjugate points,*

(3) (M, g) *is causally disconnected by a compact set.*

Proof. Suppose (M, g) satisfies (1), (2) and (3). Then (1) and (2) imply that (M, g) is strongly causal, cf. [10, p. 536], Lemma 2.10. Hence (M, g) contains an inextendible maximal nonspacelike geodesic by Theorem 2.2, which contradicts (2). ■

REMARK 3.2. This theorem implies the Hawking-Penrose Theorem for the following reason, cf. [2]. If S is a future (resp. past) trapped set, then the future (resp. past) horismos $E^+(S) = J^+(S) \setminus I^+(S)$, resp. $E^-(S) = J^-(S) \setminus I^-(S)$, causally disconnects (M, g). The assumption $\dim M \geq 3$ is necessary since no null geodesic in a two-dimensional space-time contains any conjugate points.

It is more customary to reformulate Theorem 3.1 as follows. See [9, pp. 99, 101] for the definition of the generic condition. An equivalent "invariant" formulation of this condition is given in [5, Appendix B]. Roughly speaking, this condition means that every inextendible nonspacelike geodesic runs into a point of nonzero curvature.

THEOREM 3.2. *Suppose* (M, g) *is a chronological space-time of dimension* ≥ 3 *which satisfies*

(1) $\mathrm{Ric}(v, v) \geq 0$ *for all nonspacelike* $v \in TM$,

(2) *every inextendible nonspacelike geodesic satisfies the generic condition, and*

(3) (M, g) *is causally disconnected by a compact set.*

Then (M, g) *is nonspacelike geodesically incomplete.*

Conditions (1) and (2) may also be replaced by the following condition: every inextendible nonspacelike geodesic $\gamma : (a, b) \to (M, g)$ satisfies

$$\liminf_{\substack{t_2 \to b^- \\ t_1 \to a^+}} \int_{t_1}^{t_2} \mathrm{Ric}(\gamma'(t), \gamma'(t))\, dt > 0 , \tag{4}$$

cf. [7]. Conditions (1)-(2) or (4) are used as follows. First, these conditions applied just to the null geodesics imply that the chronological space-time is either null geodesically incomplete or strongly causal, cf. [9, p. 192]. Second, conditions (1)-(2) or (4) also imply that every complete nonspacelike geodesic $\gamma : \mathbf{R} \to (M, g)$ contains a pair of conjugate points, cf. [9, pp. 98, 101], [7] respectively.

JOHN BEEM
DEPARTMENT OF MATHEMATICS
UNIVERSITY OF MISSOURI-COLUMBIA
COLUMBIA, MO 65211

PAUL EHRLICH
SCHOOL OF MATHEMATICS
INSTITUTE FOR ADVANCED STUDY
PRINCETON, NJ 08540

and

DEPARTMENT OF MATHEMATICS
UNIVERSITY OF MISSOURI-COLUMBIA
COLUMBIA, MO 65211

REFERENCES

[1] Beem, J. K., "Some examples of incomplete space-times," Gen. Rel. Grav. 7(1976), 501-509.

[2] Beem, J. K. and Ehrlich, P. E., "Constructing maximal geodesics in a strongly causal space-time," Math. Proc. Camb. Phil. Soc. 90(1981), 183-190.

[3] ———, "Cut points, conjugate points and Lorentzian comparison theorems," Math. Proc. Camb. Phil. Soc. 86(1979), 365-384.

[4] ———, "Singularities, incompleteness and the Lorentzian distance function," Math. Proc. Camb. Phil. Soc. 85(1979), 161-178.

[5] ———, *Global Lorentzian Geometry*, Marcel Dekker Monographs in Pure and Applied Math., vol. 67, 1981.

[6] ———, "The space-time cut locus," Gen. Rel. Grav. 11 (1979), 89-103.

[7] Chicone, C. and Ehrlich, P. E., "Line integration of Ricci curvature and conjugate points in Lorentzian and Riemannian manifolds," in Manuscripta Math. 31 (1980), 297-316.

[8] Geroch, R., "What is a singularity in General Relativity," Ann. Physics (New York) 48(1968), 526-540.

[9] Hawking, S. and Ellis, G., *The large scale structure of space-time*, Cambridge Monographs on Math. Physics, Cambr. U. Press, 1973.

[10] Hawking, S. and Penrose, R., "The singularities of gravitational collapse and cosmology," Proc. Roy. Soc. London Ser. A 314 (1970), 529-548.

[11] Kundt, W., "Note on the completeness of space-times," Zeitschrift für Physik 172 (1963), 488-489.

[12] Penrose, R., *Techniques of differential topology in relativity*, S.I.A.M. Regional Conference Series in Applied Mathematics, 1972.

RENORMALIZATION

Arthur Jaffe*

Two fundamental cornerstones of modern physics are Einstein's theory of special relativity, invented about 1905, and the theory of quantum mechanics which dates to the mid 1920's. Yet it is still a mathematical mystery how to unify these two basic subjects. We describe here what is known.

Physicists propose a natural form for this combination: quantum field theory. Such a theory would incorporate the usual premises of quantum theory, where the mathematical framework would be a Hilbert space $\mathcal{H}$. The vectors in $\mathcal{H}$ represent states of a physical system, and the "field operators" are linear operators on $\mathcal{H}$. The fields satisfy some system of nonlinear hyperbolic equations. These field equations are Maxwell's equations in the case of the electric or magnetic field and Dirac's equation in the case of an electron field. Recently, physicists have come to believe that somewhat more general nonlinear systems (proposed over the past thirty years by Yang, Mills, Gell-Mann, Glashow, Salam, Weinberg, and others) are the most likely candidates to unify predictions of strong, weak and electromagnetic forces of nature. (The unification of gravitational forces still has not been resolved to the satisfaction of even the most optimistic theoretical physicists.)

*Supported in part by the National Science Foundation under Grant PHY 79-16812.

Seminar on Differential Geometry
0-691-08268-5/82/000507-17$00.85/0 (cloth)
0-691-08296-0/82/000507-17$00.85/0 (paperback)
For copying information, see copyright page.

In spite of these conceptual advances, the basic mathematical puzzle remains. Can one give any example of a quantum field theory which is defined within a completely logical framework? In other words, is quantum field theory really a mathematical theory at all? This question remains unsolved for the equations which physicists believe will unify the several forces of nature mentioned above. This question remains open for any nonlinear quantum field theory in three-space plus one-time dimensions. In fact, it is only after fifteen years of work that the question has been answered affirmatively in the simpler mathematical worlds of two and three space-time dimensions. The branch of mathematical physics which led to these results is sometimes called *constructive field theory* [1].

The basic difficulty in formulating the mathematical problem is the singular nature of the nonlinear equations proposed. To explain this difficulty, consider a very elementary nonlinear wave equation in d space-time dimensions,

$$\left(\frac{\partial^2}{\partial t^2} - \sum_{j=1}^{d-1} \frac{\partial^2}{\partial x_j^{\,2}}\right) q(t,\vec{x}) = F(q(t,\vec{x})) , \tag{1.1}$$

where $F(\cdot)$ is a given (nonlinear) force function. We require, as explained below, that the solution $q(t,\vec{x})$ satisfy the constraint

$$[q(t,\vec{x}), \partial q(t,\vec{y})/\partial t] = i\delta(\vec{x}-\vec{y})I , \tag{1.2}$$

where I is the identity operator on $\mathcal{H}$ and δ is the Dirac measure. (Recall that the field $q(t,\vec{x})$ is a linear transformation on $\mathcal{H}$.) The singular constraint (1.2) ensures that fields $q(t,\vec{x})$ and $q(t',\vec{x}')$ commute whenever the space-time points $(t,\vec{x})$ and $(t',\vec{x}')$ cannot be connected by a light signal. It is because of this condition that the field $q(t,\vec{x})$ must be quite singular; it must be distribution valued. Thus the nonlinear equation (1.1) must be reinterpreted. A set of rules called *renormalization* were developed by physicists to do this.

Another assumption of quantum field theory is the existence of a unitary group $\exp(-itH)$ acting on $\mathcal{H}$, and generated by the Hamiltonian or energy operator H. This Schrödinger group defines the dynamical evolution,

$$q(t+s) = \exp(itH)q(s)\exp(-itH) .$$

The basic difficulty appears if we try to express the Hamiltonian as

$$H = H_0 + V ,$$

where H_0 is the Hamiltonian suitable for the case $F = 0$. The perturbation V appears to shift the spectrum of H by an infinite amount. Even putative small perturbations yield these apparently uncontrollable effects.

Eventually, physicists developed sets of rules of thumb to cancel these infinities and to calculate observable effects. For example, if V is a polynomial in q, then this polynomial was redefined to have infinite coefficients (diverging at some specified rates with the removal of some approximation). These rules of renormalization were remarkably accurate in producing verifiable numbers in electrodynamics. Presently 11 decimal agreement has been achieved between theory and the experimental measurement of the magnetic moment of the electron. In spite of this overwhelming practical success, a fundamental mathematical theory remained elusive. It became clear, however, that a mathematical theory needs to begin with an approximate (smoothed) equation, needs to incorporate renormalization, and finally needs to control the limit in which the smoothing was removed.

With the new attack on these mathematical problems in the period 1965-80, many results have been obtained. Examples of quantum field theories have been constructed in $d = 2, 3$. While we hope that the existence of a model in the physically relevant case $d = 4$ will soon be found, at the moment there is no concrete evidence. In fact, renormalization theory suggests that a nonabelian gauge theory should be studied. In this lecture, however, I describe methods which have been successful in simpler examples.

2. *Canonical quantization*

Consider a single non-relativistic particle with position q, moving in a potential $V = V(q)$. According to the principles of quantum theory, observables such as q, the momentum p or the energy $H = H(p, q)$ are linear, self-adjoint operators on the Hilbert space $\mathcal{H}$ of quantum states. Furthermore the trajectory $q(t)$ in time is a solution to Newton's equation

$$m \frac{d^2 q(t)}{dt^2} = F(q(t)) , \tag{2.1}$$

where $F(q) = -dV(q)/dt$. While (2.1) also holds in the classical theory, in quantum theory the initial data are chosen to satisfy the *canonical* commutation relation

$$[q(t), p(t)] = i\hbar , \tag{2.2}$$

where $p(t) = m\, dq(t)/dt$ and $\hbar$ is Planck's constant. Since the initial time is arbitrary, the solution $q(\cdot)$ should satisfy (2.2) for all t. Under mild regularity assumptions, the Lie algebra relation (2.2) can be integrated to the Weyl form of the commutation relations

$$\exp(i\alpha q(t)) \exp(i\beta p(t)) \exp(-i\alpha q(t)) \exp(-i\beta p(t)) = \exp(-i\alpha\beta) , \tag{2.3}$$

where $\alpha, \beta \in \mathbf{R}$.

For the above example, we can choose $\mathcal{H} = L_2(\mathbf{R}; dq)$ and let $q(t=0) = q$ be multiplication by the coordinate function on L_2. Then with $p(t=0) = p = -i\hbar\, d/dq$, the pair (q, p) satisfy (2.2-3) for $t = 0$. Define the Hamiltonian operator on $\mathcal{H}$ by

$$H = \frac{1}{2m} p^2 + V(q) , \tag{2.4}$$

and let $U(t) = \exp(-itH)$ be the unitary group it generates on $\mathcal{H}$. The solution to (2.1) - (2.2) is

$$q(t) = U(t)^* q U(t) , \qquad p(t) = U(t)^* p U(t) . \tag{2.5}$$

In fact, differentiating (2.5) we find

$$m\,\frac{dq(t)}{dt} = U(t)^*[iH,q]U(t) = U(t)^*pU(t) = p(t)$$

and

$$m\,\frac{d^2q(t)}{dt^2} = \frac{dp(t)}{dt} = U(t)^*[iH,p]U(t) = U(t)^*[iV(q),p]U(t)$$
$$= -U(t)^*dV(q)/dq\,U(t) = U(t)^*F(q)U(t) = F(q(t))\ .$$

Here we use the fact that $A \to U(t)^*AU(t) = A_t$ defines an automorphism of linear operators on $\mathcal{H}$, such that if $A = A(q)$, then $A_t = A(q(t))$.

3. *Quantum fields*

A wave equation, such as

$$\text{(3.1)}\qquad \frac{\partial^2}{\partial t^2}\,q(t,\vec{x}) = F(q(t,\vec{x})) + \sum_{i=1}^{d-1}\frac{\partial^2}{\partial x_i^2}\,q(t,\vec{x})\ ,$$

is a generalization of (2.1). Here $q(t)$ takes values in a space of functions of a position variable $\vec{x} \in \mathbf{R}^{d-1}$. Then $q(t,h) = \int q(t,\vec{x})h(\vec{x})\,d\vec{x}$ defines a coordinate direction in q-space. The generalization of (2.2) is

$$\text{(3.2)}\qquad [q(t,h),\,p(t,g)] = i(h,g)\ ,$$

where $p(t,h) = dq(t,h)/dt$ and (h,g) denotes a real L_2 inner product. These are the equations of canonical quantum field theory; the mathematical problem is to find a Hilbert space $\mathcal{H}$ and a set of operators $q(t,h)$ satisfying (3.1-2).

Formally, there is a solution analogous to the Hamiltonian (2.4-2.5). Namely, if $q(t,h)$, $p(t,h)$ satisfy (3.2) for $t = 0$, and if $F = -V'$, then

$$\text{(3.3)}\qquad H = \int_{t=0} \frac{1}{2}\left\{p(t,\vec{x})^2 + \sum_{i=1}^{d-1}\left(\frac{\partial q(t,\vec{x})}{\partial x_i}\right)^2 + V(q(t,\vec{x}))\right\}d\vec{x}$$

yield

$$q(t,\vec{x}) = \exp(itH)q(0,\vec{x})\exp(-itH) \tag{3.4}$$

satisfying (3.1-3.2).

The mathematical difficulty with this approach is to define H. The relation (3.2) is singular, and since $[q(t,\vec{x}), \partial_t q(t,\vec{y})] = i\delta(\vec{x}-\vec{y})$, the quantity $q(t,\vec{x})$ does not exist pointwise, but only in the sense of a distribution in the variable $\vec{x} \in R^{d-1}$. Thus one expects difficulty in defining nonlinear functions such as (3.3).

The mathematical definition of H entails solving the problem known in physics as renormalization. The Hamiltonian approach outlined above has been used successfully to construct the first examples of nonlinear quantum fields, as well as to develop the theory, mainly with $d = 2$ or 3. Basically there are three steps: (i) mollify $q(t=0)$, $p(t=0)$; (ii) introduce an approximate H in which V has certain "renormalization" parameters depending in a divergent way on the mollifier; and (iii) pass to the limit of removing the mollifier in $q(t,\vec{x})$. For this limit to exist, the renormalization parameters must be chosen so divergences cancel in perturbation theory, i.e., as physicists have proposed. Let ε denote the mollified quantity. Thus $V(q)$ is replaced by $V(q_\varepsilon) + C_\varepsilon(q_\varepsilon)$ where C_ε are the renormalization counterterms with divergent coefficients as $\varepsilon \to 0$. The coefficients are chosen according to renormalization in perturbation theory, so that the total energy H converges as $\varepsilon \to 0$. For complete details of this approach and extensive references, see [2].

4. *Quantization by functional integrals*

There is another construction of $q(t,\vec{x})$ which is based on functional integration. Here the basic problem is to define a measure $d\mu(\phi)$ on a space of functions $\phi(t)$. This measure in turn determines the Hamiltonian H as well as the space $\mathcal{H}$ on which it acts. With these objects, the solution to (3.1)-(3.2) again follows by defining $q(t)$ by (3.4).

Let $\mathcal{C}$ denote the space of continuous functions from $\mathbf{R}$ (the time) to $\mathcal{S}'(\mathbf{R}^{d-1})$. Thus a path $\phi(t) \in \mathcal{C}$ can be written as a distribution $\phi(t) = \int \phi(t, \vec{x}) f(\vec{x}) d\vec{x}$ and is a linear functional on $f \in \mathcal{S}(\mathbf{R}^{d-1})$. For $d = 1$, we suppress f. It is convenient to choose f real.

Given a countably additive Borel measure $d\mu(\phi)$ on $\mathcal{C}$, we define time translation $T(t)$ and reflection θ by lifting these actions from $\mathbf{R}$ to $\mathcal{C}$, namely

$$(T(t)\phi)(s) = \phi(s+t) , \quad (\theta\phi)(s) = \phi(-s) . \tag{4.1}$$

On functions $A(\phi)$, $(T(t)A)(\phi) = A(T(t)\phi)$, etc. We say that $d\mu$ is invariant under time translation and reflection if for all smooth, integrable A,

$$\int (T(t)A) d\mu = \int A \, d\mu = \int (\theta A) d\mu , \tag{4.2}$$

or

$$\langle T(t)A \rangle = \langle A \rangle = \langle \theta A \rangle . \tag{4.3}$$

We require an additional property, Osterwalder-Schrader positivity: Let $\mathcal{E}_+$ denote the set of finite linear combinations of exponentials

$$A(\phi) = \sum_{j=1}^{n} \xi_j \, e^{i\phi(t_j, f_h)} , \quad \xi_j \in \mathbf{C} , \quad f_j \in \mathcal{S}(\mathbf{R}^{d-1}) , \tag{4.4}$$

with the property $t_j \geq 0$, $j = 1, 2, \cdots, n$. We say that $d\mu$ is OS positive with respect to θ if

$$0 \leq \langle (\overline{\theta A}) A \rangle = \int (\overline{\theta A}) A \, d\mu \tag{4.5}$$

for all $A \in \mathcal{E}_+$.

DEFINITION 4.1. *If the Borel probability measure* $d\mu$ *is translation invariant, reflection invariant and* OS *positive, then the form* $b(A, B) =$

$<(\overline{\theta A})B>$ *on* $\mathcal{E}_+ \times \mathcal{E}_+$ *defines a quantum mechanics Hilbert space* $\mathcal{H} = (\mathcal{E}_+/\text{kernel b})^-$. *Let* $\hat{A}$ *denote the equivalence class of* $A \in \mathcal{E}_+$ *in* $\mathcal{H}$.

THEOREM 4.2. *With these hypotheses, the operators* $T(t)^\frown : \mathcal{H} \to \mathcal{H}$ *defined by*

$$T(t)^\frown \hat{A} = (T(t)A)^\frown, \qquad A \in \mathcal{E}_+, \quad t \geq 0, \tag{4.6}$$

form a self-adjoint contraction semigroup: $T(t)^\frown = \exp(-tH)$. *The generator* H *is the quantum Hamiltonian and*

$$0 \leq H = H^*, \qquad H\Omega = 0, \tag{4.7}$$

where $\Omega = \hat{1}$ *is a ground state.*

DEFINITION 4.3. *The quantum field is the operator* $q = \phi(0)^\frown$ *defined on* $\mathcal{H}$ *by*

$$q\hat{A} = (\phi(0)A)^\frown, \qquad A \in \mathcal{E}_+ .$$

The triple $\{\mathcal{H}, H, q\}$ *is the Osterwalder-Schrader quantization of* $d\mu$.

This approach is a generalization of Feynman's approach to quantum theory, and also rests on ideas of M. Kac, K. Symanzik and E. Nelson. See [1] for extensive discussion and proofs, as well as for references to related work of Guerra, Rosen, Simon, Fröhlich and Spencer. The operator

$$q(t) = e^{itH} q e^{-itH}$$

is the quantum field at time t. If $d\mu$ satisfies a simple regularity requirement, and is also Euclidean invariant, then $\{\mathcal{H}, H, q(t)\}$ satisfies the Wightman axioms for a quantum field.

Thus the construction of a field theory is reduced to the construction of $d\mu$. A candidate for $d\mu = d\mu_V$ is

$$d\mu_V = Z^{-1} e^{-\int V(\phi(x))dx} d\mu_0 \tag{4.8}$$

where $x = (t, \vec{x})$ and where $d\mu_0$ is the Gaussian measure with mean zero and covariance Δ^{-1}. Here Δ is the Laplace operator on $\mathbf{R}^d$, and Z is a normalizing constant. Formally, the quantum theory of (4.6) satisfies (3.1)-(3.2) and H is given by (3.3).

This solution is fine for $d = 1$, at least if we define (4.6) as a $T \to \infty$ limit where $\int_{-T}^{T} V(\phi(t))dt$ replaces $\int_{-\infty}^{\infty} V(\phi(t))dt$. In $d = 1$, the measure $d\mu_0$ is concentrated on continuous functions $\phi(t)$, like Wiener measure. However, for $d > 1$, the trajectories are no longer pointwise continuous; rather they take their values in some subspace of $\mathcal{S}'(\mathbf{R}^{d-1})$. Thus the nonlinear function $V(\phi(x))$ is not defined. We must mollify the paths $\phi(x)$ by a C_0^∞ approximation $\delta_\kappa(x)$ to Dirac measure:

$$\phi_\kappa(x) = (\delta_\kappa * \phi)(x)$$

where $\delta_\kappa = \kappa^d \chi(x\kappa)$ and $\int \chi(x)\,dx = 1$, $0 \le \chi \in C_0^\infty$. Thus we study

$$d\mu_V = \lim_{\Lambda \uparrow \mathbf{R}^d} \lim_{\kappa \to \infty} Z(\Lambda, \kappa)^{-1} e^{-\int_\Lambda V_\kappa(\phi_\kappa(x))\,dx} d\mu_0 . \tag{4.9}$$

The hard work of the subject goes into proving existence of the limit (4.9), and establishing properties of the limit. Here we restrict attention to V a polynomial of degree four and to $d \le 3$. Let $d\mu_{V,\kappa,\Lambda}$ denote the approximate measure.

In the following, the parameters in V will be $\lambda \in \mathbf{R}_+$, $\sigma \in \mathbf{R}$. Also α, β, γ are positive constants. Define

$$V_\kappa(\xi) = \lambda \xi^4 + a(\lambda, \kappa)\,\xi^2 \tag{4.10}$$

where

$$a(\lambda, \kappa) = \begin{cases} \sigma, & d = 1 \\ -\alpha\lambda(\ell n \kappa) + \sigma, & d = 2 \\ -\beta\lambda\kappa + \gamma\lambda^2(\ell n\kappa) + \sigma, & d = 3 . \end{cases} \tag{4.11}$$

THEOREM 4.4. *There exist* α, β, γ *such that for all* λ, σ *the* $\kappa \to \infty$, $\Lambda \uparrow \mathbf{R}^d$ *limit in (4.9) exists in the sense that*

$$\int e^{i\phi(g)} d\mu_{V,\kappa,\Lambda} \to \int e^{i\phi(g)} d\mu_V$$

for every $g \in \mathcal{S}(\mathbf{R}^d)$. *Furthermore* $d\mu$ *satisfies Euclidean invariance (including reflections) and* OS *positivity.*

This is the existence theorem for quantum fields. We now explain the $\kappa \to \infty$ limit, namely why (4.11) diverges with κ as $\kappa \to \infty$. The κ-dependent constants in (4.11) are called "renormalization constants" in physics. Without them, the $\kappa \to \infty$ limit would either not exist or would be trivial. The choice of α, β, γ is based on perturbation theory, as we explain. The proofs that the $\kappa \to \infty$ limit exists is extremely complicated for $d = 3$ [3]. For $d = 2$ it is simpler, while for $d = 1$ no renormalization divergent with $\kappa \to \infty$, is required. For the remainder of this talk, we fix Λ and investigate methods to establish uniformity in κ.

We mention, however, that Z diverges both as $\kappa \to \infty$ $(d = 3)$ or as $\Lambda \to \mathbf{R}^d$ $(d = 1, 2, 3)$. This additional (volume) divergence is studied by different methods, namely by decoupling a large volume into an ensemble of volumes of size O(1). These methods are described in detail in [1].

5. *Integration by parts*

The formula for integration by parts in a gaussian probability measure plays a useful role. Consider the case of a gaussian measure on $\mathbf{R}^N$ with mean zero and covariance C (a positive $N \times N$ matrix), namely

$$d\mu = \det(C/2\pi)^{\frac{1}{2}} \exp\left(-\frac{1}{2} \langle q, C^{-1} q \rangle\right) dq \,. \tag{5.1}$$

Here

$$\langle q, C^{-1} q \rangle = \sum_{i,j=1}^{N} q_i C_{ij}^{-1} q_j \,,$$

and dq denotes Lebesgue measure. Then integration by parts in the usual fashion yields

(5.2) $$\int_{R^N} (C^{-1}q)_i A(q)\, d\mu = \int_{R^N} \frac{\partial A}{\partial q_i}\, d\mu\,,$$

because

$$\frac{\partial}{\partial q_i} \exp\left(-\tfrac{1}{2} <q, C^{-1}q>\right) = -\sum_{j=1}^{N} (C^{-1}_{ij} q_j) \exp\left(-\tfrac{1}{2} <q, C^{-1}q>\right).$$

This formula generalizes to a gaussian probability measure $d\mu$ on $\mathcal{S}'(R^d)$ with mean zero and covariance $C : \mathcal{S} \to \mathcal{S}$, positive in the L_2 sense. Then

$$\int (C^{-1}\phi)(x)\, A(\phi)\, d\mu = \int \frac{\delta A}{\delta\phi(x)}\, d\mu\,.$$

Here $\delta A/\delta\phi(x)$ is the directional derivative in the direction $\delta_x \in \mathcal{S}'$, and the functional $A : \mathcal{S}' \to \mathbf{C}$ is assumed to be differentiable in a suitable sense.

$$\frac{\delta A(\phi)}{\delta\phi(x)} \equiv \lim_{\varepsilon \to 0} \frac{A(\phi + \varepsilon\delta_x) - A(\phi)}{\varepsilon}\,.$$

The example we generally require is $C = (\Delta + I)^{-1}$, where Δ denotes the flat-space Laplace operator $-\Sigma_i\, \partial^2/\partial x_i^2$, so the identity becomes

(5.3) $$(\Delta + I) \int \phi(x)\, A(\phi)\, d\mu = \int \frac{\delta A(\phi)}{\delta\phi(x)}\, d\mu\,.$$

An important special case is $A = B \exp(-\int V(\phi(x))\, dx)$. The factor $\exp(-V)$ can be regarded as part of the measure, defining a non-gaussian measure

$$d\mu_V = Z^{-1} \exp(-V)\, d\mu$$

where $Z = \int \exp(-V)\, d\mu$ is a normalization constant. Then the integration by parts formula in the non-gaussian measure $d\mu_V$ becomes

$$(5.4) \qquad \int \{(\Delta + I)\, \phi(x) + V'(\phi(x))\}\, A(\phi)\, d\mu_V = \int \frac{\delta A(\phi)}{\delta\phi(x)}\, d\mu_V \,.$$

For $V = 0$, (5.4) reduces to (5.3). This identity is sometimes referred to in the physics literature as the "equation of motion." In fact, for $x \notin \operatorname{suppt} A$, $\delta A/\delta\phi(x) = 0$. Thus (5.4) is the analytic continuation of (3.1) in t to the point $x_0 = -it$.

6. *Feynman diagrams*

Feynman diagrams provide a pictorial representation of the integration by parts formulas (5.3)-(5.4), or the equivalent formula

$$(6.1) \qquad \int \phi(x)\, A(\phi)\, d\mu = \iint C(x,y) \left\{\frac{\delta A}{\delta\phi(y)} - V'(\phi(y))\right\} dy\, d\mu_V \,.$$

Here $C(x,y)$ is the Green's function for $\Delta + I$, namely the kernel of the resolvent integral operator $(\Delta + I)^{-1}$.

Let us define the following diagrams:

$\phi(x) \quad = x$ 

$C(x,y) \; = x$  y .

The diagram for $\phi(x)$ is a leg attached to a vertex located at $x \in R^d$. The diagram for $C(x,y)$ is a line connecting vertices at x and y. Likewise the diagrams for products of such factors are the product diagrams. For example

(6.2) $\quad \phi(x)^2 \phi(y)\, \phi(z)^3 C(t,w)^2 =$ 

.

Finally, a vertex without a space-time label is integrated over R^d,

(6.3) $$\int_{R^d} \phi(x)^2 \phi(y)\, C(x,y)^3\, dx = $$ 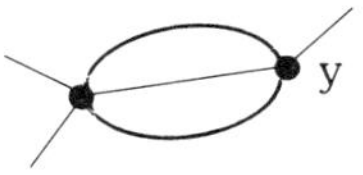

.

When we do not wish to indicate the internal structure of a diagram for a function $A(\phi)$ which is a monomial as above, we indicate the diagram by

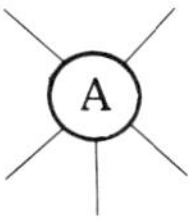

,

where the legs denote the legs of A. For example, (6.2) would have six legs and (6.3) would have one.

Let $\langle \cdot \rangle_V$ denote the expectation in the measure $d\mu_V$, namely integration with respect to that measure. Then the integration by parts formula (6.1) has a simple diagrammatic interpretation: For the case $V = 0$, integration of a ϕ leg by parts results in a sum over all possible attachments of this leg to another ϕ-leg. For example,

(6.4) 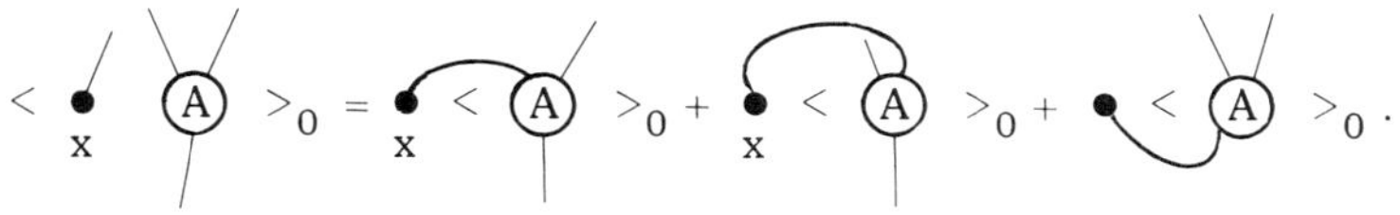

Note that one further integration by parts yields

(6.5) 

After all ϕ legs have been paired, the expectation is reduced to a sum of integrals over $R^d \times \cdots \times R^d$, i.e., to a finite dimensional integral, rather than an integral over function space.

In case $V \neq 0$, each integration by parts also introduces a V' factor, namely a V vertex with one leg attached to ϕ. Thus

(6.6) 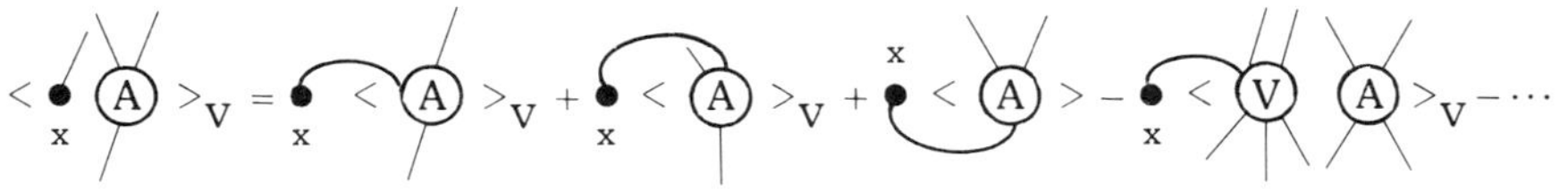

where the $\cdots$ terms indicate the sum over attachment to all possible legs of V. Consider the particular case, for example, in which A is linear and V is quadratic:

$$A = \phi(y)\,, \qquad V = V_Q = \frac{1}{2}\, a \int \phi(x)^2\, dx\,.$$

Then integration of $\phi(x)$ by parts in $\langle \phi(x)\,\phi(y)\rangle_V$ yields

(6.7) 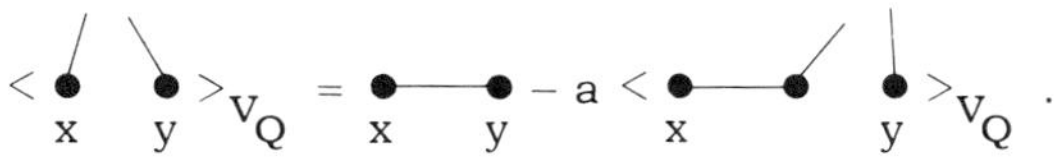

Repeated integration by parts then yields

(6.8) 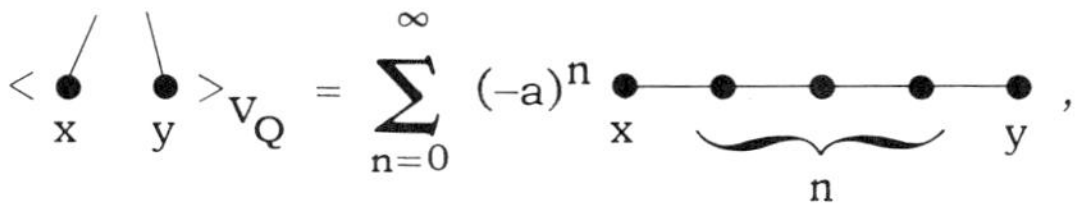

which is just the Neumann series for $(\Delta + (a+1)\,I)^{-1}(x,y)$. Alternatively, we recognize (6.7) as the integral equation for the Green's function G with kernel 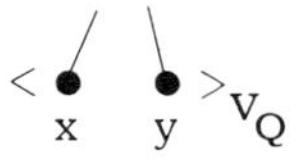 $= G(x,y)$, namely

$$(I + aC)\,G = C\,.$$

Thus

$$G = (I + aC)^{-1}C = (\Delta + (a+1)\,I)^{-1}\,.$$

7. *Perturbative renormalization*

Consider the case $V = \int \left[\lambda\phi^4 + \frac{1}{2}\, a\phi^2\right] dx$. Then integration by parts exhibits the cancellations between (the divergent part of) $a = a(\lambda, \kappa)$ and terms which arise from $\lambda\phi^4$. For example, the terms of order one in a cancel terms of order one and two in λ. For this reason, a has linear and quadratic dependence on λ. We exhibit here this cancellation in lowest order of a. The mathematical proof of the existence of $d\mu$ (i.e., of *non*perturbative renormalization) consists in giving a convergent expansion which exhibits the cancellation in all orders.

Let us first concentrate on the order one contribution to $<\phi(x)\phi(x)>_V$. Successive integration by parts yields

(7.1) $<\phi(x)\phi(y)>_V =$

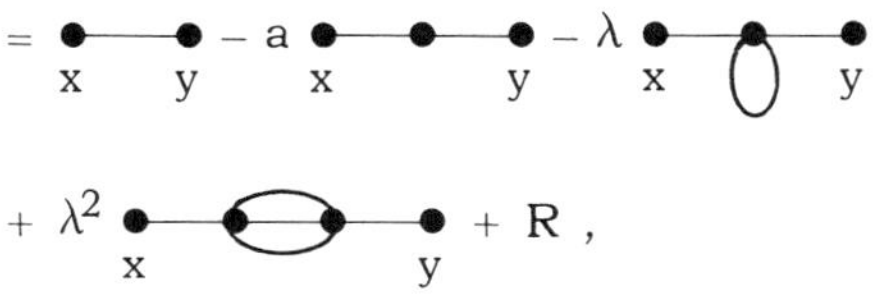

where R is either of order two in a, or else of order three in λ, as a and $\lambda \to 0$. For $d = 3$, and $C = (\Delta + I)^{-1}$,

(7.2) $$0 < C(x, y) = \frac{1}{4\pi|x-y|} e^{-|x-y|} .$$

The third diagram in (7.1) is proportional to $\lambda C(z, z)$, namely the Green's function for the Laplacian on the diagonal. With a mollifier δ_κ, this is $O(\kappa)$ as $\kappa \to \infty$. Likewise the fourth diagram in (7.1) is proportional to

(7.3) $$\iint C(z, w)^3 dz dw = O(\ell n\ \kappa) ,$$

again by the singularity of C on the diagonal. This motivates the choice of constants $\alpha, \beta > 0$ such that

(7.4) $$a(\lambda, \kappa) = -\alpha\lambda\kappa + \beta\lambda^2 \ell n\ \kappa + \sigma ;$$

they are chosen in order to cancel the divergent parts of terms two, three and four in (7.1). The remaining contribution to these terms contains the finite mass shift from σ, as well as other (finite) corrections to the mass which occur because (7.3) is not cancelled identically, but only up to a finite remainder.

In this fashion, the choice of $a(\lambda, \kappa)$ in (7.4) was motivated by the $\lambda \to 0$ dependence of the series (7.1). What is extremely surprising, is that this choice is suitable, in fact, for *all* λ. The proof is contained in the first reference of [3].

8. *Uniqueness of solutions*

The uniqueness of the measure $d\mu$ constructed by the limits $\Lambda \to \mathbf{R}^d$ (and $\kappa \to \infty$, when defined) is the question of *whether* phase transitions exist. If $d\mu$ is ergodic with respect to the translation group on $\mathbf{R}^d$, it is said to be a *pure phase*. If a given sequence has several limit points (depending, for example, on boundary conditions on $\partial\Lambda$) or if $d\mu$ is not ergodic, then phase transitions occur for V. Ergodicity of $d\mu$ is equivalent to whether or not Ω in (4.7) is the only ground state of H, i.e. whether 0 is a simple eigenvalue of H. In the case $d\mu$ is ergodic, $<\phi> = \int \phi d\mu = 0$.

For the case of V given by (4.8), it is known that the existence of phase transitions depend on σ. For $\sigma >> 0$, $<\phi> = 0$. On the other hand, for $\sigma << 0$ the value of $<\phi>$ depends on boundary conditions for the limit $\Lambda \to \mathbf{R}^d$, $d \geq 2$.

THEOREM 8.1. *For* $d = 1$ *and* σ *arbitrary, or for* $d = 2$ *or* 3 *and* $\sigma >> 0$, *the limit* $d\mu$ *is ergodic. For* $d = 2, 3$ *and* $\sigma << 0$, *the limit* $d\mu$ *has at least two ergodic components,* $d\mu_{\pm}$. *For these components*

$$<\phi>_+ = -<\phi>_- \neq 0 .$$

The general structure of phase transitions for polynomial V is only now being unravelled. One can imagine $<\phi>$ taking values at the minima of an effective potential $V_{eff}(\phi)$. In fact, the coefficients of V_{eff} are finite and the number of global minima determine the number of pure phases. For the $\phi^4_{2,3}$ models discussed here,

$V_{eff}(\phi) =$ 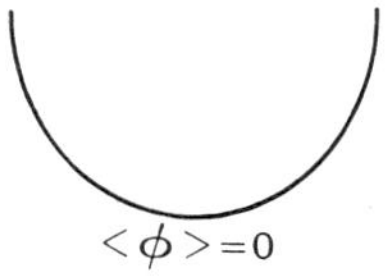

$\sigma \gg 0$,

$V_{eff}(\phi) =$ 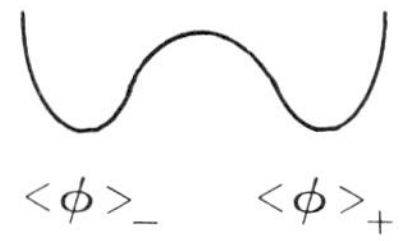

$\sigma \ll 0$,

reflecting the statements of Theorem 8.1.

HARVARD UNIVERSITY
CAMBRIDGE, MASS. 02138

REFERENCES

[1] A good general reference is J. Glimm and A. Jaffe, *Quantum Physics*, Springer Verlag, 1981. This book contains much background and general material and a complete discussion of the existence theory for $d=2$ quantum fields. It also contains details of the physical consequences of quantum fields, and how they describe particles, scattering, bound states, etc. Finally, this book contains an extensive bibliography.

[2] Field theories with fermions have also been constructed. Details of the original construction are given in Glimm and Jaffe's lectures "Field Theory Models," published in "Statistical Mechanics and Quantum Field Theory," C. DeWitt and R. Stora, editors, Gordon and Breach, 1971. More recent work is reviewed in the book described in [1].

[3] The $\kappa \to \infty$ limit of the $\lambda\phi^4$ model in $d=3$ dimensions (and for *all* λ) was originally constructed (for Λ fixed) in J. Glimm and A. Jaffe, Fort. der Physik *21*, 327-376 (1973). The extension to infinite volume by a generalization of the $d=2$ methods was done by J. Feldman and K. Osterwalder, Ann. Phys. *97*, 80-135 (1976) and J. Magnen and R. Sénéor, Ann. l'Inst. H. Poincaré *24*, 95-159 (1976). This method requires $\lambda \ll 1$. For large λ, the infinite volume limit was taken in E. Seiler and B. Simon, Ann. Phys. *97*, 470-518 (1976).

[4] Phase transitions were originally established for continuum field theories in J. Glimm, A. Jaffe and T. Spencer, Comm. Math. Phys. *45*, 203-216 (1975). A generalized method of steepest descent to estimate integrals with respect to $d\mu$ was developed in J. Glimm, A. Jaffe and T. Spencer, Ann. Phys. *101*, 610-630; 631-699 (1976). Phase transitions for $d=3$ quantum fields were established in J. Fröhlich, B. Simon and T. Spencer, Comm. Math. Phys. *50*, 79-95 (1976). Extensions to other models (Gawedzki, Sommers, Balaban and Gawedzki, Imbrie, Fröhlich and Spencer) mostly remain to be published. See [1] for further discussion of several such results.

METRICS WITH PRESCRIBED RICCI CURVATURE

Dennis M. DeTurck*

Contents

1. Introduction
2. Two-dimensional manifolds
3. Solvability of elliptic systems
4. Connections with prescribed Ricci curvature
5. The Bianchi identity and nonsolvability of the Ricci equation
6. Existence of Riemannian metrics for smooth Ricci tensors
7. Concluding remarks

1. *Introduction*

As many authors have pointed out (for example see [2] and [13]), the Ricci curvature of a differentiable manifold is an object that deserves careful investigation, although not much is known about it at the present time. One reason the Ricci curvature is important in geometry is that it can place restrictions on the topology of manifolds; in physics, it arises in Einstein's theory of general relativity. We direct the reader to [2] for a historical discussion of the study of the Ricci tensor, as well as for a summary of what is now known.

A fundamental question is to determine which symmetric covariant tensors of rank two can be Ricci tensors of Riemannian metrics. The

*Research supported in part by National Science Foundation Grant MCS79-01780.

Seminar on Differential Geometry
0-691-08268-5/82/000525-13 $00.65/0 (cloth)
0-691-08296-0/82/000525-13 $00.65/0 (paperback)

definition of Ricci curvature casts the problem of finding a metric g that realizes a given Ricci curvature R as that of solving the following nonlinear system of second-order partial differential equations:

$$\text{(1)} \qquad \operatorname{Ricc}(g) \equiv \frac{\partial \Gamma^s_{ij}}{\partial x^s} - \frac{\partial \Gamma^s_{is}}{\partial x^j} + \Gamma^s_{ij}\Gamma^t_{st} - \Gamma^s_{it}\Gamma^t_{sj}$$

where

$$\Gamma^i_{jk} = \frac{1}{2} g^{is}\left(\frac{\partial g_{js}}{\partial x^k} + \frac{\partial g_{sk}}{\partial x^j} - \frac{\partial g_{jk}}{\partial x^s}\right)$$

are the Christoffel symbols of the metric g. We will systematically write the above system as $\operatorname{Ricc}(g) = R$. In this system, there are the same number of equations as unknowns because g and R are both required to be symmetric tensors. There is a complication though, since any solution of $\operatorname{Ricc}(g) = R$ must also satisfy certain compatibility conditions imposed by the Bianchi identity. This will be discussed in §5.

Ultimately, one would like global results about existence, uniqueness and regularity—including topological obstructions—of metrics with prescribed Ricci tensors on manifolds. The first step, though, is to determine when one can solve the equation $\operatorname{Ricc}(g) = R$ locally, say, in a neighborhood of a point x_0 in $\mathbf{R}^n$. This local problem is already nontrivial and is the subject of most of this report. We present several results in complete detail, but the length of other proofs precludes their inclusion here.

In order to introduce some notation, we examine the linearization of the Ricci operator:

$$\text{(2)} \qquad \operatorname{Ricc}'(g)h \equiv \left.\frac{d}{dt}\right|_{t=0} \operatorname{Ricc}(g+th) = \frac{1}{2}\Delta_L h - \operatorname{div}^*(\operatorname{div}(Gh))$$

for $h \in S^2T^*$. This is shown, for example, in [1]. Here, Δ_L is the Lichnerowicz Laplacian:

$$\Delta_L h = -h_{ij;s}{}^{s} + R_{is}h^s_j + R_{sj}h^s_i - 2R_{isjt}h^{st} .$$

The covariant derivatives and curvature tensors that appear are those of g. The divergence operator $\operatorname{div} : S^2T^* \to T^*$ and its formal (or L^2) adjoint div^* are defined as follows for $h \in S^2T^*$ and $v \in T^*$:

$$\operatorname{div} h = -g^{st}h_{si;t}$$

$$\operatorname{div}^* v = \frac{1}{2}(v_{i;j} + v_{j;i}) .$$

Finally, G is the gravitation operator

$$Gh = h_{ij} - \frac{1}{2} g_{ij}(g^{st} h_{st}) .$$

Note that $G(\operatorname{Ricc}(g))$ is the stress-energy tensor in Einstein's theory of gravitation. The operator (2) is not elliptic since for *no* $\xi \in \mathbf{R}^n$ is the symbol of $\operatorname{Ricc}'(g)$ an isomorphism. In fact, its principal symbol is a mapping from $S^2\mathbf{R}^n$ to $S^2\mathbf{R}^n$:

$$\sigma(h)_{ij} = \frac{1}{2} g^{st}(x_0)[h_{is}\xi_j\xi_t + h_{tj}\xi_i\xi_s - h_{ij}\xi_s\xi_t - h_{st}\xi_i\xi_j] .$$

Clearly, if $h_{ij} = \xi_i\xi_j$ then $\sigma(h)_{ij} = 0$ with $h_{ij} \neq 0$. This degeneracy precludes proving local existence by applying directly the elliptic machinery developed in §3.

2. *Two-dimensional manifolds*

The problem for two-dimensional manifolds is greatly simplified by the fact that all metrics on 2-manifolds are Einstein. Thus, a necessary condition for R_{ij} to be the Ricci tensor of a Riemannian metric is that $R_{ij} = \kappa\gamma_{ij}$, where γ_{ij} is some positive definite (at each point) tensor. Locally, this condition is also sufficient, as we shall now see.

THEOREM 2.1. *Let* R_{ij} *be defined in a neighborhood of a point* p *on a 2-manifold. A metric* g_{ij} *exists so that* $\operatorname{Ricc}(g) = R$ *in a neighborhood of* p *if and only if* $R_{ij} = \kappa\gamma_{ij}$ *for some scalar function* κ *and positive definite tensor* γ_{ij}.

Proof. The metric we seek must be conformal pointwise to γ_{ij}, so we will find a function u so that $g = e^{2u}\gamma$. It is well known [15, p. 78] that, if $S_{ij} = \text{Ricc}(\gamma_{ij})$,

$$\text{Ricc}(g) = \text{Ricc}(e^{2u}\gamma) = S_{ij} - \gamma_{ij}\Delta u$$

where $\Delta u = -\gamma^{ij}u_{;ij}$ is the Laplacian operator of the γ metric. It is now clear that if we find u so that $\Delta u = \phi - \kappa$, where $S_{ij} = \phi\gamma_{ij}$, then Ricc (g) will equal R. The local solvability of the Laplacian is classical. qed.

On a compact manifold M without boundary, the range of the Laplacian is orthogonal (in $L^2(M)$) to the constant functions. Thus, $\Delta u = \phi - \kappa$ is globally solvable if and only if $\int_M (\phi-\kappa)\,dV = 0$, where dV is the volume of the γ metric. Since ϕ is the Gauss curvature of the γ metric we have the following.

COROLLARY 2.2. *Let* M *be a compact 2-manifold without boundary, and let* R_{ij} *satisfy the necessary condition* $R_{ij} = \kappa\gamma_{ij}$ *with* $\gamma > 0$. *Then* R_{ij} *is the Ricci tensor of a metric on* M *if and only if* $\int_M \kappa\,dV_\gamma = 2\pi\chi(M)$.

Along these lines, note that the Riemann curvature tensor $R^i{}_{jk\ell}(= -R^i{}_{j\ell k})$ can be considered as a 2-form with values in matrices with trace zero (since $R^s{}_{sij} = 0$). As $R^1{}_{112} = -R_{12}$, $R^1{}_{212} = -R_{22}$ and $R^2{}_{112} = R_{11}$, the condition of Theorem 2.1 yields the following.

COROLLARY 2.3. *A matrix-valued 2-form* $R^i{}_{jk\ell}$ *is locally the Riemann curvature tensor of some 2-metric if and only if its eigenvalues are purely imaginary.*

One can obtain a global result for the Riemann tensor by the same method as in Corollary 2.2.

3. *Solvability of elliptic systems*

Consider a system of p equations for q unknown functions $u = (u^1(x), \cdots, u^q(x))$ of order m

$$F_i(x, u, D^\alpha u) = 0 \qquad \begin{array}{l} i = 1, \cdots, p \\ |\alpha| \leqq m \end{array} \tag{3}$$

where the F_i are smooth in their arguments (for us, this means C^∞ in u and $D^\alpha u$, and $C^{k+\sigma}$ in x for some integer k and $0 < \sigma < 1$). The system is called *elliptic at the point* x_0 *for the function* u_0 if the following linear operator is elliptic.

$$L_i h = \sum_{\substack{|\beta| \leqq m \\ k \leqq q}} \frac{\partial F_i}{\partial(D^\beta u^k)}(x_0, u_0, D^\alpha u_0) D^\beta h^k \equiv \sum_{\substack{|\beta| \leqq m \\ k \leqq q}} c_{i\beta k} D^\beta h^k . \tag{4}$$

This last operator is called elliptic if, for every vector $\xi \in \mathbf{R}^n - \{0\}$, the matrix $[\sigma_{ik}]$, called the *principal symbol* of (3) or of (4), has maximal rank, where

$$\sigma_{ik} = (\sqrt{-1})^m \sum_{|\beta|=m} c_{i\beta k} \xi^\beta \qquad \begin{array}{l} i = 1, \cdots, p \\ k = 1, \cdots, q \end{array} .$$

The system is called *determined elliptic* (or, simply, *elliptic*) if the symbol is an isomorphism $(p = q)$, it is called *overdetermined elliptic* if the symbol is injective $(p > q)$, and it is called *underdetermined elliptic* if the symbol is surjective $(p < q)$. Finally, we say that a function $u_0(x)$ is an *infinitesimal solution* of equation (3) at a point x_0 if

$$F_i(x, u_0(x), D^\alpha u_0(x))\big|_{x=x_0} = 0 \qquad i = 1, \cdots, p .$$

THEOREM 3.1 (Local solvability). *Let* u_0 *be an infinitesimal solution of a determined or underdetermined elliptic system (3) at* x_0. *Then for* ρ *sufficiently small, there exists a* $C^{k+m+\sigma}$ *function* u *that solves (3) for* $|x - x_0| \leqq \rho$.

This theorem has been around for a while in one form or another for determined systems. The earliest place (known to the author) that it was written down is Malgrange's proof of the Newlander-Nirenberg theorem [11]. It appears subsequently in [8], [10] and [12]. The idea of the proof is as follows. For determined systems, translate so that $u_0 = 0$ and x_0 is the origin and refer to (3) collectively as $F(x, D^\alpha u) = 0$. Let $x = \rho y$ and $v = \rho^{-m}(u(\rho y))$ for v defined on $B_1(0)$ (the unit ball), and set

$$\Phi(\rho, v) = F(\rho y, \rho^{m-|\alpha|} D^\alpha v) .$$

This technique is called scaling. We then see that $\Phi(0, 0) = 0$, and that the derivative of Φ with respect to v at $(0, 0)$ turns out to be the operator (4). Since, by ellipticity, this operator admits a continuous linear right inverse, the implicit function theorem gives a solution $v = \psi(\rho)$ of $\Phi(\rho, v) = 0$ for positive ρ near zero. Unwinding the scaling operation yields a solution of (3) on $B_\rho(x_0)$. For underdetermined systems, the principal symbol σ of (3) is surjective, so $\sigma\sigma^*$ is an isomorphism. Thus, the operator LL^* is determined elliptic and so the above proof applies to the equation $F(x, D^\alpha L^* v) = 0$. The proof is completed by setting $u = L^* v$.

qed.

Even though this theorem gives local solvability for underdetermined elliptic operators, one might suspect that more is true, namely, that the scaled version of an underdetermined elliptic operator is actually a submersion. This is indeed the case, and yields the following.

THEOREM 3.2. *Let* L_0 *be the constant-coefficient part of the underdetermined elliptic equation (3) at* x_0 *and the infinitesimal solution* u_0. *Then for* ρ *sufficiently small, there is a Banach submanifold of solutions of (3) in* B_ρ *parametrized by functions* v *in the kernel of* L_0.

4. *Connections with prescribed Ricci curvature*

As an application of Theorems 3.1 and 3.2, we will prove the existence of $C^{k+1+\sigma}$ connections having prescribed $C^{k+\sigma}$ Ricci tensors, extending the work of J. Gasqui in [6] and [7]. Given an affine connection Γ^i_{jk} (not necessarily symmetric in j and k), its Ricci curvature is defined by equation (1), without the requirement that Γ^i_{jk} come from a metric. For this section, we write (1) as $\mathrm{Ricc}(\Gamma) = R$. We wish to solve this first-order equation for Γ, given a tensor R_{ij}.

THEOREM 4.1. *Suppose* R_{ij} *is a given (not necessarily symmetric)* $C^{k+\sigma}$ *tensor in a neighborhood of* x_0. *Then there is a* $C^{k+1+\sigma}$ *connection* Γ^i_{jk} *so that* $\mathrm{Ricc}(\Gamma) = R$ *in a neighborhood of* x_0.

Proof. We need only find a connection Γ_0 so that $\mathrm{Ricc}(\Gamma_0) = R$ at x_0, and show that the principal symbol of Ricc is surjective. We attack the latter problem first. For $\xi \in \mathbf{R}^n - \{0\}$, the principal symbol of Ricc operates on $a^i_{jk} \in T \otimes T^* \otimes T^*$ and maps them to $c_{ij} \in T^* \otimes T^*$ as follows:

$$\sigma(a^r_{st})_{ij} = \xi_s a^s_{ij} - \xi_j a^s_{is} = c_{ij} . \tag{5}$$

We write down the following solution of (5):

$$a^i_{jk} = \frac{1}{|\xi|^2}\left[c_{jk}\xi^i - \frac{1}{n-1} c_{j\ell}\xi^\ell \delta_{ik}\right] . \tag{6}$$

Motivated by this, it is easy to guess that an infinitesimal solution of $\mathrm{Ricc}(\Gamma) = R$ at the origin is

$$\Gamma^i_{jk} = R_{jk}x^i - \frac{1}{n-1} R_{j\ell}x^\ell \delta^i_k . \tag{7}$$

qed.

To get torsion-free connections for symmetric Ricci tensors, we must alter the Ricci operator as follows:

$$\mathrm{ricc}(\Gamma)_{ij} \equiv \frac{1}{2}[\mathrm{Ricc}(\Gamma)_{ij} + \mathrm{Ricc}(\Gamma)_{ji}] .$$

(This affects only the second term of the Ricc operator.) Although we have no snappy analogue of (6) and (7) for the symmetric case, an infinitesimal solution is found and surjectivity of the symbol of ricc is shown (under a different guise) in Lemmas 2 and 1 of [7], respectively. This yields

THEOREM 4.2. *If* R_{ij} *is a given symmetric* $C^{k+\sigma}$ *tensor in a neighborhood of* x_0, *then there is a torsion-free* $C^{k+1+\sigma}$ *connection* Γ^i_{jk} *so that* $\mathrm{ricc}(\Gamma) = R$ *in a neighborhood of* x_0.

The advantage of this proof over that given in [6] is that we do not experience the loss of derivatives that Gasqui did (he obtained a $C^{1+\sigma}$ connection for a $C^{1+\sigma}$ Ricci tensor). Of course, we do not claim that *every* solution of $\mathrm{Ricc}(\Gamma) = R$ possesses this regularity property. For example, if $\Gamma^1_{11} = x^1|x^1|$ and $\Gamma^i_{jk} = 0$ for all other choices of i, j and k, then $\mathrm{Ricc}(\Gamma) \equiv 0$. Here, the connection is $C^{1+\sigma}$ but not C^2 while R is C^∞.

5. *The Bianchi identity and nonsolvability of the Ricci equation*

We return to Riemannian geometry and consider the Bianchi identity for Ricci curvature and show that it presents obstructions to local solvability of the Ricci equation. We conclude with a proof of the fact that if R is nonsingular, there exist many metrics with respect to which R satisfies the Bianchi identity. For the remainder of this report, we assume that the dimension n is at least 3.

PROPOSITION 5.1. *Let* g *be any metric on a manifold of dimension* n *and suppose that* $\mathrm{Ricc}(g) = R$. *Then, at every point,*

$$\mathrm{Bian}(g, R) \equiv -\mathrm{div}(G(R)) = g^{st}\left(R_{sm;t} - \frac{1}{2} R_{st;m}\right) = 0 . \tag{8}$$

There are several proofs of this identity. A proof that arose from this work based solely on the fact that the Ricci operator is invariant under the group of diffeomorphisms (i.e., that $\phi^*\mathrm{Ricc}(g) = \mathrm{Ricc}(\phi^* g)$ for any diffeomorphism ϕ) can be found in [9].

Note that (8) places n conditions on the pair (g, R) that involve first derivatives of *both* g and R. The main fact to realize is that for a given R, a necessary condition for the existence of metrics that satisfy $\mathrm{Ricc}(g) = R$ is the existence of those that satisfy $\mathrm{Bian}(g, R) = 0$. This leads us to the following examples of nonsolvability of the Ricci equation.

EXAMPLE 5.2. $R = x^1 dx^1 \otimes dx^1 \pm dx^2 \otimes dx^2 \pm \cdots \pm dx^n \otimes dx^n$ is not the Ricci tensor of any Riemannian metric g near $x^1 = 0$ since there is no metric satisfying the Bianchi identity there. In fact, suppose that $\mathrm{Bian}(g, R) = 0$ for some metric g. That is to say,

$$(9) \qquad g^{st}\left(R_{sm;t} - \frac{1}{2} R_{st;m}\right) = g^{st}\left(\frac{\partial R_{sm}}{\partial x^t} - \frac{1}{2}\frac{\partial R_{st}}{\partial x^m} - \Gamma^q_{st} R_{qm}\right) = 0 .$$

If $x^1 = 0$, with R_{ij} as above, then when $m = 1$, (9) becomes $g^{11} = 0$, which is impossible for a positive definite metric. However, for $x^1 > 0$, there is a metric with this R as its Ricci tensor, namely

$$g = x^1(f(x^1))^2 dx^1 \otimes dx^1 + 2f(x^1) dx^2 \otimes dx^2 + 2f(x^1) dx^3 \otimes dx^3$$

where $f(x) = \mathrm{sech}^2((x)^{3/2}/\sqrt{18})$ (for $n=3$, arbitrary n is now easy). This example was devised by searching for metrics "of revolution", that is, diagonal metrics that depend on one variable only. This reduces the problem to one in ordinary differential equations.

EXAMPLE 5.3. $R = \sum_{i=1}^{n} (x^1 dx^i \otimes dx^i) + \frac{1}{2} \sum_{i=1}^{n} (x^i dx^i \otimes dx^1) + \frac{1}{2} \sum_{i=1}^{n} (x^i dx^1 \otimes dx^i)$ is not the Ricci tensor of any metric of any signature near the origin in $\mathbf{R}^n$. As in Example 5.2, suppose that $\mathrm{Bian}(g, R) = 0$ for some metric g. With this choice of R_{ij}, (9) implies that $g^{11} = g^{12} = \cdots = g^{1n} = 0$ at the origin, and this can never happen for a metric.

We have seen that the Bianchi identity presents a serious obstruction to local solvability for the Ricci equation. Given a "Ricci candidate" R, then, we first consider the problem of finding metrics that solve $\mathrm{Bian}(g, R) = 0$. The first step is to examine the linearization of this equation with respect to g. A modest computation yields

$$(10) \qquad \mathrm{Bian}'(g, R)h \equiv \left.\frac{d}{d\tau}\right|_{\tau=0} \mathrm{Bian}(g + \tau h, R) = R^s_m (\mathrm{div}\,(G(h)))_s - Q^{st}_m h_{st}$$

where

$$Q^{st}_m = g^{sp}g^{tq}\left(\frac{\partial R_{qm}}{\partial x^p} - \frac{1}{2}\frac{\partial R_{pq}}{\partial x^m} - \Gamma^{\ell}_{pq}R_{\ell m}\right).$$

We now show that if R is nonsingular, $\mathrm{Bian}'(g,R)$, and hence $\mathrm{Bian}(g,R)$, is an underdetermined elliptic operator. This will be true if and only if the symbol of the operator $\mathrm{div}(G\cdot)$ is surjective. The symbol maps symmetric matrices to covectors and is

$$-\xi_s a^s_m + \frac{1}{2}\xi_m a^s_s$$

for $\xi \in T^*_pM$. Thus, given $v \in T^*M$, we need to solve

(11) $$g^{st}\left[\xi_s a_{tm} - \frac{1}{2}\xi_m a_{st}\right] = v_m$$

for a. It is clear that the following specification of a solves (11):

$$a_{k\ell} = (\xi_k v_\ell + \xi_\ell v_k)/(g^{st}\xi_s\xi_t).$$

We assert that infinitesimal solutions of $\mathrm{Bian}(g,R) = 0$ exist if R is nonsingular at the point in question. These facts, along with Theorem 3.2, imply the following.

PROPOSITION 5.4. *For sufficiently small* $\rho > 0$, *the solutions of* $\mathrm{Bian}(g,R) = 0$ *on* $B_\rho(0)$ *near a given infinitesimal solution* g_0 *form a submanifold of the Banach manifold of Riemannian metrics if the inverse of* $R(0)$ *exists.*

6. *Existence of Riemannian metrics for smooth Ricci tensors*

In this section, we outline the proof of the following.

THEOREM 6.1. *If* R_{ij} *is a* $C^{k+\sigma}$ *(resp.* C^∞, *analytic) tensor field* $(k>2)$ *in a neighborhood of* x_0 *and if* $R^{-1}(x_0)$ *exists, then there is a* $C^{k+\sigma}$ $(C^\infty$, *analytic) Riemannian metric* g *such that* $\mathrm{Ricc}(g) = R$ *in a neighborhood of* x_0.

To begin, recall that in §1 it was shown that the linearization of the Ricci operator is not elliptic. However, comparing formulas (2) and (10) shows that the following is an elliptic system:

$$\text{Ricc}(g) + \text{div}^*(R^{-1}\text{Bian}(g,R)) = R \tag{12}$$

since the principal part of its linearization is simply half of the Laplacian. Unfortunately, this system is not equivalent to the original system $\text{Ricc}(g) = R$. However, combining (12) with

$$\text{div}^*(R^{-1}\text{Bian}(g,R)) = 0 \tag{13}$$

yields an overdetermined (twice as many equations as unknowns) elliptic system that is clearly equivalent to the original one. We prove local solvability for the combined system (12), (13). The details of this proof will appear in [4].

The first step is to find an infinitesimal solution for the system, which we simply assert the existence of here. Then we use Proposition 5.4 to obtain a Banach submanifold of the space of $C^{k+\sigma}$ metrics on $B_\rho(0)$ for ρ sufficiently small. For all metrics on this submanifold, equation (13) is satisfied.

Since equation (12) is elliptic, Theorem 3.1 could be applied to it. However, instead of applying the implicit function theorem directly, we intervene as follows: Recall [14, p. 59] that the implicit function theorem is commonly proved by a contracting mapping argument. This argument involves the use of an iteration procedure somewhat like Newton's method (or, more properly, a Picard method). It is here that we make the essential adjustment. In our scheme, the sequence of metrics $\{g_n\}$ is generated by a two-step procedure. The first step is to perform an ordinary "Newton-Picard" iteration for equation (12) with g_n, to obtain $\bar{g}_n$. Then, we project $\bar{g}_n$ onto the submanifold of solutions of the Bianchi identity, on which the solution of $\text{Ricc}(g) = R$ must lie, to obtain g_{n+1}.

We then demonstrate that these projection operations do not affect the convergence of the sequence and, because we pick our spaces very carefully, that the iterates actually converge to a solution of (12). Since each iterate automatically satisfies (13), we obtain in this manner the desired solution of $\mathrm{Ricc}(g) = R$.

7. *Concluding remarks*

1) In the analytic case our results can be strengthened somewhat. Using a version of Cartan-Kähler theory developed by Malgrange in [12], we can prove Theorem 6.1 to find analytic metrics of any signature (including Lorentz) for analytic nonsingular Ricci tensors.

2) For Lorentz metrics, we also have a local existence theory. In [16], we present a proof of existence of smooth solutions of the Cauchy problem for the equation $\mathrm{Ricc}(g) = R$, where R is a smooth nonsingular Ricci tensor. Existence for the Einstein equations of general relativity is also discussed there.

3) Regularity has not been discussed fully here. In [5], J. Kazdan and the author have shown that a Riemannian metric that possesses a nonsingular $C^{k+\sigma}$ Ricci tensor is also $C^{k+\sigma}$. It is also shown there that all Einstein metrics are analytic in appropriately chosen coordinate systems. A consequence of this is unique (up to diffeomorphism) continuation for Einstein metrics.

4) It may be possible to find global obstructions to the existence of metrics for certain nonsingular Ricci tensors by studying equation (12). For instance, if $\mathrm{Ricc}(g_0) = R_0$ is positive definite on a compact manifold without boundary, and if R is sufficiently near R_0, then every solution g of (12) sufficiently near g_0 is automatically a solution of $\mathrm{Ricc}(g) = R$.

5) Much of this work is contained in the author's Ph.D. thesis and has been announced in [3]. Special thanks are due Jerry Kazdan and others at the University of Pennsylvania for their encouragement and support.

REFERENCES

[1] M. Berger, *Quelques formules de variation pour une structure Riemannienne*, Ann. Scient. Éc. Norm. Sup. 4e série, t. 3 (1970), 285-294.

[2] J. P. Bourguignon, *Ricci curvature and Einstein metrics*, Global Differential Geometry/Global Analysis Proceedings, Berlin, Nov. 1979, Springer Lecture Notes, vol. 838, 42-63.

[3] D. DeTurck, *The equation of prescribed Ricci curvature*, Bull. Am. Math. Soc., 3 (1980), 701-704.

[4] ———, *Existence of metrics with prescribed Ricci tensors: Local theory*, to appear in Inventiones Math.

[5] D. DeTurck and J. Kazdan, *Some regularity theorems in Riemannian geometry*, to appear in Ann. Scient. Éc. Norm. Sup.

[6] J. Gasqui, *Connexions à courbure de Ricci donnée*, Math. Z., 168 (1979), 167-179.

[7] ———, *Sur la courbure de Ricci d'une connexion linéaire*, C. R. Acad. de Sci. Paris Sér A, 281 (1975), 283-288.

[8] ———, *Sur l'existence local d'immersions à courbure scalaire donnée*, Math. Annalen, 241 (1979), 283-288.

[9] J. Kazdan, *Another proof of Bianchi's identity in Riemannian geometry*, Proc. Am. Math. Soc., 81 (1981), 341-342.

[10] ———, *Partial Differential Equations*, lecture notes, Univ. of Pa.

[11] B. Malgrange, *Sur l'integrabilité des structures presque-complexes*, Symposia Math., vol. II (INDAM, Rome 1968), Acad. Press, 1969, 289-296.

[12] ———, *Equations de Lie II*, J. Diff. Geom., 7 (1972), 117-141.

[13] J. Milnor, *Problems of present-day mathematics* (*§XV. Differential Geometry*), Proc. Symp. Pure Math. vol. XXVIII (Mathematical Developments Arising from Hilbert Problems), Am. Math. Soc., 1976, 54-57.

[14] L. Nirenberg, *Topics in Nonlinear Functional Analysis*, Courant Institute Lecture Notes, NYU, 1974.

[15] K. Yano and S. Bochner, *Curvature and Betti Numbers*, Annals of Math Study No. 32, Princeton U. Press, 1953.

[16] D. DeTurck, *The Cauchy problem for the inhomogeneous Ricci equation*, to appear.

BLACK HOLE UNIQUENESS THEOREMS IN CLASSICAL AND QUANTUM GRAVITY

A. S. Lapedes

I. *Introduction*

A "black hole" is a concept that dates at least as far back as Laplace. The essential idea is that a black hole is an object with a gravitational field so strong that even light is dragged back into the gravitating object in much the same way that an apple is dragged back to earth if it is thrown straight up. Because no light can escape from the object it "appears black." Laplace had envisaged such an object in 1718, using the Newtonian theory of gravity. The history of the general relativistic theory of black holes starts in 1916 when K. Schwarzschild [1] published his static spherically symmetric solution of the vacuum Einstein equations describing the spacetime geometry around a nonrotating, uncharged "point mass." This was almost immediately generalized by Reissner [2], and independently by Nordstrom [3], to the electrically charged situation. However, it was to be a long forty-seven years before the stationary solution describing an electrically neutral, rotating, black hole was found by Kerr [4] in 1963, and a further two years until the solution describing an electrically charged rotating black hole was discovered by Newman [5], et al. in 1965. Although these solutions did generate considerable interest, it was generally believed that they were idealizations

Seminar on Differential Geometry
0-691-08268-5/82/000539-64$03.20/0 (cloth)
0-691-08296-0/82/000539-64$03.20/0 (paperback)

or unwarranted simplifications of the most general black hole situation, which could conceivably contain, say, arbitrary mass or charge multipole moments. Therefore, before a seminal paper of Israel [6] in 1967, it seemed as if whole classes of black hole solutions might yet be undiscovered, and in view of the slow progress made up until that point the situation looked fairly grim.

However, in 1967-1968, prospects for a full understanding of black hole theory grew much brighter. Israel, in one of the first applications of global techniques in general relativity was able to prove that any static black hole was uniquely described by two parameters, its mass M and its electrical charge Q. Furthermore, these solutions are the Schwarzschild solution if $Q = 0$ and the Reissner-Nordstrom solution if $Q \neq 0$. Thus, if a nonrotating body collapses to form a black hole, even though it may be asymmetric initially, it must lose its asymmetry as seen by an external observer, and when all the radiation and transient phenomena involved in the collapse die away, it can be described in terms of the special solutions discovered fifty-one years earlier by Schwarzschild, Reissner and Nordstrom. Israel's theorem provoked an intense investigation into general relativistic black hole theory and by the early 1970s, work of Carter, Hawking, and Robinson (and others) [7, 8] allowed one to conclude that the rotating black hole solution of Kerr was the unique solution. It has still not been proved that the charged rotating solution of Newman et al. is the unique electrically charged rotating black hole, but the problem here seems more a matter of algebraic fortitude (at least using current techniques) rather than a basic lack of understanding. One might hope that recent new techniques developed by mathematicians in proving uniqueness theorems in other nonlinear theories might have some application here [9].

As will be explained in more detail in the following sections, Einstein's theory of gravity involves a four-dimensional manifold equipped with a real metric of indefinite signature. The metric is required to satisfy one of two equations, either

$$R_{ab} = 0 \tag{1.1}$$

where R_{ab} is the Ricci tensor, or

$$R_{ab} - \frac{1}{2} g_{ab}R = \frac{8\pi G}{c^2} T_{ab} \tag{1.2}$$

where G and c are constants and T_{ab} is some prescribed tensor field which is supposed to describe "matter fields" as opposed to "gravitational fields." The matter field most commonly considered is the electromagnetic field (one really has strictly rigorous black hole uniqueness results only for the electromagnetic field) and in this case

$$T_{ab} = \frac{1}{4\pi}\left(F_{ac}F_{bd}g^{cd} - \frac{1}{4} g_{ab}F^{c}{}_{d}F_{c}{}^{d}\right) \tag{1.3}$$

where F_{ab} is an antisymmetric tensor field satisfying

$$g^{bc}\nabla_c F_{ab} = \nabla_{[c}F_{ab]} = 0 \tag{1.4}$$

where [ ] denotes antisymmetrization. For reasons of brevity we shall only consider uniqueness theorems for the situation $R_{ab} = \frac{8\pi G}{c^2} T_{ab} = 0$. The case $T_{ab} \neq 0$ is more computationally difficult than conceptually difficult, and hence, results applicable in this case will be tabulated at the end of Section III.

The spherically symmetric, static solution to $R_{ab} = 0$ discovered by Schwarzschild [1] can be written in a local coordinate chart as

$$ds^2 = -\left(1 - \frac{2m}{r}\right)dt^2 + \frac{dr^2}{1 - \frac{2m}{r}} + r^2(d\theta^2 + \sin^2\theta d\phi^2) \tag{1.5}$$

where m is a constant (the mass), t is the real line, r is a coordinate along a ray and θ and ϕ are polar and azimuthal coordinates on a two sphere. The coordinate singularity at $r = 0$ is a curvature singularity where $R_{abcd}R^{abcd}$ is unbounded ($\{a,b,c,d\} = \{1,2,3$ or $4\}$). The coordinate singularity at $r = 2m$ can be removed by introducing a new chart v', w', θ, ϕ where

$$\begin{aligned} v' &= \exp v/4m, \qquad w' = -\exp(-w/4m) \\ dv &= dt + dr/\left(1-\frac{2m}{r}\right) \\ dw &= dt - dr/\left(1-\frac{2m}{r}\right) \\ \left(1-\frac{r}{2m}\right)e^{r/2m} &= w'v' \end{aligned} \tag{1.6}$$

resulting in

$$ds^2 = -32m^3 \frac{e^{-r/2m}}{r} dv'dw' + r^2(d\theta^2 + \sin^2\theta d\phi^2) . \tag{1.7}$$

The rotating solution of Kerr [4] can be written in a local coordinate chart as

$$\begin{aligned} ds^2 &= \rho^2(dr^2/\Delta + d\theta^2) + (r^2+\theta^2)\sin^2\theta d\phi^2 - dt^2 \\ &+ \frac{2mr}{\rho^2}(a\sin^2\theta d\phi - dt)^2 \\ \rho^2 &= r^2 + a^2\cos^2\theta, \qquad \Delta = r^2 - 2mr + a^2 \end{aligned} \tag{1.8}$$

where t, r, θ, ϕ are coordinates similar to those used in the Schwarzschild solutions. $r = 0$ is again a curvature singularity and $\Delta(r) = 0$ is a removable coordinate singularity. m is again a constant (the mass) and a is related to the rotation. (1.8) reduces to (1.5) in the limit $a \to 0$.

In the following sections I will attempt to review the uniqueness theorems referred to above. In an article of reasonable length it is necessary to be ruthless in deciding what aspects of long and complicated proofs should be emphasized. Therefore only key theorems are proved in the text and subsidiary proofs left to the literature. The proof of subsidiary theorems, however, should not present any surprises as far as the techniques being used, because I have tried to include sufficient proofs in the text to familiarize the reader with common techniques. Hence one should be able to obtain a good idea of how the central theorems work from reading the text, while becoming experienced enough with causal analysis, etc. to read the literature for peripheral proofs.

"Euclidean black hole" solutions arise in Hawking's approach to quantizing gravity [10]. These solutions are again Ricci flat metrics on four-dimensional manifolds, but now the metric is positive definite and not indefinite. The metrics in equations (1.5) and (1.8) are analytic and hence one can obtain "Euclidean black holes" by analytically continuing $t \to it$ in (1.5) and $t \to it$, $a \to -i\alpha$ in (1.8). These new metrics are non-Kahler, geodesically complete, positive definite, Ricci flat metrics with topology $R^2 \times S^2$ that were apparently not known to mathematicians. The conditions under which physicists expect them to be unique are outlined in Section IV. One might hope that the uniqueness theorems of Section III would also apply to Euclidean black holes; however, we show that this is not the case. This article therefore ends with a series of conjectures concerning the uniqueness of Euclidean black hole solutions. Proof of the conjectures, while not providing the key to a yet undeveloped theory of quantum gravity, would make at least this physicist somewhat more confident of the progress made so far.

CONVENTIONS

Metric signature $-+++$. The indices a, b, c, d generally run from 1 to 4, while the indices i, j generally run from 1 to 2. Square brackets around indices denote antisymmetrization over these indices, while parentheses indicate symmetrization. Semicolons denote covariant differentiation, as does ∇. Definitions and theorems appear after their discussion or proof, and are numbered sequentially in each section.

References are also numbered sequentially in each section. No attempt has been made to reference the original proofs which are scattered throughout the literature. Instead, I often refer to chapter 9 of Reference [7] and the first two chapters of Reference [8] which also review black hole uniqueness theorems with varying degrees of rigor. Hawking's discussion contains the most detail. References to the original literature may be obtained from these master references if desired.

II. *Causal structure*

Einstein's general theory of relativity states that the physical effect of gravity is represented by a curved spacetime. Spacetime is a four-dimensional manifold, $\mathfrak{M}$, equipped with a real metric, g_{ab}, of indefinite signature which we take in this article to be $-+++$. A curved spacetime is one whose Ricci tensor, R_{ab}, is required to take a particular form: either

$$R_{ab} = 0 \tag{2.1}$$

describing empty spacetime or

$$R_{ab} - \frac{1}{2} g_{ab} R = \frac{8\pi G}{c^2} T_{ab} \tag{2.2}$$

where G is the gravitational constant, c is the speed of light (both constants determined by experiment), and T_{ab} is a prescribed tensor that is nonzero at points on the manifold where matter is present. It describes the properties of nongravitational fields. Geodesics of the spacetime represent the paths of free particles responding only to the gravitational field. The idea is that what Newton perceived as an apple affected by a force in a flat spacetime, Einstein perceives as an apple moving along a geodesic in a curved spacetime. It can be shown that general relativity subsumes Newtonian gravity theory, is logically consistent, and is in accord with those experiments which test post-Newtonian effects.

DEFINITION 2.1. *Spacetime is defined to be the pair* $\{\mathfrak{M}, g_{ab}\}$ *where* $\mathfrak{M}$ *is a connected four-dimensional Hausdorff* C^∞ *manifold and* g_{ab} *is a real metric on* $\mathfrak{M}$ *with signature* $-+++$.

The minus sign in the metric equips the manifold with a new structure, the causal structure, that may be unfamiliar to geometers who have only considered manifolds with a positive definite metric. Because black holes are *defined* in terms of the causal structure we will review in this section those concepts that culminate in the idea of "stationary regular predictable black hole spacetimes." In the following sections we will remove the "s" from "spacetimes" by showing that under certain conditions there is a unique stationary regular predictable black hole spacetime—the Kerr solution (1.8) describing a rotating black hole, which includes the Schwarzschild solution (1.5) describing a nonrotating black hole in an appropriate limit.

The existence of the indefinite metric, g, allows one to divide vectors, curves and surfaces into the following classes.

DEFINITION 2.2. *A vector* $\ell \in T_p$ *at a point* p *is timelike, null or spacelike depending on whether* $g(\ell, \ell) < 0$, $g(\ell, \ell) = 0$, $g(\ell, \ell) > 0$, *respectively.*

DEFINITION 2.3. *A curve,* γ, *with tangent vector* ℓ *at a point* p *on* γ *is a timelike, null, or spacelike curve if* $g(\ell, \ell) < 0$, $g(\ell, \ell) = 0$ *or* $g(\ell, \ell) > 0$, *respectively for all points* p *on* γ.

DEFINITION 2.4. *A surface,* S, *with normal vector* ℓ *at a point* $p \in S$ *is a timelike, null or spacelike surface if* $g(\ell, \ell) < 0$, $g(\ell, \ell) = 0$ *or* $g(\ell, \ell) > 0$, *respectively for all* $p \in S$.

If it is possible to divide nonspacelike vectors continuously into two classes: "future-directed" or "past-directed," then the spacetime is said to be "time orientable." This is analogous to space orientability; i.e., the continuous division of bases of three spacelike axes into right-handed and left-handed classes. We shall assume the existence of both time and space orientability and hence a consistent notion of future/past and right-handed/left-handed throughout spacetime. If time orientability did not hold in a spacetime then there would exist a covering manifold in which it did [11].

It is useful to separate the idea of "future" into two classes—and similarly the idea of "past." The "timelike" or "chronological" future of a point p, $I^+(p)$, is defined to be the set of all points which can be reached from p by future-directed timelike curves. The "causal" future of p, $J^+(p)$, is the union of p with the set of all points which can be reached from p by future-directed nonspacelike curves.

DEFINITION 2.5. *The timelike or chronological future of a point* p, $I^+(p)$, *is the set of all points which can be reached from* p *by future-directed timelike curves.*

DEFINITION 2.6. *The causal future of* p, $J^+(p)$, *is the union of* p *with the set of points which can be reached from* p *by future-directed nonspacelike curves.*

DEFINITION 2.7. *The timelike or chronological future of a set* S, $I^+(S)$, *is the union of* $I^+(p)$ *for all* $p \in S$. *Definitions similar to Definitions 2.5-2.6 exist for sets in an analogous fashion.*

Dual definitions for "timelike" and "causal" past exist by replacing the word "future" by "past" wherever it appears in Definitions 2.5 and 2.6. Similar definitions exist for sets.

Examples of some of these definitions are provided by Figure I. The timelike future of the origin is the interior of the future light cone. It does not include the origin. Similarly the chronological past of the origin is the interior of the past light cone. The causal future of the origin is the union of the interior of the future light cone with its boundary. Similarly for the causal past. The boundaries of the regions I^+, I^-, etc. are denoted $\dot{I}^+$, $\dot{I}^-$ and therefore, for example, the boundary of the causal future of the origin, $\dot{J}^+(0)$ is the future light cone. With these definitions it is possible to prove

THEOREM 2.1. $\dot{I}^+(S)$ *of a set* S *is a null or spacelike set.*

THEOREM 2.2. *The boundaries* $\dot{I}^+(S)$, $\dot{J}^+(S)$ *of a set* S *are generated by null geodesic segments which have past endpoints, if and only if, they intersect* S *and have future endpoints where generators intersect.*

There exist dual theorems with future replaced by "past." The proofs, although nontrivial, are not long, and are left as an exercise for the reader. They may also be found in Reference [8].

Causal structure is closely related to conformal structure. This statement can be made less Delphic by observing that the null or light cone structure (which determines the causal structure) is unchanged under conformal deformations of the metric; i.e., $g_{ab} \to \Omega^{-2} g_{ab}$ where Ω is a smooth scalar function (recall that null curves with a tangent vector ℓ^a have $g_{ab}\ell^a\ell^b = 0$). Hence conformally related metrics have identical causal structure. This is useful because one often wants to know what can be seen by an observer at infinity; e.g., are there any regions of spacetime in which light/null rays cannot escape to infinity? Conformally mapping infinity into a finite distance makes the analysis of the question more tractable and leads to the construction of "Penrose diagrams" [12]. The prototype Penrose diagram is that for the flat spacetime (Minkowski space); i.e., $\mathfrak{M} = R^4$ with the flat metric which can be written in an obvious chart as

$$ds^2 = -dt^2 + dr^2 + r^2(d\theta^2 + \sin^2\theta d\phi^2) \tag{2.3}$$

the trivial coordinate singularities at $r = 0$, $\sin\theta = 0$ can be removed by using; e.g., Cartesian coordinates. If one introduces new coordinates $\tan(p) = t+r$, $\tan(q) = t-r$, $p-q \geq 0$, $-\pi/2 < p < \pi/2$, $-\pi/2 < q < \pi/2$ then one obtains

$$ds^2 = +\sec^2 p \sec^2 q\left[-dpdq + \frac{1}{4}\sin^2(p-q)(d\theta^2 + \sin^2\theta d\phi^2)\right] . \tag{2.4}$$

Note that infinite values of $t \pm r$ have been mapped to finite values of p, q. Changing coordinates yet again to $t' = (p+q)/2$, $r' = (p-q)/2$ yields

$$ds^2 = \Omega^{-2}(t', r')\left[-dt'^2 + dr'^2 + \frac{1}{4}\sin^2 2r'(d\theta^2 + \sin^2\theta d\phi^2)\right] \tag{2.5}$$

where $\Omega^{-2} = \sec^2(t'+r')\sec^2(t'-r')$. Thus Minkowski space is conformally related to a region of "Einstein static space"

$$ds^2 = -dt'^2 + dr'^2 + \frac{1}{4}\sin^2 2r'(d\theta^2 + \sin^2\theta d\phi^2) \tag{2.6}$$

bounded by the null surfaces $t'-r' = -\pi/2$ and $t'+r' = \pi/2$. From (2.6) it's apparent that Einstein static space is a space of constant curvature with topology $R \times S^3$. The conformal structure of infinity can be represented by a drawing of the t', r' plane (Figure II) in which the t' axis is vertical, the r' axis a horizontal radial axis and each point of the diagram represents a two sphere. Null rays, for which $ds^2 = 0$, are lines at 45°. Future-directed null geodesics originate on the boundary surface labelled $\mathscr{I}^-(q = -\pi/2$, pronounced "scri minus" from "script I"), and end on the boundary $\mathscr{I}^+(p = +\pi/2)$. $\mathscr{I}^+$ and $\mathscr{I}^-$ represent "future" and "past null infinity", respectively. Future-directed timelike geodesics originate at $i^-(p, q = -\pi/2)$ and end at $i^+(p, q = +\pi/2)$. $i^\pm$ represent "future" and "past timelike infinity." Spacelike geodesics originate and end on i^0- "spacelike infinity." $i^+ i^-$ and i^0 are actually points because $\sin^2 2r'$ vanishes there. The conformal metric is regular on the null boundary surfaces $\mathscr{I}^\pm$ which have topology $R \times S^2$. It is clear from Figure II that null geodesics from any point in Minkowski space can always escape to infinity and hence an observer at infinity can "see" all of spacetime.

An example of a spacetime containing regions invisible to an observer at infinity is the Schwarzschild solution (1.1). The Kruskal extension of the original Schwarzschild chart was given in Section I. One can construct the Penrose diagram of the Schwarzschild solution by defining new coordinates that bring infinity into a finite distance as in the above. Let $v' = \sqrt{2m} \tan v''$ and $w' = \sqrt{2m} \tan w''$ so that $-\pi < v'' + w'' < \pi$ and $-\pi/2 < v'' < \pi/2$, $-\pi/2 < w'' < \pi/2$. The v'', w'' plane is drawn in Figure III. The conformal structure at infinity is similar to Minkowski space with $i^\pm$, i^0 and $\mathscr{I}^\pm$ defined for each of the two asymptotically flat regions. However, there is now a region of spacetime invisible to an observer at infinity; i.e., it is not possible to reach future null infinity, $\mathscr{I}^+$, along a future-directed timelike or null curve from any point (event) with coordinate $r < 2m$. These points are therefore not in $J^-(\mathscr{I}^+)$. The boundary of these points, $\dot{J}^-(\mathscr{I}^+)$, is called an "event horizon" and is a global concept by definition. In general, the region of spacetime not in $J^-(\mathscr{I}^+)$; i.e., that region from which it is impossible to signal to infinity with physical massive or massless particles (e.g., light rays), is called a "black hole." The black hole in the Schwarzschild solution is the spacetime region $r > 2m$. The null surface $r = 2m$ is an example of an event horizon.

DEFINITION 2.8. *A black hole is a region of spacetime from which it is impossible to escape to* $\mathcal{I}^+$ *along a future-directed nonspacelike curve.*

DEFINITION 2.9. *The event horizon bounds the region of spacetime from which it is impossible to escape to* $\mathcal{I}^+$ *along a future-directed nonspacelike curve. It is the set* $\dot{J}^-(\mathcal{I}^+)$ *where* $\cdot$ *denotes boundary.*

Penrose's conformal technique has more use than merely as a device that squashes infinity into a finite region allowing one to display the causal structure all the way out to infinity in a compact manner. It leads to a definition of "asymptotically flat" that is different (and in ways more useful) than previous definitions in terms of the rate of falloff of the metric and curvature on some embedded three-dimensional noncompact surface in spacetime. The idea is that far away from bounded objects such as stars, etc. the spacetime will asymptotically approach the flat Minkowski spacetime and thus the conformal structure near infinity will be like that of Minkowski space. In other words, one expects that a suitable definition of asymptotically flat will include the notion that one can attach a smooth boundary at infinity consisting of two disjoint null hypersurfaces, $\mathcal{I}^+$ and $\mathcal{I}^-$. Penrose has defined such spacetimes to be "asymptotically simple" [13].

DEFINITION 2.10. *A pair* $\{\mathcal{M}, g_{ab}\}$ *consisting of a four-dimensional manifold* $\mathcal{M}$ *and an indefinite metric* g_{ab} *is asymptotically simple if there exists a pair* $\{\tilde{\mathcal{M}}, \tilde{g}_{ab}\}$ *such that* $\mathcal{M}$ *can be embedded in* $\tilde{\mathcal{M}}$ *as a manifold with smooth boundary such that*

1) $\tilde{g}_{ab} = \Omega^2 g_{ab}$ *on* $\mathcal{M}$, *where* Ω *is a smooth* (C^3 *at least) function on* $\tilde{\mathcal{M}}$.
2) *on* $\partial\mathcal{M}$, $\Omega = 0$, $\nabla_a \Omega \neq 0$.

3) *each null geodesic in* $\mathfrak{M}$ *has past and future endpoints on* $\partial\mathfrak{M}$.

4) $R_{ab} = 0$ *near* $\partial\mathfrak{M}$.

Condition (2) in Definition 2.10 requires that the boundary be at infinity in the following sense: If one considers an affine parameter λ on a null geodesic in the metric g_{ab} then it is related to an affine parameter $\tilde{\lambda}$ in $\tilde{g}_{ab}$ by $d\lambda/d\tilde{\lambda} = \Omega^{-2}$. By condition (2) $\Omega = 0$ on $\partial\mathfrak{M}$ and hence $\lambda \to \infty$ on $\partial\mathfrak{M}$. Furthermore, by considering the relation of the Ricci scalars $R(g_{ab})$ and $\tilde{R}(\tilde{g}_{ab})$ it is easy to show that condition (1) implies $\partial\mathfrak{M}$ must be a null hypersurface consisting of two disjoint null hypersurfaces conventionally labelled $\mathcal{I}^+$ and $\mathcal{I}^-$. Geroch has shown [14] that $\mathcal{I}^{\pm}$ are topologically $R^1 \times S^2$. Actually, the above definition is a bit too strong because it rules out the Schwarzschild solution from being asymptotically simple by virtue of condition (3) (null geodesics can originate on $\mathcal{I}^+$ and terminate at $r = 0$ after crossing $\dot{J}^-(\mathcal{I}^+)$). By loosening condition (3) one allows spacetimes that contain regions from which it is impossible to escape to future null infinity along nonspacelike curves. Therefore, it is useful to define a "weakly asymptotically simple spacetime."

DEFINITION 2.11. *A weakly asymptotically simple spacetime* $\{\mathfrak{M}, g_{ab}\}$ *is one for which there exists an asymptotically simple spacetime* $\{\mathfrak{M}', g'_{ab}\}$ *such that a neighborhood of* $\mathcal{I}^{\pm}$ *in* $\mathfrak{M}'$ *is isometric with a similar neighborhood in* $\mathfrak{M}$.

It seems appropriate to pause at the eleventh definition and outline the remainder of this section. The Schwarzschild solution exhibits another interesting generic phenomena in general relativity, in addition to the event horizon, called a "trapped surface" by Penrose [15]. The definition of a trapped surface, however, involves an object called the "expansion" that appears in the theory of the Jacobi field for a congruence of null geodesics. It will, therefore, be necessary to digress (briefly!) and discuss null geodesic congruences and their expansion. After this excursion it will be possible to define a trapped surface and obtain an intuitive idea of the relation of trapped surfaces to singularities. Upon carefully hiding the trapped surfaces and singularities behind event horizons (which requires yet another definition of a type of spacetime) enough apparatus will have been defined into existence to actually *do* something—namely, prove two very important theorems about black holes. That will end the present section. The following section will finally get around to proving black hole uniqueness theorems.

It is time to briefly discuss null geodesic congruences. If $\gamma(v)$ is a representative null curve of the congruence parametrized by an affine parameter, v, then the tangent vector to γ obeys $k^b k^a_{;b} = 0$, where semicolon denotes covariant differentiation (a, b run from 1 to 4). One can introduce a basis, e_1, e_2, e_3, e_4 in the tangent space at a point, p, along the curve γ. It is convenient to choose e_4 equal to k^a, the tangent vector to γ at p, e_3 to be another null vector such that $g(e_3, e_3) = 0$ (e_3 is null) and $g(e_3, e_4) = -1$ (a choice of normalization); while e_1 and e_2 are chosen to be unit spacelike vectors orthogonal to each other and to e_3 and e_4: $g(e_1, e_1) = g(e_2, e_2) = 1$, $g(e_1, e_2) = g(e_1, e_3) = g(e_1, e_4) = 0$, etc. The basis can be defined at other points on $\gamma(v)$ by parallel transport. Let z^a be the tangent vector to a curve $\lambda(t)$, so that $z^a = (\partial/\partial t)_\lambda$. A family of curves, $\lambda(t, v)$ can be constructed by moving each point of $\lambda(t)$ a distance v along the flow defined by k^a. Then defining z^a as $(\partial/\partial t)_{\lambda(t,v)}$ one has that $\mathcal{L}_k z = 0$, where $\mathcal{L}$ represents the Lie derivative. z^a is the vector representing the separation of points equal distances from initial points along two neighboring curves. It obeys

$$\frac{D}{dv} z^a = z^b k^a_{;b} \tag{2.7}$$

$$\frac{D^2}{d^2v} z^a = -R^a_{bcd} z^c k^b k^d \tag{2.8}$$

where Dz^a/dv represents the covariant derivative of z^a along k. Using the basis introduced above we have that $k^a_{;4} = 0$ because k is geodesic, and also $k^3_{;c} = 0$ because $(k^a g_{ab} k^b)_{;c} = 0$. Hence $z^a k_a$ is constant along $\gamma(v)$ which means that pulses of light emitted from a single source at a time separation Δt maintain that separation in time. Factoring out this trivial behavior by restricting attention to vectors z such that $z^a k_a = 0$ (i.e., the neighboring null geodesics have purely spatial separation) and using the result $k^a_{;4} = k^3_{;c} = 0$ one obtains

$$\frac{dz^i}{dv} = k^i_{;j} z^j \tag{2.9}$$

for i, j ranging from 1 to 2. One can separate $k_{i;j}$ into three pieces: ω_{ij} = the antisymmetric part of $k_{i;j}$, called the "vorticity", θ_{ij} = the symmetric part of $k_{i;j}$, called the "rate of separation"; σ_{ij} = the trace free part of θ_{ij}, called the shear; and θ = the trace of θ_{ij}, called the expansion. Manipulation of (2.7), (2.8) leads to equations of propagation for ω_{ij}, σ_{ij} and θ

$$\frac{d}{dv}\omega_{ij} = -\theta\omega_{ij} \tag{2.10a}$$

$$\frac{d\theta}{dv} = -R_{ab}k^a k^b + 2\omega^2 - 2\sigma^2 - \frac{1}{2}\theta^2 \tag{2.10b}$$

$$\frac{d\sigma_{ij}}{dv} = -C_{i4j4} - \theta\sigma_{ij} \tag{2.10c}$$

where $C_{abcd}(a, b, c, d \varepsilon [1, 2, 3, 4])$ is the Weyl tensor, $\omega^2 = \omega^{ij}\omega_{ij}$, and $\sigma^2 = \sigma^{ij}\sigma_{ij}$.

The physical significance of these quantities is illustrated by considering a null hypersurface, S, generated by null geodesics with a tangent vector field k^a. A (infinitesimally) small area element of a spacelike two surface in S will change in area as each point of the element is moved a parameter distance δv up the null geodesics by an amount δA

$$\delta A = 2A\theta\delta v . \tag{2.11}$$

σ measures the relative rates of expansion of neighboring geodesics in the spacelike directions e_1, e_2 or, in other words, the shearing of the congruence. ω_{ab} measures the relative twist of neighboring geodesics and is zero for geodesics in three-dimensional null hypersurfaces, which are the only kind we will consider. The quantity $R_{ab}k^a k^b$ is determined by the Einstein field equation (2.2) and depends on the tensor T_{ab} describing matter. For a unit *timelike* vector at a point p one has that $V_a V_b T^{ab}$ is the local energy density of the matter as it appears to an observer moving on a timelike curve with unit tangent vector V^a at p. The requirement that the matter distribution be physical insomuch as the local energy density, $V^a V^b T_{ab}$, be nonnegative is called "the weak energy condition." By continuity the weak energy condition also requires $k^a k^b T_{ab} \geq 0$ for any null vector k^a. Thus $k^a k^b R_{ab} \geq 0$ by application of the weak energy condition and Einstein's equation (2.2). For zero vorticity this implies that once neighboring null geodesics start to converge then they are focused and intersect in finite affine parameter. Equation (2.10b) implies for $\theta = \theta_1 < 0$ at $v = v_1$ then for $v > v_1$

$$\theta \leq \frac{3}{v - (v_1 + 3/(-\theta_1))} \tag{2.12}$$

and hence θ becomes infinite for some v between v_1 and $v_1 + 3/(-\theta_1)$; i.e., there exists a focal point at v where neighboring geodesics intersect.

It is now possible to define a "trapped surface." A trapped surface is a compact two surface such that *both* families of outgoing and ingoing future-directed null geodesics orthogonal to it have negative expansion. The physical idea is that since the *outgoing* family converges (imagine, for example, a spherical beach ball S^2, with flashlights pointing radially outward covering the surface) then there is sufficient gravitational attraction so that light is getting "dragged back." Because all matter moves at speeds less than or equal to that of light the matter is dragged back also and confined to an increasingly smaller volume as the null geodesics converge. Clearly this will create a problem, typically a spacetime singularity. The idea that trapped surfaces signal "trouble" is one of the key ideas of the singularity theorems of Penrose and Hawking [16]. These theorems prove that under reasonable conditions trapped surfaces indicate that spacetime must be geodesically incomplete.

DEFINITION 2.12. *A trapped surface is a compact spacelike two surface* S, *such that the outgoing family of null geodesics orthogonal to* S *have negative expansion.*

DEFINITION 2.13. *A marginally trapped surface is a compact spacelike two surface* S *such that future-directed null geodesics orthogonal to* S *have zero expansion.*

In the Schwarzschild solution the trapped surfaces and singularities lie behind the event horizon at $r = 2m$, i.e., are not in $J^-(\mathcal{I}^+)$, and hence cannot affect and are not visible to an observer at infinity. Is this a generic feature of gravitation? Are trapped surfaces and singularities always "hidden" behind event horizons? This is a major unsolved question of classical general relativity. Penrose has proposed the "Cosmic Censorship Hypothesis": In realistic situations singularities are never naked but clothed by an event horizon. We now need to make precise the assumption that given a weakly asymptotically simple spacetime (a spacetime such that well-defined future null infinity, $\mathcal{I}^+$, and past null infinity, $\mathcal{I}^-$, exist) it is possible to predict the future, in a region called the future Cauchy development, from a suitable spacelike surface S.

DEFINITION 2.14. *The future Cauchy development of a surface* S, $D^+(S)$, *is the set of points* q *in* $\mathfrak{M}$ *such that each past-directed nonspacelike curve through* q *intersects* S *if extended far enough.*

We shall assume that the weakly asymptotically simple spacetime under consideration admits a partial Cauchy surface such that points near $\mathcal{I}^+$ are contained in $D^+(S)$. In these spacetimes there are no naked singularities in $J^+(S)$ and hence they are christened "future asymptotically predictable." Actually, it is more useful to define a "strongly future asymptotically predictable spacetime" that allows one to predict near infinity and also near the event horizon.

DEFINITION 2.15. *A strongly future asymptotically predictable spacetime is a weakly asymptotically simple spacetime containing a partial Cauchy surface* S *such that* $\mathcal{I}^+$ *is in the boundary of* $D^+(S)$ *and* $J^+(S) \cap \dot{J}^-(\mathcal{I}^+) \subset D^+(S)$.

It is known that strongly (future) asymptotically predictable spacetimes result when spherical distributions of matter, obeying physically reasonable restrictions on the stress energy tensor T_{ab}, undergoes gravitational collapse and that this feature is stable to small deviations from spherical symmetry [17, 18]. Proving that predictable spacetimes result from highly nonspherical collapse is tantamount to proving the Cosmic Censorship Conjecture (a very worthwhile endeavor!) and hence we will *assume* that this is the case. In strongly future asymptotically predictable spacetimes with S simply connected one can prove that a trapped surface T in $D^+(S)$ is really "trapped" in the sense that it is impossible to escape to $\mathcal{I}^+$ from T along a future-directed nonspacelike curve. This, in turn, implies the existence of an event horizon [7].

THEOREM 2.3. *In strongly future asymptotically predictable spacetimes the causal future of a trapped or marginally trapped surface* T, $J^+(T)$, *does not intersect the causal past of* $\mathcal{I}^+$; *i.e.,* $J^+(T) \cap J^-(\mathcal{I}^+) = \phi$. *This implies the existence of an event horizon* $\dot{J}^-(\mathcal{I}^+)$.

Enough apparatus is finally available at this point to build proofs of the two extremely important theorems concerning the event horizon mentioned earlier. The first is that the null geodesic generators of $\dot{J}^-(\mathcal{I}^+)$ may have past endpoints, but cannot have future endpoints. The proof is trivial. By definition the generators of the horizon could only have future endpoints if they intersected $\mathcal{I}^+$. Suppose some generator γ did intersect

$\mathcal{I}^+$ at a point q. Consider the generator λ of $\mathcal{I}^+$ running through q. Then one could join points on λ in $J^+(q)$ by timelike curves to points on γ in $J^-(q)$. But this contradicts the assumption that γ lies in $\dot{J}^-(\mathcal{I}^+)$ and hence generators of the horizon can have no future endpoints.

THEOREM 2.4. *Null geodesic generators of* $\dot{J}^-(\mathcal{I}^+)$ *may have past endpoints but cannot have future endpoints.*

The second theorem states that the area of a two-dimensional spacelike cross section of the horizon does not decrease towards the future. Let F be a spacelike two-surface in the event horizon lying to the future of a partial Cauchy surface S and suppose that the expansion, θ, of the null geodesic generators orthogonally intersecting F was negative at some point $p \in F$. Then one could vary F a small amount such that θ was still negative but so F intersects $J^-(\mathcal{I}^+)$. This leads to a contradiction as before because the outgoing null geodesics orthogonal to F would intersect within finite affine distance and therefore could not remain in $\dot{J}^+(F)$ all the way out to $\mathcal{I}^+$. Hence the null geodesic generators of the horizon have nonnegative expansion and since by Theorem 2.4 the generators lack future endpoints then the area of two-dimensional cross section of the event horizon cannot decrease towards the future.

THEOREM 2.5. *The area of a two-dimensional cross section of* $\dot{J}^-(\mathcal{I}^+)$ *cannot decrease towards the future.*

This behavior is reminiscent of the quantity called the "entropy" in thermodynamics. In Section IV it will be shown that a precise analogy exists between certain geometric quantities describing black holes and certain other quantities in thermodynamics such as entropy and temperature. By introducing quantum mechanics into the picture Hawking has shown that the analog is more than formal and that amazingly enough the event horizon bounding a black hole actually does have a physical entropy and temperature: $\dot{J}^-(\mathcal{I}^+)$ can burn you. Before investigating this interesting idea we shall first deal with the purely classical situation (no quantum mechanics) and show in the following section that the "final state" of a black hole is unique.

III. *The final state*

So far the causal structure of spacetime has been the topic of interest and enough subsidiary definitions have been motivated to give a precise definition of a black hole and a description of a few of its properties. We now plan to examine what is known as "the final state" of a black hole and will show that it is essentially unique. The idea of a final state arises in the scenario of a gravitational collapse of a star. When a star is sufficiently massive so that gravitational attractive forces overcome the balancing pressure forces of the constituent matter, then it begins to undergo gravitational collapse. At some point an event horizon will be formed and trapped surfaces and singularities will be behind it. Clearly things will be initially very dynamic, but at some time after the formation of the event horizon, either all the radiation and matter flying about will finally cross the event horizon or else disperse to infinity. Therefore, *outside* the event horizon (i.e., in $J^-(\mathcal{I}^+)$) spacetime quickly settles down into a stationary state that one might hope to describe with a time independent solution of Einstein's equations that approximates (arbitrarily well) initially nonsingular solution at late times.

The intuitive concept of evolution in time just used above needs to be made precise. It can be shown [19] that in strongly future asymptotically predictable spacetimes developing from a partial Cauchy surface S one can construct a time function, τ, so that there exists a family of spacelike surfaces $S(\tau)$ homeomorphic to $S = S(0)$ which cover $D^+(S) - S$ and intersect $\mathcal{I}^+$; and furthermore, $S(\tau_2) \subset J^+(S(\tau_1))$ for $\tau_2 > \tau_1$. There is therefore a rigorous notion of time evolution from one spacelike surface $S(\tau_1)$ to another one $S(\tau_2)$ in the future. As previously explained, at late times the spacetime should become time independent and hence spacetimes which

1) are strongly future asymptotically predictable from a partial Cauchy surface S
2) admit a one parameter isometry group $\phi_t : \mathfrak{M} \to \mathfrak{M}$ whose Killing vector is timelike near $\mathscr{I}^{\pm}$

are expected to be isometric at late times to a physical spacetime developing from an initially nonsingular state at some earlier time. *The time independent solution need not be initially nonsingular itself* and therefore the initial surface S may not have the topology R^3 that initially nonsingular solutions possess. However one wants the region on a spacelike surface S at a time τ outside and including the horizon, $S(\tau) \cap (J^-(\mathscr{I}^+) \cup \dot{J}^-(\mathscr{I}^+))$ to be identical to that of an initially nonsingular solution at late times. Typically this region on a spacelike surface for an initially nonsingular solution has topology $\{R^3 - \text{(a solid ball)}\}$. Hence, one requires

(α) $S \cap (J^-(\mathscr{I}^+) \cup \dot{J}^-(\mathscr{I}^+))$ is homeomorphic to R^3 minus an open set of compact closure

and it is also convenient, but not essential, to require that

(β) S is simply connected.

Furthermore, one is interested in black holes that one could fall into from infinity, i.e., those for which there exists a spacelike surface S at sufficiently large times τ so that

(γ) $S(\tau) \cap J^-(\mathscr{I}^+) \subset J^+(\mathscr{I}^-)$.

DEFINITION 3.1. *A spacetime satisfying conditions (1), (α), (β), (γ) is said to be a regular predictable spacetime.*

DEFINITION 3.2. *A spacetime satisfying conditions (1), (2), (α), (β) and (γ) is said to be a stationary regular predictable spacetime.*

At late times the closure of the region exterior to the horizon in a regular predictable spacetime will be almost isometric with a similar region in a stationary regular predictable spacetime. Muller Zum Hagen [20] has shown that in empty spacetimes ($T_{ab} = 0$) the existence of a Killing vector timelike near infinity implies the metric is analytic in that region. Elsewhere the metric is defined to be the appropriate analytic continuation.

The string of conditions and definitions above finally culminated in the idea of a "stationary regular predictable black hole spacetime." Having identified the object of interest we shall now examine its properties. However, since the proofs involved tend to be long and often subtle, but more importantly because they are also often geometric in nature and hence more accessible to geometers than causal analysis, many of them will only be sketched. More detail may be found in the references [7, 8]. After describing two theorems that apply to any regular predictable spacetime, the analysis will be split into two parts, depending on whether the Killing vector k^a is hypersurface orthogonal or not. The idea behind focussing on the Killing vector is that in the static case one can show that the *globally* defined event horizon coincides with the *locally* defined fixed point locus, or "Killing horizon," of the vector k^a. Then local analysis (differential equations, etc.) can be introduced as an additional tool to global causal analysis. The ultimate objective is to show that if k^a is hypersurface orthogonal, then the unique stationary regular predictable black hole solution to Einstein's equation, $R_{ab} = 0$, is the Schwarzschild solution (1.5) describing a nonrotating black hole. If the Killing vector k^a is not hypersurface orthogonal, then the solution can be shown to be axisymmetric, i.e., there exists an additional Killing vector, m^a, generating an action of the one parameter rotation group SO2. Then a linear combination of m^a and k^a will have a Killing horizon coincident with the event horizon, and local methods can be used again to show that the unique solution is the Kerr solution (1.8) describing a rotating black hole.

The first of the two theorems mentioned above concerning stationary regular predictable spacetimes states that the expansion θ (2.10b) and shear σ (2.10c) of the null generators of the horizon are zero. To see this, consider the action of the time translation isometry $\phi_t : \mathfrak{M} \to \mathfrak{M}$ on certain compact two surfaces F lying in the horizon. The two surfaces are constructed in a particular manner. First consider a compact spacelike two surface C in $\mathcal{I}^-$ and define from it a compact two surface F in

$\dot{J}^-(\mathcal{I}^+)$ by $\dot{J}^+(C) \cap \dot{J}^-(\mathcal{I}^+)$ i.e., the intersection of the future boundary of C with the horizon. The time translation Killing vector k^a is directed along the null generators of $\mathcal{I}^-$ and hence under the time translation ϕ_t the surface C is moved into another surface $\phi_t(C)$ lying to the future of C. Then the surface $\phi_t(F) = \dot{J}^+(\phi_t(C) \cap \dot{J}^-(\mathcal{I}^+))$ lies to the future of F in the horizon. It was shown earlier that generators of the horizon have no future endpoints and nonnegative expansion. If between F and $\phi_t(F)$ any generators had past endpoints or positive expansion, then the area of $\phi_t(F)$ would be greater than the area of F, contradicting the fact that ϕ_t is an isometry. Hence, generators of the horizon have no past endpoints and zero expansion. Examination of the propagation equations (2.10b), (2.10c) for the expansion and shear shows that the shear, the Ricci tensor term and the Weyl tensor term must all be zero on the horizon.

THEOREM 3.1. *The shear and expansion of the null geodesic generators of the horizon are zero in a stationary regular predictable spacetime.*

One can now use the fact that the shear and expansion of $\dot{J}^-(\mathcal{I}^+)$ are zero in proving the second of the two theorems mentioned previously. This theorem states that each connected component of the intersection of the event horizon with one of the spacelike hypersurfaces of constant time, $S(\tau)$, is topologically S^2. It is convenient to introduce the notion of a black hole, $B(\tau)$, on a surface $S(\tau)$ by the definition $B(\tau) = S(\tau) - J^-(\mathcal{I}^+)$. Then the boundary of B is the two surface under consideration. The plan of the proof is to consider the change in the expansion of the null geodesic normals to the compact two surface, ∂B, as the surface is deformed outward into $J^-(\mathcal{I}^+)$. It will be shown that if the initial two surface had topology other than S^2, then the slightly displaced two surface would be trapped or marginally trapped, which would contradict the previous theorem that such surfaces are bounded by the event horizon.

In line with the above plan, let n^a, ℓ^a denote the future directed ingoing and outgoing tangent vector fields to the null geodesics orthogonal to the spacelike two surface ∂B. ℓ^a is tangent to the generators of the horizon. One can choose a normalization so that $\ell^a n_a = -1$, which leaves the freedom: $\ell^a \to e^y \ell^a$, $n^a \to e^{-y} n^a$ where y is a scalar function. The induced metric on the two surface will be $h_{ab} = g_{ab} + \ell_a n_b + n_a \ell_b$. If one defines a family of two surfaces by moving each point of the original surface along a small parameter distance v along n^a, then the orthogonality of the null vectors to the surface will be maintained if

$$h_{ab}\,\ell^b{}_{;c}\,n^c = -h^b_a\, n_{c;b}\,\ell^c \quad \text{and} \quad \ell^a n_a = -1 \,. \tag{3.1}$$

It is not difficult to determine that the rate of change of the expansion of the congruence ℓ^a under the above deformation is

$$\left.\frac{d\theta}{dv}\right|_{v=0} = p_{b;d}h^{bd} - R_{ac}\,\ell^a n^c + R_{adcb}\,\ell^d n^c n^a \ell^b + p^a p_a \tag{3.2}$$

where $p^a \equiv -h^{ba} n_{c;b}\,\ell^c$ and use has been made of the fact that the horizon shear and expansion vanishes. Under the rescaling transformation $\tilde{\ell}^a = e^y \ell^a$, $\tilde{n}^a = e^{-y} n^a$ equation (3.2) becomes

$$\left.\frac{d\tilde{\theta}}{dv}\right|_{v=0} = p_{b;d}h^{bd} + y_{;bd}h^{bd} - R_{ac}\,\ell^a n^c + R_{adcb}\,\ell^d n^c n^a \ell^b + \tilde{p}^a \tilde{p}_a \tag{3.3}$$

where $\tilde{p}^a = p^a + h^{ab} y_{;b}$.

The second term is the Laplacian of y in the two surface ∂B and one can choose y such that the sum of the first four terms are a constant. The sign of this constant is determined by integrating $-R_{ac}\,\ell^a n^c + R_{adcb}\,\ell^d n^c n^a \ell^b$ over the two surface (the first two terms are divergences and therefore have zero integral). It is at this point that the topology of ∂B enters. Use of the Gauss-Codacci equation for the scalar curvature, $\hat{R}$, of an embedded two surface and application of the Gauss-Bonnet theorem, $\int \hat{R}\,dS = 2\pi\chi$ allows one to rewrite the integral of $R_{ac}\,\ell^a n^c + R_{adcb}\,\ell^d n^c n^a \ell^b$ as

$$\int_{\partial B} (-R_{ab}\,\ell^a n^b + R_{adcb}\,\ell^d n^c n^a \ell^b)\,dS = -\pi\chi + \int_{\partial B} \left(\frac{1}{2}R + R_{ab}\,\ell^a n^b\right) dS \,. \tag{3.4}$$

The field equations, $R_{ab} - \frac{1}{2} g_{ab} R = \frac{8\pi G}{c^2} T_{ab}$, show that

$$\frac{1}{2} R + R_{ab} \ell^a n^b = \frac{8\pi G}{c^2} T_{ab} \ell^a n^b \tag{3.5}$$

and for any *physical* matter distribution described by T_{ab} one has that $T_{ab} \ell^a n^b \geq 0$. This is known as the "dominant energy condition." Furthermore, the Euler number χ is $+2$ for a sphere, zero for a torus and negative for any other compact orientable two surface. Now consider the situation case by case. If χ were negative then the constant in question must be positive and thus one could choose y such that $\left.\frac{d\tilde{\theta}}{dv}\right|_{v=0} > 0$ everywhere on ∂B. But then for small negative values of v one obtains a two surface in $J^-(\mathcal{I}^+)$ such that its outgoing orthogonal null geodesics are converging—a trapped surface in $J^-(\mathcal{I}^+)$. But this contradicts the previous theorem that trapped surfaces *cannot* lie in $J^-(\mathcal{I}^+)$ and thus χ can't be negative. Similar arguments in the other cases lead to the conclusion that $\chi = 2$ is the only possibility. Therefore we have Theorem 3.2:

THEOREM 3.2. *Each connected component of* ∂B *is topologically* S^2.

As mentioned earlier, the analysis now naturally cleaves into two parts dependent on whether k^a is hypersurface orthogonal or not. The first situation is called "static" and the second "stationary."

DEFINITION 3.3. *A Killing vector is static if* $k_{[a;b}k_{c]}$, *or equivalently, if* k^a *is hypersurface orthogonal.*

By Frobenius theorem a static Killing vector can be written as $k^a = -\alpha \nabla_a \xi$ where α is a positive function. It is useful to decompose the metric into the form $g_{ab} = V^{-2} k_a k_b + h_{ab}$, where $V^2 = k^a k_a$, and h_{ab} is the induced metric on the surface $\xi = \text{constant}$. Clearly there exists a discrete isometry which maps a point on the surface ξ to a point on $-\xi$ —the metric is time symmetric. Hence if there exists a future event horizon $\dot{J}^-(\mathcal{I}^+) \cap J^+(\mathcal{I}^-)$ then there also exists a past event horizon $\dot{J}^+(\mathcal{I}^-) \cap J^-(\mathcal{I}^+)$.

Due to space restrictions (on the size of this article not on the size of the manifold) we shall assume they intersect. (It turns out that if they do not intersect then the solutions are a special limit of the case in which they do intersect.) The "exterior" region bounded by the future/past event horizons and $\mathcal{I}^{\pm}$, i.e. $J^{-}(\mathcal{I}^{+}) \cap J^{+}(\mathcal{I}^{-})$, is called the "domain of outer communication" denoted $<\mathcal{I}>$. It has the property that the trajectory, $\pi(x_0)$, of the Killing vector k^a through any point $x_0 \in <\mathcal{I}>$ will, if extended far enough, enter and remain in $I^{+}(x^0)$. The proof is left to the reader (it is not hard but somewhat finicky) [21].

DEFINITION 3.4. *The domain of outer communications,* $<\mathcal{I}>$, *is the region* $J^{-}(\mathcal{I}^{+}) \cap J^{+}(\mathcal{I}^{-})$.

THEOREM 3.3. $<\mathcal{I}>$ *is the maximally connected asymptotically flat region such that the trajectory* $\pi(x_0)$ *of* k^a *through any point* $x_0 \in <\mathcal{I}>$ *will if extended far enough, enter and remain in* $I^{+}(x^0)$.

The reasonable causality condition: "$<\mathcal{I}>$ does not contain topologically circular timelike or null curves" in conjunction with Theorem 3.3 allows the immediate conclusion that any degenerate trajectory π (any fixed point of the action generated by k^a) and any topologically circular trajectory must not lie in $<\mathcal{I}>$. Fixed points etc. can however lie on the future/past event horizons, i.e. on the boundary of the region $<\mathcal{I}>$.

Enough information is at hand to prove the coincidence of the Killing horizon and the event horizon referred to above. Let $U = -k_a k^a$ and let ζ denote the maximal connected region in which $U > 0$. It is clear from Theorem 3.3 that $\zeta \subset <\mathcal{I}>$. It is also easy to prove that the boundary, $\dot{\zeta}$, of ζ consists of null hypersurface segments except where $k^a = 0$. To see this start from Definition 3.1 of "static," i.e. $k_{[a;b}k_{c]} = 0$. By virtue of k_a being a Killing vector, $k_{(a;b)} = 0$, one has that $2k_{a;[b}k_{c]} = k_a k_{[b;c]}$. Contracting with k^a yields $U_{,[b}k_{c]} = Uk_{[b;c]}$ and hence on the boundary, $\dot{\zeta}$, where $U = 0$ the gradient of U is parallel to the Killing vector and hence null except at a fixed point locus where $k^a = 0$.

Now imagine that $<\mathcal{J}> \neq \zeta$ so that the complement of ζ in $<\mathcal{J}>$ exists. Let D be a connected component of the hypothetical complement of ζ in $<\mathcal{J}>$. Because ζ is connected so is the complement of D in $<\mathcal{J}>$. But if $<\mathcal{J}>$ is simply connected then the boundary $\dot{D}$ of D restricted to $<\mathcal{J}>$ is connected. Use of the causality condition (no closed timelike or null curves) implies k^a can't be zero in $<\mathcal{J}>$ and hence the boundary $\dot{D}$ of D consists of one connected null hypersurface. Now the outgoing normal to this hypersurface will be everywhere future directed or everywhere past directed. If it is future directed, then future directed timelike curves in $<\mathcal{J}>$ could not enter D. If it is past directed, then future directed timelike curves in $<\mathcal{J}>$ could not enter ζ from D. Either way, the only manner in which D could lie in $<\mathcal{J}>$ is if D is empty. Therefore we have the following theorem.

THEOREM 3.4. *A static regular predictable spacetime with a simply connected domain of outer communication,* $<\mathcal{J}>$, *containing no closed timelike or null curves has* $U = -k^a k_a > 0$ *on* $<\mathcal{J}>$. *In such spacetimes the event horizon coincides with the Killing horizon.*

The restriction that $<\mathcal{J}>$ be simply connected is not objectionable because the topology of $<\mathcal{J}>$ after any reasonable gravitational collapse will be $R \times \{R^3 - \text{a solid ball}\}$ —which is necessarily simply connected (see Definition 3.1). Physicists would tend to frown on further artificial identifications etc. by mathematicians, as they consider such behavior nonphysical.

It is now possible to appeal to a theorem of Israel [6], which historically preceded the above theorems but is here transposed for reasons of logical clarity, to prove that the unique static, regular predictable, Ricci flat black hole spacetime is the Schwarzschild solution (1.5).

THEOREM 3.5. *A static, regular predictable spacetime must be the Schwarzschild solution if*

1) *the staticity Killing vector* k^a *has nonzero gradient everywhere in* $<\mathcal{J}>$.

2) $R_{ab} = 0$.
3) $\dot{J}^+(\mathcal{I}^-) \cap \dot{J}^-(\mathcal{I}^+)$ *is a compact two surface* F.

Requirement (1) is a nontrivial restriction. Requirement (3) plus Theorem (3.2) states that F is topologically S^2.

The idea of Israel's proof is to show that a static regular predictable spacetime satisfying conditions (1), (2) and (3) must be spherically symmetric. It then immediately follows that such a spherically symmetric spacetime must be isometric to a piece of the Schwarzschild solution (2.5) by applying the well-known Birkhoff theorem [22]. The proof of spherical symmetry makes heavy use of the Gauss-Codazzi equations to recast the Einstein equation $R_{ab} = 0$ into a form moulded to the geometrically special ξ = constant, V = constant surfaces. First introduce a coordinate chart $\{x^1, x^2, x^3, x^4\}$ so that locally the metric can be written in a form where the Killing vector is manifest: $k^a = V^2 \nabla^a(x^4)$, $k^a \nabla_a(x^\alpha) = 0$ where $\alpha = 1, 2, 3$. In this chart the metric can be written as

$$ds^2 = -V^2(dx^4)^2 + g_{\alpha\beta} dx^\alpha dx^\beta, \quad V = V(x^\alpha), \quad g_{\alpha\beta} = g_{\alpha\beta}(x^\alpha) . \tag{3.6}$$

The Gauss-Codazzi equations allow the condition of Ricci flatness to be rewritten in terms of geometrical structures of the x^4 = constant hypersurfaces as follows

$$g^{\alpha\beta} R_{\alpha\beta} = 0, \qquad R_{\alpha\beta} + V^{-1} V_{;\alpha;\beta} = 0 .$$

This implies that V is harmonic $g^{\alpha\beta} V_{;\alpha;\beta} = 0$. (Note that the extrinsic curvature of the x^4 = constant surface is absent in the equation above because it vanishes.) The Gauss-Codazzi equations can be used again to rewrite these equations in terms of geometrical structures of the equipotential $V = C$ = constant two-surfaces embedded in the x^4 = constant hypersurface. Algebraic manipulating of this nature leads to the equation

$$\frac{\partial}{\partial V} (\sqrt{g}/\rho) = 0, \quad \rho = \{|\nabla V|^2\}^{-1/2} \tag{3.7}$$

where g is the 2×2 determinant of the metric g_{ij} on the two surface $x^4 = \text{constant}$, $V = C$ $(i=1,2;\ j=1,2)$. One can also obtain

$$\frac{1}{8} R_{abcd}R^{abcd} = (V\rho)^{-2}\left[k_{ij}k^{ij} + 2\rho^{-2}\rho_{;j}\rho^{;j} + \rho^{-4}\left(\frac{\partial\rho}{\partial V}\right)^2\right] \tag{3.8}$$

where $a, b, c, d \in [1,4]$ and k_{ij} is the second fundamental form of the surface $x^4 = \text{constant}$, $V = C$. Integration of equation (3.7) over the $x^4 =$ constant surface subject to the boundedness of (3.8) (a condition that the manifold be regular) leads to

$$S_0/\rho_0 = 4\pi m \tag{3.9}$$

where

m = a constant

S_0 = the area of the two surface: $x^4 = \text{constant}$, $\lim_{C\to 0} V = C$

$\rho_0 = \lim_{C\to 0} \rho$.

On the other hand one can also use other Gauss-Codazzi equations to eventually prove that

$$\begin{aligned} \rho_0 &\geq 4m \\ S_0 &\geq \pi\rho_0^2 \end{aligned} \tag{3.10}$$

with equality if and only if

$$\partial_i\rho = 0 = \rho\left(k_{ij} - \frac{1}{2} g_{ij}k\right) . \tag{3.11}$$

Equation (3.10) is inconsistent with (3.9) unless equality holds, and then spherical symmetry follows immediately from (3.11).

Birkhoff's theorem [22] that "spherical symmetry implies Schwarzschild" is trivial. A spherically symmetric spacetime admits an isometry group SO(3) with group orbits spacelike two spheres. It is not hard to show that locally the metric for a spherically symmetric spacetime can be written

$$ds^2 = -V^2(r)\,dt^2 + \frac{dr^2}{V^2(r)} + r^2(d\theta^2 + \sin^2\theta + d\phi^2) \qquad (3.12)$$

where r is a coordinate along a ray and $d\theta^2 + \sin^2\theta d\phi^2$ is the standard metric on a two sphere. Explicit integration of the coupled equations obtained by substituting (3.12) in $R_{ab} = 0$ leads to $V^2 = 1 - \frac{2m}{r}$ where m is a constant. Therefore one is finally able to conclude that the unique, static, asymptotically flat, regular spacetime containing an event horizon is the Schwarzschild solution (1.5). Robinson et al.[23] have removed condition (1) from Theorem 3.5.

To complete the uniqueness proofs it is now necessary to consider the case where k^a is not hypersurface orthogonal. In this case k^a can become spacelike in $<\mathcal{J}>$ and hence Theorem 3.4 (that $k_a k^a < 0$ in $<\mathcal{J}>$) does not hold. The region on which k^a is spacelike is called the ergosphere. It is sufficient for the purpose of this article to note that ergospheres *can* exist when k^a is not hypersurface orthogonal by noticing that the Kerr solution exemplifies this behavior. It is possible to prove that an ergosphere *must* exist when k^a is not hypersurface orthogonal [24], but we will not do so here. Given the existence of an ergosurface then there are two possibilities: either it intersects the horizon or it does not. The best one can do (at least so far) is to give a physical plausability argument that rules out the second possibility. The argument uses a mechanism proposed by Penrose [25] to extract energy from a black hole with an ergosphere by sending particles into the ergosphere from infinity. Let $p_1^{\ a} = m_1 V_1^{\ a}$ be the momentum vector for a small particle of mass m_1 moving on a curve with $V_1^{\ a}$ the unit future directed tangent vector to the curve. Then if the motion is geodesic the energy $E_1 = p_1^{\ a} k_a$ will be constant. Suppose the particle fell into the region where k^a was spacelike (the ergosphere) and then blew itself apart into two particles with momenta $p_2^{\ a}$ and $p_3^{\ a}$. Conservation of momenta and energy requires $p_1^{\ a} = p_2^{\ a} + p_3^{\ a}$ and therefore one could arrange $E_2 = p_2^{\ a} k_a$ to be negative since k_a is spacelike. Thus $E_3 > E_1$ and particle three could escape to infinity where its total energy would be greater than that of the original particle. Particle two must remain in the ergosphere and hence could not escape to infinity nor could it fall into the black hole if the ergosphere was disjoint from the horizon. One could repeat this procedure and extract an unlimited amount of energy if the ergosphere remained disjoint. This solution to the energy crisis seems physically implausible. On the other hand, one would expect the solution to change gradually as the energy was extracted and since the ergosphere cannot disappear (because the particles left behind have to exist somewhere) then presumably it would move to touch the horizon. But it will be proven next that if the ergosphere does intersect then a stationary solution must also be axisymmetric and hence the black hole would have to spontaneously change to axisymmetry. Either possibility: unlimited energy extraction or spontaneous symmetry changes, indicate an unstable initial state and therefore one can conclude that in any physically realistic situation the ergosphere will intersect the horizon. It would be very nice to have a rigorous proof of this however.

DEFINITION 3.5. *The ergosphere is the region of* $<\mathcal{J}>$ *where the stationarity Killing vector is spacelike*: $k^a k_a > 0$.

ASSUMPTION 3.1. *Let* $\{\mathcal{M}, g_{ab}\}$ *be a stationary (non-static) regular predictable spacetime. Then the ergosphere intersects the horizon.*

The physical significance of the intersection of ergosphere and horizon can be seen by considering a connected component of the horizon, $\mathcal{C}$. By Theorem 3.2 the quotient of a connected component $\mathcal{C}$ by its generators is topologically S^2. The isometry $\phi_t : \mathcal{M} \to \mathcal{M}$ generated by the Killing vector k^a maps generators into generators and can be regarded as an isometry group on $\mathcal{C}$. If the ergosphere intersects the horizon then k^a will be spacelike somewhere on the horizon and the action of the group will be a nontrivial rotation. A particle moving along a generator of the horizon would therefore appear to be *rotating* with respect to the stationary frame at infinity defined by k^a. Furthermore a physicist would expect a rotating black hole to be axisymmetric. This is because a rotating, non-axisymmetric black hole would gravitationally radiate away its asymmetries and would eventually become an axisymmetric black hole rotating at a slower angular velocity than it had in its initial non-axisymmetric state. The actual proof [26] of "stationarity implies axisymmetry" (subject to Assumption 3.1) is quite long and is therefore only sketched below.

The idea of the proof is to show that there exists an axial Killing vector intrinsic to the geometry of the horizon that can actually be extended off the horizon to be an axial Killing vector of the full spacetime. The proof involves considering the Cauchy problem to the past of the intersection of the horizon with an ingoing null hypersurface (see Figure IV). The Cauchy data will be shown to remain unchanged as the spacelike two surface in which the two null surfaces intersect is moved down the generators of the horizon. It follows from the uniqueness of the Cauchy problem that there exists a Killing vector $\hat{k}^a$ in the region to the past of the intersection of the two null surfaces which coincides on the horizon with a generator of the horizon, ℓ^a. In other words, the vector $\ell^a - k^a$, where ℓ^a is a generator of the horizon and k^a is the stationarity Killing vector, can actually be continued off the horizon to be the Killing vector $\hat{k}^a - k^a$ generating an isometry of the full spacetime. The orbits of $\hat{k}^a - k^a$ will be shown to be closed spacelike curves corresponding to a rotation about an axis of symmetry.

To begin the proof, consider a stationary, regular predictable spacetime containing a black hole (subject to Assumption 3.1) that is rotating with period t_1. This means that the orbit of the Killing vector k^a generating the isometry $\phi_t : \mathcal{M} \to \mathcal{M}$ will be spirals which repeatedly intersect null geodesic generators of the horizon (see Figure IV). Consider a point p on a generator λ, then $\phi_{t_1}(p)$ is also on λ. One can choose a parameter on λ such that the future directed null vector tangent to λ satisfies

$$\ell^b \ell^a_{;b} = 2\varepsilon \ell^a \tag{3.13}$$

where ε is constant on λ, and the difference in the values of the parameter at p and $\phi_t(p)$ is t_1. The vector field ℓ^a defined in this way satisfies $\mathcal{L}_k \ell = 0$ where $\mathcal{L}$ denotes Lie derivative (i.e. it is invariant under the action of the isometry generated by k^a). A spacelike vector field m^a in a connected component of the horizon can be defined by $m^a = \frac{t_1}{2\pi}(k^a - \ell^a)$. m^a satisfies $\mathcal{L}_k m = \mathcal{L}_\ell m = 0$ and its orbits will be closed spacelike curves. Now choose a spacelike surface F in a connected component of the horizon, C, tangent to m^a and consider the family of two surfaces obtained by moving each point of F an equal parameter distance down the generators of the horizon. Let n^a be the null vector tangent to a null surface N orthogonal to F and ℓ^a, and normalized so that $n^a \ell_a = -1$. Let $\tilde{m}^a$ be a second spacelike vector tangent to F satisfying $\mathcal{L}_k \tilde{m} = \mathcal{L}_\ell \tilde{m} = 0$. $\tilde{m}$ will be orthogonal to ℓ, n and m.

The idea now is to consider the Cauchy problem for the region to the past of the horizon and the null surface N. If one introduces the useful notation of Newman and Penrose

$$Z^a = \frac{1}{\sqrt{2}} \left\{ \frac{m^a}{\sqrt{m^a m_a}} + i \frac{\tilde{m}^a}{\sqrt{\tilde{m}^a \tilde{m}_a}} \right\} \tag{3.14}$$

where the two real vectors m^a, $\tilde{m}^a$ are combined into one complex null vector then it turns out that the Cauchy data for the empty space Einstein equations consists of

$\psi_0 \equiv C_{abcd} \ell^a Z^b n^c Z^d$ on the horizon

$\psi_4 \equiv C_{abcd} n^a \bar{Z}^b n^c \bar{Z}^d$ on the surface N (where $\bar{Z}$ denotes Z complex conjugate)

and

$$\left.\begin{aligned} \mu &\equiv -n_{a;b}\bar{Z}^a Z^b \\ \psi_2 &\equiv C_{abcd}(\ell^a n^b \ell^c n^d - \ell^a n^b m^c \bar{m}^d) \\ \rho &\equiv \ell_{a;b} Z^a \bar{Z}^b \end{aligned}\right\} \text{ on the surface } F . \tag{3.15}$$

It can be shown that ρ, ψ_0 and μ are zero for a stationary event horizon and furthermore one can prove that ψ_2 is a constant along the generators of the horizon ℓ^a. Hence the only nontrivial data is on N. One wants to show that the data remains unchanged when moving N towards the past by moving each point of the two surface F an equal parameter distance down generators of the horizon. To do this, it is easiest to assume that the solution is analytic. Then data on N can be represented by their partial derivatives on F in the direction along N. One can then evaluate the change in these quantities as F is moved down the horizon by calculating their derivatives along a generator of the horizon. By clever manipulation one can always obtain expressions for the derivatives of these quantities along the generator in the form

$$\frac{dx}{dv} = ax + b \tag{3.16}$$

where v is a parameter along the generator, a and b are constant along the generator and x is the quantity in question.

Equation (3.16) implies x must be constant. To see this consider displacing F a distance t_1 to the past along the generators of the horizon where t_1 is the period of rotation. This is equivalent to the isometry ϕ_{-t_1} which implies x must be the same at F and at the displaced F. But since x satisfies (3.16) x must be the constant $-b/a$. One can proceed in this manner to show that all derivatives at the horizon of the Cauchy data on N are constant along the generators of the horizon. Then, by the uniqueness of the Cauchy problem, it follows that there exists a Killing vector $\hat{k}^a$ which coincides with ℓ^a on the horizon. If one forms the quantity $\tilde{k}^a = \left(\frac{t_1}{2\pi}\right)(\hat{k}^a - k^a)$ then $\tilde{k}^a$ will be a Killing vector whose orbits are closed curves since they are closed on the horizon. By the causality condition the curves must be spacelike. Therefore there exists a Killing vector in the full spacetime that coincides with the horizon generator ℓ^a on the horizon (and hence is null there) which generates rotations about an axis of symmetry.

THEOREM 3.6. *In a stationary, nonstatic, regular predictable spacetime, $\{\mathfrak{M}, g_{ab}\}$ subject to Assumption 3.1, there exists a one-parameter cyclic isometry group of $\{\mathfrak{M}, g_{ab}\}$ that commutes with the stationarity isometry group.*

Although there is considerable work left to do in proving the uniqueness of stationary axisymmetric black holes, no step along the way will be as difficult as the last theorem. Equipped with the two Killing vectors of Theorem 3.6, the plan will now be to use the Killing vectors to end up with a local problem in a somewhat analogous manner to the procedure used in the static situation. Recall that the static Killing vector was hypersurface orthogonal. Analogously the surface of transitivity of the two parameter group action generated by k^a and $\tilde{k}^a$ are everywhere orthogonal to another family of two surfaces, or in other words, the plane of the Killing vectors k^a and $\tilde{k}^a$ are orthogonal to another two surface family. To see this form the bivector $k_{[a}\tilde{k}_{b]} = \omega_{ab}$ in terms of which the orthogonal transitivity condition becomes $\omega_{[ab;c}\omega_{d]e} = 0$. This is equivalent to the vanishing of the scalars χ^1 and χ^2 where

$$\begin{aligned} \chi^1 &= k_{a;b}\omega_{cd}n^{abcd} \\ \chi^2 &= \tilde{k}_{a;b}\omega_{cd}n^{abcd}. \end{aligned} \qquad n^{abcd} = \text{the completely antisymmetric tensor.} \tag{3.17}$$

Now consider the expression $n^{abcd}\chi^1_{;d}$, which upon straightforward evaluation expands out to nine terms. All terms except one either vanish by virtue of k^a being a Killing vector or else by the fact that k^a and $\tilde{k}^a$

commute. The surviving term yields

$$n^{abcd}\chi^1_{;d} = -12k^{[a}\tilde{k}^b k^{c];d}{}_{;d} \tag{3.18}$$

which becomes

$$n^{abcd}\chi^1_{;d} = 12k^{[a}\tilde{k}^b R^{c]}{}_d k^d \tag{3.19}$$

because $k^{a;d}{}_{;d} = -R^a{}_b k^b$ for any Killing vector k^a. The right-hand side vanishes if $R^c d$ is that for a sourceless electromagnetic field or if $R^c d = 0$. Hence χ^1, and by similar manipulation χ^2, are constants. By (3.17) χ is proportional to ω_{ab} which vanishes on the rotation axis where $\tilde{k}^a = 0$. Hence we have the following theorem

THEOREM 3.7. *Let* $\{\mathcal{M}, g_{ab}\}$ *be a regular predictable spacetime with a two parameter Abelian isometry group with Killing vectors* k *and* $\tilde{k}$. *If* T *is a subdomain of* $\mathcal{M}$ *which intersects the rotation axis,* $\tilde{k}^a = 0$, *and if* $k^{[a}\tilde{k}^b R^{c]}{}_d k^d = 0$ *in* T *then the surfaces of transitivity of the two parameter isometry group are orthogonal to another family of two surfaces i.e.* $\omega_{[ab;c}\omega_{d]e} = 0$ *where* $\omega_{ab} = k_{[a}\tilde{k}_{b]}$. *Equivalently* $k_{[a;b}\omega_{cd]} = 0 = \tilde{k}_{[a;b}\omega_{cd]}$.

Theorem 3.7 implies that except where ω_{ab} is null or degenerate, then it is possible to choose a chart $\{t, \phi, x_1, x_2\}$ such that $k^a = \frac{\partial}{\partial t}$, $\tilde{k}^a = \frac{\partial}{\partial \phi}$ and $k^a x^i{}_{;a} = \tilde{k}^a x^i{}_{;a} = 0$ for $i = 1, 2$. Impose the normalization $t_{,a}k^a = \phi_{,a}\tilde{k}^a = 1$ and write the metric on the region T of Theorem 3.7 in the form

$$dS^2 = -Udt^2 + 2Wd\phi dt + Xd\phi^2 + g_{ij}dx^i dx^j \tag{3.20}$$

where U, W, X and g_{ij} are functions only of x^i, $i = 1$ or 2. If T is simply connected then t can be taken to be a globally well-behaved function on T while ϕ can be a well-defined angular variable defined modulo 2π.

It is perhaps obvious that the next step is to prove that the domain of outer communications lies in T and that, in analogy to the static case, the event horizon is also a Killing horizon. The remainder of the proof will then consist of gleaning more information about g_{ij}, substituting the new local expression for the metric in the equation $R_{ab} = 0$, and proving that the solution of the resulting set of coupled nonlinear partial differential equations is unique if it is subject to physically realistic boundary conditions. The uniqueness proof utilizes a miraculous identity solved by the metric components when restricted to be Ricci flat which was kindly supplied by Robinson after an algebraic tour de force.

The discussion of the domain of outer communications in the stationary, axisymmetric case above is facilitated by introducing yet more taxonomy for the geometric objects one finds. As before, let ζ denote the maximal connected region in which k^a is timelike so that $U > 0$ in ζ. Recall that $\zeta \subset \langle \mathfrak{I} \rangle$. Now define σ as $\sigma = -\frac{1}{2}\omega_{ab}\omega^{ab}$ so that $\sigma > 0$ is the region in which the surfaces of transitivity of the two parameter Abelian isometry group is timelike. The maximal connected asymptotically flat region $\mathfrak{W}$ in which $\sigma > 0$ is called the stationary axisymmetric domain of $\mathfrak{M}$. Examination of the metric (3.20) shows that

$$\begin{aligned} U &\equiv -k^a k_a \\ X &\equiv \tilde{k}^a \tilde{k}_a \\ W &\equiv k^a \tilde{k}_a \end{aligned} \tag{3.21}$$

and therefore $\sigma = UX + W^2$. Clearly $\zeta \subset \mathfrak{W}$ (recall $X > 0$ by the causality condition of no closed nonspacelike curves in $\langle \mathfrak{I} \rangle$). Now the trajectory of the action of k^a through a point, x_0, on one of the cylindrical timelike two surfaces of transitivity in $\mathfrak{W}$ will enter the chronological future of x_0 defined in relation to the locally intrinsically flat geometry of the cylinder and hence also enter $I^+(x_0)$ in the four dimensional geometry of $\mathfrak{M}$. By applying Theorem 3.3 we finally obtain $\zeta \subset \mathfrak{W} \subset \langle \mathfrak{I} \rangle$.

DEFINITION 3.6. *The stationary axisymmetric domain* $\mathfrak{W}$ *of* $\mathfrak{M}$ *is the maximal connected asymptotically flat region of* $\mathfrak{M}$ *in which the two surface of transitivity are timelike i.e.* $\sigma > 0$.

DEFINITION 3.7. *The boundary* $\dot{\mathfrak{W}}$ *of* $\mathfrak{W}$ *is the rotosurface.*

In analogy to the analysis in the static case, it is now useful to show $\dot{\mathfrak{W}}$ consists of null hypersurface segments except at degenerate points where $\omega^{ab} = 0$. To see this start with the orthogonal transitivity condition of Theorem 3.7 $k_{[a;b}\omega_{cd]} = 0 = \tilde{k}_{[a;b}\omega_{cd]}$ and use the Killing antisymmetry condition $k_{[a;b]} = k_{a;b}$ to obtain after a little manipulation the result $2\omega_{ae;[b}\omega_{cd]} = \omega_{ae}\omega_{[cd;b]}$. Contraction with ω_{ab} yields $\sigma_{,[b}\omega_{cd]} = \sigma\omega_{[cd;b]}$. This states that (except in the degenerate case $\sigma_{,b} = 0$) that the normal to $\dot{\mathfrak{W}}$ (which is parallel to $\sigma_{,b}$) lies in the plane of ω_{cd}. This is only possible if the normal is null; i.e., the rotosurface $\dot{\mathfrak{W}}$ is null. With more care it is possible to deduce that $\dot{\mathfrak{W}}$ is null even in the degenerate case $\sigma_{,b} = 0$ except on lower dimensional surfaces of degeneracy such as the rotation axis. To continue the parallel to the case where k^a is hypersurface orthogonal let D be a connected component of the complement of $\mathfrak{W}$ in $\langle \mathfrak{J} \rangle$. As before, the boundary $\dot{D}$ of D as restricted to $\langle \mathfrak{J} \rangle$ is connected. Now the causality condition of no nonspacelike closed curves in $\langle \mathfrak{J} \rangle$ implies that ω_{ab} is nowhere zero in $\langle \mathfrak{J} \rangle$ except on the rotation axis where $\tilde{k}^a = 0$. This is easily seen to be true, for if $\omega_{ab} = 0$ in $\langle \mathfrak{J} \rangle$ then k^a parallels $\tilde{k}^a$ which gives circular trajectories of k^a that violate causality. Now the fact that $\mathfrak{W}$ consists of null hypersurface segments implies that the boundary $\dot{D}$ of D restricted to $\langle \mathfrak{J} \rangle$ consists of null hypersurface segments, except perhaps at points on the rotation axis. The outgoing normal to the boundary will be everywhere future-directed or everywhere past-directed as before, despite the problem on the rotation axis because this axis must be a timelike two-surface everywhere and therefore couldn't form the boundary of a null surface. The conclusion is that D must be empty in this case, just as it was in the static case, and hence the rotosurface $\dot{\mathfrak{W}}$ coincides with the hole boundary $\langle \dot{\mathfrak{J}} \rangle$.

THEOREM 3.8. *Let* $\{\mathfrak{M}, g_{ab}\}$ *be a regular predictable, stationary, axisymmetric spacetime with a simply connected domain of communication* $<\mathcal{I}>$ *subject to the causality condition and the orthogonal transitivity condition. Then* $<\mathcal{I}> = \mathfrak{W}$ *and hence the rotosurface, the boundary of the region* $\sigma > 0$, *coincides with the event horizon.*

Theorem 3.8 shows that the globally defined event horizon that bounds $<\mathcal{I}>$ actually coincides with the locally defined rotosurface and therefore the analysis from here on is tractable local analysis. Theorem 3.8 also shows that the metric (3.20)

$$ds^2 = -Udt^2 + 2Wd\phi dt + Xd\phi^2 + g_{ij}dx^i dx^j \tag{3.22}$$

where $U = U(x^i)$, $W = W(x^i)$, $X = X(x^i)$, $g_{ij}(x^i)$ for $i = 1, 2$ is expressed in a globally good chart in $<\mathcal{I}>$ apart from degeneracies on the rotation axis and the horizon. For a metric of the form (3.22) it turns out that Ricci flatness implies that the projection of the Ricci tensor into the surfaces of transitivity must have zero trace, which, in turn, implies that the scalar, ρ, defined as the nonnegative root of $\rho^2 = \sigma$ must satisfy $\nabla^2 \rho = 0$ where ∇^2 is the Laplacian in the metric g_{ij}. Now ρ is strictly greater than zero in $<\mathcal{I}>$ where $\sigma > 0$, apart from the rotation axis, and is zero on the horizon (by application of Theorem 3.8) while at infinity the asymptotic flatness condition implies ρ behaves like an ordinary cylindrical radial coordinate. Carter [8] has constructed a simple argument using Morse theory of harmonic functions to show that under these boundary conditions the harmonic function ρ has no critical points in $<\mathcal{I}>$ and therefore ρ can be used as a globally good coordinate in $<\mathcal{I}>$ except on the rotation axis. One can also choose a globally well-behaved scalar z such that $z =$ constant curves intersect $\rho =$ constant curves orthogonally and can then write the two-dimensional metric, g_{ij}, in the form

$$ds_{II}^2 = g_{ij}dx^i dx^j = \Sigma\,(d\rho^2 + dz^2) \tag{3.23}$$

where Σ is a strictly positive function in $<\mathcal{I}>$ (see Figure Va).

Clearly a globally good chart in $<\mathcal{J}>$ is desired. Carter [8] after a fairly long and tedious analysis, has shown that the domain of outer communications can be covered globally by a manifestly stationary and axisymmetric ellipsoidal coordinate system (Figure Vb) $\{\lambda, \mu, \phi, t\}$ with ϕ, t being ignorable coordinates such that the metric takes the form

$$ds^2 = -Udt^2 + 2Wd\phi dt + Xd\phi^2 + \sum \left\{\frac{d\lambda^2}{\lambda^2 - C^2} + \frac{d\mu^2}{1-\mu^2}\right\} \qquad (3.24)$$

where λ ranges from the constant, C, to infinity while μ ranges from 1 to -1. $\lambda = C$ is the horizon and $\mu = \pm 1$ are the north and south poles of the symmetry axis. Ernst has shown that the Einstein equation $R_{ab} = 0$ neatly reduces to just two equations in terms of the background metric

$$\frac{d\lambda^2}{\lambda^2 - C^2} + \frac{d\mu^2}{1-\mu^2}$$

$$\begin{aligned} \nabla \left\{\frac{X\nabla W - W\nabla X}{\rho}\right\} &= 0 \\ \nabla \left\{\frac{\rho \nabla X}{X}\right\} &+ \frac{|X\nabla W - W\nabla X|^2}{\rho X^2} \end{aligned} \qquad (3.25)$$

where $\rho^2 = (\lambda^2 - C^2)(1-\mu^2)$, U and $\sum$ are determined in terms of X and W by quadrature and the covariant derivatives ∇ are in the two-dimensional metric

$$ds^2_{II} = \frac{d\lambda^2}{\lambda^2 - C^2} + \frac{d\mu^2}{1-\mu^2}. \qquad (3.26)$$

It is convenient (actually "essential" will turn out to be a better word) to introduce the "twist potential", Y, by requiring

$$\begin{aligned} (1-\mu^2)\, Y_{,\mu} &= XW_{,\lambda} - WX_{,\lambda} \\ -(\lambda^2 - C^2)\, Y_{,\lambda} &= XW_{,\mu} - WX_{,\mu} \end{aligned} \qquad (3.27)$$

where comma denotes partial differentiation and the integrability condition for Y is equation (3.25). Equations (3.25) become the expressions $E(X,Y) = 0$, $F(X,Y) = 0$ where:

$$E(X,Y) = \nabla \cdot (\rho X^{-2} \nabla X) + \rho X^{-3}(|\nabla X|^2 + |\nabla Y|^2) = 0$$
$$F(X,Y) = \nabla \cdot (\rho X^{-2} \nabla Y) = 0 \qquad (3.28)$$

and ρ and ∇ are defined the same way as before.

It is, of course, necessary to supply boundary conditions for the coupled equations (3.24). Carter [8] has determined that the requirements of asymptotic flatness plus regularity conditions on the horizon and rotation axis lead to certain conditions on X and Y. These conditions are: as $\mu \to \pm 1$ X and Y are well-behaved functions of λ and μ with

$$X = 0(1-\mu^2)$$
$$X^{-1}X_{,\mu} = -2\mu(1-\mu^2)^{-1} + 0(1) \qquad (3.29)$$
$$Y_{,\lambda} = 0((1-\mu^2)^2); \quad Y_{,\mu} = 0(1-\mu^2)$$

and as $\lambda \to C$, X and Y are well-behaved functions with

$$X = 0(1); \qquad X^{-1} = 0(1)$$
$$Y_{,\lambda} = 0(1); \qquad Y_{,\mu} = 0(1) . \qquad (3.30)$$

Asymptotic flatness requires that as $\lambda^{-1} \to 0$, Y and $\lambda^{-2}X$ are well-behaved functions of λ^{-1} and μ with

$$\lambda^{-2}X = (1-\mu^2)[1 + 0(\lambda^{-1})]$$
$$Y = 2J\mu(3-\mu^2) + 0(\lambda^{-1}) \qquad (3.31)$$

where J is a constant that will turn out to be the angular momentum measured in the asymptotically flat region and 0 stands for "on the order of."

The problem of proving the uniqueness of stationary (nonstatic) regular predictable spacetimes subject to Assumption 3.1 and Theorems 3.6, 3.7, 3.8 is therefore equivalent to proving the uniqueness of the set of coupled equations 3.28 in the two-dimensional background metric (3.26) subject to the boundary conditions (3.29), (3.30), and (3.31). This proof has been supplied by Robinson [27b]. The key part of Robinson's proof is the identity

$$\begin{aligned}
&(X_1X_2)^{-1}(Y_2-Y_1)[X_1^2F(X_1,Y_1)-X_2^2F(X_2,Y_2)] + \tfrac{1}{2}X_2^{-1}[(Y_2-Y_1)^2 + \\
&X_2^2-X_1^2]E(X_1,Y_1) + \tfrac{1}{2}X_1^{-1}[(Y_2-Y_1)^2 + X_1^2-X_2^2]E(X_2,Y_2) + \\
&\qquad \frac{1}{2}\nabla\left[\rho\nabla\left(\frac{(X_2-X_1)^2+(Y_2-Y_1)^2}{X_2X_1}\right)\right] \qquad (3.32)\\
&= \rho(2X_2X_1)^{-1}|X_1^{-1}(Y_2-Y_1)\nabla Y_1 - \nabla X_1 + X_2^{-1}X_1\nabla X_2|^2 \\
&+ \rho(2X_2X_1)^{-1}|X_2^{-1}(Y_2-Y_1)\nabla Y_2 + \nabla X_2 - X_1^{-1}X_2\nabla X_1|^2 \\
&+ \rho(4X_1X_2)^{-1}|(X_2+X_1)(X_2^{-1}\nabla Y_2 - X_1^{-1}\nabla Y_1) - (Y_2-Y_1)(X_2^{-1}\nabla X_2 + X_1^{-1}\nabla X_1)|^2 \\
&+ \rho(4X_1X_2)^{-1}|(X_2-X_1)(X_1^{-1}\nabla Y_1 + X_2^{-1}\nabla Y_2) - (Y_2-Y_1)(X_1^{-1}\nabla X_1 + X_2^{-1}\nabla X_2)|^2 .
\end{aligned}$$

For fixed parameters c and J there is an associated Kerr solution (1.4) with $c^2 = m^2 - a^2$ and $J = am$. Suppose that (X_1, Y_1) corresponds to this Kerr solution and (X_2, Y_2) corresponds to a hypothetical second black hole solution satisfying the boundary conditions. Integration of (3.32) over the two-dimensional manifold (3.26) leads to a boundary integral on the left-hand side of the identity which vanishes by the boundary conditions (3.29), (3.30), (3.31). The integrand of the right-hand side is a sum of four positive definite terms each of which must now necessarily vanish. Simple manipulation of the resulting first order partial differential equations soon leads to the conclusion that $Y_2 = Y_1$ and $X_2 = X_1$, i.e. that the Kerr solution (1.8) is the unique stationary, axisymmetric solution satisfying the boundary conditions.

THEOREM 3.9. *The unique stationary (nonstatic) regular predictable Ricci flat spacetime subject to Assumption 3.1 and Theorems 3.6-3.8 is the Kerr solution (1.8).*

Non-vacuum theorems

In the non-vacuum case ($T_{ab} \neq 0$) it has not been possible so far to prove the uniqueness of the rotating electrically charged black hole solution of Kerr-Newman [5]. However, Robinson [27] has shown that continuous variations of this solution are fully determined by continuous variations of the constants: m = mass, J = angular momentum, Q = electric charge, by using a linearized version of the identity (3.32) extended to the electromagnetic case. Such a result is colloquially known as a "no-hair" theorem. Israel [6a] has proved a uniqueness theorem for the electrically charged, nonrotating black hole solution of Reissner-Nordstrom [2, 3]. Various results have been obtained for other fields. Hawking [28] has shown that no regular solution to the non-vacuum equations exists for a scalar (spin 0) field, Hartle similarly for the Fermi (spin 1/2) field, and Beckenstein [29] has shown no regular solutions exist for massive scalar (spin 0), massive vector (spin 1) and massive (spin 2) fields.

Perhaps a word is in order concerning "multi-solutions", e.g. multi-Schwarzschild, multi-Kerr, etc. "multi" means here that there is more than one connected component of the horizon and hence the above theorems are inapplicable. Physically, the idea is that one is considering more than one black hole. Although physical arguments yield some information about such configurations it might be nice to have a rigorous proof that, say, no nonsingular multi-Schwarzschild solution exists.

IV. *Classical solutions in quantum gravity*

It was pointed out in Section II that the property that the area of a two-dimensional spatial cross section of the horizon never decreases towards the future was analogous to the property of entropy in thermodynamics: Entropy never decreases towards the future. In fact, it is possible to prove that each of the Four Laws of Thermodynamics (i.e., four fundamental equations defining thermodynamics) have an analogy in black hole theory where thermodynamic quantities are replaced by geometric quantities as in the substitution "area" for "entropy." The relevant geometric quantities are:

(i) the scalar ε defined by $\ell^b \ell^a_{;b} = 2\varepsilon \ell^a$ where ℓ^a is a generator of the horizon $\ell^a \equiv \frac{2\pi}{t_1} \tilde{k}^a + k^a$ by Theorem 3.6). It is conventional to redefine ε and t_1 as $\kappa = 2\varepsilon$, $\Omega = \frac{2\pi}{t_1}$.

(ii) the mass, M

(iii) the area A of a two-dimensional cross section of the horizon

(iv) the "angular velocity" $\Omega = 2\pi/t_1$.

The relevant thermodynamic quantities are:

(i) the temperature, T

(ii) the entropy, S

(iii) the pressure, P

(iv) the volume, V.

The Four Laws of Thermodynamics are:

(0) The temperature T is a constant for a system in equilibrium.

(1) In a change from one equilibrium state to another characterized by changes in E, S, and V then

$$dE = TdS + PdV .$$

(2) In any process in which a thermally isolated system goes from one state to another $dS \geq 0$.

(3) It is impossible to reduce the temperature T to absolute zero by a finite sequence of steps.

The Four Laws of Black Hole Mechanics are:

(0) The scalar κ is a constant on the horizon

(1) In a change from one black hole equilibrium to another

$$dM = \frac{\kappa dA}{8\pi} + \Omega dJ .$$

(2) In any change in a black hole state

$$dA \geq 0 .$$

(3) It is impossible to reduce κ to zero by a finite sequence of steps. Comparison of the Four Laws leads to the *formal* equations: $T = \kappa/2\pi$ and $S = A/4$ and the temptation to include black holes in thermodynamics. Of course, classical black holes do not really have a temperature because nothing can ever escape to $\mathcal{I}^+$ once it crosses the horizon and hence a classical black hole could not stay in equilibrium with a heat bath. However, in 1975 Hawking [30] was able to prove using a semiclassical formalism that if one treats the matter fields using quantum mechanics, instead of classical mechanics, then particles can escape to $\mathcal{I}^+$ from behind the horizon and furthermore a black hole *emits* particles as if it were a hot body with temperature $\kappa/2\pi$ and entropy $A/4$.

Those remarkable results on the thermal quantum properties of black holes can also be recovered using the Euclidean path integral approach to quantum gravity [10]. This approach has a strong geometrical content that might appeal to differential geometers. In this approach "Euclidean black hole" solutions play an important role.

"Euclidean" or "Euclidean section" will mean that the metric on a four-dimensional manifold is of positive definite signature. "Solution" will mean that the metric is Ricci flat. For example, the Euclidean Schwarzschild solution can be written in a local coordinate chart as:

$$ds^2 = \left(1 - \frac{2m}{r}\right) d\tau^2 + dr^2 / \left(1 - \frac{2m}{r}\right) + r^2(d\theta^2 + \sin^2\theta \, d\phi^2) . \tag{4.1}$$

It can be obtained from the Lorentzian Schwarzschild solution describing a nonrotating black hole of mass m

$$ds^2 = -\left(1 - \frac{2m}{r}\right) dt^2 + dr^2 / \left(1 - \frac{2m}{r}\right) + r^2(d\theta^2 + \sin^2\theta \, d\phi^2) . \tag{4.2}$$

by $t \to i\tau$. τ must be identified with period $8\pi m$ for the Euclidean section to be regular. θ and ϕ are the usual polar and azimuthal coordinates on a 2-sphere and $r \in [2m, \infty)$. The manifold is geodesically complete and has topology $R^2 \times S^2$. The Euclidean Kerr solution

$$ds^2 = [d\tau - \alpha \sin^2\theta d\phi]^2 \Delta/\rho^2 + [(r^2 - \alpha^2) d\phi - \alpha d\tau]^2 \sin^2\theta/\rho^2 + \rho^2 dr^2/\Delta + \rho^2 d\theta^2 \tag{4.3}$$

$$\Delta = r^2 - 2mr - \alpha^2 \qquad \rho^2 = r^2 - \alpha^2 \cos^2\theta$$

can be obtained from the Lorentzian Kerr solution describing a rotating black hole of mass m and angular momentum ma

$$ds^2 = -[dt - a \sin^2\theta d\phi]^2 \Delta/\rho^2 + [(r^2 + a^2) d\phi - a dt]^2 \sin^2\theta/\rho^2 + \rho^2 dr^2/\Delta + \rho^2 d\theta^2 \tag{4.4}$$

$$\Delta = r^2 - 2mr + a^2 \qquad \rho^2 = r^2 + a^2 \cos^2\theta$$

by $\tau \to i\tau$, $a \to -i\alpha$. The $\{\tau, \phi\}$ plane must be identified as $\{\tau, \phi\} = \{\tau + \beta, \phi + \beta\Omega_H\}$ where $\beta = 4\pi m(m + (m^2 + \alpha^2)^{1/2})/(m^2 + \alpha^2)^{1/2}$ and $\Omega_H = \alpha[2(m^2 + m(m^2 + \alpha^2)^{1/2})]^{-1}$. θ and ϕ are again the usual 2-sphere coordinates and $r \in [m + (m^2 + \alpha^2)^{1/2}, \infty)$. The manifold has topology $R^2 \times S^2$ and is geodesically complete with the metric given above.

We will now briefly review the Euclidean path integral approach to quantum gravity following the analysis given in reference [10]. The essential idea is that the partition function for a system of temperature $1/\beta$ can be represented as a functional integral over fields periodic with period β in Euclidean time:

$$Z = \int_C d[\phi] e^{-I[\phi]}. \tag{4.5}$$

Here Z is the partition function, $d[\phi]$ denotes functional integration over fields ϕ (indices to be appropriately added for spinor, vector, tensor), $I[\phi]$ is the classical action functional for ϕ on the Euclidean section, while the subscript C on the integral denotes the class of fields to be integrated, e.g. periodic in imaginary time with Dirichlet boundary conditions.

The appropriate action for gravity is

$$I = \frac{1}{16\pi G} \int R \sqrt{g} d^4x + \frac{1}{8\pi G} \int K \sqrt{h} d^3x + C_0 \tag{4.6}$$

where G is Newton's constant in natural units, R is the Ricci scalar, h is the determinant of the induced metric h_{ab} on the boundary, K is the trace of the second fundamental form of the boundary, and C_0 is a constant adjusted to make the action of flat space vanish. The integral is over all asymptotically flat metrics, periodic in Euclidean time, which fill in a $S^2 \times S^1$ boundary at infinity. The $S^2 \times S^1$ boundary is chosen to represent a large spherical "box", S^2, bounding three space; cross the periodically identified Euclidean time axis, S^1.

It is impossible to perform the functional exactly and hence a steepest descents approximation is employed. That is, one expands the action about a classical solution of the field equations, $\left.\frac{\delta I}{\delta g_{ab}}\right|_{g_{ab}^{\text{classical}}} = 0$ and integrates over fluctuations away from this solution. Hence

$$g_{ab} = g_{ab}^{\text{classical}} + \tilde{g}_{ab} \tag{4.7}$$

and

$$I[g] = I_0[g^{\text{classical}}] + I_2[\tilde{g}_{ab}] + \cdots \tag{4.8}$$

$I_2[\tilde{g}_{ab}]$ is quadratic in the fluctuation $\tilde{g}_{ab}$ and has the form $\int \tilde{g}_{ab}\, 0^{abcd}\, \tilde{g}_{cd} \sqrt{g} d^4x$ where 0^{abcd} is a second order differential operator in the "background" metric g_{ab}. Truncation of the expansion at quadratic order is called the "one loop expansion" and leads to an expression for $\log Z$ of the form:

$$\log Z = -I[g_{ab}^{\text{classical}}] + \log\left\{\int d[\tilde{g}_{ab}]e^{-I_2[\tilde{g}_{ab}]}\right\}, \tag{4.9}$$

where I_0 is the contribution of classical background fields to $\log Z$ and the second term (the "one loop" term) represents the effect of quantum fluctuations about the background fields. Evaluation of the second term involves the determinant of the operator 0^{abcd}. A convenient definition of $\det 0^{abcd}$ is the zeta function definition of Singer [32]. Hawking [33] has employed this definition to calculate one loop terms. Gibbons and Perry [34] have investigated the one loop term in detail. It should be noted that more than one background field (classical solution) may satisfy the boundary conditions, and in this event there are contributions to $\log Z$ of the form (4.9) for each classical background field.

One background field satisfying the boundary conditions of asymptotic flatness, $S^2 \times S^1$ boundary, and periodicity β in Euclidean time is flat space

$$ds^2 = d\tau^2 + dr^2 + r^2(d\theta^2 + \sin^2\theta d\phi^2) \tag{4.10}$$

with τ identified with period β. The action, (4.6) of flat space is zero. In the limit of a very large "spherical box", S^2, with radius r_0 tending to infinity, the one loop term can be evaluated exactly [33] as $\dfrac{4\pi r_0^3}{135\beta^3}$.

The interpretation is that this is the contribution to the partition function for thermal gravitons on a flat space background. Another background field satisfying the boundary conditions is the Euclidean Schwarzschild solution.

$$ds^2 = \left(1 - \frac{2m}{r}\right) d\tau^2 + \frac{dr^2}{1 - \frac{2m}{r}} + r^2(d\theta^2 + \sin^2\theta d\phi^2) \tag{4.11}$$

where regularity requires $\tau = \tau + \beta$, $\beta = 8\pi m$. This has action $I = 4\pi m^2$ and a one loop term [34] $\dfrac{106}{45} \log \dfrac{(\beta)}{\beta_0} \dfrac{4\pi r_0^3}{135\beta^3}$ for $r_0 >> \beta$ i.e. for a box size large compared to the black hole. β_0 is related to the one loop renormalization parameter.

Given the partition function one can evaluate relevant thermodynamic quantities such as energy and entropy in the usual fashion

$$\langle E \rangle = -\frac{\partial}{\partial \beta} \log Z \tag{4.12a}$$

$$S = \beta \langle E \rangle + \log Z . \tag{4.12b}$$

Applying this to the contribution to $\log Z$ from the classical action of the Schwarzschild solution yields

$$S = 4\pi m^2 = A/4 \tag{4.13}$$

where A is the area of the "event horizon", $r = 2m$. Hence the classical background contribution to the partition function yields a temperature, $T = \frac{1}{\beta} = \frac{1}{8\pi m}$, and an entropy, $S = 4\pi m^2$. These are precisely the expressions for the temperature and entropy of a nonrotating black hole that Hawking first obtained in 1975 by completely different methods.

One can calculate the (unstable) equilibrium states of a black hole and thermal gravitons in a large box by including the one loop terms in the expression for $\log Z$. Maximization of the entropy with fixed energy leads to the conclusion that if the volume, V, of the box satisfies

$$E^5 < \frac{\pi^2}{15}(8354.5)\, V \tag{4.14}$$

then the most probable state of the system is flat space with thermal gravitons, while if the inequality is not satisfied the most probable state is a black hole (Schwarzschild solution) in equilibrium with thermal gravitons.

One can also consider the partition function for grand canonical ensembles in which a chemical potential is associated with a conserved quantity. For example one can consider a system at a temperature $T = 1/\beta$

and a given (conserved) angular momentum J with associated chemical potential, Ω, where Ω is the angular velocity. The partition function would then be given by a functional integral over all fields with $(t, r, \theta, \phi) = (t+\beta, r, \theta, \phi+\beta\Omega)$. The Euclidean Kerr solution (4.3) would then be a classical background solution around which one could expand the action in a one loop calculation analogous to the above.

It is clear from the analysis reviewed above that the Euclidean black hole solutions, both Schwarzschild and Kerr, play a key role in approximating the functional integrals occurring in quantum gravity, and connect in a fundamental way to the thermal properties of black holes discovered by Hawking [30] and summarized earlier. The claim has been made [31] that the Lorentzian black hole theorems apply to the Euclidean section. It is straightforward to show that Israel's theorem [6], which in essence proves that (4.2) is the unique, static (hypersurface orthogonal Killing vector), asymptotically flat solution to Einstein's equations with a regular fixed point surface of the staticity Killing vector, can be taken over to the Euclidean section essentially line for line. However, in the next section it will be shown that Robinson's theorem, proving the uniqueness of the Lorentzian Kerr solution no longer works on the Euclidean section.

If the Euclidean black hole solutions are not unique then there exists at least one other Euclidean solution, satisfying the conditions above, which would necessarily have to be included in the steepest descents approximation of the functional integral. This would mean there exists the possibility of a third phase, in addition to the Euclidean black hole solutions and flat space, contributing to the analysis of the possible states of a gravitational field in a box. One might call such a solution a new "Euclidean black hole" solution. This new Euclidean black hole solution would either not admit a Lorentzian section, or if a Lorentzian section exists, it would violate the conditions of a Lorentzian black hole solution by being, for example, singular or perhaps not asymptotically flat. Hence the new Euclidean black hole solution would play a role somewhat analogous to the instantons of Yang Mills theory, insomuch as the Lorentzian sections of such solutions are not physical objects, although they do have a physical effect by making a large contribution to the functional integral in the quantization of the theory.

V. *Euclidean black hole uniqueness theorems* [44]

The first part of the classical black hole uniqueness theorems described in previous sections, that which assumes a locally timelike Killing vector and utilizes spacetime causal structure, is clearly inapplicable to the Euclidean section for two reasons. First, there is no reason for assuming the existence of a Killing vector as one wishes to include in the functional integral all positive definite metrics satisfying

(i) asymptotic flatness

(ii) an $S^2 \times S^1$ boundary at infinity

(iii) an identification of the metric

$$(t, r, \theta, \phi) = (t+\beta, r, \theta, \phi)$$

or

$$(t, r, \theta, \phi) = (t+\beta, r, \theta, \phi+\Omega\beta)$$

depending on the physical situation chosen and hence the extremal metric need not ab initio have a Killing vector. Secondly, there is no causal structure on the Euclidean section. However, one might hope that the second part of the classical uniqueness theorems, the Israel [6] and Robinson [27] theorems, would allow one to draw a more restricted conclusion concerning the extremal metric in the class of metrics satisfying conditions (i), (ii), and (iii) and furthermore possessing either a hypersurface orthogonal Killing vector (Euclidean analogue of staticity); or a nonhypersurface orthogonal Killing vector (Euclidean analogue of stationarity) that commutes with a second Killing vector generating the action of SO(2) (Euclidean analogue of axisymmetry).

A positive definite metric possessing a hypersurface orthogonal Killing vector, ∂_t, can be obtained from (3.6) by $X^0 \to it$

$$ds^2 = V^2 dt^2 + g_{\alpha\beta}(X^1, X^2, X^3)\, dX^\alpha dX^\beta, \quad V = V(X^1, X^2, X^3). \tag{5.1}$$

It is clear that Israel's theorem can be transcribed to the Euclidean section essentially line for line because, as described in Section III, much of the analysis involves the two geometry $V = \text{constant}$, $t = \text{constant}$. The part explicitly involving the four geometry and hence the metric signature, for example equation (3.8), remains unchanged independent of whether the signature is $+2$ or $+4$. The surface $V = 0^+$ is the fixed point locus of the Killing vector ∂_t or a "bolt" in the parlance of Reference [31], and therefore the manifold has an Euler characteristic, $\chi = 2$, by the fixed point theorems. The Euclidean version of Israel's theorem therefore proves that the unique, nonsingular, Ricci flat, positive definite metric satisfying the conditions of

(i) asymptotic flatness

(ii) an $S^2 \times S^1$ boundary at infinity

(iii) an identification of the metric $(t, r, \theta, \phi) = (t+\beta, r, \theta, \phi)$ on boundary

(iv) two dimensional fixed point locus of hypersurface orthogonal Killing vector (staticity + nontrivial topology)

is the Euclidean Schwarzschild solution (4.1) where $\beta = 8\pi m$.

It is natural to expect a similar Euclidean analogue of Robinson's theorem, however we will now show that there are grave difficulties with the analogy. A positive definite, axisymmetric, "stationary" (nonhypersurface orthogonal Killing vector) metric is obtained from (3.24) by $t \to it$ and $W \to -iW$. This procedure was used in going from the Lorentzian Kerr metric (4.4) to the Euclidean Kerr metric (4.3), i.e. $t \to it$ and $a \to -i\alpha$. It is important to realize that one should not merely put $U \to -U$ in (3.24). Equation (3.27) implies that $Y \to -iY$ and similarly in (3.28) and (3.32). Therefore the Euclidean Robinson identity (3.32) has a sum of two positive definite and two negative definite terms on the right-hand side. Hence when one integrates the Euclidean Robinson identity over the manifold it is no longer possible to conclude that each term on the right-hand side must separately equal zero. Therefore one cannot conclude from this analysis that the Euclidean Kerr solution is unique.

One can introduce a new set of variables for which there exists a Robinson identity with the right-hand side being positive definite [35]. We start from the Lorentzian field equations in terms of the metric quantities W and X, as given, e.g. by Carter [8].

$$\begin{aligned} &\nabla\left(\frac{X\nabla W - W\nabla X}{\rho}\right) = 0 \\ &\nabla\left(\frac{\rho\nabla X}{X}\right) + \frac{|X\nabla W - W\nabla X|^2}{\rho X^2} = 0\,. \end{aligned} \tag{5.3}$$

The Euclidean equations $(W \to -iW)$ are therefore

$$\begin{aligned} &\nabla\left(\frac{X\nabla W - W\nabla X}{\rho}\right) = 0 \\ &\nabla\left(\frac{\rho\nabla X}{X}\right) - \frac{|X\nabla W - W\nabla X|^2}{\rho X^2} = 0\,. \end{aligned} \tag{5.4}$$

Introduction of the quantities $\tilde{X} = \rho/X$ and $\tilde{Y} = W/X$ leads to

$$\begin{aligned} &\nabla\left(\frac{\rho\nabla\tilde{X}}{\tilde{X}}\right) + \frac{\rho|\nabla\tilde{Y}|^2}{(\tilde{X})^2} = 0 \\ &\nabla\left(\frac{\rho\nabla\tilde{Y}}{\tilde{X}^2}\right) = 0\,. \end{aligned} \tag{5.5}$$

These equations for the Euclidean variables $\tilde{X}$, $\tilde{Y}$ are identical to Equations (3.28) for the Lorentzian variables X and Y. Therefore the Robinson identity (3.32) exists on the Euclidean section in terms of the Euclidean variables $\tilde{X}$, $\tilde{Y}$. Integration of the twiddled identity over the manifold leads to a sum of four positive terms on the right-hand side as desired. However, the twiddled divergence on the left-hand side does not integrate up to a boundary term that vanishes, in fact it diverges on the "horizon", i.e. the two dimensional fixed point locus (bolt) of the Killing vector ∂_t. Once again it is impossible to prove the uniqueness of the Euclidean Kerr black hole using a Euclidean Robinson theorem. Next we try (and fail) to disprove uniqueness by searching for possible counter-examples.

The failure of the Euclidean Robinson theorem discussed above suggests that perhaps another Euclidean solution satisfying the boundary conditions exists. One manner in which stationary, axisymmetric Euclidean solutions may be found is by analytically continuing stationary, axisymmetric Lorentzian solutions to the Euclidean section. Clearly all Lorentzian solutions, apart from Kerr, will be pathological in some sense since the Lorentzian Robinson uniqueness theorem works. The idea would be that the pathologies would not be present on the Euclidean section. Some Euclidean solutions cannot be obtained by analytic continuation of Lorentzian ones. A sufficient, but not necessary, condition for this is that the curvature be (anti) self dual. In this section we consider examples from both categories.

Apart from the Lorentzian Kerr solution, the only other stationary, axisymmetric, asymptotically flat solution for which the metric is explicitly known is the Lorentzian Tomimatsu-Sato solution [36, 37]. There is actually a family of such solutions, characterized by a parameter, δ, taking integer values with $\delta = 1$ being the Kerr solution. The complexity of the metric grows rapidly with δ. The Tomimatsu-Sato solutions contain event horizons for odd δ, however they are not black hole solutions because curvature singularities exist outside the horizon. The Euclidean section of the T–S solutions may be defined in analogy with the Euclidean section of the Kerr solution (4.3) and the singularity outside the horizon disappears (viz. the disappearance of the $r = 0$ singularity in Kerr). However, new singularities appear at the north and south poles of the horizon, so the Euclidean T–S solution is not a counterexample to the conjectured uniqueness of the Euclidean Kerr solution.

A class of Euclidean solutions which cannot be obtained from Lorentzian solutions are those with (anti) self dual curvature. A reasonable physical requirement to impose on any Euclidean solution is that the manifold admit spin structure. Gibbons and Pope [39] have constructed an argument proving that self dual, asymptotically Euclidean solutions (i.e. the curvature falls off to zero at infinity in the four dimensional sense) with spin structure cannot exist. Their argument applies equally well to the asymptotically flat situation (curvature falls off to zero in the three dimensional sense) under consideration here. The argument proceeds as follows. The index of the Dirac operator, $\gamma^a \nabla_a$ for a manifold with boundary is given by

$$\text{index}\,[\gamma^a \nabla_a] = \frac{1}{192\pi^2}\int R_a{}^b \wedge R^a{}_b\, d(\text{vol}) - \frac{1}{192\pi^2}\int_{\partial m} \theta^a{}_b \wedge R^b{}_a\, d(\text{surf}) - [\eta_D(0)] \tag{5.6}$$

where $R_a{}^b$ is the curvature 2-form in an orthonormal basis, $\theta^a{}_b$ is the second fundamental form of the boundary, and $\eta_D(0)$ is the expression

$$\eta_{D(s)}\Big|_{s=0} = \sum_n \text{sign}(\lambda_n)|\lambda_n|^{-s}\Big|_{s=0} \tag{5.7}$$

where the eigenvalues λ_n are eigenvalues of the Dirac operator restricted to the boundary. h is the dimension of the kernel which is zero for $S^1 \times S^2$. $\eta_D(0)$ measures the "handedness" of the manifold and vanishes if the boundary of the manifold admits an orientation reversing isometry as does the boundary $S^2 \times S^1$ under consideration here, and also the S^3 boundary considered by Gibbons and Pope. The second term in the index (5.6) vanishes by virtue of asymptotic flatness while the first vanishes by the condition of (anti) self duality. Hence an asymptotically flat, self dual solution, if it exists, should admit at least one normalizable spinor. However, Lichnerowicz's theorem [40] proves that spinors on manifolds with $R \geq 0$ are covariantly constant and therefore not normalizable if the manifold is noncompact. Hence one must conclude that asymptotically flat, self dual solutions do not exist.

Despite the failure of the Euclidean Robinson Theorem one can prove a Euclidean "No Hair" theorem. The phrase "No Hair" theorem usually refers to the Lorentzian theorem of Carter [8]: stationary, axisymmetric spacetimes satisfying the usual black hole boundary conditions fall into families depending on at most two parameters, the mass m and the angular momentum $J = ma$; and that continuous variations of these solutions are uniquely determined by continuous variations of m and J. Hence the only regular perturbations of the Lorentzian Kerr solution are the "trivial" perturbations in m and J. A corollary is that the Kerr solution is the unique family with a regular zero angular momentum $(J = 0)$ limit. The method of proof involves a linearized version of the Robinson identity (3.32), where "linearized" means X_1, Y_1 differs from X_2, Y_2 by quantities of the first order. Clearly this theorem will have the same difficulties on the Euclidean section as the Robinson uniqueness theorem. Teukolsky [41, 42] and Wald [43] have employed a different method to show that no bifurcations occur off the Kerr sequence. The idea behind their method is to explicitly solve the Teukolsky [41] master equation for perturbations off the Kerr background solution and thereby show that the only stationary, regular perturbations are the trivial perturbations $m \to m + \delta m$, $J \to J + \delta J$. This method also works on the Euclidean section [44], when combined with recent results of Lapedes and Perry [45].

VI. *Uniqueness conjectures*

The Euclidean Schwarzschild and Euclidean Kerr solutions (4.1), (4.3) are nonsingular, non-Kähler, four dimensional, positive definite, Ricci flat metrics. In Section IV the importance of the uniqueness of these solutions was outlined and a rough statement was formulated of the conditions under which the solutions are suspected to be unique. In this section we make these conjectures precise.

CONJECTURE I.

Let the pair $\{\mathcal{M}, g_{ab}\}$ represent a noncompact four dimensional manifold with an associated positive definite metric. The Euclidean Schwarzschild solution $\{R^2 \times S^2, g_{ab}\}$ with g_{ab} given by (4.1) is the unique solution that satisfies the following conditions

(i) Ricci flat

(ii) geodesically complete

(iii) asymptotically flat, i.e. the induced metric $g_{\alpha\beta}$ on a regular noncompact embedded three dimensional hypersurface satisfies

$$\lim_{r\to\infty} g_{\alpha\beta} = \delta_{\alpha\beta} + \mathcal{O}(r^{-1}) \qquad \lim_{r\to\infty} \partial_\gamma g_{\alpha\beta} = \mathcal{O}(r^{-2})$$

where $r^2 = \delta_{\alpha\beta} X^\alpha X^\beta$ in suitable coordinates

(iv) an $S^2 \times S^1$ boundary at infinity such that in a suitable chart

$$ds^2 = d\tau^2 + dr^2 + r^2(d\theta^2 + \sin^2\theta\, d\phi^2) + \mathcal{O}(1/r)$$

where τ is identified with period $8\pi m$. r is a coordinate along a ray and θ, ϕ are the usual polar and azimuthal angles on S^2.

(v) nontrivial topology.

Condition (v) excludes suitably identified flat space from being a counterexample.

Note that if one further requires that the metric admit a hypersurface orthogonal Killing vector then the Euclidean version of Israel's theorem (Section V) proves this more restricted conjecture.

CONJECTURE II.

Let the pair $\{\mathcal{M}, g_{ab}\}$ represent a noncompact, four dimensional manifold with an associated positive definite metric as before. The Euclidean Kerr solution (4.3) is the unique solution satisfying conditions (i), (ii), (iii) and (v) (above) which has an $S^2 \times S^1$ boundary at infinity such that in a suitable chart

$$ds^2 = d\tau^2 + dr^2 + r^2(d\theta^2 + \sin^2\theta\, d\phi^2) + \mathcal{O}(1/r)$$

where the pair $\{\tau, \phi\}$ is identified with $\{\tau+\beta, \phi+\beta\Omega\}$, r is a coordinate along a ray, θ and ϕ are the usual polar and azimuthal angles on S^2, and β and Ω are constants defined in Section IV.

Note that if one further requires that the metric admit two commuting Killing vectors, one of which is nonhypersurface orthogonal, and the other is a generator of SO(2) (the Euclidean analogue of stationarity and axisymmetry) then the Ernst, Carter, Robinson formalism of Section V does *not* prove this more restricted theorem. The formalism does provide a restatement of the more restrictive problem as follows.

CONJECTURE IIa.

Subject to the following conditions, the unique solution X, Y, to the coupled set

$$\nabla \cdot (\rho X^{-2} \nabla X) + \rho X^{-3}(|\nabla X|^2 - |\nabla Y|^2) = 0$$

$$\nabla \cdot (\rho X^{-2} \nabla Y) = 0$$

in the background metric

$$ds^2 = d\lambda^2/(\lambda^2 - c^2) + d\mu^2/(1-\mu^2)$$

where

$$\rho^2 = (\lambda^2 - c^2)(1-\mu^2)$$

is

$$X = (1-\mu^2)\{(\lambda+m)^2 - a^2 - a^2(1-\mu^2)2mr/(r^2 - a^2\mu^2)\}$$

$$Y = 2ma\mu(3-\mu^2) + 2a^3\mu m(1-\mu^2)^3/[(\lambda+m)^2 - a^2\mu^2] .$$

The conditions are

(i) In the limit $\mu \to \pm 1$ X and Y are well-behaved functions of λ and μ with

$$X = \mathcal{O}(1-\mu^2)$$

$$X^{-1}X_{,\mu} = -2\mu(1-\mu^2)^{-1} + \mathcal{O}(1)$$

$$Y_{,\lambda} = \mathcal{O}((1-\mu^2)^2)\,; \quad Y_{,\mu} = \mathcal{O}(1-\mu^2)\,.$$

(ii) In the limit $\lambda \to c$, X and Y are well-behaved functions with

$$X = \mathcal{O}(1)\,; \quad X^{-1} = \mathcal{O}(1)$$

$$Y_{,\lambda} = \mathcal{O}(1)\,; \quad Y_{,\mu} = \mathcal{O}(1)\,.$$

(iii) In the limit $\lambda^{-1} \to 0$, Y and $\lambda^{-2}X$ are well-behaved functions of λ^{-1} and μ with

$$\lambda^{-2}X = (1-\mu^2)[1+\mathcal{O}(\lambda^{-1})]$$

$$Y = 2ma\mu(3-\mu^2) + \mathcal{O}(\lambda^{-1})$$

m and a are constants.

Proofs of the conjectures above are left as a challenge to mathematicians.

SCHOOL OF NATURAL SCIENCES
THE INSTITUTE FOR ADVANCED STUDY
PRINCETON, NEW JERSEY 08540, U.S.A.

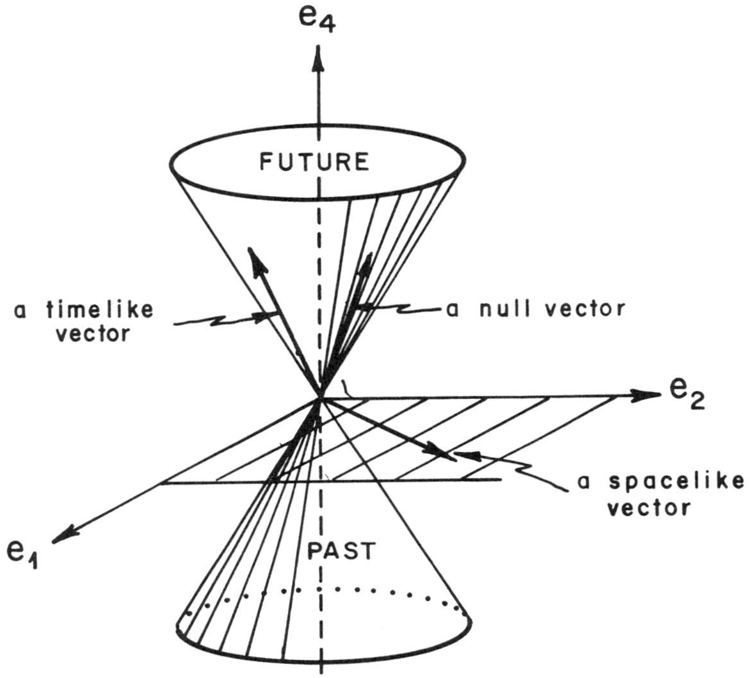

Figure I: The null cone separates timelike from spacelike vectors.

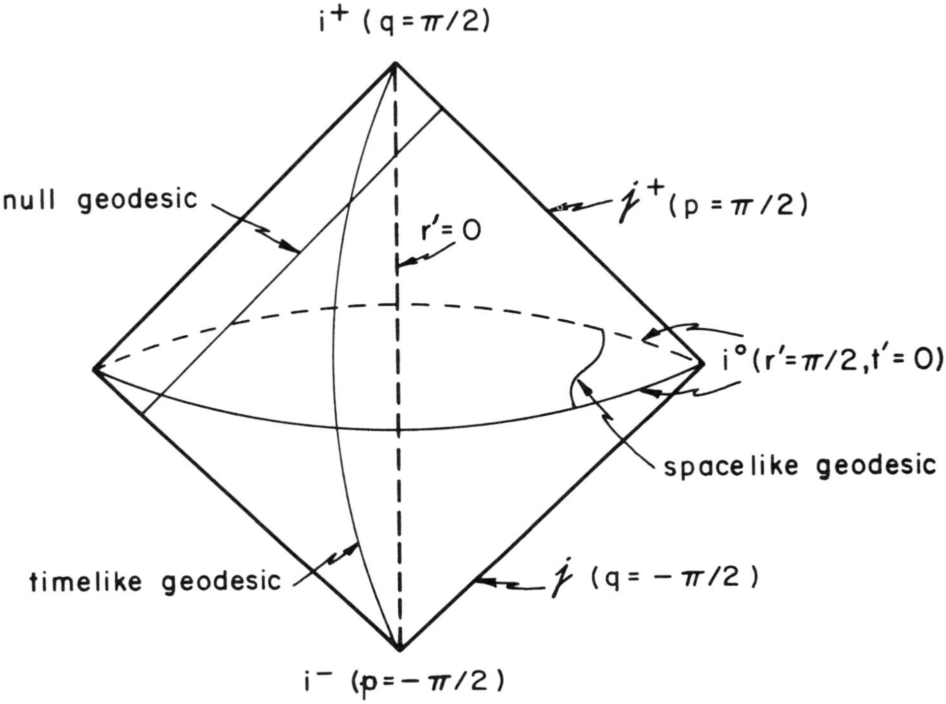

Figure II: Penrose diagram of Minkowski spacetime. Null lines are at 45°. $\mathcal{J}^+$ and $\mathcal{J}^-$ are at $p = -\pi/2$, $q = -\pi/2$, respectively.

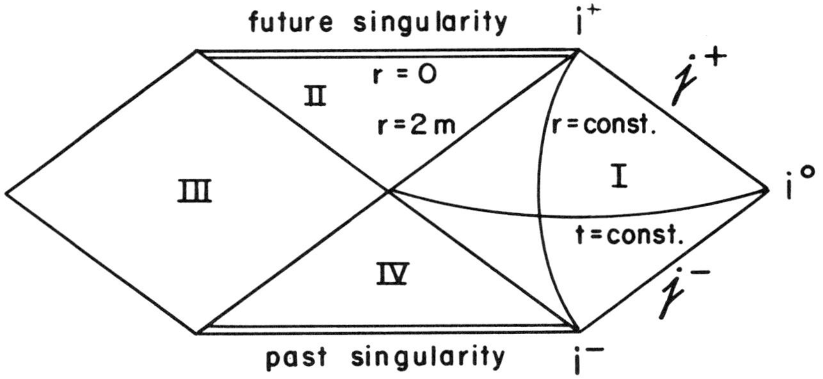

Figure III: Penrose diagram of the Schwarzschild solution. The diagram is reflection symmetric for regions I-III and II-IV. Null lines are at 45°. The double lines are curvature singularities at $r = 0$. A representative $r =$ constant timelike geodesic starts at i^- and ends at i^+. A representative $t =$ constant spacelike surface is also drawn.

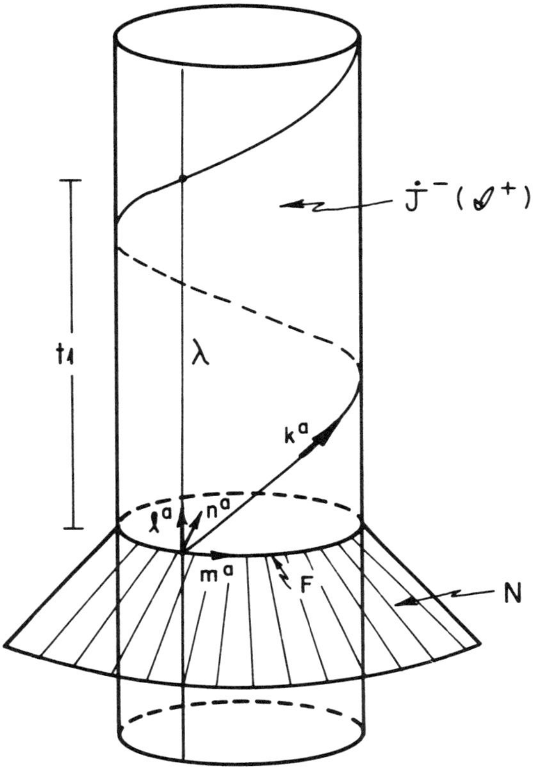

Figure IV: The event horizon is represented by a cylinder with ℓ^a a future-directed null geodesic generator of $\dot{J}^{-}(\mathscr{I}^{+})$. n^a is a null vector orthogonal to ℓ^a, m^a, and m^a are mutually orthogonal spacelike vectors with $m^a = k^a - \ell^a$. N is a null surface orthogonal to $\dot{J}^{-}(\mathscr{I}^{+})$. F is a spacelike two surface in $\dot{J}^{-}(\mathscr{I}^{+})$. t_1 is the period of rotation of the black hole.

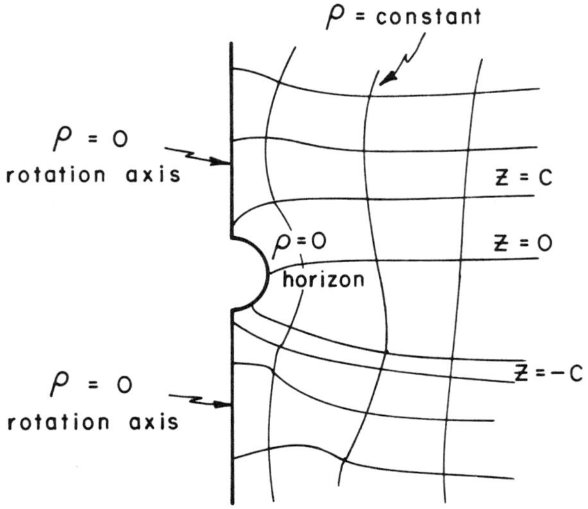

Figure Va: The ρz plane. $ds^2 = \Sigma(d\rho^2 + dz^2)$.

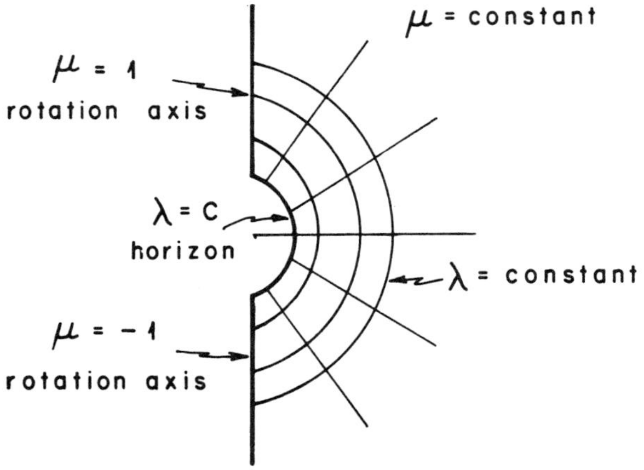

Figure Vb: Ellipsoidal coordinates $ds^2 = \dfrac{d\lambda^2}{\lambda^2 - c^2} + \dfrac{d\mu^2}{1-\mu^2}$.

REFERENCES

[1] K. Schwarzschild, Sitzber. Deut. Akad. Wiss. Berlin Kl. Math. Phys. Tech., 189(1916).

[2] H. Reissner, Ann. Phys. (Germany) *50*, 106(1916).

[3] G. Nordstrom, Proc. Kon. Ned. Akad. Wet. *20*, 1238(1918).

[4] R. Kerr, Phys. Rev. Lett. *11*, 237(1963).

[5] E. Newman, J. Math. Phys. *6*, 918(1965).

[6] W. Israel, Phys. Rev. *164*, 1776(1967).

[6a] ________, Comm. Math. Phys. *8*, 245(1968).

[7] Reviewed in S. W. Hawking, G. F. Ellis, *Large Scale Structure of Spacetime*, Cambridge University Press, 1973, chapter 9.

[8] Reviewed in B. Carter, "Black Hole Equilibrium States" in *Black Holes*, C. DeWitt and B. DeWitt, eds., Gordon and Breach Publishers, New York, 1973.

[9] B. Gidas, et al., Commun. Math. Phys. *68*, 209(1979).

[10] Reviewed in S. W. Hawking, "The Path Integral Approach to Quantum Gravity," in *General Relativity: An Einstein Centenary Survey*, S. W. Hawking and W. Israel, eds., Cambridge University Press, Cambridge, England, 1979.

[11] L. Markus, Ann. Math. *62*, 411 (1955).

[12] R. Penrose, "Structure of Spacetime," in *Batelle Rencontre*, C. DeWitt and J. Wheeler, eds., W. A. Benjamin Co., New York, 1968.

[13] ________, Proc. Roy. Soc. A*284*, 159(1965).

[14] R. Geroch, "Spacetime Structure from a Global Viewpoint," in *General Relativity and Cosmology*, Proceedings of the International School in Physics 'Enrico Fermi', course XLVII, R. K. Sachs, ed., Academic Press, New York, 1971.

[15] R. Penrose, Phys. Rev. Lett. *14*, 57(1965).

[16] Reference [7], chapter 8.

[17] A. Doroshkevich, et al., Sov. Phys. J.E.T.P. *22*, 122(1966).

[18] R. Price, Phys. Rev. *5*, 2419(1972).

[19] Reference [7], chapter 9.

[20] H. Muller zum Hagen, Proc. Camb. Phil. Soc. *68*, 199(1970).

[21] The proof may be found in Reference [8].

[22] See Reference [7], Appendix V.

[23] D. Robinson, Gen. Relat. Grav. *8*, 695(1977).

[24] Reference [7], chapter 9.

[25] R. Penrose, R. Floyd, Nature *229*, 177 (1971).

[26] Reference [7], chapter 9.

[27] D. Robinson, Phys. Rev. D*10*, 458 (1974).

[27b] ———, Phys. Rev. Lett., *34*, 905 (1975).

[28] S. W. Hawking, Commun. Math. Phys. *25*, 167 (1972).

[29] J. Beckenstein, Phys. Rev. D*5*, 1239, 2403 (1972).

[30] S. W. Hawking, Commun. Math. Phys. *43*, 199 (1975).

[31] G. W. Gibbons, S. W. Hawking, Commun. Math. Phys. *66*, 291 (1979).

[32] M. McKean, I. Singer, J. Diff. Geom. *1*, 43 (1967).

[33] S. W. Hawking, Commun. Math. Phys. *55*, 133 (1977).

[34] G. W. Gibbons, M. J. Perry, Nucl. Phys. B*146*, 90 (1978).

[35] D. Robinson, private communication.

[36] S. Tomimatsu, H. Sato, Phys. Rev. Lett. *29*, 1344 (1972).

[37] ———, Prog. Theor. Phys. *50*, 95 (1973).

[38] G. W. Gibbons, Phys. Rev. Lett. *30*, 398 (1973).

[39] ———, G. N. Pope, Commun. Math. Phys. *66*, 267 (1979).

[40] A. Lichnerowicz, Comptes Rendue *257*, 5 (1968).

[41] S. Teukolsky, Phys. Rev. Lett. *29*, 1114 (1972).

[42] ———, Ap. J. *185*, 639 (1973).

[43] R. Wald, J. Math. Phys. *14*, 1453 (1973).

[44] A. S. Lapedes, Phys. Rev. D*22*, 1837 (1980).

[45] A. S. Lapedes, M. J. Perry, Phys. Rev. D*24*, 1478, (1981).

GRAVITATIONAL INSTANTONS

Malcolm J. Perry*

This work reviews the overall nature of gravitational instantons. I discuss their introduction from the viewpoint of covariant quantum gravity. I then discuss their general topological classification, and finally list those known to date, together with their properties.

§1. *Introduction*

The first application of differential geometry to physics was made by Einstein, and culminated in the general theory of relativity in 1915 [1]. General relativity is a theory of gravitation and of spacetime where the spacetime metric g_{ab} has Lorentz signature $(-+++)$, and is determined through the Einstein equations

$$R_{ab} - \frac{1}{2} R g_{ab} + \Lambda g_{ab} = 8\pi G T_{ab} . \qquad (1.1)$$

R_{ab} is the Ricci tensor of g_{ab}, R is the Ricci scalar, Λ the cosmological constant, G is Newton's constant, and T_{ab} is the energy-momentum tensor of matter. This theory is at present entirely classical. This means that the theory is completely deterministic and does not really fit into the fundamental conceptual framework of physics as viewed in the

*Supported by the National Science Foundation under Grant PHY-78-01221.

Seminar on Differential Geometry
0-691-08268-5/82/000603-28$01.40/0 (cloth)
0-691-08296-0/82/000603-28$01.40/0 (paperback)

second half of the twentieth century. It is believed that all theories must be essentially quantum mechanical. One can think of four areas (at least) where classical general relativity must break down and be replaced by some sort of quantum mechanical counterpart.

1) Under a wide range of plausible physical circumstances, general relativity generates spacetime singularities. The fact that such singularities occur are predicted by a series of theorems of Hawking and Penrose [2-4]. In these examples, spacetime is shown to be necessarily geodesically incomplete. Physically this corresponds to paths of observers in free fall terminating after a finite proper time.

2) Numerous spacetimes admit the possibility of causality violation. That is, there are curves through a given point p in spacetime which are timelike and closed. The existence of such things gives rise to existential problems of an imponderable nature [5].

3) Gravitational radiation is now an observational fact [6]. All radiation must have a quantum nature which accounts for how it propagates energy and momentum [7] and how it is emitted and absorbed [7].

4) In classical relativity, an event horizon (the boundary of a black hole) is a surface which cannot be seen from outside the black hole. Such a surface can absorb things, but not emit them, thus acts thermodynamically as a surface at a temperature of absolute zero. If this existed it could behave like a perpetual motion machine, in contradiction to the third law of thermodynamics.

All of these problems should mysteriously solve themselves if one has a sensible quantum theory of gravity. Indeed, some progress has been made toward understanding 3) and 4) within the context of covariant approaches to quantum gravity [7, 8].

One possible approach to the quantization of gravity is to adopt the functional integral approach. Here one starts with the classical action, I.

The action I is defined in such a way that extremization of I with respect to the metric tensor g_{ab} on a spacetime region M yields the Einstein equations on M subject to ∂M being fixed. Thus, the metric tensor on ∂M, h_{ab}, is fixed up to coordinate transformations. The action I is thus [9, 10]

$$I = \frac{1}{16\pi G}\int_M (R-2\Lambda)(-g)^{1/2} d^4x \pm \frac{1}{8\pi G}\int_{\partial M} \mathrm{Tr}K\, d\Sigma + C\,. \tag{1.2}$$

K is the trace of the second fundamental form in ∂M. The second fundamental form is defined in terms of the unit normal to ∂M, n_a.

$$n^a n_a = \pm 1 \tag{1.3}$$

$$K_{ab} = \nabla_{(c} n_{d)} h^c_a h^d_b \tag{1.4}$$

$$h_{ab} = g_{ab} \mp n_a n_b \tag{1.5}$$

$$K = h^{ab} K_{ab}\,. \tag{1.6}$$

The $\pm$ signs refer to a spacelike (timelike) ∂M. C is an arbitrary constant (possibly infinite) which is usually adjusted so that the action of flat spacetime I is zero. Extremization of I yields the vacuum Einstein equations

$$R_{ab} = \Lambda g_{ab}\,. \tag{1.7}$$

Although there is virtually no evidence that $\Lambda \neq 0$, we include the cosmological term since it is of importance in spacetime foam (see Section 2).

In the functional approach to any quantum field theory describing a field ϕ one constructs amplitudes $\langle \phi'', t_2 | \phi', t_1 \rangle$ to pass from a given field configuration ϕ' at time t_1 to ϕ'' at time t_2 by writing

$$\langle \phi'', t_2 | \phi', t_1 \rangle = \int_B D[\phi] \exp(-iI[\phi] \tag{1.8}$$

where $D[\phi]$ is some measure on the space of all fields ϕ [11] and $I[\phi]$

is the action for the quantum field ϕ. B is the space of all fields which interpolate between ϕ' and ϕ''.

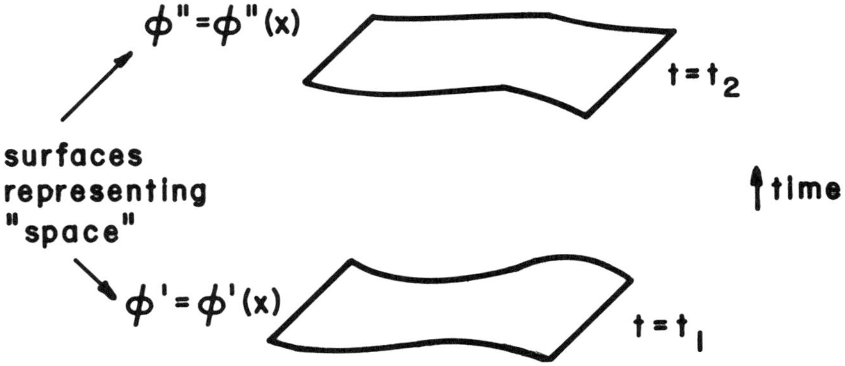

Figure 1

In gravitation, one cannot guarantee that such a split into space and time exists in a unique fashion since the theory itself determines the nature of spacetime. However, most attempts to quantize the theory boldly assert that something like this can be done. Thus, we consider the amplitude

$$\langle h'', t_2 | h', t_1 \rangle = \int_B D[g] \exp(-iI[g]) \tag{1.9}$$

where h is the induced metric on the spacelike 3-surfaces, and g is a 4-metric which interpolates between them. B is now the space of all metrics modulo coordinate transformations. Clearly a prerequisite of this type of approach is that

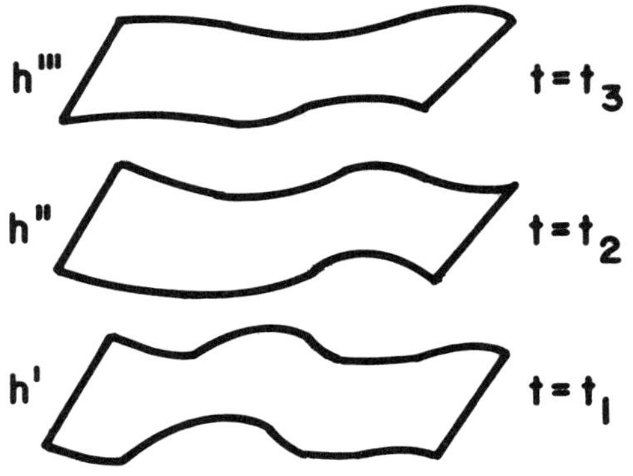

Figure 2

$$<h''', t_3 | h', t_1> = \sum_{h''} <h''', t_3 | h'', t_2> <h'', t_2 | h', t_1> . \tag{1.10}$$

This is essentially a statement of conservation of probability. To be more concrete, if one arrives at the state h''' from h', one must have evolved via some state h''. The action I satisfies this condition; however it should be noted that if we omitted the surface term in the action, Equation 1.10 could not hold [12].

Such integrals, Feynman integrals are not well defined owing to the Lorentzian signature of spacetime. One can overcome this obstacle by an analytic continuation to integrate over spaces of Euclidean signature $(++++)$. One calculates all one's results in "Euclidean space" and then analytically continues results back to a Lorentzian spacetime. This type of construction is useful in certain situations, for example the calculation of thermal partition functions, or the behavior of the vacuum in gravity. One does not expect these techniques to be of much use in the computation of dynamic processes such as the evolution of the universe, or the final stages of black hole evaporation [13]. One may regard this as the analog in gravitation of the Euclidicity postulate for regular field theories [14], or of the construction of the vacuum state in quantum chromodynamics [15].

In Euclidean space then, we wish to consider expressions of the form

$$Z \sim \int_{B'} D[g] \exp(-I[g]) \tag{1.11}$$

where $I[g]$ is the Euclidean gravitational action

$$I[g] = \frac{-1}{16\pi} \int_M (R-2\Lambda)\, d(\mathrm{Vol}) - \frac{1}{8\pi} \int_{\partial M} \mathrm{Tr}K \, d(\mathrm{Surface}) + C . \tag{1.12}$$

B' is the space of all metrics with appropriate boundary conditions (see Section 2) modulo coordinate transformations. To restrict ourselves to integrate over each space once and only once, we will use the Faddeev-Popov prescription [16].

The first unfortunate problem that we encounter is that the Euclidean action $I[g]$ is unbounded below [17]. This could lead to various problems. To see this, consider a conformal transformation of the metric

$$\hat{g}_{ab} = \Omega^2 g_{ab} \tag{1.13}$$

with $\Omega = 1$ on M. This induces a change in the action, taking to

$$I[\hat{g}] = \frac{-1}{16\pi} \int \Omega^2 R + 6\, \nabla_a \Omega \nabla^a \Omega \; d(\mathrm{Vol}) \tag{1.14}$$

$$+ \text{ surface terms.}$$

Clearly, since $\Omega^2 \times \text{const}$ does not bound $|\nabla\Omega|^2$, this is unbounded below.

The only practical way of evaluating (1.12) is by the saddle point method [18]. Here one picks all saddle points of the action, with metric $g^{(i)}$, on a manifold $M^{(i)}$. $g^{(i)}$ are objects which obey the Einstein equations everywhere and are geodesically complete. (These form a pair $(M^{(i)}, g^{(i)})$.) Each pair is known colloquially as a gravitational instanton. One then proceeds to evaluate quantum corrections to these instantons by means of a perturbation expansion $g_{ab} = g^{(i)}_{ab} + \phi_{ab}$ about each instanton.

Thus

$$Z \sim \sum_i \exp(-I[g^{(i)}]) \int^i D[\phi] \exp(-I[\phi]) \,. \tag{1.15}$$

The remainder of this paper concerns itself exclusively with a study of gravitational instantons. Further details of how to deal with the quantum fluctuations can be found in [19, 20].

§2. *Topology*

Our problem now has become one of finding all Einstein 4-manifolds subject to various boundary conditions. The boundary conditions emerge from physical consideration so we will not dwell on the matter here, but summarize those boundary conditions which are interesting.

One interesting class is compact instantons. These have to do with the spacetime foam description of gravitational physics [12, 21-3]. In this class we are interested in Einstein spaces. These will be discussed in detail in Section 4.

The other interesting classes comprise of non-compact instantons. In these classes we are only interested in Ricci-flat spaces. These fall into four distinct classes.

(1) Asymptotically Euclidean (AE)

A space is asymptotically Euclidean if outside a compact set the metric approaches the flat metric on R^4. One may thus regard the boundary of such objects as an S^3 at infinity. Thus, it is possible to write the metric in this asymptotic region as $g = \text{flat} + 0(\rho^{-2})$ where ρ is a circumferential coordinate on S^3. In order to reconstruct physics from such an instanton, one would wish to map this boundary into the boundary of Minkowski space. To see how this is done, introduce a set of coordinates on R^4 which are related to the Cartesian coordinates x, y, z, τ by

$$\begin{aligned} x+iy &= \rho\cos\frac{1}{2}\theta\exp\left(\frac{1}{2}i(\phi-\psi)\right) \\ z+i\tau &= \rho\sin\frac{1}{2}\theta\exp\left(\frac{1}{2}i(\phi+\psi)\right) \end{aligned} \tag{2.1}$$

The flat metric on R^4 is now

$$ds^2 = d\rho^2 + \frac{\rho^2}{4}\left((d\psi+\cos\theta d\phi)^2 + d\theta^2 + \sin^2\theta d\phi^2\right) . \tag{2.2}$$

θ, ϕ and ψ are the Euler angles on S^3 and thus run from 0 to π, 2π and 4π respectively. Alternatively, the metric may be written as

$$ds^2 = d\rho^2 + \frac{\rho^2}{4}(\sigma_1^2+\sigma_2^2+\sigma_3^2) \tag{2.3}$$

where σ_i are a basis of 1-forms on S^3 which are left-invariant under the action of SU(2) on S^3. This space can be analytically continued to Minkowski space by the transformations

$$\begin{aligned} \tau &= it \\ \phi - \psi &= \hat{\phi} - \hat{\psi} \\ \phi + \psi &= i(\hat{\phi} + \hat{\psi}) \end{aligned} \tag{2.4}$$

then

$$\begin{aligned} x + iy &= \hat{\rho} \cos \tfrac{1}{2} \hat{\theta} \exp\left(\tfrac{1}{2} i(\hat{\phi} - \hat{\psi})\right) \\ z + t &= \hat{\rho} \sin \tfrac{1}{2} \hat{\theta} \exp\left(\tfrac{1}{2} (\hat{\phi} + \hat{\psi})\right) \\ z - t &= \hat{\rho} \sin \tfrac{1}{2} \hat{\theta} \exp\left(-\tfrac{1}{2} (\hat{\phi} + \hat{\psi})\right) . \end{aligned} \tag{2.5}$$

The metric on Minkowski space is

$$ds^2 = d\hat{\rho}^2 + \frac{\hat{\rho}^2}{4} (d\hat{\theta}^2 + \cos \hat{\theta}(d\hat{\phi}^2 + d\hat{\psi}^2) - 2d\hat{\phi}\, d\hat{\psi}) . \tag{2.6}$$

Physically useful quantities, for example Green's functions, are computed as functions of the coordinates in the instanton background, $f(\rho, \theta, \phi, \psi; \rho', \theta', \phi', \psi')$. The amplitude in Minkowski space is then obtained by analytic continuation,

$$\begin{aligned} \hat{\theta} &= \theta \\ \hat{\phi} &= \frac{1}{\sqrt{2}} (e^{i\pi/4}\phi - e^{-i\pi/4}\psi) \\ \hat{\psi} &= \frac{1}{\sqrt{2}} (-e^{-i\pi/4}\phi + e^{i\pi/4}\psi) \end{aligned} \tag{2.7}$$

or equivalently

$$\begin{aligned} \theta &= \hat{\theta} \\ \phi &= \frac{1}{\sqrt{2}}\left(e^{-i\pi/4}\hat{\phi} - e^{i\pi/4}\hat{\psi}\right) \\ \psi &= \frac{1}{\sqrt{2}}\left(e^{-i\pi/4}\hat{\psi} - e^{i\pi/4}\hat{\phi}\right) . \end{aligned} \tag{2.8}$$

These coordinates (θ, ϕ, ψ) may be introduced on the sphere at infinity in any AE space, and so this continuation can always be defined.

Physically, the AE spaces correspond to zero temperature, or vacuum instantons.

(2) Asymptotically Locally Euclidean (ALE)

A space is asymptotically locally Euclidean if outside a compact set the metric approaches the flat metric on R^4/Γ. Γ is a discrete subgroup of SO(4) which acts freely on the S^3 at infinity. Thus the boundary of such spaces is S^3/Γ. If Γ is the identity, this class degenerates to AE. As an example, Γ may be Z_2 (as in the Eguchi-Hanson instanton, see Section 3). In this case we can introduce a coordinate system on $S^3/Z_2 = P^3$ at infinity exactly as before. Now however the Euler angles θ, ϕ and ψ are identified with period π, 2π and 2π respectively. It is not presently clear whether ALE spaces are physically interesting. However, the fact that the sphere at infinity has been identified under the action of Γ indicates that any Green's function f will have far greater symmetry properties than usually expected as a direct consequence of (2.7). This leads to problems of physical interpretation [24].

(3) Asymptotically Flat (AF)

A space is asymptotically flat if outside a compact set the metric approaches the flat metric on $R^3 \times S^1$. One may thus regard the boundary at infinity of such objects as $S^2 \times S^1$. Thus, at infinity, the metric can be written as $g = \text{flat} + 0(r^{-1})$ where r is a circumferential coordinate on S^2. Metrically, this means that the line element can be written as

$$ds^2 = dr^2 + r^2 d\Omega^2 + d\psi^2 \tag{2.9}$$

where $d\Omega$ is the solid angle element on S^2, and ψ is a cyclic coordinate on S^1. These spaces correspond physically to finite temperature boundary conditions with

$$\text{Period}(\psi) = (\text{Temperature})^{-1} . \tag{2.10}$$

(4) Asymptotically Locally Flat (ALF)

A space is asymptotically locally flat if outside a compact set the metric approaches the metric

$$ds^2 = dr^2 + r^2(\sigma_1^2 + \sigma_2^2) + \sigma_3^2 \tag{2.11}$$

where σ_i are the left-invariant 1-forms on S^3/Γ where Γ is an isometry of this $U(1) \otimes SU(2)$ invariant metric. Since these configurations are periodic in ψ, they presumably represent finite temperature objects, although the precise physical meaning of ALF boundary conditions is still rather opaque.

One way to classify Gravitational Instantons is to introduce Euler characteristic χ and Hirzebruch signature τ for these manifolds. Let $b_p (p = 0, 1, 2, 3, 4)$ be the pth Betti number of the manifold M with metric g_{ab}. b_p is the number of closed p-surfaces in M that are not boundaries of a (p+1)-surface. Thus, b_p can be thought of as the number of "p-dimensional holes" in M. By Hodge's theorem, b_p is equal to the number of linearly independent L^2 harmonic p-forms on a compact (M, g_{ab}). Since we are working in four dimensions it is useful to decompose b_2 into two pieces b_2^+ and b_2^-.

$$b_2 = b_2^+ + b_2^- \tag{2.12}$$

$b_2^{+(-)}$ is the number of linearly independent L^2 self-dual (anti-self-dual) harmonic 2-forms on (M, g_{ab}). Poincaré duality guarantees further that

$$b_1 = b_3 . \tag{2.13}$$

For compact manifolds without boundary $b_0 = b_4 = 1$. For manifolds with boundary $b_0 = 1$, $b_4 = 0$. If the manifold is simply connected, then $b_1 = 0$. The Euler characteristic is then given by

$$\chi = \sum_{p=0}^{4} (-)^p b_p . \tag{2.14}$$

The Hirzebruch signature is

$$\tau = b_2^+ - b_2^- . \tag{2.15}$$

It is perhaps useful to note that χ and τ classify compact simply-connected orientable 4-manifolds with spin structure up to homotopy [25]. One might think that since some physical fields are spinor fields that it was essential that a manifold have a spin structure. This is not really the case since it is possible to define a spin^c structure on various manifolds. This allows one to have a combination of Yang-Mills fields and spinor fields, even though a spin structure does not exist, [26, 27]. A necessary condition for a compact 4-manifold to have a spin structure is that $\tau = 0$ mod 16.

Now we introduce the curvature tensor $R^a{}_{bcd}$ of the metric g_{ab} on M, and the second fundamental form K_{ab} on ∂M, where the unit outward normal to ∂M is n^a, the induced metric on ∂M is

$$h_{ab} = g_{ab} - n_a n_b \tag{2.16}$$

and then

$$K_{ab} = \nabla_{(c} n_{d)} h_a^c h_b^d . \tag{2.17}$$

We can now use the Chern integral formula [28] for the Euler character

$$\chi = \frac{1}{128\pi^2} \int_M \varepsilon_{abef}\, \varepsilon^{abcd} R^{efgh} R_{cdgh}\, g^{1/2} d^4x$$
$$+ \frac{1}{128\pi^2} \int_{\partial M} R_{abcd} K^{ac} n^b n^d + 64 \det\{K^a_b\} h^{1/2} d^3x \,. \tag{2.18}$$

Also, we can use the Atiyah-Patodi-Singer formula for τ [29]

$$\tau = \frac{1}{96\pi^2} \int_M \varepsilon_{abcd} R^{abef} R^{cd}{}_{ef}\, g^{1/2} d^4x$$
$$- \frac{1}{48\pi^2} \int_{\partial M} R_{abcd}\, \varepsilon^{cdef} n^a n_e K^b_f\, h^{1/2} d^3x \tag{2.19}$$
$$- \eta(0) \,.$$

This is a remarkable formula since it involves the non-local contribution $\eta(0)$. This is derived from the η-function on ∂M. The η-function is defined as

$$\eta(s) = \sum_n |\lambda_n|^{-s} (\text{sign } \lambda_n) \tag{2.20}$$

where λ_n's are the eigenvalues of the graded operator $(-)^p(*d-d*)$ acting on 2p-forms in the boundary, ∂M. ($*$ is the Hodge dual operator.) $\eta(s)$ is guaranteed to be convergent for $\text{Re } s > -\frac{1}{2}$.

The boundary terms that appear in (2.18) and (2.19) have been explicitly calculated for some of the boundary conditions that we have mentioned [30, 31]. For ALE boundaries, it seems that the discrete subgroups of SU(2), namely the cyclic groups Z_k, the binary dihedral groups of order $4k$, D^*_k, the binary tetrahedral group, T^*, the binary octahedral group 0^*, and the binary icosahedral group I^*, can all be associated with self-dual instantons [32, 33]. For ALF spaces, there are instantons with $\Gamma = Z_k$, and it seems likely that $\Gamma = D_k^*$ examples also exist (see Section 3 for further details). These boundary terms are tabulated below.

		Boundary contribution to χ	Boundary contribution to τ
AE		1	0
AF		0	0
ALE	$\Gamma = Z_k$	$1/k$	$\frac{(k-2)(k-1)}{3k}$
	D_k^*	$1/4k$	$\frac{2k^2+1}{6k}$
	T^*	$1/24$	$\frac{49}{36}$
	0^*	$1/48$	$\frac{121}{72}$
	I^*	$1/120$	$\frac{361}{180}$
ALF	$\Gamma = Z_k$	0	$\frac{k}{3}-1$
	D_k^*	0	$\frac{k}{3}$

Since most of the instantons we know about have symmetries, expressed in terms of Killing vectors, it is useful to classify them in terms of the fixed point sets of these Killing vectors [34]. Let k^a be a Killing vector which generates an isometry group G. Then $\mu_t : M \to M$ is the action of the group where t is the group parameter. This is related to k^a by

$$K = k^a \frac{\partial}{\partial x^a} = \frac{\partial}{\partial t} . \tag{2.21}$$

The group G has a fixed point where $K = 0$. At a fixed point, $\mu_t{}^* : T_p(M) \to T_p(M)$ where $T_p(M)$ is the tangent space at p to M, and $\mu_t{}^*$ is generated by the antisymmetric tensor $\nabla_a k_b$. This tensor can have rank two or four. The rank is the codimension of the fixed point set.

If $\nabla_a k_b$ has rank two, the fixed point set is termed a "Bolt." We can express the components of $\mu_t{}^*$ in an orthonormal frame at p as

$$\begin{pmatrix} 1 & 0 & 0 & 0 \\ 0 & 1 & 0 & 0 \\ 0 & 0 & \cos \kappa t & \sin \kappa t \\ 0 & 0 & -\sin \kappa t & \cos \kappa t \end{pmatrix}.$$

κ is the non-trivial skew eigenvalue of $\nabla_a k_b$. κ is sometimes termed the surface gravity of the bolt. The existence of a bolt implies that t must be periodically identified with period $2\pi\kappa^{-1}$. The two dimensional submanifold, or bolt, is usually compact. It thus may be assigned an Euler characteristic, $\hat{\chi}$. For a spherical bolt, as found in many of our subsequent examples, $\hat{\chi} = 2$. However M need not be simply connected so that $\hat{\chi} = 2(1-g)$ where g is the genus of the bolt. M need not be orientable [35], so the bolt could even be a Klein bottle. Also associated with each bolt is a self-intersection number Y.

If $\nabla_a k_b$ has rank four, then the fixed-point set is a point or "nut." The components of $\mu_t{}^*$ can then be written as

$$\begin{pmatrix} \cos \alpha t & -\sin \alpha t & 0 & 0 \\ \sin \alpha t & \cos \alpha t & 0 & 0 \\ 0 & 0 & \cos \beta t & -\sin \beta t \\ 0 & 0 & \sin \beta t & \cos \beta t \end{pmatrix}$$

in an orthonormal frame at p.

α, β are the skew eigenvalues of $\nabla_a k_b$. If $\alpha\beta < 0$ this is sometimes called an antinut as opposed to $\alpha\beta > 0$, a nut. ($\alpha\beta = 0$ is a bolt.) If α/β is rational, then there are a pair of coprime integers (p,q) $\alpha/\beta = p/q$. This is called a nut of type (p,q). If α/β is irrational, then one has a pair of independent isometries generated by a pair of Killing vectors ℓ, m. Linear combinations of this pair lead to regular nuts or bolts.

$\nabla_a k_b$ can be split into its self-dual and anti-self-dual pieces K^+_{ab} and K^-_{ab}. A nut is self-dual if $K^-_{ab} = 0$, then $p = q = 1$. It is anti-self-dual if $K^+_{ab} = 0$ when $p = -q = 1$. If $K^+_{ab} = K^-_{ab}$ at a fixed point, the fixed point is a bolt.

Using the index theorems for isometries we can write

$$\chi = \text{No. of nuts} + \text{No. of antinuts} + \sum_{\text{bolts}} \hat{\chi} \tag{2.22}$$

and

$$\tau = -\sum_{\substack{\text{nuts}\\ \text{antinuts}}} \cot p\theta \cot q\theta + \sum_{\text{bolts}} Y/\sin^2\theta + \eta(0,\theta) \tag{2.23}$$

where $\eta(0,\theta)$ is determined by the nature of the boundary. For simple examples, AE and AF, $\eta(0,\theta) = 0$. For ALE and ALF with $\gamma = Z_k$, $\eta(0,\theta) = k\operatorname{cosec}^2\theta - 1$.

Finally, there is a set of inequalities, called the Hitchin inequalities which are useful for seeing what types of spaces are possible. Hitchin showed that

$$3\chi \geq 2|\tau| + \text{boundary terms} \tag{2.24}$$

for Einstein spaces [36], with equality being attained iff M admits a metric with an (anti) self-dual curvature tensor. We can thus immediately see, for example, that $S^1 \times S^3$ does not admit an Einstein metric since then $\chi = 0$ which implies that $\tau = 0$. Thus the metric must be self-dual if it exists. However, Yau has shown that all such metrics must be isometric to K3, or T^4 or coverings thereof [37]. These identities, in combination with 2.22 and 2.23 are useful in ruling out many generic possibilities.

§3. *Non-compact instantons*

This section lists all presently explicitly known non-compact instantons. We begin with

1) Asymptotically Euclidean

The positive action theorem states that any AE manifold with $R = 0$ has positive action, and that $I = 0$ iff the metric is flat [17, 37]. Suppose that an instanton has action I_0. A constant scale transformation takes the metric g_{ab} into λg_{ab}. Then $I_0 \to \lambda I_0$. However, this will map AE solutions into AE solutions and so the action must be invariant. Thus the action I_0 must be zero for an AE instanton. Since the action is zero only for flat space, the unique AE instanton is flat space with topology R^4.

2) Asymptotically Locally Euclidean

The generalized positive action conjecture states that any ALE manifold with $R = 0$ has positive action, and that $I = 0$ iff the metric has (anti)-self-dual curvature [38]. Hitchin [32, 33] has shown that there exist self-dual ALE instantons associated with the groups $\Gamma = Z_k, D_k^*, T^*, 0^*$, and I^*. However, only the Z_k solutions are known explicitly [39, 40]. One can write the metric as

$$ds^2 = \frac{1}{V}(dt + \vec{\omega}\cdot d\vec{x})^2 + V\, d\vec{x}\cdot d\vec{x} \tag{3.1}$$

and

$$V = 2\lambda \sum_{i=1}^{k} \frac{1}{|\vec{x} - \vec{x}_i|} \tag{3.2}$$

The vectors here are three vectors in flat Euclidean space. $\vec{x}_i$ are arbitrary three vectors. $\vec{x}$ is the position three vector. $\vec{\omega}$ is determined (although not uniquely) by

$$\vec{\nabla} V = \pm \vec{\nabla} \times \vec{\omega} \tag{3.3}$$

the $\pm$ being chosen depending on whether we wish the metric to be self-dual or anti-self-dual. The locations $\vec{x} = \vec{x}_i$ are nuts with respect to the Killing vector $\partial/\partial t$. These are self-dual (or anti-self-dual) nuts. The skew eigenvalues of $\nabla_a K_b$ are $\pm \frac{1}{4\lambda}$. $\vec{\omega}$ has string singularities in it.

These singularities can be eliminated by taking a pair of coordinate patches $\{t_n, \vec{x}\}$ and $\{t_{n+1}, \vec{x}\}$ close to the nuts $\vec{x} = \vec{x}_n$. If we construct ϕ_n, the azimuthal angle defined around the line from $\vec{x}_n$ to $\vec{x}_{n+1}$, and make identifications

$$t_{n+1} = t_n + 4|\lambda|\phi_n \tag{3.4}$$

the string singularities are resolved. This identification is compatible with the resolution of the singularities of the nuts [40]. The result is a non-singular manifold the boundary of which is S^3/Z_k. The Euler character χ is k, and the signature is $\tau = \pm(k-1)$. The action for these solutions is zero.

If we choose $k = 1$, the metric (3.1) turns out to be that of flat space. If we choose $k = 2$, we get the Eguchi-Hanson space [39]. This was first discovered in a different coordinate system, the metric then taking the form

$$ds^2 = \frac{1}{f} dr^2 + \frac{1}{4} r^2(\sigma_1^2 + \sigma_2^2 + f\sigma_3^2) \tag{3.5}$$

where

$$f(r) = 1 - a^4/r^4 . \tag{3.6}$$

r is a radial coordinate and σ_i are a basis of 1-forms on S^3.

$$\begin{aligned} \sigma_1^2 + \sigma_2^2 &= d\theta^2 + \sin^2\theta\, \phi^2 \\ \sigma_3 &= d\psi + \cos\theta d\phi . \end{aligned} \tag{3.7}$$

θ, ϕ and ψ are Euler coordinates on S^3.

In order for $r = a$ not to be a conical singularity, ψ must be identified with period 2π. This displays the ALE character of this metric explicitly, the boundary being P^3. One could, for example, choose as a Killing vector the vector $\partial/\partial\psi$. This has a fixed point at $r = a$, a spherical bolt. Alternatively, we could have chosen $\partial/\partial\phi$, which has 2 fixed points, both self-dual nuts located at the north and south poles of the 2-sphere $r = a$. Of course, the nut and bolt decomposition is not unique, however the results of applying 2.22 and 2.23 are of course unique. The explicit transformation from (3.1) to (3.5) has been given in [41]. It should be mentioned that this metric admits a Kähler structure [42].

3) Asymptotically Flat

The obvious AF instanton is flat space with topology $R^3 \times S^1$. This instanton is of great physical significance since it corresponds to finite temperature field theory in flat space. If we write the metric as

$$ds^2 = dt^2 + dx^2 + dy^2 + dz^2 \tag{3.8}$$

the Killing vector $\partial/\partial t$ has no fixed points, therefore $\chi = \tau = 0$. The action for flat space is zero.

The only other explicitly known AF gravitational instanton is the Kerr instanton [10, 12, 43]. In Boyer-Lindquist coordinates, the metric is given by

$$ds^2 = (r^2-a^2\cos^2\theta)\,\frac{dr^2}{\Delta} + d\theta^2 + \frac{1}{(r^2-a^2\cos^2\theta)} - [\Delta(dt+a\sin^2\theta\, d\phi)^2 + \sin^2\theta((r^2-a^2)d\phi - a\,dt)^2] \tag{3.9}$$

where

$$\Delta = r^2 - 2mr - a^2 . \tag{3.10}$$

The region

$$\infty \geq r \geq M + (a^2 + M^2)^{1/2} \tag{3.11}$$

is the region we are interested in where the metric is positive definite. θ and ϕ are to be taken as spherical polar coordinates. The Killing vector $\partial/\partial t$ has a spherical bolt at $r = M + (a^2+M^2)^{1/2}$. To be free of conical singularities, we must identify

$$(r, t, \theta, \phi) - (r, t, \theta, \phi + 2\pi)$$

and

$$(r, t, \theta, \phi) - (r, t + 2\pi/\kappa, \theta, \phi + 2\pi\,\Omega/\kappa)$$

where

$$\kappa = \frac{(M^2+a^2)^{1/2}}{2M(M + (M^2+a^2)^{1/2})}$$

and

$$\Omega = \frac{1}{2} a \frac{1}{M^2 + M(M^2+a^2)^{1/2}} .$$

Thus, the instanton has topology of $R^2 \times S^2$.

The Kerr solution is a 2-parameter family of solutions, the only restriction on M and a is that $M > 0$. We regard M as, physically, a mass parameter and a as a rotation parameter. If we let $a \to \infty$, the metric becomes flat. If we set $M = 0$, we obtain flat space, but the topology of $R^2 \times S^2$ must then be abandoned.

If we take the $\partial/\partial t$ Killing vector, it has a single bolt of self-intersection number zero, and Euler character 2. Thus using 2.22 and 2.23 we find that $\chi = 2$ and $\tau = 0$. The action of this instanton is $I = \pi M/\kappa$.

There are probably other instantons which consist of a series of Kerr type bolts. We call these the Multi-Kerr solutions. Such things would have $\chi = 2N$ and $\tau = 0$ where N is the number of bolts. For each bolt κ and Ω must be the same. The action would be $I = \frac{1}{2} \pi N M/\kappa$ [44, 45, 46]. It has been shown that there are no multi-instanton solutions with zero rotation [47].

4) ALF Instantons

First, we will deal with $\Gamma = 1$. There is a class of metrics called the Kerr-Taub-NUT metrics [47, 48, 49]. These form a 3-parameter family of Ricci flat metrics. The metric, in Boyer-Lindquist type coordinates is

$$ds^2 = \Xi\left(\frac{dr^2}{\Delta} + d\theta^2\right) + \frac{\sin^2\theta}{\Xi}(adt + P(r)d\phi)^2 + \frac{\Delta}{\Xi}(dt + P(\theta)d\phi)^2 \qquad (3.14)$$

where

$$\begin{aligned} \Delta &= r^2 - 2Mr + N^2 - a^2 \\ \Xi &= r^2 - (N + a\cos\theta)^2 \\ P(r) &= r^2 - a^2 - N^4(N^2-a^2)^{-1} \\ P(\theta) &= 2N\cos\theta - a\sin^2\theta - \frac{aN^2}{N^2-a^2} . \end{aligned} \qquad (3.15)$$

θ and ϕ are polar coordinates on S^2. If $N = 0$, this degenerates to the Kerr metric, which is AF rather than ALF. We will assume that $N \neq 0$. In this metric, there are potential string singularities at $\theta = 0, \pi$. These are avoided if t is identified with period $8\pi|N|$.

$$M + (M^2 - N^2 - a^2) \leq r \leq \infty . \tag{3.16}$$

$\partial/\partial t$ is a Killing vector, and has a fixed point where $\Delta = 0$, at $r = M + (M^2 - N^2 - a^2)^{1/2}$. If $M = |N|$, the metric is self-dual or anti-self-dual. If $a = 0$ also, then the point where $\Delta = 0$, $r = N$ is a self-dual (or anti-self-dual) nut [49]. Then we have the self-dual, or anti-self-dual Taub-NUT instanton. It has $\chi = 1$ and $\tau = 0$, and action $I = 4\pi N^2$. Alternatively $\Delta = 0$ can be a spherical bolt of self-intersection number $Y = -1$. For the bolt to be non-singular

$$\frac{1}{4|N|} = \frac{(M^2 - N^2 + a^2)}{2M^2 - N^2 - N^2(N^2 - a^2)^{-1} + 2M(M^2 - N^2 + a^2)} . \tag{3.17}$$

This condition arises from the requirement that the identification of t at the bolt is consistent with the removal of the string singularities. The metric (3.14) also has curvature singularities. These lie in the hyperbolic region, $r < M + (M^2 - N^2 + a^2)^{1/2}$ provided that $M > |N|$. Thus provided (3.17) is satisfied and $M > |N|$ we have an m-complete space. These spaces have $\chi = 2$, $\tau = 1$, and action $I = 4\pi|N|M$.

For $a = 0$, we find that $M = \frac{5}{4}|N|$, a solution first discovered by Page and christened the Taub-Bolt solution [48]. One can increase a up to $a \simeq 0.693|N|$. There is then a range of a, $0.693|N| < a < (1 + \sqrt{17})|N|/4$ for which there are no instantons. For $a \geq (1 + \sqrt{17})|N|/4$, $M \geq |N|$ there is another family of solutions. As $a \to \infty$, $M \to 2|N|$, and the metric tends to locally flat space. This is summarized diagrammatically in Figure 3.

The only other explicitly known ALF instantons have a boundary with $\Gamma = Z_s$. These are the self-dual multi-Taub-NUT family. These have metric

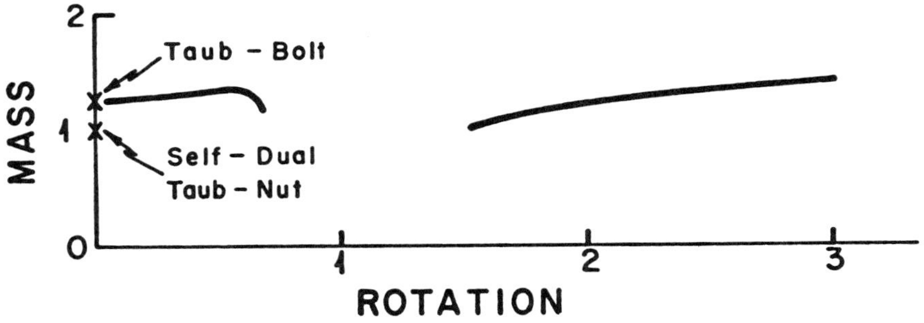

Figure 3. This shows the allowed portions of the M, a, N plane for the Kerr-Taub-NUT metric. The axes are in units of $|N|$.

$$ds^2 = \frac{1}{V}(dt + \vec{\omega}\cdot d\vec{x})^2 + V\, d\vec{x}\cdot d\vec{x} \tag{3.18}$$

where

$$V = 1 + 2N \sum_{i=1}^{s} \frac{1}{|\vec{x} - \vec{x}_i|} . \tag{3.19}$$

This metric is identical to metric (3.1) and (3.2) apart from the constant term in V. Again $\vec{\omega}$ is determined by the relation

$$\vec{\nabla} V = \pm \vec{\nabla} \times \vec{\omega} . \tag{3.19}$$

$\partial/\partial t$ is a Killing vector, and has fixed points at $\vec{x} = \vec{x}_i$. $\vec{\omega}$ has string singularities in it running each of the s nuts to infinity unless t is identified with period $8\pi|N|$. Then, these fixed points are regular (anti)-self-dual nuts. Hence $\chi = s$, $\tau = s-1$ and the action $I = 4\pi s N^2$. If $s = 1$, we have the self-dual version of 3.9.

There seems to exist a series of ALF instantons with $\Gamma = D_s^*$. However, we do not yet know the metrics for these spaces explicitly [33].

§4. *Compact instantons*

We turn now to the question of compact instantons. We regard these a solution of the Einstein equation

$$R_{ab} = \Lambda g_{ab} \, . \tag{4.1}$$

First, we observe that for $\Lambda < 0$, these instantons cannot have any Killing vectors that are globally well defined. This follows from Yano and Nagano's theorem [50]. Let k^a be a Killing vector, and (4.1) be satisfied, then

$$\int k^a \Box k_a + \Lambda k^a k_a \, d(\mathrm{Vol}) = 0 \tag{4.2}$$

is an identity. Then since $\Box$ is a negative semi-definite operator, there can only be Killing vectors if $\Lambda \geq 0$.

The only explicitly known metrics that have Killing vectors are the Einstein metric on S^4 ("de Sitter space"), the Einstein metric on $S^2 \otimes S^2$, the Fubini-Study metric on CP^2 [51-4], and the Page metric on $CP^2 \# \overline{CP}^2$ [55]. The first two examples possess a spin structure, the second two do not. The only non-trivial example is the Page metric. It was found by taking a singular limit of the Kerr-Taub-NUT-de Sitter metric. Its metric is

$$ds^2 = \frac{3}{\Lambda}(1+\alpha^2)\left\{\frac{(1-\alpha^2 x^2)}{3-\alpha^2-\alpha^2(1+\alpha^2)x^2}\,\frac{dx^2}{1-x^2} + \frac{1-\alpha^2 x^2}{3+6\alpha^2-\alpha^4}\,(d\theta^2+\sin^2\theta\, d\phi^2) + \right.$$
$$\left. + \frac{3-\alpha^2-\alpha^2(1+\alpha^2)x^2}{4(3+\alpha^2)^2(1-\alpha^2 x^2)}\,(1-x^2)\,(d\psi+\cos\theta\, d\phi)^2\right\} \tag{4.3}$$

where $-1 \leq x \leq 1$, $0 \leq \theta \leq \pi$, $0 \leq \phi \leq 2\pi$ and $0 \leq \psi \leq 4\pi$, and α is

$$a = -1 - (2+a-b)^{1/2} + (4-a+b+8(a-b)^{-1/2}(a+b)^{-1})^{1/2}$$
$$a = (2^{1/2}+1)^{1/3} \tag{4.4}$$
$$a = (2^{1/2}-1)^{1/3} \, .$$

Hence $\alpha \simeq 0.2817$.

This has action $I \simeq -0.9553 \times 2\pi/\Lambda$, $\chi = 4$, $\tau = 0$. One should contrast this with $S^2 \times S^2$ which has action $I = -2\pi/\Lambda$, and identical χ

and τ. To complete the list, S^4 has $\chi = 2$, $\tau = 0$ and $I = -3\pi/M$, and CP^2 $\chi = 3$, $\tau = \pm 1$ and $I = -9\pi/4\Lambda$. One knows furthermore that S^4 is the space with the most negative action for fixed Λ that is a solution of 4.1.

If $\Lambda = 0$, then only two explicit examples are known. One is the 4-torus, $S^1 \times S^1 \times S^1 \times S^1$ with flat metric. The other is the Einstein-Kähler metric on K3 [56, 7, 8]. This is not known explicitly, although in References [57] and [58] an approximate metric is constructed.

For $\Lambda < 0$, a number of spaces are known to exist although they are not known explicitly. One may, for example, take a space of constant curvature and factor out by some discrete group [59, 60].

One can have spaces of constant holomorphic sectional curvature, formed by taking CP^2, making Λ negative and factoring by a discrete group [60]. One can take direct products of a pair of two dimensional spaces of constant curvature. A two dimensional space is specified by its genus, thus, if these genera are g_1 and g_2, we find a space where $\chi = 4(1-g_1)(1-g_2)$. One can take complex hypersurfaces of degree κ in CP^n, [56] to yield Kähler metrics. If $n = 3$, $\kappa = 4$ we obtain K3.

We now consider some general inequalities [12]. If we define the volume of an instanton as V,

$$V = \int g^{1/2} d^4x \tag{4.5}$$

then the action is

$$I = -\frac{\Lambda V}{8\pi}. \tag{4.6}$$

The Weyl tensor obeys

$$C_{abcd}C^{abcd} \geq \frac{1}{2} \left| C_{abcd}C^{edef} \varepsilon^{ab}{}_{ef} \right| \tag{4.7}$$

with equality iff the metric is conformally self-dual. Examples of such spaces are CP^2 and K3. It should be remembered that if the Riemann tensor is self-dual, the Weyl tensor is self-dual, although not vice versa. Integration of 4.7 yields

$$2\chi - 3|\tau| \geq \Lambda^2 V/6\pi^2 \tag{4.8}$$

again with equality iff the Weyl tensor is self-dual.

If the metric is Kähler, then we can relate χ and τ to the first Chern number c_1, by

$$2\chi + 3\tau = c_1^2 .$$

Thus, if in addition the metric is Einstein then

$$c_1^4 = \frac{V\Lambda^2}{4\pi^4} \tag{4.9}$$

and so

$$\chi \geq \frac{V\Lambda^2}{6\pi^2} \tag{4.10}$$

with equality iff the Weyl tensor is self-dual.

For algebraic subvarieties of CP^n [61], then

$$\chi \leq \frac{\Lambda^2 V}{2\pi^2} \leq 2\chi . \tag{4.11}$$

The lower bound is attained for hypersurfaces generated by curves of degree κ where

$$\begin{aligned} \chi &= \kappa(\kappa^2 - 8\kappa + 22) \\ \tau &= \frac{1}{3}\kappa(\kappa+2)(\kappa-3) . \end{aligned} \tag{4.12}$$

The upper bound is attained for products of pairs of two-dimensional spaces of constant curvature.

JOSEPH HENRY LABORATORIES
PRINCETON UNIVERSITY
PRINCETON, NEW JERSEY 08540

REFERENCES

[1] A. Einstein (1915), "Die Feldgleichungen der Gravitation," Preuss. Akad. Wiss. Berlin, Sitzber, 844-847.

[2] R. Penrose (1965), "Gravitational Collapse and Space-time Singularities," Phys, Rev. Lett. *14,* 57-59.

[3] S.W. Hawking (1967), "The Occurrence of Singularities in Cosmology: Causality and Singularities," Proc. Roy. Soc. A*300,* 187-201.

[4] S.W. Hawking and R. Penrose (1970), "The Singularities of Gravitational Collapse and Cosmology," Proc. Roy. Soc. A*314,* 529-548.

[5] See for example S.W. Hawking and G.F.R. Ellis (1973), "The Large Scale Structure of Spacetime," pp. 189-201, Cambridge University Press, London.

[6] J.H. Taylor and P.M. McCulloch (1980), "Evidence for the Existence of Gravitational Radiation from Measurements of the Binary Pulsar PSR1913+16," Annals of the New York Academy of Sciences, *336,* 442-446.

[7] S. Weinberg (1972), "Gravitation and Cosmology," Wiley, New York.

[8] S.W. Hawking (1975), "Particle Creation by Black Holes," pp. 219-267 in "Quantum Gravity," eds. C.J. Isham, R. Penrose and D.W. Sciama, Oxford University Press, Oxford.

[9] J.W. York (1972), "Role of Conformal Three Geometry in the Dynamics of Gravitation," Phys. Rev. Lett. *28,* 1082-5.

[10] G.W. Gibbons and S.W. Hawking (1977), "Action Integrals and Partition Functions in Quantum Gravity," Phys. Rev. D*15,* 2752-2756.

[11] R.P. Feynman and A.R. Hibbs (1965), "Quantum Mechanics and Path Integrals," McGraw-Hill, New York.

[12] S.W. Hawking (1979), "The Path-Integral Approach to Quantum Gravity," pp. 746-789 in "General Relativity: An Einstein Centenary Survey," eds. S.W. Hawking and W. Israel, Cambridge University Press, London.

[13] D. Gross, M.J. Perry and L. Yaffe (1980), "On the Instability of Quantum Gravity," Princeton preprint.

[14] B. Simon (1979), "Functional Integration and Quantum Physics," Academic Press, New York.

[15] C. Callan, R. Dashen and D. Gross (1978), "Toward a Theory of the Strong Interactions," Phys. Rev. D*17,* 2717-2763.

[16] L. Faddeev and V.N. Popov (1967), "Feynmann Diagrams for the Yang-Mills Field," Phys. Lett. *25B,* 29-30.
R.P. Feynmann (1963), "The Quantum Theory of Gravitation," Acta Physica Polonica *24,* 697-722.
B.S. DeWitt (1964), "The Dynamical Theory of Groups and Fields," Blackie, London.

[17] G.W. Gibbons, S.W. Hawking and M.J. Perry (1978), "Path Integrals and the Indefiniteness of the Gravitational Action," Nucl. Phys. B*138*, 141-150.

[18] See Reference [15] for example.

[19] G.W. Gibbons and M.J. Perry (1978), "Quantizing Gravitational Instantons," Nucl. Phys. B*146*, 90-108.

[20] M.J. Perry (1980), "The Decomposition Theorem," in preparation.

[21] S.W. Hawking (1978), "Spacetime Foam," Nucl. Phys. B*144*, 349-362.

[22] S.W. Hawking, D.N. Page and C.N. Pope (1979), "The Propagation of Particles in Spacetime Foam," Phys. Lett. *86*B, 175-178.

[23] ________ (1979), "Quantum Gravitational Bubbles," University of Cambridge preprint.

[24] M.J. Perry (1980), "ALE Instantons," in preparation.

[25] J.H.C. Whitehead (1949, "On Simply Connected 4-Dimensional Polyhedra," Comm. Math. Helv. *22*, 48-92.

[26] S.W. Hawking and C.N. Pope (1978), "Generalized Spin Structures in Quantum Gravity," Phys. Lett. *73B*, 42-44.

[27] A. Back, P.G.O. Freund and M. Forger (1978), "New Gravitational Instantons and Universal Spin Structures," Phys. Lett. *77B*, 181-184.

[28] S.S. Chern (1945), "On the Curvatura Integra in a Riemannian Manifold," Ann. Math. 46, 674-684.

[29] M.F. Atiyah, V.K. Patodi and I.M. Singer (1975), "Spectral Asymmetry and Riemannian Geometry," I, Proc. Camb. Phil. Soc. *77*, 43-69; II, *ibid. 78*, 405-432; III, *ibid. 79*, 71-99.

[30] H. Romer (1979), "G-Index theorem for natural operators on ALE gravitational Instanton Spaces," pp. 293-300 in "Supergravity," ed. D.Z. Freedman and P. van Nieuwenhuizen.

[31] G.W. Gibbons, C.N. Pope and H. Romer (1979), "Index Theorem Boundary Terms for Gravitational Instantons," Nucl. Phys. B*157*, 377-386.

[32] N. Hitchin (1979), "Polygons and Gravitons," Math. Proc. Camb. Phil. Soc. *85*, 465-476.

[33] ________ (1980), unpublished.

[34] G.W. Gibbons and S.W. Hawking (1979), "Classification of Gravitational Instanton Symmetries," Comm. Math. Phys. *66*, 291-310.

[35] C.J. Isham (1978), "Twisted Quantum Fields in a Curved Space-Time," Proc. Roy. Soc. A*362*, 383-404.

[36] N. Hitchin (1974), "Compact 4-dimensional Einstein Manifolds," J. Diff. Geom. *9*, 435-441.

[37] R.M. Schoen and S.-T. Yau (1979), "Proof of the Positive Action Conjecture in Quantum Relativity," Phys. Rev. Lett. *42,* 547-548.

[38] G.W. Gibbons and C.N. Pope (1979), "The Positive Action Conjecture and Asymptotically Euclidean Metrics in Quantum Gravity," Comm. Math. Phys. *66,* 267-290.

[39] T. Eguchi and A.J. Hanson (1978), "Asymptotically Flat Self-Dual Solutions to Euclidean Quantum Gravity," Phys. Lett. *74B,* 249-251. T. Eguchi and A.J. Hanson (1979), "Self-Dual Solutions to Euclidean Gravity," Ann. Phys. *120,* 82-106.

[40] G.W. Gibbons and S.W. Hawking (1978), "Gravitational Multi-Instantons," Phys. Lett. *78B,* 430-432.

[41] M.K. Prasad (1979), "Equivalence of Eguchi-Hanson Metric to Two Center Gibbons-Hawking Metric," Phys. Lett. *83B,* 310.

[42] E. Calabi (1979), "Métriques Kähleriennes et fibrés Holomorphes," Ann. Scient. Ec. Norm. Sup. 4e series *4,* 269-294.

[43] G.W. Gibbons and M.J. Perry (1978), "Black Holes and Thermal Green Functions," Proc. Roy. Soc. *A358,* 467-494.

[44] M.J. Perry (1980), "The Multi-Kerr Configurations," in preparation.

[45] G. Neugebauer (1980), "A General Integral of the Axially Symmetric Stationary Einstein Equations," J. Phys. A*13,* L19-L21.

[46] D. Kramer and G. Neugebauer (1980), "The Superposition of Two Kerr Solutions," Phys. Lett. *75A,* 259-261.

[47] G.W. Gibbons and M.J. Perry (1980), "New Gravitational Instantons and their Interactions," Phys. Rev. to be published.

[48] D.N. Page (1978), "Taub-NUT Instanton with an Horizon," Phys. Lett. *78B,* 249-251.

[49] S.W. Hawking (1977), "Gravitational Instantons," Phys. Lett. *60A,* 81-3.

[50] K. Yano and T. Nagano (1959), "Einstein Spaces Admitting a One-par-Family of Conformal Transformations," Ann. Math. *69,* 451-461.

[51] T. Eguchi and P.G.O. Freund (1976), "Quantum Gravity and World Topology," Phys. Rev. Lett. *37,* 1251-1254.

[52] G.W. Gibbons and C.N. Pope (1978), "CP2 as a Gravitational Instanton," Comm. Math. Phys. *61,* 239-248.

[53] A. Trautman (1977), "Solutions of the Maxwell and Yang-Mills Equations Associated with Hopf Fibrings," Int. J. Theor. Phys. *16,* 561-565.

[54] S. Kobayashi and K. Nomizu (1963), "Foundations of Differential Geometry," Vol. II, pp. 159. Interscience, New York.

[55] D.N. Page (1978), "A Compact Rotating Gravitational Instanton," Phys. Lett. *79B*, 235-238.

[56] S.-T. Yau (1971), "Calabi's Conjecture and Some New Results in Algebraic Geometry," Proc. Natl. Acad. Sci. *74*, 1798-1799.

[57] See Reference [52].

[58] D.N. Page (1978), "A Physical Picture of the K3 Gravitational Instanton," Phys. Lett. *80B*, 55-57.

[59] J. Wolf (1967), "Spaces of Constant Curvature," McGraw-Hill, New York.

[60] G.W. Gibbons (1979), "Gravitational Instantons." Talk presented at International Conference on Mathematical Physics, Lausanne, August 1979.

[61] N. Hitchin, unpublished.

SOME UNSOLVED PROBLEMS IN CLASSICAL GENERAL RELATIVITY

R. Penrose

Although a great deal of the research into general relativity carried out at present is concerned with its relation to quantum theory, there are many interesting outstanding problems in the original classical theory which have remained unsolved for many years. In view of the important recent work of Schoen and Yau (1979, 1980), which establishes the positive-definiteness of mass in general relativity when measured at spatial infinity, it would seem reasonable to hope that others of these classical problems of general relativity may also be resolved in the not-too-distant future. It should be pointed out, moreover, that general relativity has become, in recent years, an experimentally well-tested theory. So any mathematical results of importance to the classical theory will be assured a permanent place in physics. But the same cannot at all be said of results relating to current work in the quantum theory of gravity, since that remains a highly speculative collection of physical and mathematical ideas which cannot yet be properly said to constitute a theory at all, let alone a well-tested one.

I shall make no attempt here to be at all comprehensive in the classical relativity problems I propose to discuss. They are mainly related to the lines of work that I have myself chosen to pursue. There are undoubtedly many other unresolved problems of equal interest to others.

Seminar on Differential Geometry
0-691-08268-5/82/000631-38$01.90/0 (cloth)
0-691-08296-0/82/000631-38$01.90/0 (paperback)

The conventions adopted here are: the metric tensor g_{ab} (and its inverse g^{ab}) have signature $(+,-,-,-)$; lower case latin letters denote space-time indices; the symbol ∇ (or ∇_a) stands for (Christoffel) covariant derivative; Riemann tensor sign conventions are fixed by $(\nabla_a\nabla_b - \nabla_b\nabla_a)q^d = R_{abc}{}^d q^c$, with $R_{ab} = R_{acb}{}^c$; physical units are taken so that $G = c = 1$.

The result of Schoen and Yau refers to space-times which are appropriately asymptotically flat at spatial infinity. Asymptotically flat space-times are interesting, not because they are thought to be realistic models for the entire universe, but because they describe the gravitational fields of isolated systems, and because it is only with asymptotic flatness that general relativity begins to relate in a clear way to many of the important aspects of the rest of physics, such as energy, momentum, radiation, etc. Asymptotic flatness is thus a highly significant physical idealization, and it is gratifying that certain apparently appropriate conditions for a space-time to be asymptotically flat can be put in an elegant geometrical form, —though exactly *how* appropriate these conditions are is itself an unresolved issue.

In detail, the asymptotic conditions that one chooses to impose may depend on the nature of the problem at hand. For the Schoen-Yau theorem, suitable conditions need to be placed at spatial infinity, and this is adequate for the discussion of total mass-energy of a system, where one is not concerned with questions of radiation; such as mass carried away from a system by gravitational waves. To discuss radiation problems it is more appropriate to impose conditions at *null* infinity.

A glance at Figure 1 should explain why this is so. The intuitive meaning of the mass defined at spatial infinity is that it is the total integrated mass-energy that is intercepted by some spacelike hypersurface approaching a flat spacelike hyperplane at infinity. The mass carried away by outgoing radiation (whether electromagnetic, gravitational, or of any other massless field) travels outwards from the source essentially with the speed of light, and this would always be incorporated in the total

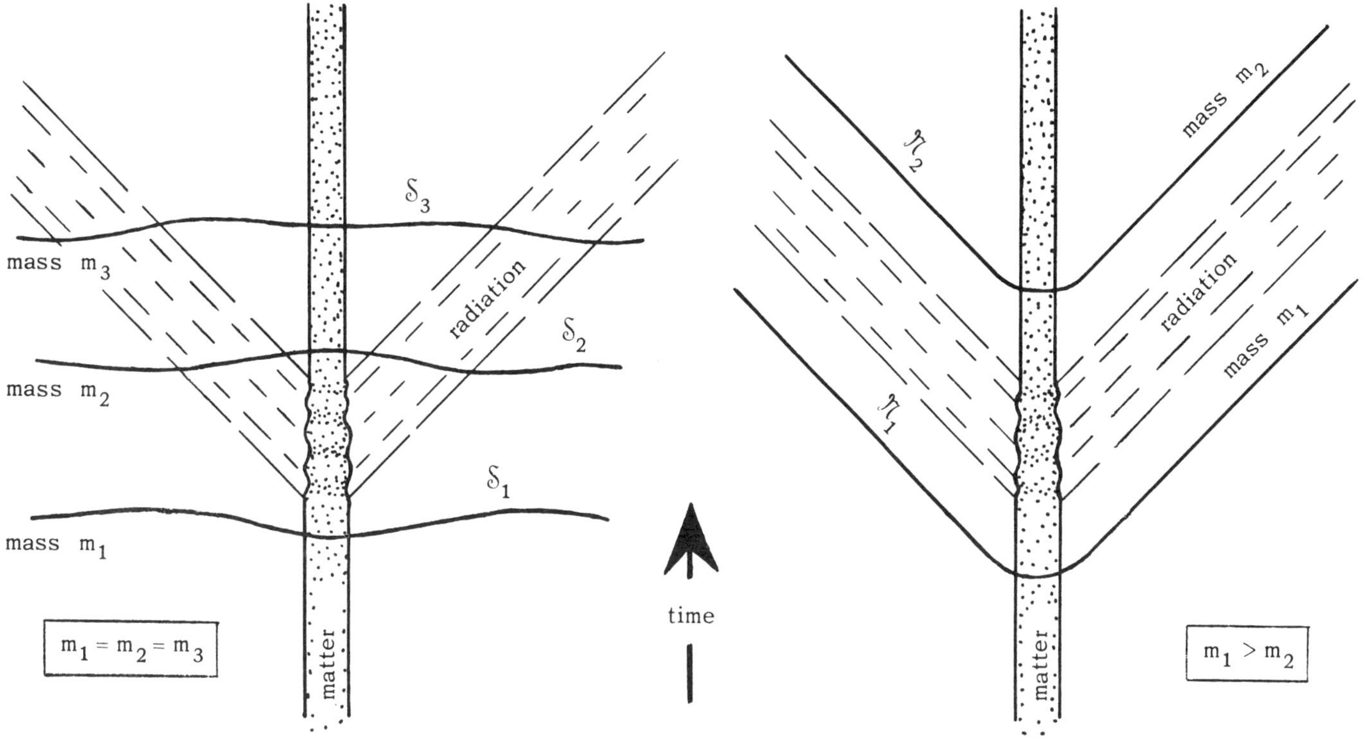

Figure 1

(i) the total mass-energy intercepted by spacelike hypersurfaces $\mathcal{S}_1$, $\mathcal{S}_2$, $\mathcal{S}_3$ remains constant and one cannot ascertain the portion of the mass carried away by radiation.

(ii) by using (asymptotically) *null* hypersurfaces $\mathcal{N}_1$, $\mathcal{N}_2$ we can infer the mass-loss due to radiation.

mass measure no matter how far into the future the spacelike hypersurface is chosen. If the *total* mass-energy is the only quantity that we have, then its measure on each spacelike hypersurface will stay the same and will not enable us to distinguish the part residing in the radiation from that part which remains in the source. However if a *null* or asymptotically null hypersurface is used, which opens out into the future, then radiation can escape between two such hypersurfaces, the difference between the total mass intercepted by the earlier hypersurface and that intercepted by the later one being a measure of the mass carried by the escaping radiation.

It should be pointed out that the reason we are forced into considerations of this kind is that there is no *local* definition of mass-energy in general relativity which takes into account the contribution coming from the gravitational field itself. (That the gravitational field itself must have mass follows from the fact that the potential energy of a pair of gravitating particles depends upon their distance apart, so that *total* energy, and therefore the total mass, of the system particle+particle+field must depend upon this distance and therefore on the gravitational field configuration.) To appreciate the essentially non-local nature of gravitational energy, we recall that in *special* relativity we have, in the flat Minkowski space-time background, a local energy-momentum tensor T_{ab} which is symmetric and which satisfies a vanishing divergence law

$$\nabla^a T_{ab} = 0 . \tag{1}$$

Taking the four components of this equation with respect to a standard constant basis frame, we have four independent equations of the type

$$\nabla^a J_a = 0 , \tag{2}$$

where J_a is each of T_{a0}, T_{a1}, T_{a2}, T_{a3} in turn.

Equation (2) is, of course, simply $dJ^* = 0$ where the 3-form J^* is the Hodge dual of the covector J, so J provides us with a conserved current. In the case of electromagnetic theory we have one such J, providing the familiar charge conservation law: the total electric charge entering a compact space-time 4-volume equals the total charge leaving it—and the total charge intercepting any hypersurface is measured simply by the integral of J^* over it. Here we have four such conservation laws, providing the four components of the conserved energy-momentum 4-vector.

One of the cornerstones of *general* relativity is that the equation (1) should remain true in that theory. But since we do not now have constant basis frames, we cannot obtain a globally conserved energy-momentum simply by integrating components of T_{ab}. From the physical point of view this is reasonable. For in Einstein's equation

$$R_{ab} - \frac{1}{2} R g_{ab} = -8\pi T_{ab} \tag{3}$$

(where, usually, the gravitational constant G is inserted after "8π"; but we have adopted units with $G = 1$), the T_{ab} on the right-hand side satisfies (1) and describes the energy-momentum of all the matter fields (i.e. everything except gravity itself). The contribution to the energy-momentum from the *gravitational* field does not appear in T_{ab}, so it is not to be expected that T_{ab} alone should provide us with a conserved energy-momentum. Instead, the gravitational contribution enters non-locally, showing up in the total energy-momentum as measured at infinity but not showing up in the components of any locally defined tensor quantity.

It may be, however, that some sort of *quasi-local* definition of energy-momentum in general relativity exists, where one does not need to go "all the way to infinity" in order for the concept to be meaningfully defined, but where, for example, a total energy-momentum might be assigned to any 2-surface, $\partial\mathcal{H}$ bounding a compact portion $\mathcal{H}$ of a spacelike hypersurface, the physical interpretation of this energy-momentum being the total energy-momentum (gravitational plus that due to matter fields) intercepted by $\mathcal{H}$. To be useful, such a quasi-local energy-momentum concept should be subject at least to certain inequalities expressing positivity and semi-additivity requirements. Thus, we have, for our first problem:

PROBLEM 1. Find a suitable quasi-local definition of energy-momentum in general relativity.

In special relativity, such a quasi-local definition exists, as we have seen, the gravitational contribution being zero, and the answer being obtained simply by integrating the appropriate components of T_{ab} over $\mathcal{H}$. Another way of thinking of these components is that they are given by currents of the form

$$J_a = T_{ab} k^b$$

where k is a Killing vector generating a translational symmetry of the Minkowski space-time. But J still satisfies (2) when k is any Killing vector whatever, such as that generating a rotational motion of the space-time. In this way we obtain *ten* conserved currents rather than just four, the six new ones providing the components of the total relativistic angular momentum (including those describing the motion of the mass-centre). Thus, there is a more ambitious successor to Problem 1:

PROBLEM 2. Find a suitable quasi-local definition of angular momentum in general relativity.

To expect a positive solution to Problems 1 and 2 may well be felt to be excessively optimistic. But the vagueness of their statements allows for a certain latitude in their interpretations. In fact a suggested approach to these problems was provided some years ago, namely the "linkages" of Tamburino and Winicour (1966, cf. Winicour 1980), but the resulting definitions involve some ambiguities and lack adequate useful inequalities. In the absence of satisfactory solutions to Problems 1 and 2, therefore, we appear to be forced into the asymptotic considerations that we opened with.

One of the important landmarks of general relativity in the post-Einstein period was the work of Bondi (1960, Bondi *et al.* 1962) and its generalization due to Sachs (1962a) which showed that, with a suitable definition at null infinity, the mass carried by gravitational waves is necessarily positive-definite. Thus, in Figure 1 we have, with the Bondi-Sachs definition, $m_1 \geq m_2$ provided that $\mathfrak{N}_2$ lies entirely to the future of $\mathfrak{N}_1$. For this result, it is required that an appropriate condition of approach to flatness at infinity is imposed in future-null directions. In the initial work of Bondi and Sachs this was done in terms of the existence of a certain type of coordinate system in which a certain coordinate u called "retarded time" takes constant values on outgoing null hypersurfaces (this being valid outside a certain spatially compact region—or world-tube—which contains all the matter). Thus $\mathfrak{N}_1$ could be given by $u = u_1$ and $\mathfrak{N}_2$ by $u = u_2$. In fact the coordinate conditions imposed by Bondi and Sachs imply a certain restriction on the relation between the hypersurfaces $\mathfrak{N}_1$ and $\mathfrak{N}_2$, namely that one should be obtained from the other by a so-called "translation" rather than by a more general "supertranslation," but this restriction is actually not necessary for the mass-loss inequality $m_1 \geq m_2$ (Penrose (1964), (1967)). The meaning of the terms "translation" and "supertranslation" will be given shortly. These concepts are best understood in terms of a formalism somewhat different from that of Bondi and Sachs, and this will be described next.

The idea is to borrow from projective geometry or, more appropriately from complex function theory, the concept of "points at infinity." We envisage a conformal factor Ω which rescales the space-time metric g_{ab} to a new one

$$\hat{g}_{ab} = \Omega^2 g_{ab} ,$$

the scalar field Ω approaching zero, at space-time infinity, at such a rate that $\hat{g}_{ab}$ smoothly attains finite values there. The possibility of introducing such a conformal factor is, in effect, equivalent to the Bondi-Sachs asymptotic flatness condition.

The fix idea a little more clearly, let us see how this works in the case of Minkowski space. The standard form of the metric is

$$ds^2 = dt^2 - dx^2 - dy^2 - dz^2$$

but it is convenient first to choose new coordinates u, v, θ, ϕ where

$$u = t - r, \quad v = t + r, \quad \text{with} \quad r = (x^2 + y^2 + z^2)^{1/2},$$

and where

$$z = r \cos\theta, \quad x = r \sin\theta \cos\phi, \quad y = \sin\theta \sin\phi$$

so the metric form becomes

$$ds^2 = dudv - \frac{1}{4}(v-u)^2 (d\theta^2 + \sin^2\theta d\phi^2).$$

The coordinate u is, in fact, an example of a Bondi-Sachs retarded time parameter (and v is a Bondi-Sachs advanced time parameter), so $u =$ const. would provide examples of the outgoing null hypersurfaces referred to earlier. Any suitable conformal factor Ω has to have, as it turns out, an asymptotic behavior that is like the reciprocal of an affine parameter on null geodesics. Thus, along each null line $u, \theta, \phi =$ const. we expect $\Omega \sim v^{-1}$ in future directions, and similarly $\Omega \sim u^{-1}$ in past directions, along $v, \theta, \phi =$ const. This suggests the choice

$$\Omega = \{(1+u^2)(1+v^2)\}^{-1/2}$$

(although there is much arbitrariness in Ω).

The further coordinate change to

$$p = \tan^{-1} v, \quad q = \tan^{-1} u$$

provides us with a regular coordinate system p, q, θ, ϕ "at infinity" (where $\Omega = 0$), the rescaled metric $d\hat{s} = \Omega ds$ being now given by

$$ds^2 = dpdq - \frac{1}{4}\sin^2(p-q)(d\theta^2 + \sin^2\theta d\phi^2).$$

The original Minkowski space corresponds to the range $-\frac{1}{2}\pi < q \le p < \frac{1}{2}\pi$

(where $\Omega > 0$), but we can widen this to include the boundary points, at which $q = -\frac{1}{2}\pi$ or $p = \frac{1}{2}\pi$. The relevant coordinate range is indicated in Figure 2.

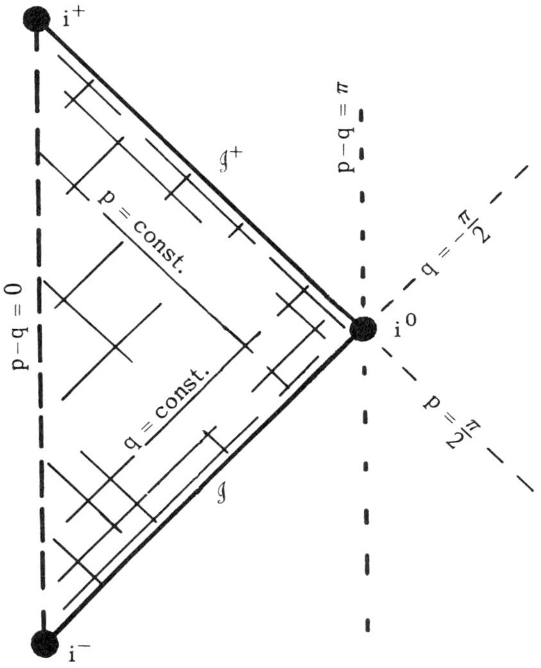

Figure 2

The interior of the triangular region (together with the left-hand symmetry axis) traces out the region of the Einstein universe that is conformal to Minkowski space.

Each point of the triangular (p,q)-region describes a 2-sphere whose radius (in the $d\hat{s}$ metric) is $\frac{1}{2}\sin(p-q)$ —except that each point of the "symmetry axes" $p = q$ and $p - q = \pi$ represents a single point. If we extend the coordinate range of p, q to the whole of that lying between these axes (i.e. to the range $0 \leq p - q \leq \pi$), then we obtain the metric of the "Einstein universe," which is topologically $S^3 \times \mathbf{R}$. (This is a static model which is spatially an exact 3-sphere.) The region lying between the null hypersurfaces $\mathscr{I}^-$ (given by $q = -\frac{1}{2}\pi$, $-\frac{1}{2}\pi < p < \frac{1}{2}\pi$)

and $\mathcal{I}^+$ (given by $p = \frac{1}{2}\pi$, $-\frac{1}{2}\pi < q < \frac{1}{2}\pi$) is the part conformal to Minkowski space, but the ds metric is regular (with $\Omega = 0$) also at all points of $\mathcal{I}^-$ and $\mathcal{I}^+$. We refer to $\mathcal{I}^-$ as past null infinity and $\mathcal{I}^+$ as future null infinity for Minkowski space. Any null geodesic acquires a past end-point on $\mathcal{I}^-$ and a future end-point on $\mathcal{I}^+$. (We recall that the concept of a null geodesic is conformally invariant—it is the same with respect to g_{ab} as it is with respect to $\hat{g}_{ab}$.)

The well-known Schwarzschild and Kerr metrics (cf. Hawking and Ellis (1973) are other examples for which conformally regular boundary hypersurfaces $\mathcal{I}^-$ and $\mathcal{I}^+$ can be introduced. These metrics satisfy Einstein's vacuum equations (i.e. are Ricci-flat), but they possess singularities in the interior regions. ("Singularities" may be recognized in terms of geodesic incompleteness, if desired.) The singularities may be removed in the interior by replacing the metric there by another one in which the vacuum equations are violated (i.e. $R_{ab} \neq 0$), and the resulting Ricci tensor is interpreted, according to Einstein's equations, as referring to a mass-energy (matter) region which acts as a source for the external gravitational field. The space-time with its boundary $\mathcal{I}^- \cup \mathcal{I}^+$ has a structure like that depicted in Figure 3 (with one spatial dimension being suppressed). Again every null geodesic acquires an end-point on each of $\mathcal{I}^-$ and $\mathcal{I}^+$. The manifold-with-boundary $\bar{\mathfrak{M}}$ that we obtain (and whose interior is conformal to the physical space-time $\mathfrak{M}$) is not quite compact. Roughly speaking, it has "three points missing." These would be the points $i^-(p = q = -\frac{1}{2}\pi)$, $i^0(p = -q = \frac{1}{2}\pi)$ and $i^+(p = q = \frac{1}{2}\pi)$ in the Minkowski case (representing past-timelike, spacelike and future-timelike infinities, respectively), but they would generally be singular points for the conformal metric of $\bar{\mathfrak{M}}$, when $\mathfrak{M}$ is asymptotically flat but not flat, and so must be excluded.

This discussion motivates the definition of a space-time $\mathfrak{M}$, with metric $\hat{g}$, as being k-*asymptotically simple* (Penrose 1965b) if there exists a manifold-with-boundary $\bar{\mathfrak{M}}$, with C^k metric g $(k \geq 2)$ and boundary $\mathcal{I}$, such that

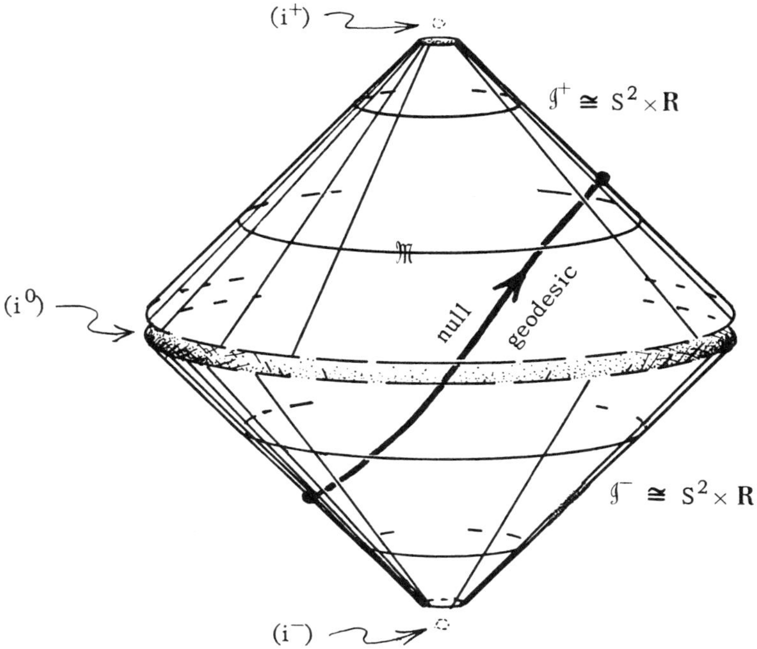

Figure 3

$\bar{\mathcal{M}}$ for asymptotically simple space-time $\mathcal{M}$.

(i) $\bar{\mathcal{M}} - \mathscr{I}$ is conformal to $\mathcal{M}$, with $\hat{g} = \Omega^2 g$,

(ii) Ω is C^k throughout $\bar{\mathcal{M}}$, while $\Omega = 0$ and $d\Omega \neq 0$ on $\mathscr{I}$,

(iii) every null geodesic in $\mathcal{M}$ acquires two end-points on $\mathscr{I}$.

If Einstein's vacuum equations hold for g in the neighborhood of $\mathscr{I}$, then it follows that $\mathscr{I}$ is *null* and consists of two disjoint pieces $\mathscr{I}^-$ and $\mathscr{I}^+$ each having topology $S^2 \times \mathbf{R}$, just as in the Minkowski, Schwarzschild and Kerr cases (Penrose 1965b, Geroch 1971). Furthermore $\mathcal{M} \cong \mathbf{R}^4$ (Geroch 1971).

Asymptotic simplicity has the appearance of being a very reasonable condition to impose on a space-time which describes an isolated system in general relativity and in which all light rays escape to infinity. And one would anticipate that there are many asymptotically flat Ricci-flat space-times representing imploding-exploding gravitational waves (where the intensity and localization is insufficient to produce collapse to a black hole). Yet the following is at present an unsolved problem:

PROBLEM 3. Find an asymptotically simple Ricci-flat space-time which is not flat—or at least prove that such space-times exist.

A related problem is:

PROBLEM 4. Are there restrictions on k for non-stationary k-asymptotically simple space-times, with non-zero mass, which are vacuum near $\mathscr{I}$?

(A stationary space-time is one possessing a timelike Killing vector field; "vacuum near $\mathscr{I}$" means that the g-metric is Ricci-flat in some neighborhood of $\mathscr{I}$ in $\tilde{\mathfrak{M}}$.)

Some work by Schmidt and Stewart (1979) and by Porrill and Stewart (1980) rather suggests that an affirmative answer to Problem 4 may perhaps be the case, with k having a rather small value. If this turns out to be so it would make the asymptotic analysis of isolated gravitating systems much less geometrically pleasant than one had originally hoped. The difficulty seems to be a possible incompatibility between a high degree of differentiability on $\mathscr{I}^-$ and on $\mathscr{I}^+$ whenever the total mass is non-zero. In effect, i^0 is a singular point off which the gravitational (curvature) field can "scatter," possibly leading to a loss of differentiability on $\mathscr{I}^+$ or $\mathscr{I}^-$. However, the calculations which have been carried out so far depend on a linear perturbation analysis and it is not really clear what the rigorous results should actually be.

One consequence of Einstein's vacuum equations holding (for g) near $\mathscr{I}$ is that the generators of $\mathscr{I}$ (i.e. the null geodesic curves lying on $\mathscr{I}^{\pm}$, these curves constituting a fibration of $\mathscr{I}^{\pm}$) are free of shear (Penrose 1965b). This has the effect that any two spacelike cross-sections of $\mathscr{I}^+$ (or $\mathscr{I}^-$) are mapped *conformally* to one another by these generators. Since these cross-sections are all topologically 2-spheres, it follows that the induced $\hat{g}$-metric on each can, with suitable choice of Ω, be chosen to be (minus) that of a unit sphere, so the metric on $\mathcal{I}^+$ (and similarly $\mathcal{I}^-$) becomes

$$-d\hat{s}^2 = d\theta^2 + \sin^2\theta d\phi^2 + 0 \cdot du^2 . \tag{4}$$

The final term is included as a reminder that $\mathcal{I}^+$ is *three*-dimensional, the third coordinate u not actually contributing to the form of the metric. The $d\hat{s}$ metric is in fact degenerate (vanishing determinant) with signature $(-, -, 0)$.

The conformal $d\hat{s}$ metric (4) is considered to be part of the universal intrinsic structure of $\mathcal{I}^+$ (universal, in the sense that *any* space-time which is asymptotically simple and vacuum near $\mathcal{I}$ has a $\mathcal{I}^+$ metric—and similarly a $\mathcal{I}^-$ metric—which is conformal to (4)). In fact there is a somewhat stronger (universal) intrinsic structure (Penrose 1964, 1974) that can be assigned to $\mathcal{I}^\pm$, sometimes referred to as its *strong conformal geometry*. One way of describing this structure is to say that in addition to the conformal $d\hat{s}$ metric (4), the ratio

$$d\hat{s} : du \tag{5}$$

is taken as invariant. Another way to phrase this is to say that a tensor of the form

$$\hat{g}_{\alpha\beta}\, \hat{n}^\gamma\, \hat{n}^\delta \tag{6}$$

is specified on $\mathcal{I}^\pm$, where Greek indices refer to intrinsic directions within $\mathcal{I}^\pm$, $\hat{g}_{\alpha\beta}$ being the metric tensor defined by $d\hat{s}^2$, and $\hat{n}^\gamma$ pointing along the generators of $\mathcal{I}^\pm$ (in the future direction) with $\hat{n} = \partial/\partial u$. We have

$$\hat{n}^\alpha\, \hat{g}_{\alpha\beta} = 0$$

expressing the relation between $\hat{n}$ and the degenerate direction of $\hat{g}_{\alpha\beta}$. The particular strong conformal geometry required for $\mathcal{I}^\pm$ is determined by the condition that, on $\mathcal{I}^\pm$,

$$\mp\, \hat{n}^a = \hat{g}^{ab}\, \hat{\nabla}_b\, \Omega . \tag{7}$$

An important condition on $\mathscr{I}^{\pm}$ is that its generators be *infinitely long* in the sense that the parameter u, as defined above in relation to the structure (5), (6), should cover the entire range $(-\infty, \infty)$ for each θ, ϕ (Geroch and Horowitz 1978). This suggests:

PROBLEM 5. Find conditions on the Ricci tensor R_{ab} throughout $\mathfrak{M}$ which ensure that the generators of $\mathscr{I}$ are infinitely long.

A possible such condition might be the *null convergence condition*

$$R_{ab}\ell^a\ell^b \leq 0 \quad \text{whenever} \quad \ell^a\ell_a = 0 \tag{8}$$

which, with Einstein's equation (3), is the physically reasonable *weak energy condition*

$$T_{ab}\ell^a\ell^b \geq 0 \quad \text{whenever} \quad \ell^a\ell_a = 0 .$$

(Sign conventions have been chosen so that (8) indeed expresses convergence rather than divergence.) With asymptotic simplicity holding, (8) implies that every null geodesic in $\mathfrak{M}$ either possesses a pair of conjugate points or else only just fails to do so in the sense that its end-points on $\mathscr{I}^+$ and $\mathscr{I}^-$ are conjugate. To say that the generators of $\mathscr{I}$ are infinitely long is to say that these generators also only just fail to possess pairs of conjugate points in an appropriate corresponding sense.

The generators of $\mathscr{I}^+$ being assumed infinitely long, we obtain the Bondi-Metzner-Sachs group (BMS group) as the intrinsic symmetry group of either $\mathscr{I}^+$ or $\mathscr{I}^-$, in which the strong conformal geometry (and orientations) are preserved. The BMS group consists of transformations of the form

$$\theta \mapsto \Theta(\theta, \phi)$$
$$\phi \mapsto \Phi(\theta, \phi)$$
$$u \mapsto K(\theta, \phi)\{u + A(\theta, \phi)\}$$

where

$$d\hat{s}^2 \mapsto \{K(\theta, \phi)\}^2 d\hat{s}^2$$

the functions Θ and Φ defining a (non-reflective) conformal motion of S^2 (so the function K is restricted in having a specific form, with three free real parameters) and $A(\theta, \phi)$ being an *arbitrary* (appropriately smooth) real-valued function on S^2.

The subgroup of the BMS group for which the θ, ϕ transformation is the identity on S^2 is the group of *supertranslations* of $\mathscr{I}^{\pm}$. A general supertranslation is given by $\theta \mapsto \theta$, $\phi \mapsto \phi$, $u \mapsto u + A(\theta, \phi)$ where A is arbitrary. The particular cases given when A is composed only of spherical harmonics of zeroth and first order are called *translations*. The reason for this name is that in the case of Minkowski space, any translation in the ordinary sense does in fact induce a translation on $\mathscr{I}^{\pm}$ in the sense just described. In fact, the translation subgroup of the BMS group is uniquely singled out by its group-theoretic properties, as the only 4-parameter normal subgroup of the BMS group (Sachs 1962b).

In the case of Minkowski space, with the specific u-parameter that was introduced earlier, we may identify the restricted* Lorentz group as the subgroup of the (restricted) BMS group $\mathbf{B}$ for which $A(\theta, \phi) \equiv 0$, i.e. for which the cross-section $u = 0$ of $\mathscr{I}^{+}$ is left invariant. In fact $u = 0$ is the future light cone of the (t, x, y, z)-origin O, which is invariant under standard Lorentz Transformations. The restricted Poincaré (i.e. inhomogeneous Lorentz) group is generated by these transformations together with the translations—and so may be identified as a subgroup $\mathbf{P}$ of $\mathbf{B}$. There are however, many subgroups of $\mathbf{B}$ which are isomorphic with the restricted Poincaré group. Specifically, if we take any supertranslation S and conjugate $\mathbf{P}$ with respect to it we get a distinct such subgroup $\mathbf{P}' = S\mathbf{P}S^{-1}$. As far as the group structure of $\mathbf{B}$ is concerned—and indeed, as far as the geometric structure of $\mathscr{I}^{+}$ is concerned, $\mathbf{P}$ and $\mathbf{P}'$ are completely equivalent to one another. Indeed, it is one of the characteristic difficulties of gravitational radiation theory that one must come to terms with this fact. The Poincaré group is the symmetry group of flat space-time,

---

*Here the "restricted ..." is to be interpreted as "connected component of the ...."

so it might have been thought that a suitably *asymptotically* flat space-time should, in some appropriate sense, have the Poincaré group as an asymptotic symmetry group. Instead, it turns out that in general we seem only to obtain the BMS group (which has the unpleasant feature of being an infinite-parameter group) as the asymptotic symmetry group of an asymptotically flat space-time.

The group $\mathbf{P}$, for Minkowski space $\mathfrak{M}$, is the subgroup of $\mathbf{B}$ which leaves invariant not only the strong conformal geometry of $\mathcal{I}^+$ but also a family of cross-sections of $\mathcal{I}^+$ referred to as *good cuts*. The good cuts are characterized by the fact that they are the intersections with $\mathcal{I}^+$ of *future light cones* of points of $\mathfrak{M}$. They are obtained from the particular good cut $u = 0$ by translations, and their equations are given, generally, by setting u equal to a function of θ and ϕ which is composed only of zeroth and first order spherical harmonics. The difficulty, in the case of a general asymptotically simple $\mathfrak{M}$, lies in the fact that there seems to be no suitable family of cross-sections of $\mathcal{I}^+$ that can properly take over the role of the Minkowskian good cuts. This means that although the translation elements of $\mathbf{B}$ are canonically singled out, there is no canonical concept of a "supertranslation-free Lorentz rotation." These Lorentz rotations, in the Minkowski case, are distinguished as BMS transformations leaving some good cut invariant.

The physical concepts of energy and momentum are associated with the translations of Minkowski space, while the concept of angular momentum is associated with Lorentz rotations of Minkowski space. (Recall the discussion of the role of a Killing vector k in the discussion preceding Problem 2.) In accordance with this, it turns out that for an asymptotically flat space-time a good concept of energy-momentum exists at $\mathcal{I}$, namely the Bondi-Sachs definition, whereas the concept of angular momentum remains somewhat problematical.

The main ingredient of the Bondi-Sachs energy-momentum definition is a certain curvature quantity, often denoted Ψ_2, which describes a part of the (conformal) curvature whose fall-off at large distances along a null geodesic is like r^{-3}, where r is an affine parameter on the geodesic. In terms of $\mathcal{I}$, say $\mathcal{I}^+$, we choose a convenient $\hat{g}$ metric for which the metric of $\mathcal{I}^+$ takes the form (4), with u scaled compatibly with the strong conformal geometry, cf. (5), (6), (7). This last may be achieved by taking the physical g-metric in the form

$$\begin{aligned} ds^2 &= 2du dr - r^2(d\theta^2 + \sin^2\theta d\phi^2) \\ &+ O(r^{-2})dr^2 + O(r^{-1})\cdot dr(\times \text{ differentials not involving } dr) \\ &+ O(r)\,(\times \text{ products of differentials not involving } dr) \end{aligned} \tag{9}$$

where, near $\mathcal{I}^+$, the hypersurfaces $u = \text{const.}$ are taken to be null with r as an affine parameter on each null geodesic generator, and where $\Omega = r^{-1}$ near $\mathcal{I}^+$. Coordinates for which the metric takes the form (9) are called *Bondi-Sachs coordinates* (although usually a slightly different condition, irrelevant for our purposes, is placed on the coordinate r).

It is also convenient, near $\mathcal{I}^+$, to choose a *null tetrad* (with respect to $\hat{g}$) of vectors $\hat{\ell}^a$, $\hat{m}^a$, $\hat{\bar{m}}^a$, $\hat{n}^a$ at points of $\bar{\mathfrak{M}}$, and, at points of $\mathfrak{M}$, also the null tetrad (with respect to g) defined by

$$\ell^a = \Omega^2\hat{\ell}^a, \quad m^a = \Omega\hat{m}^a, \quad n^a = \hat{n}^a . \tag{10}$$

(The vectors ℓ^a, n^a are real null; and m^a is complex null—in the sense $g_{ab}m^am^b = 0$ —with complex conjugate $\bar{m}^a$. The normalization conditions are $g_{ab}\ell^an^b = 1 = -g_{ab}m^a\bar{m}^b$, the remaining scalar products all vanishing. The conditions on the "hatted" tetrad are the same, but now with respect to $\hat{g}$.) The vector $\hat{n}^a$ is given by (7) on $\mathcal{I}^+$ and

$$g_{ab}\ell^b = \ell_a = \nabla_a u = \hat{\ell}_a = \hat{g}_{ab}\hat{\ell}^b \tag{11}$$

in $\mathfrak{M}$ (near $\mathcal{I}^+$). With these conventions, we define the complex tetrad components

$$\begin{aligned} \Psi_0 &= C_{abcd}\ell^am^b\ell^cm^d \\ \Psi_1 &= C_{abcd}\ell^am^b\ell^cn^d \\ \Psi_2 &= C_{abcd}\ell^am^b\bar{m}^cn^d \\ \Psi_3 &= C_{abcd}\ell^an^b\bar{m}^cn^d \\ \Psi_4 &= C_{abcd}\bar{m}^an^b\bar{m}^cn^d \end{aligned}$$

of the Weyl curvature tensor, the corresponding definitions for the "hatted" components also holding, whence, by (10)

$$\Psi_i = \Omega^{i-6}\hat{\Psi}_i \qquad (i = 0, \cdots, 4) \ . \tag{12}$$

It can be shown (Penrose 1965b) that for a k-asymptotically simple space-time $(k \geq 3)$ which is vacuum near $\mathcal{I}^+$, all the quantities $\hat{\Psi}_i$ have smoothly vanishing limits at $\mathcal{I}^+$ and so

$$\Psi_i = O(r^{i-5}) \qquad (i = 0, \cdots, 4) \tag{13}$$

which is the "peeling property" of Sachs (1962a, cf. also Newman and Penrose 1962, Penrose 1965b).

We note, from (13), that the quantity Ψ_4 behaves as r^{-1} and so dominates at large distances. It is therefore often referred to as the *gravitational radiation field*. Also the quantity Ψ_2 behaves as r^{-3}, as was alluded to earlier.

We are still not quite in a position to define the Bondi-Sachs mass. For this, we need the further quantity

$$\sigma = m^a m^b \nabla_a \ell_b \tag{14}$$

which defines the (complex) *shear* of the null geodesic generators of each null hypersurface $u = \text{const.}$, i.e. of the system of integral curves of ℓ^a. Under conformal rescaling we have

$$\hat{\sigma} = \Omega^{-2}\sigma \tag{15}$$

which implies that

$$\sigma = O(r^{-2}) \ ,$$

since $\hat{\sigma}$ must be continuous at $\mathcal{I}^+$. A particular cut $u = \text{const.}$ is then defined to be a good cut iff $\hat{\sigma} = 0$ all over it (at $\mathcal{I}$). In Minkowski space, this definition agrees with the one given earlier, but it also allows us to extend the concept of a "good cut" to general asymptotically simple space-times.

In fact, for this purpose, the coordinate system that we have introduced is largely an irrelevance. All we need to do is to set up $\hat{\ell}^a$ and $\hat{m}^a$ at each point of the cut (some general cross-section of $\mathcal{J}^+$), where $\hat{m}^a$ is complex null, with real and imaginary parts both tangent to the cut and where $\hat{\ell}^a$ is real null and orthogonal to $\hat{m}^a$ (but not tangent to $\mathcal{J}^+$). The cut is a good cut iff $\hat{\sigma} = 0$, with $\hat{\sigma} = \hat{m}^a \hat{m}^b \hat{\nabla}_a \hat{\ell}_b$ as in (14). In fact σ is the complex shear, at $\mathcal{J}^+$, of that particular null hypersurface in $\bar{\mathfrak{M}}$ which is generated by the null geodesics (other than those in $\mathcal{J}^+$) meeting the cut orthogonally. This is the unique null hypersurface meeting $\mathcal{J}^+$ in the cut in question. In Minkowski space, this null hypersurface is a null cone iff the cut is a good cut.

Another way to think of $\hat{\sigma}$, at points of some general cut $\mathcal{C}$ of $\mathcal{J}^+$, is as a measure of the trace-free part of the extrinsic curvature of $\mathcal{C}$, as it is imbedded in $\mathcal{J}^+$. The modulus of $\hat{\sigma}$ defines the magnitude of this curvature and the argument of $\hat{\sigma}$, the directions of maximum extrinsic curvature (in relation to $\mathrm{Re}(\hat{m}^a)$ and $\mathrm{Im}(\hat{m}^a)$).

Merely having a *definition* of a good cut for general asymptotically flat space-times does not remove the difficulties of defining "supertranslation-free Lorentz rotations" however. For it turns out that when outgoing gravitational radiation is present ($\hat{\Psi}_4 \neq 0$ on $\mathcal{J}^+$), good cuts do not generally exist at all on $\mathcal{J}^+$. The reason for this, roughly speaking, is that $\hat{\sigma}$, being complex, represents two real numbers per point of the cut, whereas to define a particular cut, we have available only one real number per point of the cut (i.e. the "distance" along each generator of $\mathcal{J}^+$). Furthermore, in any Bondi coordinate system the rate of change of $\hat{\sigma}$ (on $\mathcal{J}$), with respect to u, for the family of cuts u = const. is given by

$$\frac{\partial}{\partial u}\hat{\sigma} = -\bar{N} \tag{16}$$

where N is the Bondi-Sachs (complex) *news function* on $\mathcal{J}^+$, satisfying

$$\frac{\partial}{\partial u} N = \hat{\Psi}_4 \tag{17}$$

showing that the presence of radiation is an obstruction to the Bondi family $u = \text{const.}$ being all good cuts. It may be remarked, also, that N has an additional geometrical interpretation:

$$\bar{N} = -\frac{1}{2}\hat{R}_{ab}\hat{m}^a\hat{m}^b$$

(and, of course, $\hat{R}_{ab}$ need not vanish near $\mathscr{I}^+$ even though R_{ab} does).

The news function N is important also in that it features in the definition of the Bondi-Sachs mass at a cut $\mathcal{C}$ of $\mathscr{I}^+$:

$$m = \frac{1}{4\pi}\int(\hat{\sigma}N - \hat{\Psi}_2)\,d\mathcal{C} \tag{18}$$

(which is always real, even though the integrand is complex), the integral being taken over $\mathcal{C}$, where $d\mathcal{C}$ is the element of surface area

$$d\mathcal{C} = \sin\theta\, d\theta \wedge d\phi\ .$$

The original Bondi-Sachs mass-loss formula (Bondi 1960, Bondi et al. 1962, Sachs 1962) can be stated as

$$\frac{dm}{du} = -\frac{1}{4\pi}\int N\bar{N}\,d\mathcal{C} \leq 0$$

where the cuts are taken as the $u = \text{const.}$ surfaces in a Bondi-Sachs coordinate system. This entails that each cut is obtainable from each other cut by a *translation*—in fact by a translation generated by $\partial/\partial u$. But the more general mass-loss formula

$$m_2 - m_1 = -\frac{1}{4\pi}\int_{\mathcal{C}_1}^{\mathcal{C}_2} N\bar{N}\,d\mathcal{C} \wedge du \leq 0 \tag{19}$$

also holds (cf. Penrose 1964, 1967) where m_1 and m_2 simply refer to two arbitrary cuts $\mathcal{C}_1$ and $\mathcal{C}_2$ of $\mathscr{I}^+$, respectively, where $\mathcal{C}_2$ lies entirely to the future of $\mathcal{C}_1$, the same formula (18) being used in each case. The integral in (19) is taken over the region of $\mathscr{I}^+$ lying between $\mathcal{C}_1$ and $\mathcal{C}_2$, the equation (19) being a consequence of $R_{ab} = 0$ holding in the immediate neighborhood of this region. There is also a generalization of this result when merely the weak energy condition holds in place of Einstein's vacuum equations. An additional term is added to $N\bar{N}$ in the integrand in (19) which is necessarily non-negative whenever (8) holds, so the same inequality applies as before.

It should be clear that the reason we get a mass loss rather than a mass gain is simply that the above discussion has been applied to $\mathscr{I}^+$ rather than $\mathscr{I}^-$. Exactly analogous results to the above would be obtained, as applied now to the Bondi-Sachs mass defined on cuts of $\mathscr{I}^-$, but where now the mass is non-decreasing with time rather than non-increasing. (This assumes that the appropriate regularity and energy conditions are now assumed to hold at $\mathscr{I}^-$ rather than at $\mathscr{I}^+$.) The reason that for physically interesting space-times one is normally concerned with the mass defined at $\mathscr{I}^+$, rather than at $\mathscr{I}^-$, is that physically realistic models would usually involve retarded (i.e. outgoing) radiation but no advanced (i.e. incoming) radiation. The radiation field on $\mathscr{I}^-$ is usually set to zero, in accordance with this, and the Bondi-Sachs mass defined on $\mathscr{I}^-$ is therefore simply a constant. In such circumstances the study of quantities defined at $\mathscr{I}^-$ is less interesting than of the corresponding quantities defined at $\mathscr{I}^+$.

We are now in a position to state several more problems:

PROBLEM 6. Show that if a cut $\mathcal{C}$ of $\mathscr{I}^+$ (or $\mathscr{I}^-$) can be spanned by a spacelike hypersurface along which an appropriate energy condition holds, then the Bondi-Sachs mass defined at $\mathcal{C}$ is non-negative.

The usual energy condition on the energy tensor T_{ab} that is imposed for this sort of problem is the *dominant energy condition* (cf. Synge 1956, Hawking and Ellis 1973):

$$T_{ab}t^a u^b \geq 0 \text{ whenever } t^a \text{ and } u^a \text{ are future-timelike}$$

which is to be taken in conjunction with Einstein's equation (3). An affirmative resolution of Problem 6 would be a null-infinity version of the Schoen-Yau theorem (1979, 1980). However it is as yet not even completely clear that the two different mass concepts involved here are, in an appropriate sense, the same. If $\mathfrak{M}$ admits a spacelike hypersurface which is sufficiently asymptotically flat that a mass can be defined at spacelike infinity according to one of the standard definitions (e.g. Einstein 1918, Arnowitt *et al.* 1962, Landau and Lifshitz 1962) then we ask:

PROBLEM 7. Does the Bondi-Sachs mass defined on cuts of $\mathcal{I}^+$ have a well-defined limit as the cuts recede into the past along $\mathcal{I}^+$, this limit agreeing with the mass defined at spacelike infinity?

The statement of Problem 7 is left deliberately a little vague, it being not quite clear what the appropriate definitions at spatial infinity should be (but cf. Ashtekar and Magnon-Ashtekar 1979). It is necessary, however, that one feature of our mass definitions that I have so far glossed over be now properly taken into account. This is that the mass is not actually a scalar quantity but is one component of an "asymptotic 4-vector," the other three components defining the spatial momentum. The way that the non-scalar nature of the Bondi-Sachs mass, as defined by (18) manifests itself is that there is a certain non-uniqueness in the choice of unit sphere metric that arises in (4), compatible with the given conformal structure. This non-uniqueness shows up in the 3-parameter freedom in the choice of the function $K(\theta, \phi)$ that arises in the BMS transformations. The different choices of $K(\theta, \phi)$ correspond to the different possible choices of "asymptotic time-axis."

If we rescale the unit sphere metric in this way, we obtain a new weighting in the integrand in (18). The 4-dimensional linear space spanned by the factors that so arise provides the asymptotic 4-vector-space for the Bondi-Sachs energy-momentum. (In fact, for each spacelike component, the resulting weighting factor is negative in places and does not itself arise from a rescaling of the sphere metric.) In a convenient orthonormal basis frame for the asymptotic space, the Bondi-Sachs 4-momentum has components

$$p_a = \frac{1}{4\pi}\int W_a(\hat{\sigma}N - \hat{\Psi}_2)\,d\mathcal{C} \tag{20}$$

with

$$W_0 = 1\,,\quad W_1 = \cos\theta\,,\quad W_2 = \sin\theta\sin\phi\,,\quad W_3 = \sin\theta\cos\phi\,.$$

In fact, an affirmative solution to Problem 6 would imply that p_a is future-causal (i.e. future-timelike, future-null or zero) in the sense

$$p_0 \geq 0\,,\quad (p_0)^2 - (p_1)^2 - (p_2)^2 - (p_3)^2 \geq 0\,. \tag{21}$$

But one would anticipate that, except in the cases when $\mathcal{C}$ spans a spacelike hypersurface entirely lying in flat space-time, the *strict* inequality holds in (21) (assuming satisfaction of the dominant energy condition), i.e. that p_a is future-timelike. Indeed, it follows from Bondi-Sachs mass loss that p_a is future-timelike if it is already so for some later cut of $\mathscr{I}^+$. As far as I am aware, the expected relation (21) at spacelike infinity is not known either, with or without strict inequality.

PROBLEM 8. Show that if the dominant energy condition holds, then the Bondi-Sachs energy-momentum, and also the energy-momentum defined at spacelike infinity, are future-timelike, the space-time being assumed not to be flat everywhere in the region of an appropriate spacelike hypersurface.

The various weighting factors that occur in the definition (20) may be thought of as providing the generators of *translations* of the BMS group. There are other ways of expressing the Bondi-Sachs energy-momentum which makes its relationship to such asymptotic Killing vectors more explicit (Tamburino and Winicour 1966, cf. Winicour 1980). In a corresponding way, one may define a concept of angular momentum at cuts of $\mathscr{I}^+$

whenever a generator of a "Lorentz rotation" belonging to the BMS group has been provided. However, this would require a suitable definition of a "good cut," or something similar, in order to distinguish the six "genuine" components of angular momentum from the infinite-dimensional space of "spurious" components that involve parts arising from generators of supertranslations.

The problem is particularly blatant in the case of (asymptotically simple) space-times for which the outgoing radiation tails off suitably in both directions along $\mathscr{I}^+$, so that "near i^0" and "near i^+" good cuts exist, but where the "good cuts near i^0" and the "good cuts near i^+" are non-trivial supertranslations of one another. Let us say that i^0 *and* i^+ *are non-trivially related* in such space-times. It would seem that the very concept of angular momentum gets "shifted" with time whenever i^0 and i^+ are non-trivially related, so it is hard to see in these circumstances how one can rigorously discuss such questions as the angular momentum carried away by gravitational radiation. Moreover, it would seem that i^0 and i^+ are likely to be non-trivially related in the general case:

PROBLEM 9. In an asymptotically simple space-time which is vacuum near $\mathscr{I}$ and for which outgoing radiation is present and falls off suitably near i^0 and i^+, is it necessarily the case that i^0 and i^+ are non-trivially related? (At least, are there some examples in which i^0 and i^+ are non-trivially related?)

It might be that a somewhat different approach to the problem of angular momentum will evade this problem. Thus we have:

PROBLEM 10. Find a good definition of angular momentum for asymptotically simple space-times.

In this connection, it is worth mentioning an idea due to Newman (1976, cf. Hansen et al. 1978), although a discussion of its many intriguing ramifications would take us too far from our essential purposes. The idea is to allow *complex* cuts of $\mathscr{I}^+$ —or, more accurately, cuts of the complexification $\mathbb{C}\mathscr{I}^+$ of $\mathscr{I}^+$. For this purpose one must assume that the space-time $\mathfrak{M}$ has an appropriately analytic metric at $\mathscr{I}^+$ (with analytic conformal factor). Then the quantity $\hat{\sigma}$ has a holomorphic extension to these complex cuts, and we can define a complex good cut to be one for which this holomorphically extended $\hat{\sigma}$ vanishes. It can then be shown, assuming that the metric of $\mathfrak{M}$ is, in an appropriate sense, "close" enough to that of Minkowski space, that there is a four-complex-dimensional family of such complex good cuts, and, rather remarkably, that the four-complex-dimensional space whose points represent these cuts can, in a natural way, be assigned a complex metric which is automatically Ricci-flat with a self-dual conformal curvature (Hansen et al. 1978, cf. also Penrose 1976). In fact, these complex "space-times" provide *generic* such self-dual solutions of Einstein's vacuum equations—and for this, the vacuum equations need not be assumed to hold for the original metric g_{ab}. Newman's idea leads us into the theory of curve-space *twistors* (Penrose 1976, Penrose and Ward 1980) which provides, among other things, methods of constructing self-dual solutions of Einstein's vacuum equations. A major problem of that theory is to find an appropriate way of extending these techniques which will enable the self-dual restriction to be removed, (cf. Penrose 1979b).

Most of these later problems have been concerned only with outgoing gravitational radiation (or, by time-reflection, only with incoming radiation). There are various problems which are concerned with the interrelations between incoming and outgoing radiation. Let me state, as a mathematical problem, just one of these:

PROBLEM 11. If there is no incoming radiation and no outgoing radiation and the space-time $\mathfrak{M}$ is vacuum near $\mathscr{I}$ and (in some suitable sense) near i^0, is $\mathfrak{M}$ necessarily stationary near $\mathscr{I}$?

One would anticipate an affirmitive answer to Problem 11. The result of Papapetrou (1957) which asserts the non-existence of asymptotically flat non-stationary periodic space-times which are Ricci-flat near spatial infinity may be regarded as a step towards a solution to this problem. (A periodic asymptotically simple $\mathfrak{M}$ would necessarily be free of both outgoing and incoming radiation, because of (17) and the mass-loss formula (19).) Problem 11 should be considered in conjunction with Problem 4.

There are many other outstanding problems in general relativity involving gravitational radiation, but often they are hard to state simply as mathematical problems. Most notable among these, perhaps, is to obtain an entirely rigorous derivation of Einstein's formula for the radiation emitted by a system of varying quadrupole moment (cf. Landau and Lifshitz 1962). At one level, the study of the structure of $\mathcal{I}^+$ can be said to give a clear-cut derivation (cf. Newman and Unti 1962, Janis and Newman 1965). But the quadrupole moment is defined in terms of quantities that may not be easy to relate directly to local measurable quantities (e.g. star orbits). In the view of some physicists (cf. Ehlers 1980), there could even be some significant doubt about the validity of the Einstein formula. A more substantial problem even than this, however, is to obtain a general understanding of how gravitational waves scatter one another. But this question is rather open-ended. Another problem is to understand the significance (either mathematical or physical) of a certain curious set of ten exactly conserved asymptotically defined quantities (Newman and Penrose 1968, cf. Exton et al. 1969).

I have been discussing problems, for the most part, which refer to asymptotically simple space-times. Roughly, these are space-times in which every light ray escapes (both in past and future directions) to an asymptotically flat region. There are, however, interesting asymptotically flat space-time models in which light rays can get trapped. Most notably, there are the black hole models and models representing dynamical collapse to a black hole, as in Figure 4. In the collapse models there is an initial Cauchy hypersurface $\mathcal{S}$ along which a non-singular space-time geometry is specified. The space-time near $\mathcal{S}$ is taken to be Ricci-flat outside some compact region of $\mathcal{S}$, and inside this region has a Ricci

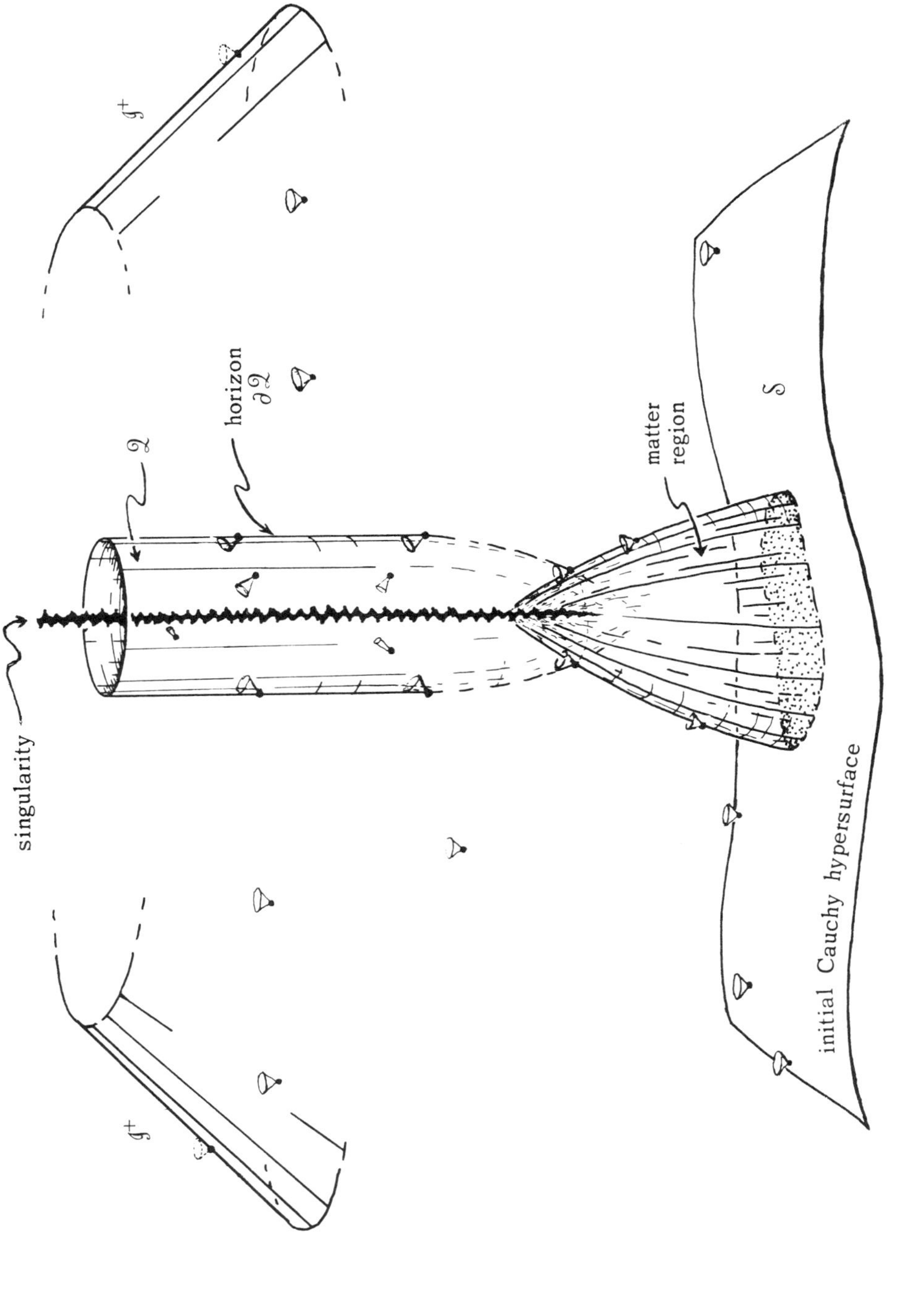

Figure 4. Dynamical collapse to a black hole.

tensor subject to some physically reasonable equation of state (for an appropriate matter distribution). These equations are evolved into the future away from $\mathcal{S}$ and, for certain sets of initial conditions in which the matter region is sufficiently concentrated or falling rapidly enough inwards, one finds that a space-time singularity arises. In this context, by a "space-time singularity arising," we mean that the maximal evolution of the field equations provides a space-time that is not geodesically complete in future-causal (or past-causal) directions—and, indeed, cannot be extended to such a complete space-time even if the field equations are abandoned in the extension. In normal circumstances, the singularity arises because curvature scalars (e.g. $R_{abcd}R^{abcd}$) diverge to infinity at finite affine distance.

It is not hard to produce explicit models of collapse that exhibit this behavior in which spherical symmetry is assumed (cf. Oppenheimer and Snyder 1939). At one time many people believed that the singularity in these models was physically spurious, arising merely because the spherical symmetry implied an exact focussing towards a central point. It was argued that perhaps the introduction of deviations from spherical symmetry would remove this focussing effect, so that infinite curvature would not arise in the central regions. It turned out, however, that this is not the case (Penrose 1965a, Hawking and Penrose 1970, Hawking and Ellis 1973), the formation of such singularities being, in a clear sense, a *stable* property of the initial data sets. Any small-enough (but finite) perturbation of the initial data for spherically symmetrical collapse will also lead to a space-time singularity arising, it being essentially only necessary to assume that the equations of state are such that the *strong energy condition*

$$2\,T_{ab}t^a t^b \geq T^a{}_a t^b t_b \quad \text{whenever} \quad t^a t_a \geq 0$$

always holds, i.e. (with Einstein's equation (3)), we have the *timelike convergence condition*:

$$R_{ab}t^a t^b \leq 0 \quad \text{whenever} \quad t^a t_a \geq 0\,. \tag{22}$$

If it assumed that the hypersurface $\mathcal{S}$ remains a Cauchy surface for the entire space-time $\mathfrak{M}$ no matter how far $\mathfrak{M}$ is extended, then we need only assume null convergence condition (8). A "Cauchy surface" here means a hypersurface with the property that it intersects every maximally extended timelike curve in exactly one point. The hypersurface $\mathcal{S}$, here, will necessarily be a Cauchy surface for the space-time which represents the maximal evolution (cf. Choquet-Bruhat and Geroch 1969) of the initial data on $\mathcal{S}$, but it is conceivable that the space-time could be extended beyond this maximal evolution—to result in a larger space-time for which $\mathcal{S}$ is not a Cauchy hypersurface.

In the spherically symmetrical collapse models there is a certain region $\mathcal{Q}$ composed of those points from which it is not possible to draw a timelike curve of infinite length into the future. Any timelike curve in $\mathcal{Q}$, if extended maximally into the future will encounter the space-time singularity. The boundary $\partial\mathcal{Q}$ of $\mathcal{Q}$ will be referred to here as the *absolute event horizon,* or simply the *horizon,* and the region $\mathcal{Q}$ as a *black hole.** One anticipates that if the initial data is perturbed slightly away from spherical symmetry, then a picture arises, qualitatively similar to the one just described. In particular, one does not anticipate that timelike curves can escape, in the future directions from the close vicinity of the singularity, to reach the exterior infinity. If the initial data is perturbed by a *large* amount away from spherical symmetry, but in such a way that a singularity still arises, then it is not nearly so clear that the picture will remain essentially unchanged. Perhaps the initial data can evolve into a so-called *naked singularity* from which signals (that is, causal curves) can escape in the future direction to external infinity. This possibility is a disturbing one for astrophysicists, since one has no clear way to make predictions once a naked singularity arises. The normal belief seems to be that naked singularities will not in fact arise—at least if one

*Often a slightly different definition of these concepts is given, where future-timelike curves in $\mathcal{Q}$ do not reach $\mathcal{J}^+$ (see Hawking and Ellis 1973).

assumes that the initial setup is, in some appropriate sense "generic," and that the evolution can be treated as taking place according to standard classical physics. If this is the case, then we say that *Cosmic Censorship* holds (cf. Penrose 1969, 1978, Tipler *et al.* 1980). The proving (or disproving) of an appropriate form of Cosmic Censorship in classical general relativity is perhaps the most important unsolved problem in that theory. It is particularly important because, unlike many of the other problems that I have been discussing, it has genuine astrophysical significance and, furthermore, it is by no means absolutely clear on physical grounds that Cosmic Censorship should always hold true. Thus, we would certainly like to know the answer to:

PROBLEM 12. Is Cosmic Censorship a valid principle in classical general relativity?

The statement of this problem has been left deliberately rather vague. In fact, it is a problem in itself to find a satisfactory mathematical formulation of what is physically intended. For myself, I prefer a rather strong "local" version of Cosmic Censorship which makes little reference to infinity (Penrose 1978, 1979a). The question concerns the structure of the singularities that can arise in the evolution of suitably "generic" initial data on a suitable spacelike hypersurface. Suppose the local structure of these singularities is always such that they provide a *future* boundary to the space-time (for the forward-evolved data, that is; for the backward-evolved data we require the singularities to provide a *past* boundary). Then (in the forward region) there would be no timelike curves going into the future from the vicinity of the singularity to reach external infinity, so we could say that a form of Cosmic Censorship holds. In effect, we can phrase this version of Cosmic Censorship roughly in the following way: Let $\mathcal{S}$ be a complete spacelike initial data surface whose metric approaches that of Euclidean 3-space at infinity and on which are specified initial data (satisfying the necessary constraints) for the gravitational field and for appropriate matter fields. Suppose that these data are evolved into the future (and past) to obtain the space-time of maximal evolution. The question is then whether this space-time can be extended further, in a way consistent with the field equations. If—at least for "almost all" initial data sets—it is not possible to extend the maximal evolution in this way, then we say that a form of Cosmic Censorship holds. The connection between this kind of extension and the existence of naked singularities is illustrated in Figure 5.

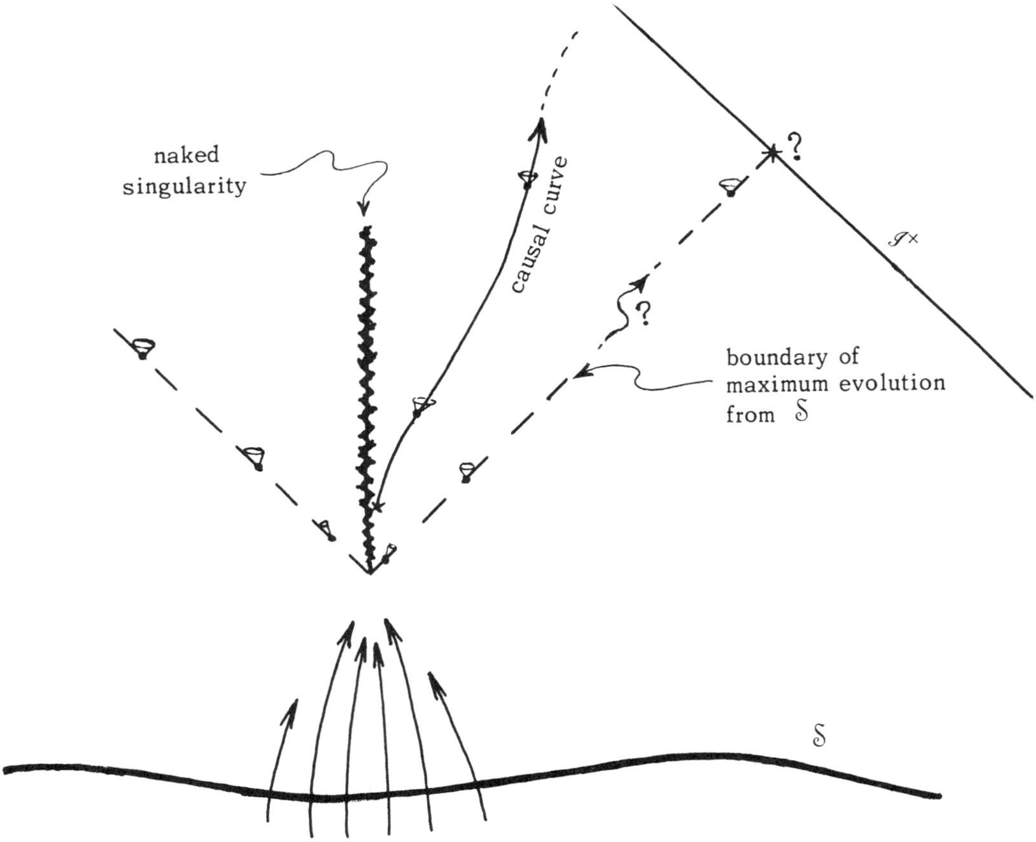

Figure 5. The development of a naked singularity

It may be felt, however, that this version of Cosmic Censorship does not capture all of its physical aspects. One might require, in addition, that provided the initial data tail off appropriately at infinity, the evolution leads to a suitable $\mathcal{I}$ with "infinitely long" generators (cf. Tipler *et al.* 1980). Furthermore, one must be somewhat careful about one's choice of equation of state for the matter region. (For example, it is known that the equations of a pressureless fluid or "dust" will lead to spurious "shell crossing" naked singularities, cf. Yodzis *et al.* 1973.) An important preliminary version of Cosmic Censorship would be the case in which all matter fields are absent, so we are simply exploring the structure of Ricci-flat space-times. The suggestion, then, is that "generic" maximally extended Ricci-flat space-times (which contain asymptotically flat spacelike hypersurfaces) should be *globally hyperbolic* in the sense of Leray (1952). For according to a theorem of Geroch (1967), global hyperbolicity is equivalent to the existence of a Cauchy surface. (See also Penrose 1972.)

There are various curious inequalities that can, in effect, be derived from a suitable assumption of Cosmic Censorship. One of the criteria that can be used for the occurrence of a black hole is the existence of a *trapped surface*, which is a compact spacelike 2-surface $\mathcal{T}$ (normally $\cong S^2$) having the property that both systems of null normals to $\mathcal{T}$ are converging in future directions. (For an ordinary spacelike 2-sphere in Minkowski space, only the ingoing null normals, representing a light flash going inwards from the sphere, are converging. The outgoing null normals represent a light flash coming outwards from the sphere and are diverging.) The essential property of a trapped surface is that the "outgoing" null normals are actually converging. This occurs inside the horizon in the (extended) Schwarzschild space-time and it also occurs inside the horizon of a general black hole (cf. Hawking and Ellis 1973), the singularity in the central region being a consequence of this light-trapping property (and energy conditions, etc.).

Now, with a suitable assumption of Cosmic Censorship, together with various mild other physical assumptions, it is possible to prove a certain sequence of inequalities (Penrose 1973, cf. Hawking 1972). First, the area A of $\mathcal{T}$ will be less than the area of the horizon $\partial\mathcal{Q}$, at the intersection of $\partial\mathcal{Q}$ with the null hypersurface generated by the geodesics tangent to the ingoing null normals at $\mathcal{T}$. Second, the various cross-sections of $\partial\mathcal{Q}$ (which is a null hypersurface, where non-singular) are non-decreasing in area as they proceed into the future. Third, in accordance with the so-called "no-hair" theorem of Israel-Carter-Hawking-Robinson (cf. Carter 1979) one anticipates (problem!) that the black hole will settle down into a Kerr space-time (cf. Hawking and Ellis 1973), with perhaps other regions containing matter and further black holes (moving away). The area of the horizon for a Kerr hole of mass m and angular momentum am is

$$8\pi m(m+(m^2-a^2)^{1/2})$$

which is not greater than

$$16\pi m^2$$

(the result for the Schwarzschild metric). The mass measured at infinity at late times, i.e. the Bondi-Sachs mass (18) measured at cuts of $\mathcal{I}^+$ in the future limit $u \to +\infty$ should, on physical grounds, be not less than m (since any other matter should contribute positively). Finally, the mass measured at any earlier cut should, by the mass-loss formula (19), be not less than the future limiting value. Combining these inequalities together (and redefining m), we anticipate an affirmative answer to:

PROBLEM 13. Let $\mathcal{S}$ be a spacelike hypersurface in $\overline{\mathcal{M}}$ which is compact with boundary, the boundary consisting of a cut $\mathcal{C}$ of $\mathcal{I}^+$ together with a trapped surface $\mathcal{T}$. Let m be the Bondi-Sachs mass evaluated at $\mathcal{C}$ and let A be the area of $\mathcal{T}$. Show that

$$A \leq 16\pi m^2$$

provided that the dominant energy condition holds throughout some neighborhood of $\mathcal{S}$.

It should be remarked that when black holes are present, the space-time $\mathcal{M}$ will not actually be asymptotically simple, since condition (3) of asymptotic simplicity will fail (e.g. for the generators of $\partial\mathcal{Q}$). However this should cause no difficulty, since $\mathcal{I}^+$ can still exist for various weakened versions of asymptotic simplicity (e.g. Penrose 1974, Geroch and Horowitz 1978). All that is required is that conditions (i) and (ii) should hold near $\mathcal{I}^+$, with $\mathcal{C} \cong S^2$. Furthermore, the surface $\mathcal{T}$ need not be assumed to be connected in Problem 13; the effects of the various separate black holes combine together appropriately (cf. Hawking 1972). It may be pointed out, also, that an affirmative answer to Problem 13 will have the consequence that the apparently stronger inequality

$$A \leq 16\pi(m^2 - (p_1)^2 - (p_2)^2 - (p_3)^2)$$

must also hold, where p_a are as in (20), because the inequality in Problem 13 must hold for all allowed scalings for the metric on $\mathcal{I}^+$. Note that Problem 13 generalizes Problem 6. There is also a version of Problem 13 which applies at spacelike infinity (cf. Problem 7) and this has been considered by Schoen and Yau.

Problem 13 may be considered as a sort of consistency check on the various physical assumptions involved. The most doubtful of these is presumably Cosmic Censorship, so inequalities of this nature have been regarded as giving some sort of test of that hypothesis. However, they provide probably a rather weak test, as it seems to me that the Cosmic Censorship hypothesis has been appealed to in only a rather peripheral way. It would be nice to understand more about this, however.

In particular classes of models it is sometimes possible to check the inequality in Problem 13 directly (cf. Penrose 1973, Gibbons 1972, Jang and Wald 1977). Indeed, Gibbons has shown that under certain circumstances, the inequality to be proved follows from Minkowski's inequality for convex surfaces in Euclidean 3-space. Another closely related special case, for a positive function f defined on the (θ,ϕ)-sphere, is

$$4\pi \int f^2 d\mathcal{C} \leq \left\{ \int f(1 - \nabla^2 \log f) d\mathcal{C} \right\}^2 , \tag{23}$$

where $d\mathcal{C} = \sin\theta d\theta \wedge d\phi$ as before, the integrals being taken over the sphere, with Laplacian ∇^2, and where

$$\nabla^2 \log f < 1$$

(cf. Penrose 1973). Gibbons showed that with suitable convexity assumptions, (23) follows from Minkowski's inequality. Perhaps (23) is a well-known result?

There are many other problems involving black holes which have not yet been solved. I shall not dwell on these, but mention one:

PROBLEM 14. Show that there is no vacuum equilibrium configuration involving more than one black hole.

To negate Problem 14, one would require a stationary solution of the vacuum equations that is complete up to its horizon $\partial\mathcal{Q}$, this horizon consisting of two or more disconnected pieces. Work by Gibbons, D.C. Robinson, Lindenblom and others has established Problem 14 affirmatively in special cases, but the general result remains unknown. If instead of the vacuum equations we allow a matter density given by a Maxwell field (Einstein-Maxwell equations) then we obtain an analogue of Problem 14 which is not true if we allow certain "degenerate" charged black holes for which electrostatic repulsion just balances the gravitational attraction (Israel and Wilson 1972, Hartle and Hawking 1972). It is not known whether there is a purely gravitational analogue of this, where spin takes over the role of electric charge. This seems rather unlikely. But generally results for the Einstein-Maxwell equations follow closely those for the pure Einstein vacuum equations. Thus Einstein-Maxwell analogues exist for the various other problems that I have stated only in the Einstein vacuum case, namely Problems 3, 4, 9 and 11.

I am grateful to the Institute for Advanced Study, Princeton, for their hospitality and to the National Science Foundation for support under Contract MCS 79-12938.

REFERENCES

Arnowitt, R., Deser, S. and Misner, C. W., (1962), in *Gravitation, and Introduction to Current Research* (ed. L. Witten, Wiley, New York).

Ashtekar, A. and Magnon-Ashtekar, A., (1979), *Phys. Rev. Lett.* 43, 181.

Bondi, H. (1960), *Nature*, Lond. 186, 535.

Bondi, H., van der Burg, M.G.J. and Metzner, A.W.K. (1962), Proc. Roy. Soc. (Lond.) A269, 21.

Carter, B. (1979), in *General Relativity, an Einstein Centenary Survey* (eds. S.W. Hawking and W. Israel, Camb. Univ. Press).

Choquet-Bruhat, Y. and Geroch, R. (1969), Commun. Math. Phys. 14, 329.

Ehlers, J. (1980), *Ann. N.Y. Acad. Sci.*

Einstein, A. (1918), *S. B. preuss Akad. Wiss.* 448.

Exton, A. R., Newman, E. T. and Penrose, R. (1969), *J. Math. Phys.* 10, 1566.

Geroch, R. (1967), *J. Math. Phys.* 8, 782.

__________, (1971), in *General Relativity and Cosmology*, International School of Phys. "Enrico Fermi" 47 (Ed. R. K. Sachs, Acad. Press).

Geroch, R. and Horowitz, G. (1978), *Phys. Rev. Lett.* 40, 203; 40, 483.

Gibbons, G.W. (1972), *Commun. Math. Phys.* 27, 87.

__________, (1974), *Commun. Math. Phys.* 35, 13.

Hartle and Hawking (1972), *Commun. Math. Phys.* 26, 87.

Hansen, R. O., Newman, E. T., Penrose, R. and Tod, K. P. (1978), *Proc. Roy. Soc.* (Lond.) A363, 445.

Hawking, S.W. (1972), *Commun. Math. Phys.* 25, 152.

Hawking, S.W. and Ellis, G.F.R. (1973), *The Large Scale Structure of Space-Time* (Camb. Univ. Press).

Hawking, S.W. and Penrose, R. (1970), *Proc. Roy. Soc. Lond.* A314, 529.

Israel, W. and Wilson, G. A. (1972), *J. Math. Phys.* 13, 865.

Jang, P. S. and Wald, R. (1977), *J. Math. Phys.* 18, *41*.

Janis, A. I. and Newman, E. (1965), *J. Math. Phys.* 6, 902.

Landau and Lifshitz (1962), *The Classical Theory of Fields* (Trans. M. Hamermesh, Pergamon, Oxford).

Leray, J. (1952), *Hyperbolic Differential Equations* (Princeton, I.A.S.).

Newman, E. T. (1976), *Gen. Rel. Grav.* 7, 107.

Newman, E. T. and Unti, T.W.J. (1962), *J. Math. Phys.* 3, 891.

Newman, E. T. and Penrose, R. (1962), *J. Math. Phys.* 3, 566; 4, 998.

———, (1968), *Proc. Roy. Soc.* **A305**, 175.

Oppenheimer, J. R. and Snyder, H. (1939), *Phys. Rev.* **56**, 455.

Papapetrou, A. (1957), *Ann. Phys.* **20**, 399.

Penrose, R. (1964), in *Relativity Groups and Topology: the 1963 Les Houches lectures* (eds. C. M. deWitt and B. S. deWitt, Gordon and Breach, New York).

———, (1965a), *Phys. Rev. Lett.* **14**, 57.

———, (1965b), *Proc. Roy. Soc.* **A284**, 159.

———, (1967), in *Relativity Theory and Astrophysics I* (ed. J. Ehlers, Amer. Math. Soc.).

———, (1969), *Rivista del Nuovo Cim. Numero Speciale* **1** (ser. 1), 252.

———, (1972), *Techniques of Differential Topology in Relativity* (S.I.A.M., Philadelphia).

———, (1973), *Ann. N.Y. Acad.* **224**, 125.

———, (1974), in *Group theory in non-linear problems* (ed. A. O. Barut, Reidel, Dordrecht).

———, (1976), *Gen. Rel. Grav.* 7, 31.

———, (1978), in *Theoretical Principles in Astrophysics and Relativity* (eds. N. R. Lebovitz, W. H. Reid and P. O. Vandervoort, Univ. of Chicago Press).

———, (1979a), in *General Relativity, and Einstein Centenary Survey* (eds. S. W. Hawking and W. Israel, Camb. Univ. Press).

———, (1979b), in *Advances in Twistor Theory* (eds. L. P. Hughston and R. S. Ward), Pitman, San Francisco.

Penrose, R. and Ward, R. S. (1980), in *General Relativity and Gravitation*, Vol. 2 (ed. A. Held, Plenum, New York).

Porrill, J. and Stewart, J. M. (1980), *Proc. Roy. Soc.* (Lond.) **A**

Sachs, R. K. (1962a), *Proc. Roy. Soc.* (Lond.) **A270**, 103.

———, (1962b), *Phys. Rev.* **128**, 2851.

Schmidt, B. G. and Stewart, J. M. (1979), *Proc. Roy. Soc.* (Lond.) **A367**, 503.

Schoen, R. and Yau, S.-T. (1979), *Commun. Math. Phys.* **65**, 45.

________, (1981), *Commun. Math. Phys.* **79**, 231-260.

Synge, J. L. (1956), *Relativity, the Special Theory* (North-Holland, Amsterdam).

Tamburino, L. and Winicour, J. (1966), *Phys. Rev.* **150**, 1039.

Tipler, F. J., Clarke, C.J.S. and Ellis, G.F.R. (1980), in *General Relativity and Gravitation Vol. 2* (ed. A. Held, Plenum, New York).

Winicour, J. (1980), in *General Relativity and Gravitation, Vol. 2* (ed. A. Held, Plenum, New York).

Yodzis, P., Seifert, H.-J. and Muller zum Hagen, H. (1973), *Commun. Math. Phys.* **34**, 135.

Added in proof:

A new line of argument, due to E. Witten, has provided a greatly simplified proof of the Schoen-Yau theorem. It seems that Problems 6 and 8 are also likely to succumb to this line of argument.

PROBLEM SECTION

Shing-Tung Yau

During the last part of the geometry year in the Institute, there were a lot of requests for a lecture on open problems. After discussions with the following mathematicians: Bourguignon, Calabi, Cheng, Kazdan, Li, Schoen, Simon, Treibergs and Uhlenbeck, I made up a list of sixty problems and gave two lectures on it. Later I was encouraged to enlarge the problem set somewhat.

I should emphasize that the problem set is by no means exhaustive. The choice of the topics depends a lot on the author's personal taste. Apart from a few exceptions, I do not attempt to mention problems in the other subjects which are closely related to differential geometry or can probably be solved by geometric means.

The difficulty of the problems ranges from "elementary" to "deep." Whereas "deep" problems may be solved by a beginning student within a few months and elementary problems can be open for a long time, I do hope that this problem set will provide a condensed overview of the subject for beginning students.

Most, if not all, of the problems are well known. If a problem has an explicit reference or the problem is explicitly suggested by a single person, it will be mentioned. Otherwise the reader can assume that the problem is well known.

Seminar on Differential Geometry
0-691-08268-5/82/000669-38$01.90/0 (cloth)
0-691-08296-0/82/000669-38$01.90/0 (paperback)

Finally, I would like to thank all the mathematicians who read the original draft and suggested corrections, further discussion and references, including F. J. Almgren, Jr., M. Berger, A. Borel, E. Calabi, J. Cheeger, M. Gromov, and H. B. Lawson. I am also grateful to James Mackraz for his valuable help in organizing the problems and compiling the bibliography, and to Neola Crimmins and Kathy Lunetta for a wonderful job of typing this work.

I. Curvature and the Topology of Manifolds

A. Sectional curvature

1. (The Hopf Conjecture.) Does $S^2 \times S^2$ admit a metric with positive sectional curvature?

The only progress on this problem is due to Bourguignon and the others [BDS], improving a result of Berger [Br1]. They proved that in a neighborhood of the product metric of $S^2 \times S^2$ there is no metric with positive curvature.

In general, one does not know any example of a compact, simply-connected manifold of nonnegative sectional curvature which does not admit a metric of strictly positive curvature. It would be nice to know whether a compact simply-connected symmetric space of rank > 1 admits a metric with positive curvature or not. Eventually, one should be able to classify four-dimensional manifolds of positive curvature. (At this time, only S^4 and $\mathbb{C}P^2$ are known examples.)

2. Is there any metric with positive curvature on exotic spheres? Gromoll and Meyer [GM1] found a metric of nonnegative curvature on a Milnor seven-sphere which was of strictly positive curvature outside a set of large codimension. In [H1], N. Hitchin also proved that spheres which are "very exotic" cannot even admit metrics with positive scalar curvature.

3. Let M be an N-dimensional manifold with nonnegative sectional curvature. Is it true that the ith Betti number of M is not greater than the ith Betti number of T^N, the N-torus?

Recently, Gromov, in an as yet unpublished work, proved that there is an upper bound of the ith Betti member depending only on i and N. Hence if M is the connected sum of many copies of CP^2, M does not admit a metric with nonnegative sectional curvature.

4. Let M be a compact manifold with positive curvature. Does M admit a smooth, effective S^1 action?

This question is motivated by the fact that all known examples of manifolds with positive curvature have a lot of symmetry.

5. Is there an example of a compact, simply-connected manifold with nonnegative Ricci curvature that does not admit a metric with nonnegative sectional curvature?

Most likely, the answer is "yes" and one might try the connected sum of N copies of CP^2.

6. Do all vector bundles over a manifold with positive curvature admit a complete metric with nonnegative sectional curvature?

This is an attempt to understand the converse of the theorem of Cheeger and Gromoll [CG2] that asserts that every complete, nonnegatively curved manifold is diffeomorphic to a vector bundle over a totally geodesic, compact, nonnegatively curved manifold. There are works done by J. Nash in [Na] where he considered the analogous situation for Ricci curvature. There is also work in this direction by A. Rigas [Ri].

7. (Chern). Let M be a compact, positively curved manifold. Is it true that every abelian subgroup of $\pi_1(M)$ is cyclic?

This was proposed by S. S. Chern in the Kyoto Conference in Differential Geometry. He based his conjecture on the theorem of Preissmann [P] and the solution of the space-form problem (see Wolf [W]) that the conclusion is true if the curvature is either negative or equal to a positive constant. It is possible that for a nonnegatively curved compact manifold, the rank of an abelian subgroup of $\pi_1(M)$ is dominated by the rank of the curvature tensor of the manifold if we define the latter suitably. Recently G. Carlsson was able to prove that if an abelian group acts freely on the product of k copies of the sphere, then the rank of the abelian group is not greater than k.

8. For an even-dimensional compact manifold with positive sectional curvature, prove that the Euler number is positive.

Affecting one approach is an example of a six-dimensional open manifold with a (noncomplete) metric of positive curvature and a negative Gauss-Bonnet-Chern integrand provided by Geroch [G]. For the details of the conjecture, see Chern [Ch1].

9. Characterize the groups which can appear as the fundamental group of some compact manifold with negative curvature.

By the Cartan-Hadamard theorem one knows that the manifold is a $K(\pi, 1)$ which gives certain conditions on the group, e.g. the group must be torsion free. The Preissmann theorem [P] asserts that every abelian subgroup must be cyclic. Milnor [M1] showed that it must have exponential growth.

In fact, using a result of Margulis [Ma], one can show that the number of conjugate classes of a cyclic subgroup grows at least exponentially. Eberlein [E] also showed that the group contains a nontrivial free subgroup. If the manifold is also Kähler, it is not known if a finite cover of such a manifold has nonzero first Betti number. The method of Margulis should give more fruitful results.

If the manifold is an irreducible locally symmetric manifold of dimension greater than two, a theorem of A. Borel [Bor] (which is also a consequence of the later strong rigidity theorem of Mostow) tells us that the outer automorphism group of the fundamental group is finite. It is not known that the same statement is true for general manifolds with negative curvature. Note that the example of Mostow and Siu [MS] is a negatively curved manifold which is not homotopic to any locally symmetric space. Millson [Miℓ] and Vinberg [Vi] have constructed examples of hyperbolic manifolds with nonzero first Betti number.

10. As a continuation of the previous problem, let M^{2N} be a compact manifold with negative curvature. Is it true that $(-1)^N \chi(M) > 0$?

This is part of the Hopf conjecture, and is known for $N = 2$ (see Chern [Ch2]). Singer has proposed to settle this problem by looking at the universal cover of M. He points out that if the L^2 harmonic forms on the universal cover are all zero except in the middle dimension, then one can apply the index theorem for coverings (see Atiyah [At1]) to prove the statement in the affirmative.

11. The Cohn-Vossen inequality says that the total curvature of a complete surface is dominated by its Euler number. Finn [Fi] and Huber [Hu] studied the difference of these numbers in terms of geometric quantities. The question is how to generalize this inequality to higher dimensions. If M is a complete manifold with finite volume and bounded curvature, when is the Euler number equal to the Gauss-Bonnet-Chern integral? If the manifold is locally symmetric, this is true and due to Harder [Ha]. It can be proved that the assertion is true if the curvature of M is bounded between two negative constants. In a private conversation, Gromov claimed that the Gauss-Bonnet-Chern integral is an integer and is the Euler number if one merely assumes that the curvature is nonpositive and the metric is real analytic.

If M is complete and has nonnegative curvature, Poor [Po], R. Walter [Wa] and Greene and Wu (Theorem 9 of [GWu]) proved that the Cohn-Vossen inequality holds true if $\dim M = 4$. What is the geometric constraint on M if the equality holds? What happens when $\dim M = 2n$ with $n > 2$?

12. Let M_1 and M_2 each have negative curvature. If $\pi_1(M_1) = \pi_1(M_2)$, prove that M_1 is diffeomorphic to M_2.

There is some progress due to Cheeger, Gromov [Gr1], Farrell and Hsiang [FH]. Cheeger proved that $\pi_1(M)$ determines the second Stiefel bundle of M. Gromov proved that $\pi_1(M)$ determines the unit tangent bundle of M. Farrell-Hsiang proved $\pi_1(M)$ determines $M \times R^3$. Farrel-Hsiang have only to assume that one of the manifolds has negative curvature.

13. Let M_1 and M_2 be compact Einstein manifolds with negative curvature. Suppose $\pi_1(M_1) \cong \pi_1(M_2)$ and $\dim M_1 \geq 3$. Is M_1 isometric to M_2?

If the manifolds are both locally symmetric this is Mostow's rigidity theorem.

14. Let M be a compact manifold of dimension N. Can one find a positive constant δ_N depending only on N so that M is diffeomorphic to a manifold with constant negative curvature whenever the curvature of M lies between $-1-\delta_N$ and -1 ?

Let M be a compact manifold whose sectional curvature lies between -4 and -1. If M is not diffeomorphic to a manifold with constant negative curvature, is M locally symmetric?

15. Develop a useful notion of curvature for p.l. manifolds so that one could obtain appropriate pinching theorems and formulas for characteristic classes.

One would like, for example, some kind of p.l. approximations for positively curved manifolds which have positive curvature in this sense.

Of the progress in developing analogues for geometric quantities for p.l. manifolds are the works of Banchoff on a p.l. Gauss-Bonnet formula, Regge's proposal for scalar curvature, and Cheeger's studies of several curvature invariants (see Cheeger [C3], [C4]).

16. What can one say about the Pontrjagin classes and the Stiefel-Whitney classes of a compact manifold with negative curvature? For example, is it true that a finite covering of such a manifold is spin?

17. Prove that the Stiefel-Whitney numbers of a flat manifold are zero.

Not much progress has been made on this well-known question. See Auslander and Szczarba [ASz].

18. Given M a complete, noncompact manifold with sectional curvature $K \geq 0$; if, at some point x, $K(x) > 0$, prove that M is diffeomorphic to R^N.

This conjecture appears in [CG2]. Further, can a metric with curvature positive everywhere except perhaps zero on a set of low dimension be deformed to have positive curvature everywhere? (See Gromoll-Meyer [GM1].)

19. Let Ω be a closed 4k-form defined on a compact manifold M which represents some Pontryagin class of M. Can one find a metric on

M so that Ω is represented by the curvature form according to Chern [Ch3]? If there are other topological obstructions, what are they? It is rather clear that such obstructions should exist. The problem is to find a sufficient condition.

One can ask the same question for Chern classes of Kähler manifolds. For the first Chern class, this was conjectured by Calabi and solved in [Y].

A resolution of this question will give a deep understanding of the curvature tensor.

B. Ricci curvature

20. Find necessary and sufficient conditions on a symmetric tensor T_{ij} on a compact manifold so that one can find a metric g_{ij} to satisfy

$$R_{ij} - R/2\, g_{ij} = T_{ij}$$

whence R_{ij} is the Ricci tensor and R is the scalar curvature of g_{ij}. If g_{ij} is the Lorentz metric on a four-dimensional manifold, this is simply the Einstein field equation. If M has boundary, what are suitable boundary conditions to impose?

21. Let M be a complete manifold with positive Ricci curvature. Can M be deformed to a compact manifold with boundary?

22. Characterize the fundamental group of a complete manifold with positive Ricci curvature.

If the manifold is compact, the splitting theorem of Cheeger-Gromoll [CG1] provides a rather satisfactory answer.

The case of noncompact manifolds is more complicated. Recently, P. Nabonnand [Nab], under the direction of Bérard-Bergery, provided an example of a noncompact, complete manifold with positive Ricci curvature whose fundamental group is infinite cyclic. This example has dimension ≥ 4.

In the case of three dimensions, Schoen and Yau [SY1] proved that such manifolds are diffeomorphic to $\mathbb{R}^3$. It may be possible that the fundamental group of the manifold is always a finite extension of a polycyclic group.

23. Construct an explicit metric with zero Ricci curvature on the K-3 surface. The existence of such a metric was proved in Yau [Y1]. Is there any four-dimensional manifold with zero Ricci curvature which is not covered by a torus or a K-3 surface? A simpler unsolved question is whether such a manifold can be diffeomorphic to S^4 or $S^2 \times S^2$.

24. Can every manifold with dimension ≥ 3 admit a metric with negative Ricci curvature?

It is very hard to suggest what the right answer to this question is. One does not know if S^3 or T^3 admit such a metric. However, there are many examples of simple-connected Kähler manifolds with negative Ricci curvature [Y1].

Perhaps a compact manifold with nonpositive Ricci curvature does not admit an SU(2) action. This conjecture is partly motivated by Bochner's theorem that SU(2) cannot act isometrically and partly motivated by the theorem of Lawson and Yau [LY1] stating that a manifold with effective SU(2) action admits a metric of positive scalar curvature.

If this last conjecture is true, then one can prove that the complex structures over $S^2 \times S^2$ are given by the standard Hirzebruch surfaces.

25. Classify four-dimensional compact Einstein manifolds with negative Ricci curvature. Can S^4 admit such a metric?

The Thorpe-Hitchin inequality [H2] gives some relation on the Euler number and the index of these manifolds.

26. Find for each N constants $c_N < C_N$ so that if the Ricci curvature of a compact manifold satisfies $c_N \delta_{ij} \leq R_{ij} \leq C_N \delta_{ij}$ then the manifold admits an Einstein metric.

C. Scalar curvature

27. Classify complete, three-dimensional manifolds with nonnegative scalar curvature.

This is considerably interesting for general relativity because the "universe" tends to have such metrics. In fact, under physically reasonable assumptions, Schoen and Yau [SY2] did prove that such metrics always exist on the universe.

Schoen-Yau [SY3] also proved that the fundamental group of such a manifold does not contain a subgroup which is isomorphic to the fundamental group of a compact surface of genus ≥ 1. In the case of a compact manifold, this was proven in [SY4]. For dimension exceeding three, the problem was considered in Schoen-Yau [SY5] and Gromov-Lawson [GL].

28. Classify all compact, four-dimensional, Einstein manifolds with positive scalar curvature.

29. Prove that a compact manifold with nonnegative scalar curvature is a $K(\pi, 1)$ if and only if it is flat.

30. Prove that a compact, simply-connected, three-dimensional manifold with positive scalar curvature is homeomorphic to the sphere. It is proved in Meeks-Simon-Yau [MSY] that the connected sum of two fake three-spheres does not admit a metric with positive Ricci curvature.

31. Classify compact hypersurfaces in R^{N+1} which have constant scalar curvature. Are they isometric to S^N ? If they are convex, then the answer is yes and was proved by Cheng-Yau [CY1].

32. (Yamabe). Prove that any metric on a compact manifold can be conformally deformed to a metric of constant scalar curvature. Yamabe published a proof [Yam], but N. Trudinger [Tr] found a gap in the work after Yamabe's death. Nonetheless, Yamabe's original proof can be pushed to cover a large class of metrics, as was made clear by Trudinger (see also Eliasson [Eℓ]).

Pushing further, Aubin [Au] solved the problem for an even broader class, in dimension $s \geq 6$.

However, even for surfaces of genus zero it is nontrivial to find a proof without use of a complex analysis.

One can formulate a similar conjecture in the class of complete noncompact manifolds. Progress has been made recently by W.M. Ni [Ni].

II. Curvature and Complex Structure

33. Let M be a compact Kähler manifold with nonnegative bisectional curvature. Prove that M is biholomorphic to a locally symmetric Kähler manifold, at least when the Ricci curvature is positive.

If the bisectional curvature is strictly positive, the manifold is in fact biholomorphic to CP^N, as was conjectured by Frankel and proved in Mori [Mo] and Siu-Yau [SiY1].

34. Let M be a complete, noncompact, Kähler manifold with positive bisectional curvature. Prove that M is biholomorphic to C^N.

It is not even known if this manifold is Stein. If the sectional curvature is positive, then M is Stein as was observed by Greene and Wu [GWu]. For geometric conditions which guarantee manifolds to be C^N, see Siu-Yau [SiY2].

35. Let M be a complete, simply-connected, Kähler manifold with negative bisectional curvature. Prove that M is Stein.

It is not even known that M must be noncompact. What are the examples of compact surfaces with negative tangent bundle? Are they non-simply-connected? B. Wong observed that one can reduce the higher dimensional problem to the surfaces.

36. If M is complete, Kähler, of finite volume, and has bounded curvature, is M a Zasiski open set of some projective manifold? If M has negative bisectional curvature, does M have a finite automorphism group?

Recently, Siu and Yau [SiY3] proved that if the sectional curvature is bounded between two negative constants, then the first question is affirmative.

For the second question see [LY2], [Ko].

37. Let M be a compact Kähler manifold with negative sectional curvature. Prove that if $\dim_{C} M > 1$, then M is rigid, i.e., there is only one complex structure over M.

When M is covered by the complex two-dimensional ball, this was proved by Yau [Y1] using the Kähler-Einstein metric and Mostow's theorem. Under the constraint that M is "strongly negative," Siu [Si] proved the most general form of the statement.

38. Given M a simply-connected, complete, Kähler manifold with sectional curvature less than or equal to -1, prove there exists a bounded holomorphic function on M.

One would like to even prove that there is a branched immersion of M onto a bounded domain in $\mathbb{C}^N$.

39. Let M be a compact Kähler manifold with positive first Chern class. Suppose M admits no holomorphic vector field. Prove that M admits a Kähler-Einstein metric. This was conjectured by Calabi [Cal].

40. Let M be a complete Kähler manifold with zero Ricci curvature. Prove that M is a Zariski open set of some compact Kähler manifold. If this is true, we shall have algebraic means to classify these manifolds.

41. Classify all compact, two-dimensional Kähler surfaces with zero scalar curvature. (See [Y2].)

42. Let M be a compact simply-connected symplectic manifold. Does M admit a Kähler structure? M. Berger says that Serre indicated a counter-example to him in 1955, where $\pi_1(M) \neq 0$. See [Bs].

For any symplectic structure over a manifold, one can define an almost-complex structure. Conversely, it may be true that for any almost-complex structure, one can also find an associated symplectic structure. Is it true that the almost-complex structure determines the symplectic structure up to conjugation by a diffeomorphism? One does not know the answer of this last question even for $\mathbb{C}P^N$. However, Moser [Mos] has proved that all elements of a one-parameter family of symplectic structures are mutually conjugate by diffeomorphisms.

43. Let Ω be a bounded, pseudoconvex domain in $\mathbb{C}^N$. Cheng and Yau [CY2] have established the existence of a canonical Kähler-Einstein metric on Ω. Under general conditions, e.g. $\partial\Omega \in C^2$, that metric is complete. Is the metric always complete?*

44. Describe the Kähler-Einstein metric constructed by Cheng-Yau [CY2] on the Teichmüller space. What is its relation to the Bergmann metric? In general, if a domain is not biholomorphic to a product domain

*Moh-Yau have recently shown that a bounded domain admits a complete Kähler-Einstein metric iff the domain is pseudoconvex. However, it is still desirable to study the boundary behavior of this metric.

and covers a compact Kähler manifold, is the Kähler-Einstein metric equal to the Bergmann metric?

45. Let M be a compact Kähler-Einstein manifold of complex dimension N, with negative scalar curvature. Yau [Y1] proved then that $(-1)^N \frac{2(N+1)}{N} C_1^{N-2} C_2 \geq (-1)^N C_1^N$. Are there any other inequalities of this sort among the Chern numbers of M ? When $N = 4$, Bourguignon asked whether or not $C_4(M)$ is positive.

46. (Calabi). Let u be a real-valued function defined on C^N so that $\det \frac{\partial^2 u}{\partial z_i \partial \bar{z}_j} = 1$ and $\sum_{i,j} \frac{\partial^2 u}{\partial z_i \partial \bar{z}_j} dz^i dz^j$ defines a complex metric. Prove that this metric is flat (see [CA2]).

The difficulty lies in the fact that the automorphism group of C^N is very large.

47. Let M be a compact Kähler manifold with positive holomorphic sectional curvature or positive Ricci curvature. Prove that M is rationally connected, i.e. any two points of M can be joined by a chain of rational curves.

48. Let M be a compact Kähler manifold with negative sectional curvature. Prove that M is covered by a bounded domain of C^N. One might prove a weaker assertion that the universal cover of M has an abundacy of bounded holomorphic functions. (See the example of Mostow-Siu [MS].)

49. Let M_t be a holomorphic family of Kähler manifolds. Let ds_t^2 be the canonical Kähler-Einstein metric on M_t. What is the behavior of ds_t^2 where the family M_t degenerates?

III. Submanifolds

50. Prove that a compact surface in R^3 is rigid, i.e. one cannot find a continuous family of surfaces in R^3 which are isometric to each other and are not obtained from each other by a rigid motion.

This is a very long-standing problem. If we consider polyhedral surfaces, there is a counterexample due to R. Connelly [Co]. Sullivan asked if the (signed) volume enclosed by the surfaces is invariant under isometric deformation.

For the smooth case, Cohn-Vossen proved the rigidity for convex surfaces. In an attempt to generalize Cohn-Vossen's result, L. Nirenberg [Nir] studied the surfaces with $\int K^+ = 4\pi$. He generalized Cohn-Vossen's result assuming the nonexistence of more than one closed asymptotic line. The real analytic case was first established by A.D. Alexandrov [Aℓ].

51. Let M be the space of immersions of a fixed compact surface into R^3. Prove that the subspace of M which consists of infinitesimally rigid immersions is "generic" in M. How can we describe its complement? Study the same problem in the category of surfaces of rotation.

52. The Nash embedding theorem insures that every manifold can be isometrically embedded into some Euclidean space, but it does not give us geometric properties of the embedding. For example, one hopes to show that a complete manifold with bounded Ricci curvature and positive injectivity radius can be embedded with bounded mean curvature in a higher dimensional Euclidean space.

53. Can one generalize Weyl's embedding problem to higher dimensions?

This would be to prove that a compact, N-dimensional manifold with positive sectional curvature can be isometrically immersed into the Euclidean space of $\frac{N(N+1)}{2}$ dimensions.

One difficulty comes from the lack of understanding of the nonuniqueness of the immersion. P. Griffiths has recently obtained some new insight into this problem.

54. Given a smooth metric in a neighborhood of a point p in a 2-dimensional manifold, can one find a neighborhood of p that embeds isometrically into R^3?

The cases where the metric is either C^ω or of strictly positive or negative curvature are well known. See Pogorelov [Pg] for a possible counterexample in the smooth category.

55. Suppose one defines an isometric embedding of a manifold into R^N to be elliptic if the second fundamental form corresponding to each normal has at least two nonzero eigenvalues of the same sign (see Tanaka [Ta]). Suppose, then, that we have two elliptic isometric embeddings of a fixed compact manifold. Are they congruent to each other? What is the correct generalization of the Cohn-Vossen rigidity theorem to higher dimensions? If M is a complete immersed surface in R^3 with finite area and if K is bounded and non-positive, then is M rigid?

56. The famous Efimov theorem [Ef] states that no complete surface with curvature ≤ -1 exists in R^3. One may ask whether or not a complete hypersurface with Ricci curvature less than -1 can exist in R^N. This was asked in [Y3] and [R].

One may also try to generalize Hilbert's theorem and ask if the hyperbolic space form of dimension N can be isometrically embedded in R^{2N-1}.

Another problem is the nature of the singularities of a surface with $K = -1$ in R^3. (See Hopf [Ho].) Can one give a good definition of weak solution for the $K = -1$ embedding equations, analogous to minimal currents for the zero mean curvature equation? Possibly it would be useful to consider objects in the frame bundle.

57. Find nontrivial sufficient conditions for a complete, negatively curved surface to embed isometrically in R^3. Such a condition might be a rate of decay of the curvature. Related to this is the Dirichlet problem for prescribed Gauss curvature.

58. Recall that a Weingarten surface is a surface where the mean curvature H and the Gaussian curvature K satisfy a suitable functional relation of the form $\phi(K, H) = 0$ where ϕ is a nonsingular function defined on the plane. It would be interesting to know if the ellipsoid of rotation is characterized among compact surfaces by $\lambda_1 = c\lambda_2^3$ where λ_i are the principal curvatures and c is a constant. In general, Voss was able to establish that a compact real analytic Weingarten surface of genus zero is a surface of revolution (see Hopf [Ho]). What are the compact, real

analytic Weingarten surfaces of higher genus? Must they have genus = 1 and be either a tube surface or a surface of revolution?

Hopf proved that for a closed real analytic Weingarten surface of genus zero, the number $\frac{dk_2}{dk_1}$ (where k_1 and k_2 are the principle curvatures) must take the following discrete values at the umbilical point: $0, -1, (2k+1)^{\pm 1}$ for $k \geq 1$ and ∞. Is the same statement valid for compact smooth Weingarten surfaces of genus zero?

Another problem for surfaces in R^3 is to give an intrinsic characterization of compact surfaces defined by a real algebraic polynomial. How does one express the degree of the polynomial in terms of the invariants of the metric?

59. Let h be a real-valued function on R^3. Find (reasonable) conditions on h to insure that one can find a closed surface with prescribed genus in R^3 whose mean curvature (or curvature) is given by h.

F. Almgren made the following comments:

For "suitable" h one can obtain a compact smooth submanifold ∂A in R^3 having mean curvature h by maximizing over bounded open sets $A \subseteq R^3$ the quantity

$$F(A) = \int_A h\, d\mathcal{L}^3 - \text{Area}(\partial A) .$$

A function h would be suitable, for example, in case it were continuous, bounded, and $\mathcal{L}^3$ summable, and $\sup F > 0$. However, the relation between h and the genus of the resulting extreme ∂A is not clear.

In fact, the problem in this context is a special case of a variety of minimal partitioning problems. One can see [Alm2] for this type of problem, and there is one of interest in the work of Sir W. Thomson (Lord Kelvin) [Th]. With a suitable restriction on h, Bakel'man [Ba] and Treibergs-Wei [TW] have found solutions of this problem for the closed surface of genus zero.

60. (Willmore [Wi]). Let M be a compact, two-dimensional torus embedded in R^3. Let H be its mean curvature. Is it true that $\int_M H^2 \geq 2\pi^2$ with equality implying that M is obtained from the circular torus by a Möbius transformation? Recently, Li-Yau [LY2] defined the concept of conformal area for a conformal structure on a surface. They prove that $\int_M H^2$ is not less than this area. Using this, they show that $\int_{RP^2} H^2 \geq 6\pi$ and $\int_M H^2 \geq 2\pi^2$ if M is conformally equivalent to the square torus.

61. (Alexandrov [Aℓ2]). Let S be the boundary surface of a convex body in R^3. If the intrinsic radius of S is bounded by 1, what is the largest surface area of S possible?

62. (Milnor [KO]). Let Σ be a complete noncompact surface immersed in R^3, and let λ_1, λ_2 be its principal curvatures. Prove that either $|\lambda_1 - \lambda_2|$ is not bounded away from zero on Σ, or K changes sign, or $K \equiv 0$.

63. (Hopf). Prove that a closed surface Σ immersed in R^3 with constant mean curvature is isometric to S^2.

Hopf proved this in the case that Σ is homeomorphic to S^2. Alexandrov [Aℓ] accomplished the proof under the assumption that Σ is embedded. (See Hopf [Ho].) Reilly [R] gave another proof of this case recently.

64. Prove the Caratheodory conjecture that every closed convex surface in R^3 has at least two umbilical points. In the real analytic case, this was asserted by Bol [Bℓ] and Hamburger [Ham], but doubts were later expressed about these papers—see Klotz [K] for corrections.

65. Can one define the rank of a compact C^ω-manifold M with nonpositive curvature so that if M is a locally symmetric space the definition agrees with the standard one? Suppose there is a totally geodesic, immersed, flat 2-plane in M. Can one find an immersed totally geodesic torus in M? (See Gromoll and Wolf [GW], Lawson and Yau [LY2].) If the "rank" of M is greater than one, one expects that M is very rigid metrically. How do we describe this rigidity?

66. (Kuiper). Let M be the surface obtained by attaching a handle to RP^2. Can M be immersed into R^3 with a "two-piece" property, i.e. every plane that cuts the surface divides the image into exactly two components?

See Kuiper's survey paper which is to appear in the Chern Symposium volume.

IV. The Spectrum

67. Let Ω_1 and Ω_2 be two bounded smooth domains in R^2 so that the eigenvalues (counting multiplicity) of the Laplacian acting on functions defined on Ω_1 and Ω_2 having zero boundary conditions are the same. Is Ω_1 isometric to Ω_2 ?

This is an old problem. For closed manifolds one can formulate an analogous problem, however, the answer is negative. This is by virtue of examples of Milnor [M2] and Vigneras [V], the latter providing a two-dimensional counterexample with negative curvature.

68. In problem 67, suppose the spectrums of Ω_1 and Ω_2 are equal except for a finite number of exceptions. Are the two spectrums in fact identical? One can ask a similar question if the set of exceptions is infinite but has density zero.

69. Let g(t) be a one-parameter family of metrics on a compact manifold with the same spectrum for the Laplacian. Prove that the metrics g(t) are isometric to each other.

Guillemin and Kazhdan [GK] proved that this is the case if the manifold is a surface of negative curvature, or if the manifold is suitably negatively pinched when the dimension of the manifold is greater than two.

70. Let Ω be a bounded domain in R^2. Let λ_i be the spectrum of the Laplacian acting on functions with zero boundary data (again, and henceforth, counting with multiplicity). Prove that

$$\lambda_i \geq \frac{4\pi i}{\text{area}(\Omega)} .$$

This was conjectured by Pólya [Poℓ] and was proved by him in the case that Ω can tile $\mathbf{R}^2$. One can formulate a similar question for eigenvalues of the Neumann problem (with the inequality in the opposite direction).

71. Let M be a two-dimensional, closed, compact surface. Can one find a universal constant C so that

$$\frac{\lambda_i}{i} \leq \frac{C(g+1)}{\text{area}(M)} \ ?$$

Here, g is the genus of M. If M is diffeomorphic to S^2, is $\lambda_i(M)$ not greater than $\lambda_i(S^2)$ where S^2 is equipped with a metric of curvature $= \frac{4\pi}{\text{area}(M)}$?

In the case $i = 1$, this is known to be true. The case when M is diffeomorphic to S^2 was proved by Hersch [He]. For M orientable and $g > 0$, this was proved by Yang and Yau [YY]. Recently, P. Li and Yau were able to find similar bounds for nonorientable surfaces.

72. Study the discrete spectrum of a complete manifold whose curvature is bounded and negative and whose volume is finite. When is it nonempty? What is its asymptotic behavior and relation to the closed geodesics?

Let $M = \{(x,y) \in \mathbf{R}^2 | y > 0\}/\Gamma$, where Γ is a congruent subgroup of $SL(2, \mathbf{Z})$. It is an old conjecture that λ_1 for M is at least $\frac{1}{4}$. Selberg [Se] proved that $\lambda_1 \geq \frac{3}{16}$.

It will also be important to study the continuous spectrum of a general complete manifold with finite volume. Hopefully, one can obtain some kind of L^2 index theorem for elliptic operators for these manifolds.

73. The behavior of the spectrum of a compact manifold of negative curvature is quite different for dimension two and dimension three. For example, R. Schoen [Sch] proved that for a three-dimensional hyperbolic space form (with curvature $= -1$), $\lambda_1 \geq \frac{c}{\text{vol}(M)^2}$ where c is a universal constant. This is certainly false for surfaces (see Schoen-Wolpert-Yau [SWY]).

Is it true that $\lambda_1 \operatorname{vol}(M)^2$ has an upper bound if M is a three-dimensional hyperbolic space.

74. Let M be a compact surface. Let $\lambda_1 \leq \lambda_2 \leq \cdots$ be the spectrum of M and $\{\phi_i\}$ be the corresponding eigenfunctions. For each i, the set $\{x \mid \phi_i(x) = 0\}$ is a one-dimensional rectifiable simplicial complex. Let L_i be the length of such a set. It is not difficult to prove that $\liminf_{i \to \infty} \sqrt{\lambda_i}^{-1}(L_i)$ has a positive lower bound depending only on the area of M. (This was independently observed by Bruning [B].)

It seems more difficult to find an upper bound of $\limsup_{i \to \infty} \sqrt{\lambda_i}^{-1}(L_i)$.

75. S. Y. Cheng [Cn] proved that for a compact surface, the multiplicity of λ_i has an upper bound depending only on the genus of the surface. Can one generalize this to higher dimensions? Most likely this is not true without modification. What is the correct statement? For a compact surface with fixed genus g, can one exhibit a metric (explicitly) with highest multiplicity in λ_i ?

76. Let M be a compact manifold and denote by f_i, $i = 1, 2, \cdots$, the eigenfunctions for the Laplacian on M. Show that the number of critical points of f_i is increasing with i.

77. Let Ω be a bounded domain in $\mathbf{R}^2$. Denote by $\lambda_1(\Omega)$ and $\lambda_2(\Omega)$ the first and second (nonzero) eigenvalues of the Laplacian for functions with zero boundary values. Show

$$\frac{\lambda_2(\Omega)}{\lambda_1(\Omega)} \leq \frac{\lambda_2(D)}{\lambda_1(D)}$$

where D is the disk in $\mathbf{R}^2$, and that equality implies Ω is a disk.

This will mean that one can determine whether the drum is circular or not by knowing the first two tones of the drum. For more details, see [PPW].

78. Let Ω be a bounded convex domain in $\mathbf{R}^2$. Let f_2 be the second eigenfunction for the Laplacian with zero boundary conditions. Show that the nodal line of f_2 cannot enclose a compact subregion of Ω.

In general, one likes to know the qualitative behavior of the nodal line. This conjecture has been around for a long time.

79. Let M be a compact manifold without boundary. Then we can define the eigenvalues for the Laplacian acting on the differential forms. How can we estimate the first nonzero eigenvalue in terms of computable geometric quantities? See Li [Li], Li-Yau [LiY1] and the recent work of Gromov [Gr3] where the estimates of λ_n on functions depend on the diameter of M and a lower bound for the Ricci curvature. See also the paper of Uhlenbrock [U].

80. (The Schiffer conjecture, or Pompieu problem). Let Ω be a smooth, compact bounded domain in $\mathbf{R}^2$. Suppose there exists an f which is an eigenfunction for the Laplacian with Neumann boundary conditions. If also f is constant on the boundary of Ω, prove Ω is a disk.

This problem is relevant to the following classical problem:

Given a function f defined on $\mathbf{R}^2$, and a bounded domain Ω, if one knows the value of the integral of f over all images of Ω under Euclidean motions of the plane, can the function f be recovered?

V. Problems Related to Geodesics

81. Prove that every compact manifold M has an infinite number of closed geodesics.

This is an old problem. Klingenberg has studied this extensively, and has obtained many deep results. See his book [Kℓ] for the case where $\pi_1(M)$ is finite.

82. Let M be a compact manifold without conjugate point. If M is homotopically equivalent to the torus, prove that M is flat.

This was conjectured by E. Hopf and proved by him for two-dimensional M. L. Green [Ge] has proved that the total scalar curvature of M must be non-positive, and is zero only if M is flat. It is believable that the fundamental group of a compact manifold without conjugate point has exponential growth unless the manifold is flat.

83. Prove the Blaschke conjecture for other symmetric spaces of rank one besides S^N. For the sphere, this was established through the efforts of Green, Weinstein, Berger, Kazdan, and Yang (see [Bs] for the precise history of the problem).

84. Prove that a compact harmonic manifold is symmetric.

A manifold is defined to be harmonic if geodesic spheres of small radius have constant mean curvature (see [Bs]).

85. Let M be a compact manifold with finite fundamental group. Can one find a non-hyperbolic closed geodesic? For the details of this problem see [Kℓ], and the paper of Ballman-Thorbergsson-Ziller in these proceedings.

86. If M is diffeomorphic to the N-sphere, give a lower estimate on the number of embedded closed geodesics.

It is well known that Lusternik-Schnirelmann have proven the existence of three distinct embedded closed geodesics if $N = 2$. (See Lusternik-Schnirelmann [LS].) For contributions to this type of problem, see [Kℓ].

87. Generalize Loewner and Pu's inequality to higher dimensions. The Loewner inequality says that for the two-torus,

$$\frac{A}{\ell^2} \geq C$$

where ℓ is the length of the shortest closed homotopically nontrivial loop and C is a universal constant. In this regard, consult the work of Berger [Br2] and Gromov [Gr4].

VI. Minimal Submanifolds

88. Prove that any three-dimensional manifold must contain an infinite number of immersed minimal surfaces. Sacks and Uhlenbeck [SU] proved the existence of a minimal sphere in any compact manifold which is not covered by a contractible space. Sacks-Uhlenbeck and Schoen-Yau [SY4] independently proved that any incompressible surface can be deformed into a minimal surface. When the ambient manifold is three-dimensional, an argument of Osserman shows that they are immersed. In most cases, they are in fact embedded by the results of Meeks and Yau [MY] and more recent work of Freedman, Hass and Scott. The work of

Meeks-Simon-Yau also showed that, starting from any compact surface in a compact three-dimensional manifold, one can minimize its area "within its isotopy class."

For a general three-dimensional manifold, Pitts [Pi] proved the existence of one such minimal surface. However, one does not know the genus of the surface from his method.

89. Prove that there are four distinct embedded minimal spheres in any manifold diffeomorphic to S^3. One should study the work of Sacks and Uhlenbeck [SU] in this regard.

90. Is it true that every compact differentiable manifold can be minimally embedded into S^N for some N ?

Recently (in a yet unpublished work), W. Y. Hsiang and W. T. Hsiang studied the problem of minimally embedding some exotic spheres in S^M.

91. Is there any complete minimal surface of R^3 which is a subset of the unit ball?

This was asked by Calabi [Ca3]. There is an example of a complete minimally immersed surface between two planes due to Jorge and Xavier [JX]. Calabi has also shown that such an example exists in R^4. (One takes an algebraic curve in a compact complex surface covered by the ball and lifts it up.)

92. What are the complete, embedded, minimal surfaces (with finite genus) in R^3 ?

The only known examples are the catenoid and the helicoid. It is possible to prove that any such surface is standardly embedded, in the topological sense.

93. Prove that every smooth, regular Jordan curve in R^3 can bound only a finite number of stable minimal surfaces.

If the Jordan curve is real analytic, Tomi [To] proved that it can bound only a finite collection of locally minimal disks. Tomi's argument is quite general, and the basic point that he requires to generalize the theorem to the smooth case is the proof of the absence of boundary branch points for stable minimal surfaces whose boundary is a smooth regular curve. To date, this last assertion is unestablished.

If we assume that the minimal surface has least area in the strong sense, Hardt-Simon [HS] established the absence of boundary branch points, thus proving the finiteness in this case.

There are various uniqueness theorems after suitable perturbation of the boundary. These were due to Böhme, Morgan, Tomi, Tromba and others.

94. Given a single smooth, regular, Jordan curve, can one find a nontrivial, continuous family of minimal disks bounded by this curve?

There is a classical example due to P. Levy [Le] and Courant [Cou] of a rectifiable Jordan curve which is smooth except at one point and which bounds an uncountable number of minimal disks. (A proof of the validity of this example depends on the "bridge principle" which was first established by Kruskal [Kr]. A more rigorous proof of the bridge principle was independently established by Almgren-Solomon [AS], and Meeks-Yau [MY].) Morgan [Mor] has found an example of continuous family of minimal surfaces whose boundary consists of four disjoint circles.

95. Let σ be a smooth Jordan curve in S^3 which bounds an embedded disk in the unit ball of R^4. Prove that there is a curve σ' isotopic to σ in S^3 which bounds an embedded minimal disk in the unit ball of R^4. An application of this would be the proof that the sliced knot is a ribbon knot.

96. What is the structure of the space of minimal surfaces of a fixed genus in S^3? Lawson [L1] has proved that, besides RP^2, any closed surface can be minimally embedded in S^3. Which conformal structures can be realized in such a way? What happens if we replace S^3 by S^N with $N > 3$?

97. (Lawson). Is the only embedded minimal torus in S^3 the Clifford torus?

There are many minimal torus in S^3 which are not Clifford tori, but they are not embedded.

98. (Lawson). Let M be an embedded minimal surface in S^3. Prove that the two domains in S^3 divided by M have equal volume.

This is a delicate form of the Gauss-Bonnet theorem. Indeed, if $M^{2N-1} \subseteq S^{2N}$ is a compact connected hypersurface such that all the odd elementary symmetric functions of the second fundamental form are zero, then the general Gauss-Bonnet theorem proves that the two components of $S^{2N} - M^{2N-1}$ have equal volume provided they have the same Euler characteristic. For a minimal surface in S^3, these two components are always diffeomorphic (cf. Lawson [L3]).

In the general case of S^N for $N > 3$, the conjecture fails. C. L. Terng, for example, shows $S^P((P/N)^{1/2}) \times S^{N-P}(((N-P)/N)^{1/2})$ does not divide S^{N+1} into two equivolume pieces unless $P = N-P$.

99. (Chern). Prove that the only embedded minimal hypersurface in S^{N+1} which is diffeomorphic to S^N is the totally geodesic sphere.

An affirmative answer will be interesting even for a special case where we assume the cone over the hypersurface is stable in R^{N+2}. Hopefully this would mean that an area-minimizing hypersurface which is a topological manifold is smooth. Under the assumption of stability of the cone the conjecture is true for $N = 2, 3, 4, 5$, see [Sim]. For higher codimension the conjecture is false; see [LO].

100. Is it true that the first eigenvalue for the Laplace-Beltrami operator on an embedded minimal hypersurface of S^{N+1} is N ?

This is not known even for $N = 2$. An affirmative answer will imply that the area of embedded minimal surfaces in S^3 will have an upper bound depending only on the genus. This is a consequence of the theorem of Yang-Yau [YY].

101. Is there a closed minimal surface in S^N with negative curvature?

102. As a generalization of Bernstein's theorem, Schoen and Fischer-Colbrie [F-CS], and do Carmo and Peng [DP], proved that any complete stable minimal surface in R^3 is linear. Can one generalize this statement to the case of a complete stable hypersurface in R^N for $N \leq 8$?

103. If u is an entire solution of the minimal surface equation on R^N, does u have polynomial growth?

One should read the paper of Bombieri and Giusti [BG]. Bombieri also suggested that it may have some connections with the first eigenvalue of minimal hypersurfaces in S^N. See also Allard and Almgren [AA].

104. Classify the topological type of the seven-dimensional area minimizing cones in R^8.

It was observed by Lawson that the space of diffeomorphism classes of these cones is finite and that explicit bounds should be obtainable. For example, merely from the assumption of stability, Simons [Sim] deduces an explicit L^2-estimate on the second fundamental form of the minimal hypersurface $M^6 \subseteq S^7$ corresponding to such a cone. Similar bounds on the L^p-norms for $p = 2, \cdots, n$ would give a priori bounds on the sum of the Betti numbers.

105. (Chern). Consider the set of all compact minimal hypersurfaces in S^N with constant scalar curvature. Think of the scalar curvature as a function on this set. Is the image of this function a discrete set of positive numbers?

There have been the works of Simons [Sim], Chern-doCarmo-Kobayashi [CDK], Lawson [L2] and Yau [Y3]. More recently, Terng and Peng [TP] made a breakthrough on this problem.

106. Let M be a compact, three-dimensional manifold with curvature $= -1$. Let Σ be a surface of genus g so that there is some continuous $f : \Sigma \to M$ with $f_* : \pi_1(\Sigma) \to \pi_1(M)$ an injection. It was known ([SY4], [SU]) that such a map can be deformed to be a minimal immersion. Is it true that for most M, the resulting immersion would be unique?

107. Let M be a complete minimal surface in R^3. Osserman [O1] has proved that the Gauss map of M cannot omit a set of positive capacity in S^2 and he conjectured that it in fact could not omit more than four points in S^2. Recently, Xavier [X] proved that it cannot omit more than eleven points. Based on the method of Xavier, Bombieri [Bo] improved the number to seven. Can one improve it to four? Can one generalize these assertions to three-dimensional minimal hypersurfaces?

108. Suppose H is an area minimizing hypersurface in a manifold M. Prove that by a perturbation of the metric on M, the singularities of H may be eliminated, retaining the $N-1$ dimensional homology class represented by H. For some related problems, see B. White [Wh].

Is it true that the support of a codimension one area minimizing current has a p.l. structure? For a general high codimensional minimal current, it is not true that the support is a real analytic variety (see Milani [Mi]). For the present state of the problem, see Almgren [Alm2].

109. Let Ω be a k-dimensional, compact, minimal submanifold of $\mathbf{R}^N$. Prove the isoperimetric inequality

$$\mathrm{Vol}(\Omega)^{k-1} \leq c_k \mathrm{Vol}(\partial\Omega)^k$$

where c_k is given by

$$c_k = \frac{\mathrm{Vol}(B(1))^{k-1}}{\mathrm{Vol}(\partial B(1))^k}$$

with B(1) signifying the unit ball in $\mathbf{R}^k$ It is true if $k = N$, that is, domains in $\mathbf{R}^N$.

This inequality with c_k greater than the above is known to be true. (See [FF], [Alm1], [Aℓℓ], [MiS], and [BDG].)

If $k = 2$ and Ω is simply connected, this result is classical and is due to Carleman (see Osserman [O2]). If $k = 2$ and Ω is doubly connected, it is again true, and due to Osserman and Schiffer [OS] and J. Feinberg [F].

An approach is to show that the extremal case for the inequality can be realized as a stationary integral varifold which one might be able to show is a flat k disk. That flat k disks are indeed extreme among nearby non-parametric surfaces has been studied by B. White [Wh].

110. Let Σ be a compact surface, and let $f: \Sigma \to M$ be a minimal immersion into a three-dimensional manifold that f has least area among all the maps homotopic to f. (If Σ has boundary, we consider only immersions which are embeddings on $\partial\Sigma$ and we fix the image of $f(\partial\Sigma)$ also.) If Σ is S^2 or a planar domain, Meeks-Yau [MY] prove that f is

an embedding. When Σ has higher genus, Freedman-Hass-Scott prove that f is an embedding assuming that f is homotopic to an embedding. Without this latter assumption, f need not be an embedding. However, it is believed that f tends to minimize the complexity of the self-intersection set. It is of basic interest for the topologists to estimate the number of triple points of f.

111. Let $f: M_1 \to M_2$ be a diffeomorphism between two compact manifolds with negative curvature. If $h: M_1 \to M_2$ is a harmonic map which is homotopic to f, is h a univalent map? For $n = 2$, this was proved in [SY6] and [Sa].

For $n > 2$, Calabi has a counterexample if we do not impose conditions on M_1 and M_2. (In Calabi's example, M_2 is a torus.)

112. Prove that $\pi_i(S^N)$ can be represented by harmonic maps. What happens if we replace S^N by a compact manifold with finite fundamental group?

One should refer to the paper of R. T. Smith [S].

113. (Affine geometry).

(a) (Chern) Establish Bernstein's theorem for affine geometry: Any convex graph over affine space which is an affine maximal hypersurface must be a paraboloid.

(b) Classify 3-dimensional compact affine-flat manifolds. In general, it is not known whether any compact affine flat manifold has zero Euler number (see Milnor [M3], Kostant-Sullivan [KS], Sullivan [Su1], [Su2], and Wood [Wo1]).

VII. General Relativity and the Yang-Mills Equation

114. This is the problem of "cosmic censorship," as coined by Penrose.

Let M be a 3-dimensional manifold equipped with a metric g_{ij} and a symmetric tensor h_{ij}. Assume that g_{ij} and h_{ij} satisfy the compatibility requirement necessary for them to represent the induced metric and second fundamental form, respectively, that M would inherit as a spacelike hypersurface of 4-dimensional asymptotically flat space-time satisfying the (vacuum) Einstein field equations.

In the study of global solutions to the vacuum field equation with Cauchy data g_{ij} and h_{ij} on M, one wishes to know the nature of the singularities of the solution obtained. Perhaps the most important open problem in general relativity is this:

Is it true that generically a singularity will have a horizon? (Is there no "naked" singularity?)

This question amounts to asking if the future can be theoretically predicted.

One should consult the book of Hawking and Ellis [HE] for background of this problem.

115. The splitting theorem of Cheeger and Gromoll [CG1] says that if a Riemannian manifold M of nonnegative Ricci curvature contains a line γ (i.e., an absolutely minimizing geodesic), then M decomposes isometrically as a cross product $\mathbf{R} \times N$, the first factor being represented by γ.

It would be of interest in studying the structure of space-time to prove that a geodesically complete Lorentzian 4-manifold of nonnegative Ricci curvature in the timelike direction which contains an absolutely maximizing timelike geodesic is isometrically the cross product of that geodesic and a spacelike hypersurface.

116. Prove that a static stellar model is isometric to a sphere. See Lindbloom [Lin] for the case when the model has uniform density.

S. Hawking demonstrated that a static black hole is axially symmetric, but his argument is based in part on physical reasoning.

From the work of Israel, Hawking, Carter, and Robinson, one knows that a stationary, rotating black hole must be the Kerr black hole (see [Ro]). Can one make a similar statement about a charged stationary black hole? If the metric is Riemannian, there are similar questions. Lapedes (these proceedings) points out that Robinson's method does not apply, but Israel's approach still works, in the static case.

117. Prove that any Yang-Mills fields on S^4 is either self-dual or antiself-dual.

See the paper of Bourguignon and Lawson [BL] in this volume.

Atiyah, Drinfield, Hitchen, and Manin [AHDM] have classified the self- and antiself-dual solutions.

118. Prove that the moduli space of the self-dual fields on S^4 with a fixed Pontryagin number is connected. Prove that the Pontryagin number of a L^2-integrable gauge field on R^4 is an integer.

Both problems 117 and 118 are very well known. See the excellent article of Atiyah [At2].

119. Physicists have a notion of asymptotically flat manifolds (see [SY7], for example). The definition depends heavily on the choice of coordinate system and is not intrinsic. If one replaces the definition by requiring the curvature to decay suitably, do we obtain an equivalent condition?

120. Given an asymptotically flat space, can one give a good definition of total angular momentum? What would the relationship be with total mass? (See [Pe].)

BIBLIOGRAPHY

[Aℓ] A. D. Alexandrov, "On a class of closed surfaces," *Recuil Math.* (Moscou) *4*(1938), 69-77.

[Aℓ2] ———, "Die innere geometrie des convexen flachen," Akad. Verlag, Berlin, 1955.

[Aℓ3] ———, "Uniqueness theorems for surfaces in the large," Vestnik Leningrad *11*(1956), AMS Translation 21 (1962), 341-353.

[Aℓℓ] W. K. Allard, "On the first variation of a varifold," *Ann. Math.*, *95*(1972), 417-491.

[AA] W. K. Allard and F. Almgren, "On the radial behavior of minimal surfaces and the uniqueness of their tangent cones," *Ann. Math.* (1981).

[Alm1] F. Almgren, "Some interior regularity theorems for minimal surfaces and an extension of Bernstein's theorem," *Ann. Math., 84* (1966), 277-292.

[Alm2] ———, "Multiple valued functions minimizing Dirichlet's integral and the regularity of mass minimizing integral currents," preprint.

[AS] F. Almgren and B. Solomon, "How to connect minimal surfaces by bridges," preprint.

[At1] M. Atiyah, "Elliptic operators, discrete groups and von Neumann algebras," *Société Mathématique de France, Asterisque, 32, 33* (1976), 43-72.

[At2] ————, "Geometrical aspects of group theories," *Proc. International Congress Math.*, Helsinki, (1973), 881-885.

[AHDM] M. Atiyah, N. Hitchin, U. Drinfeld, and Yu. Manin, "Construction of instantons," *Physics Letters, 65A* (1978), 185-187.

[Au] T. Aubin, "The scalar curvature," in *Differential Geometry and Relativity*, Holland, 1976, 5-18.

[ASz] L. Auslander and R. H. Szczarba, "Characteristic classes of compact solvmanifolds," *Ann. Math. 76*(1962), 1-8.

[Ba] I. Ja. Bakel'man, these proceedings.

[B-B] L. Bérard-Bergery, to appear.

[Br1] M. Berger, "Trois remarques sur les varietés riemanniennes à courbure positive," *C. R. Acad. Sci. Paris*, Sér A-B *263*(1966), A76-A78.

[Br2] ————, "Du Coté de chez Pu," *École Norm. Sup.*, (4) *5*(1972).

[Bs] A. Besse, *Manifolds All of Whose Geodesics are Closed*, Springer-Verlag, Berlin-Heidelberg-New York, 1978.

[Bℓ] G. Bol, "Über Mabelpunkte auf einer Eifläche," Math. Zeit. *49* (1943/44), 389-410.

[Bo] E. Bombieri, to appear.

[BDG] E. Bombieri, E. DiGiorgi, and E. Guisti, "Minimal cones and the Bernstein problem," *Inv. Math., 7*(1969), 243-268.

[BG] E. Bombieri and E. Guisti, "Harnack's inequality for elliptic differential equations on minimal surfaces," *Inv. Math., 15*(1972), 24-46.

[Bor] A. Borel, "On the automorphisms of certain subgroups of semi-simple Lie groups," *Proc. Bombay Colloquium on Alg. Geometry*, 1968, 43-73.

[BDS] J. P. Bourguignon, A. Deschamps, and P. Sentenac, "Conjecture de H. Hopf sur les produits de varietés," *École Norm. Sup.*, (4) *5*(1972), fasc. 2.

[BL] J. P. Bourguignon and B. Lawson, this volume.

[B] J. Bruning, "Über knoten von eigenfunktionen des Laplace-Beltrami operators," *Math. Z., 158*(1978), 15-21.

[Cal] E. Calabi, "The space of Kähler metrics," *Proc. Internat. Congress Math.*, Amsterdam, *2*(1954), 206-207.

[Ca2] E. Calabi, "Examples of Bernstein problems for some nonlinear equations," *Proc. AMS Symposium on Global Analysis*, XV, 223-230.

[Ca3] ———, see page 170 of *Proceedings of the United States-Japan Seminar in Differential Geometry*, Kyoto, 1965, Nippon Hyoronsha Co., Tokyo, 1966.

[C1] J. Cheeger, "Some examples of manifolds of nonnegative curvature," *J. Diff. Geom.*, *8*(1972), 623-628.

[C2] ———, "Finiteness theorems for Riemannian manifolds," *Am. J. Math.*, *92*(1970), 61-74.

[C3] ———, "On the spectral geometry of spaces with cone-like singularities," *Proc. Nat. Acad. Sci. USA*, *76*(1979), 2103-2106.

[C4] ———, "On the Hodge theory of Riemannian pseudo-manifolds," in *AMS Symposium on the Geometry of the Laplace Operator*, University of Hawaii at Manoa, 1979, 91-146.

[CG1] J. Cheeger and D. Gromoll, "The splitting theorem for manifolds of nonnegative Ricci curvature," *J. Diff. Geom.*, *6*(1971), 119-128.

[CG2] ———, "On the structure of complete manifolds of nonnegative curvature," *Ann. Math.*, *96*(1972), 413-443.

[Cn] S. Y. Cheng, "Eigenfunctions and nodal sets," *Comm. Math. Helv.*, *51*(1976), 43-55.

[CY1] S. Y. Cheng and S. T. Yau, "Hypersurfaces with constant scalar curvature," *Math. Ann.*, *225*(1977), 195-204.

[CY2] ———, "On the existence of complete Kähler metric on non-compact complex manifolds and the regularity of Fefferman's equation," *Comm. Pure Appl. Math.*, to appear.

[Ch1] S. S. Chern, "On the curvatura integra in a Riemannian manifold," *Ann. Math.*, *46*(1945), 964-971.

[Ch2] ———, "The geometry of G-structures," *Bull. Am. Math. Soc.*, *72*(1966), 167-219.

[Ch3] ———, "On curvature and characteristic classes of a Riemannian manifold," *Abh. Math. Sem. Univ. Hamburg*, *20*(1955), 117-126.

[CDK] S. S. Chern, M. do Carmo, and S. Kobayashi, "Minimal submanifolds on the sphere with second fundamental form of constant length," in *Functional Analysis and Related Essays*, Springer-Verlag, Berlin-Heidelberg-New York, 59-75.

[Co] R. Connelly, "An immersed polyhedra which flexes," *Indiana Univ. J. Math.*, *25*(1976), 965-972.

[Cou] R. Courant, *Dirichlet's Principle*, Interscience Publishers, Inc., New York, 1950, pp. 119-122.

[DP] M. do Carmo and C. K. Peng, "Stable complete minimal surfaces in $\mathbf{R}^3$ are planes," *Bull. Am. Math. Soc., 1* (1979), 903-905.

[E] P. Eberlein, "Some properties of the fundamental group of a Fuchsian manifolds," *Inv. Math., 19*(1973), 5-13.

[EF] M. V. Efimov, "Hyperbolic problems in the theory of surfaces," *Proc. Inter. Congress of Mathematicians*, Moscow (1966), AMS Translation *70*(1968), 26-38.

[Eℓ] H. Eliasson, "On variations of metrics," *Math. Scand., 29*(1971), 317-327.

[FH] T. Farrell and W. C. Hsiang, to appear.

[FF] H. Federer and W. Fleming, "Normal and integral currents," *Ann. Math., 72*(1960), 458-520.

[F] J. Feinberg, "The isoperimetric inequality for doubly-connected minimal surfaces in $\mathbf{R}^N$," *J. d'Analyse Math., 32*(1977), 249-278.

[Fi] R. Finn, "On a class of conformal metrics, with applications to differential geometry in the large," *Comment. Math. Helv., 40* (1965), 1-30.

[F-CS] D. Fischer-Colbrie and R. Schoen, "The structure of complete stable minimal surfaces in 3-manifolds of nonnegative scalar curvature," *Comm. Pure Appl. Math., 33*(1980), 199-211.

[G] R. Geroch, "Positive sectional curvature does not imply positive Gauss-Bonnet integrand," *Proc. Am. Math. Soc., 54*(1976), 267-270.

[Ge] L. Green, "A theorem of E. Hopf," *Michigan Math. J., 5*(1958), 31-34.

[GWu] R. Greene and H. Wu, "C^∞ convex functions and manifolds of positive curvature," *Acta Math., 137*(1976), 209-245.

[GM1] D. Gromoll and W. Meyer, "An exotic sphere with nonnegative sectional curvature," *Ann. Math., 100*(1974), 401-406.

[GW] D. Gromoll and J. Wolf, "Some relations between the metric structure and the algebraic structure of the fundamental group in manifolds of nonpositive curvature," *Bull. Am. Math. Soc., 77* (1971).

[Gr1] M. Gromov, "Three remarks on geodesic dynamics and the fundamental groups," unpublished.

[Gr2] ________, "Manifolds of negative curvature," *J. Diff. Geom., 13* (1978), 223-230.

[Gr3] ________, "Paul Levy's isoperimetric inequality," I.H.E.S. preprint, 1980.

[Gr4] ________, "Structures métriques pour les variétés riemanniennes," Textes Mathematique *1*, edited by J. Lafontaine and P. Panou, Cedic-Nathan.

[GL] M. Gromov and B. Lawson, "The classification of simply-connected manifolds of positive scalar curvature," *Ann. Math.*, *111* (1980), 423-434.

[GK] V. Guillemin and D. Kazhdan, "Some inverse spectral results for negatively curved n-manifolds," *AMS Symposium on the Geometry of the Laplace Operator*, University of Hawaii at Manoa, 1979, 153-180.

[Ham] H. Hamburger, "Beweis einer Carathéodoryschen Vermutung I," *Ann. Math.*, *41* (1940), 63-68, II, III, *Acta. Math.*, *73* (1941), 174-332.

[Ha] G. Harder, "A Gauss-Bonnet formula for discrete arithmetically defined groups," *École Norm. Sup.*, *9* (1971), 409-455.

[HS] R. Hardt and L. Simon, "Boundary regularity and embedded solutions for the oriented Plateau problem," *Ann. Math.*, *110* (1979), 439-486.

[HL] R. Harvey and H. B. Lawson, Jr., "A constellation of minimal varieties defined over the group G_2," to appear.

[HE] S. W. Hawking and G. F. R. Ellis, *The Large Scale Structure of Space-Time*, Cambridge University Press, Cambridge, Mass., 1973.

[He] J. Hersch, "Sur la fréquence fondementale d'une membrane vibrante: évaluations par défaut et principé de maximum," *Z. Angw. Math. Phys.*, Vol. XI, Fasc. 5 (1960), 387-413.

[H1] N. Hitchin, "Harmonic spinors," *Advances in Math.*, *14* (1974), 1-55.

[H2] ———, "Compact four-dimensional Einstein manifolds," *J. Diff. Geom.*, *9* (1974), 435-441.

[Ho] H. Hopf, *Lectures on Differential Geometry in the Large*, mimeographed lecture notes, Stanford University, 1956-57.

[Hu] A. Huber, "Konforme and metricsche kreise auf vollständigen flächen," *Comment. Math. Helv.*, *47* (1972), 409-420.

[JX] L. P. M. Jorge and F. Xavier, "A complete minimal surface in R^3 between two parallel planes," *Ann. Math.*, *112* (1980), 203-206.

[Kℓ] W. Klingenberg, *Lectures on Closed Geodesics*, Springer-Verlag, Berlin-Heidelberg-New York, 1978.

[K] T. Klotz, "On G. Bol's proof of Caratheodory's conjecture," *Comm. Pure and Appl. Math.*, *12* (1959), 277-311.

[KO] T. Klotz and R. Osserman, "Complete surfaces in E^3 with constant mean curvature," *Comm. Math. Helv.* *41* (1966-67), 313-318.

[Ko] S. Kobayashi, in *Differential Geometry and Relativity*, Cahan and Flato, eds., Holland, 1976.

[KS] B. Kostant and D. Sullivan, "The Euler characteristic of an affine space form is zero," *Bull. AMS* *81* (1975), 937-938.

[Kr] M. Kruskal, "The bridge theorem for minimal surfaces," *Comm. Pure Appl. Math.*, *7*(1954), 297-316.

[L1] H.B. Lawson, Jr., "Complete minimal surfaces in S^3," *Ann. Math.*, *92*(1970), 335-374.

[L2] ________, "Local rigidity theorems for minimal hypersurfaces," *Ann. Math.*, *89*(1969), 187-197.

[L3] ________, "The unknottedness of minimal embeddings," *Inv. Math.*, *11*(1970), 183-187.

[LO] H.B. Lawson, Jr. and R. Osserman, "Non-existence, non-uniqueness and irregularity of solutions to the minimal surface system," *Acta Math.*, *139*(1977), 1-17.

[LY1] H.B. Lawson, Jr. and S.T. Yau, "Scalar curvature, non-abelian group actions, and the degree of symmetry of exotic spheres," *Comm. Math. Helv.*, *49*(1974), 232-244.

[LY2] ________, "Compact manifolds of non-positive curvature," *J. Diff. Geom.*, *7*(1972), 211-228.

[Le] P. Levy, *Lecons d'analyse fonctionelle*, Gauthier-Villars, Paris, 1922.

[Li] P. Li, "A lower bound for the first eigenvalue of the Laplacian on a compact manifold," *Indiana U. Math. J.*, *28*(1979), 1013-1019.

[LiY1] P. Li and S.T. Yau, "Estimates of eigenvalues of a compact Riemannian manifold," *AMS Symposium on the Geometry of the Laplace Operator*, University of Hawaii at Manoa, 1979, 205-239.

[LiY2] ________, to appear.

[Lin] L. Lindbloom, "Static, uniform density stellar models must be spherical," preprint.

[LS] L. Lusternik and L. Schnirelmann, *Méthodes Topologiques dans les Problèmes Variationnels*, Hermann and Co., Paris, 1934.

[Ma] R.A. Margulis, "On connections between metric and topological properties of manifolds of non-positive curvature," *Proc., Sixth Topological Conference, Tbilsi, USSR*, 1922, p. 83 (in Russian).

[MY] W. Meeks, III and S.T. Yau, "Topology of three dimensional manifolds and the embedding problem in minimal surface theory," *Ann. Math.*, to appear.

[MSY] W. Meeks, III, L. Simon, and S.T. Yau, to appear.

[MiS] J. Michael and L. Simon, "Sobolev and mean value inequalities on generalized submanifolds of R^n," *Comm. Pure Appl. Math.*, XXVI (1973), 361-379.

[Mi] A. Milani, "Non-analytical minimal varieties," Instituto Matematico "Leonida Tonelli," (1977), 77-13.

[M1] J. Milnor, "A note on curvature and fundamental group," *J. Diff. Geom.*, *2*(1968), 1-7.

[M2] ———, "Eigenvalues of the Laplace operator of certain manifolds," *Proc. Nat. Acad. Sci. USA. 51*(1964), 542.

[M3] ———, "On the existence of a connection with curvature zero," *Comm. Math. Helv. 32*(1958), 215-223.

[Miℓ] J. Millson, "On the first Betti number of a constant negatively curved manifold," *Ann. Math.*, *104*(1976), 235-247.

[Mo] S. Mori, "Projective manifolds with ample tangent bundles," *Ann. Math.*, *110*(1979), 593-606.

[Mor] F. Morgan, "A smooth curve in $\mathbf{R}^3$ bounding a continuum of minimal surfaces," preprint.

[Mos] J. Moser, "On the volume elements of a manifold," *Trans. Am. Math. Soc.*, *120*(1965), 286-294.

[MS] G. D. Mostow and Y. T. Siu, "A compact Kähler surface of negative curvature not covered by the ball," *Ann. Math.*, *112*(1980), 321-360.

[Nab] P. Nabonnand, Thesis, University of Nancy, 1980.

[Na] J. Nash, Ph.D. thesis, Stanford Univ. 1975. Also *J. Diff. Geom.*, to appear.

[Ni] W. M. Ni, this volume.

[Nir] L. Nirenberg, in *Nonlinear Problems*, edited by R. E. Langer, Univ. of Wisconsin Press, Madison, 1963, 177-193.

[O1] R. Osserman, "Global properties of minimal surfaces of E^3 and E^N," *Ann. Math.*, *80*(1964), 340-364.

[O2] ———, "The isoperimetric inequality," *Bull. Am. Math. Soc.*, *84*(1978), 1182-1238.

[OS] R. Osserman and M. Schiffer, "Doubly-connected minimal surfaces," *Arch. Rational Mech. Anal.*, *58*(1975), 285-307.

[PPW] L. Payne, G. Polya, and H. Weinberger, "On the ratio of consecutive eigenvalues," *J. Math. Phys.*, *35*(1956), 289-298.

[Pe] R. Penrose, "Some unsolved problems in classical general relativity," in this volume.

[Pi] J. Pitts, "Existence and regularity of minimal surfaces on Riemannian manifolds," *Mathematical Notes*, Princeton University Press.

[Pg] A. V. Pogorelov, "An example of a two-dimensional Riemannian metric admitting no local realization in E^3," *DAN 198*(1971), *Soviet Math. Dok. 12*(1971), 729-730.

[Po] W. A. Poor, Jr., "Some results on nonnegatively curved manifolds," *J. Diff. Geom.*, *9*(1974), 583-600.

[P] A. Preissman, "Quelques propriétés globales des espaces de Riemann," *Comment. Math. Helv.*, *15*(1942-43), 175-216.

[R] R. Reilly, "Applications of the Hessian operator in a Riemannian manifold," *Mich. Math. J.*, *26*(1977), 457-472.

[Ri] A. Rigas, "Geodesic spheres as generators of the homotopy groups of O, BO," *J. Diff. Geom.*, *13*(1978), 527-545.

[Ro] D. C. Robinson, "Uniqueness of the Kerr black hole," *Phys. Rev. Lett.*, *34*(1975), 905-906.

[SU] J. Sacks and K. Uhlenbeck, "The existence of minimal immersion of 2-spheres," *Ann. Math.*, to appear.

[Sa] J. Sampson, "Some properties and applications of harmonic mappings," *Ann. Scient. Ec. Norm. Sup.*, *11*(1978), 211-228.

[Sch] R. Schoen, to appear.

[SS] R. Schoen and L. Simon, "Regularity of stable minimal hypersurfaces," preprint.

[SWY] R. Schoen, S. Wolpert, and S. T. Yau, "Geometric bounds on the low eigenvalues of a compact surface," *AMS Symposium on the Geometry of the Laplace Operator*, University of Hawaii at Manoa, 1979, 279-285.

[SY1] R. Schoen and S. T. Yau, "Complete three-dimensional manifolds with positive Ricci curvature and scalar curvature," in this volume.

[SY2] ———, "Positivity of the total mass of a general space-time," *Physical Review Letters*, *43*(1979), 1457-1459.

[SY3] ———, to appear.

[SY4] ———, "Existence of incompressible minimal surfaces and the topology of three-dimensional manifolds with non-negative scalar curvature," *Ann. Math.*, *110*(1979), 127-142.

[SY5] ———, "On the structure of manifolds with positive scalar curvature," *Manuscripta Math.*, *28*(1979), 159-183.

[SY6] ———, "On univalent harmonic maps between surfaces," *Inv. Math.*, *44*(1978), 265-278.

[SY7] ———, *Commun. Math. Phys.*, *65*(1979), 45-76.

[Se] A. Selberg, "On the estimation of Fourier coefficients of modular forms," *AMS Symposium on Number Theory*, California Inst. of Technology, Pasadena, 1963.

[Sim] J. Simons, "Minimal varieties in Riemannian manifolds," *Ann. Math.*, *88*(1968), 62-105.

[Si] Y. T. Siu, ''Complex analyticity of harmonic maps and the strong rigidity of compact Kähler manifolds,'' *Ann. Math., 112*(1980), 73-111.

[SiY1] Y. T. Siu and S. T. Yau, ''Compact Kähler manifolds of positive curvature,'' *Inv. Math.*, to appear.

[SiY2] ————, ''Complete Kähler manifolds with non-positive curvature of faster than quadratic decay,'' *Ann. Math., 105*(1977), 225-264.

[SiY3] ————, ''Compactifications of negatively curved Kähler manifolds of finite volume,'' in this volume.

[S] R. T. Smith, ''Harmonic mappings of spheres,'' *Am. J. Math., 97* (1975), 364-385.

[Su1] D. Sullivan, ''A generalization of Milnor's inequality concerning affine foliations and affine manifolds,'' *Comm. Math. Helv. 51* (1976), 183-189.

[Su2] ————, ''Cycles for the dynamical study of foliated manifolds and complex manifolds,'' *Inv. Math. 36*(1976), 225-255.

[Ta] N. Tanaka, ''Rigidity for elliptic isometric imbeddings,'' *Nagoya Math. J., 51* (1973), 137-160.

[TP] C. L. Terng and C. K. Peng, this volume.

[Th] Sir W. Thomson (Lord Kelvin), ''On the division of space with minimal partitional area,'' *Phil. Mag.* (5), *24*(Dec. 1887), 503-514.

[To] F. Tomi, ''On the local uniqueness of the problem of least area,'' *Arch. Rat. Mech. Anal., 52*(1973), 312-318.

[TW] A. Treibergs and W. Wei, preprint.

[Tr] N. Trudinger, ''Remarks concerning the conformal deformation of Riemannian structures on compact manifolds,'' *Ann. Scuola Norm. Sup. Pisa. 3*(1968), 265-274.

[U] D. A. Uhlenbrock, ''Spectral distributions of Laplacians and Kato's inequality,'' *AMS Symposium on the Geometry of the Laplace Operation,* Iniversity of Hawaii at Manoa, 1979, 293-299.

[V] M. F. Vignéras, ''Variétés Riemannianes isospectrales er non isométriques,'' *Ann. Math., 112*(1980), 21-32.

[Vi] E. B. Vinberg, ''Some examples of crystallographic groups in Lobacevski space,'' *Math. Sbornik, 78*(1969), 633-639, trans. *7* (1969), 617-622.

[Wa] R. Walter, ''A generalized Allendorfer-Weil formula and an inequality of the Cohn-Vossen type,'' *J. Diff. Geom., 10*(1975), 167-180.

[Wh] B. White, ''Perturbations of singularities and uniqueness of tangent cones for minimal surfaces, Thesis, Princeton University, 1981.

[Wi] T. J. Willmore, "Note on embedded surfaces," *An. Şti. Univ. "Al. I. Cuza," Iaşi Şect. Ia Mat.* *11*(1965), 493-496.

[W] J. Wolf, *Spaces of Constant Curvature*, McGraw-Hill, New York, 1966.

[Wol] J. Wood, "Bundles with totally disconnected structure group," *Comm. Math. Helv.* *46*(1971), 257-273.

[X] F. Xavier, "The gauss map of a complete non-flat minimal surface cannot omit 11 points of the sphere," preprint.

[Yam] H. Yamabe, "On a deformation of Riemannian structures on compact manifolds," *Osaka Math. J.*, *12*(1960), 21-37.

[YY] P. Yang and S. T. Yau, "Eigenvalues of the Laplacian of compact Riemannian surfaces and minimal submanifolds," *Ann. Scuola Norm. Sup. Pisa*, *7*(1980), 55-63.

[Y] S. T. Yau, "On the Ricci curvature of a compact Kähler manifold and the complex Monge-Ampère equation, I," *Comm. Pure and Appl. Math.*, XXXI(1978), 339-411.

[Y1] ————, "On Calabi's conjecture and some new results in algebraic geometry," *Proc. Nat. Acad. Sci. USA*, *74*(1977), 1798-1799.

[Y2] ————, "On the curvature of compact Hermitian manifolds," *Inv. Math.*, *25*(1974), 213-239.

[Y3] ————, "Submanifolds with constant mean curvature II," *Am. J. Math.*, *97*(1975), 76-100.

ication Data

2)
eminars in
d by the Institute
ial Science

-Addresses, essays,
equations, Partial—
I. Yau, S.-T.
II. Series.
Q 516.36 81-8631
ISBN 0- AACR2
ISBN 0-691- pbk.)

Shing-Tung Yau is Professor of Mathematics at the Institute for Advanced Study in Princeton, New Jersey.